THE CHEMIST'S
READY REFERENCE
HANDBOOK

Other McGraw-Hill Books of Interest

AIP • AMERICAN INSTITUTE OF PHYSICS HANDBOOK
Dean • LANGE'S HANDBOOK OF CHEMISTRY
Grant • HACKH'S CHEMICAL DICTIONARY
Lapedes • McGRAW-HILL DICTIONARY OF PHYSICS AND MATHEMATICS
Parker • McGRAW-HILL DICTIONARY OF SCIENTIFIC AND TECHNICAL TERMS
Parker • McGRAW-HILL ENCYCLOPEDIA OF CHEMISTRY
Perry and Green • PERRY'S CHEMICAL ENGINEERS' HANDBOOK
Shugar, Shugar, Bauman, and Bauman • CHEMICAL TECHNICIANS' READY
REFERENCE HANDBOOK

For more information about other McGraw-Hill materials,
call 1-800-2-MCGRAW in the United States. In other
countries, call your nearest McGraw-Hill office.

THE CHEMIST'S READY REFERENCE HANDBOOK

Gershon J. Shugar, B.S., M.A., Ph.D.

Professor, Essex County College
Newark, N.J.

John A. Dean, B.S. in Chem., M.S., Ph.D.

Professor Emeritus of Chemistry
University of Tennessee
Knoxville, Tenn.

CONSULTING EDITORS

Ronald A. Shugar, B.S., M.D.

Medical Associates
Edison, N.J.

Lawrence Bauman, B.S., D.D.S.

Former Professor, NYU School of Dentistry
Fanwood, N.J.

Rose Shugar Bauman, B.S.

Science Writer
Watchung, N.J.

Peggy A. Dean, C.P.S.

Office of the President
University of Tennessee
Knoxville, Tenn.

McGRAW-HILL PUBLISHING COMPANY

New York   St. Louis   San Francisco   Auckland   Bogotá
Caracas   Hamburg   Lisbon   London   Madrid   Mexico
Milan   Montreal   New Delhi   Oklahoma City
Paris   San Juan   São Paulo   Singapore
Sydney   Tokyo   Toronto

Library of Congress Cataloging-in-Publication Data
Shugar, Gershon J., date.
    The chemist's ready reference handbook / Gershon J. Shugar, John
A. Dean.
    p. cm.
    Includes index.
    ISBN 0-07-057178-3
    1. Chemistry—Handbooks, manuals, etc. I. Dean, John Aurie,
    date.   II. Title.
QD65.S537   1989   89-8166
543—dc20   CIP

34567890   DOC/DOC   9543210

ISBN 0-07-057178-3

*The editors for this book were Harold B. Crawford and David E. Fogarty
and the production supervisor was Suzanne W. Babeuf. This book was set
in Times Roman. It was composed by the McGraw-Hill Publishing Company Professional & Reference Divisi*

Printed and bound by R. R. Donnell

Information contained in this work has been obtained by McGraw-
Hill, Inc., from sources believed to be reliable. However, neither
McGraw-Hill nor its authors guarantees the accuracy or completeness
of any information published herein and neither McGraw-Hill nor its
authors shall be responsible for any errors, omissions, or damages
arising out of use of this information. This work is published with the
understanding that McGraw-Hill and its authors are supplying information but are not attempting to render engineering or other
professional services. If such services are required, the assistance of
an appropriate professional should be sought.

*For more information about other McGraw-Hill materials,
call 1-800-2-MCGRAW in the United States. In other
countries, call your nearest McGraw-Hill office.*

CONTENTS

PREFACE

This Handbook is designed to be an "omnibook" for the practicing chemist and for the college student who is taking laboratory courses; also for the graduate student who needs a ready source of information which meaningfully relates to actual laboratory practice. This timely Handbook differs from all others in the field by treating theory in the briefest of terms as needed background. Instead it covers the "real-world" questions and problems faced in the chemical laboratory. The busy professional will welcome this collection of data, procedures, precautions, and troubleshooting hints for all important areas, both instrumental methods and wet chemistry, in the chemist's day-to-day work.

Profusely illustrated throughout, the Handbook. . .

- Describes a wide variety of laboratory instruments and their use
- Details a variety of analytical procedures
- Offers precautions and safety procedures
- Provides practical checklists and troubleshooting hints

Relevant material was derived from Shugar's *Chemical Technician's Ready Reference Handbook*, Second Edition. A major portion is entirely new, however, and includes sections on. . .

- Laboratory computers and automation
- Gas, liquid, and planar chromatography
- Ultraviolet, visible, infrared, fluorescence and Raman spectrophotometry
- Flame emission and atomic absorption
- X-ray and atomic emission and induction-coupled plasma spectroscopy
- NMR, radiochemistry, and mass spectrometry
- Electroanalytical methods
- Thermal methods
- Automated analyses
- Determination of physical properties
- Preliminary operations of analysis, including the balance, distillation, moisture and drying, extraction, filtration, and pressure and vacuum
- Volumetric analysis
- Practical laboratory information such as heating, cooling, and temperature measurement, statistical treatment of data, and chemical resistance of polymers and rubbers.

The desire was to produce a compilation of laboratory methods within the limits set by the economy of available space. One difficulty always faced by the authors of such a book is that they must decide which information is to be excluded in order to keep the volume from becoming unwieldy in size and too expensive for an individual to purchase.

It is hoped that users of this Handbook will offer suggestions of material that might be included, or even excluded, in future editions.

Gershon J. Shugar
John A. Dean

THE CHEMIST'S READY REFERENCE HANDBOOK

CHAPTER 1

LABORATORY COMPUTERS AND AUTOMATION

1.1 INTRODUCTION

Chemical laboratory instrumentation and procedures are undergoing a dramatic change. The thrust of the technical articles and the manufacturers' advertisements for the products are all oriented, wherever possible, to computer-assisted laboratory instrumentation and automation. Even the balance, the most basic analytical instrument, is coupled to a computer. Chemists no longer have to record data by hand. In the computer-assisted laboratory the combination of the microcomputer and microprocessor relieves chemists of many manual manipulations involved in the procedure and instantaneously presents the results of the analysis both as hard copy (from a printer) and on screen [from the cathode-ray tube (CRT)]. The trend is to expedite and improve the productivity of the laboratory and, at the same time, to minimize experimental errors and the amount of calculation required.

To understand the role of a computer in a specific instrumental method, the interactions among the instrument, the computer, and the analyst or user must be considered. Several combinations of interactions (data processing) are considered next.

1.1.1 Batch or Off-Line

The batch (or off-line) mode of processing uses computer systems which incorporate mainframe central processing units. Data are collected from the instrument and stored on an external input medium, such as magnetic tape, punched cards, or a floppy disk. Then, along with a user-written program, the operator (user) schedules everything for input to the computer as a single package. Computer programs are written in an analyst-oriented language, such as Fortran or Basic. This type of processing tends to optimize the use of the central processing unit because it allows all data and software to be available concurrently when needed. The output from a run is stored on a peripheral device for subsequent printing or display (Fig. 1.1).

Batch processing is useful for complex calculations or for manipulation of large amounts of data. Since there is no direct communication in a batch environment between the computer and the instrument or between the computer and

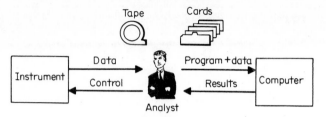

FIGURE 1.1   Batch or off-line computer configuration.

the operator, this configuration tends to be unresponsive to the immediate needs of instrumentation or analysis. Jobs are processed in the order submitted.

1.1.2   Real Time or On-Line

In real-time systems the computer is connected directly to one or more instruments through an electronic interface (Fig. 1.2). Data from one experiment can be fed directly into the computer system where they are processed and made available immediately to the operator. The analyst interacts with both the computer and the instrument to obtain and process data, control instrument operation, and retrieve results. Further processing of data can be requested. In this mode the computer responds instantaneously to data acquired from the instrument and thus can be used to control or modify the conditions of the experiment. All the data generated can be stored for future reference.

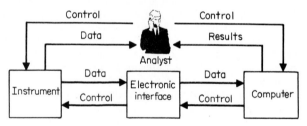

FIGURE 1.2   Real time or on-line computer configuration.

In a real-time system the computer and the instrument form a symbiotic relationship which provides a dynamic approach to experimentation and analysis. Methods that require the rapid execution of complex mathematical transformation functions, such as Fourier transform nuclear magnetic resonance and infrared spectroscopy, would be impossible without on-line computers.

1.1.3   In-Line or Integrated Systems

When the computer becomes an integral, dedicated part of the packaged instrument, the configuration is known as an in-line or integrated system (Fig. 1.3). The computer supervises instrument operation by prompting the analyst for input parameters, by monitoring and actively controlling the instrument operation, and by

Instrument -computer system

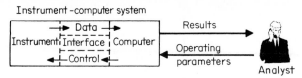

FIGURE 1.3 In-line or integrated computer configuration.

processing the data and outputting it in predefined formats. In these systems electrical and mechanical components are replaced by instructional statements (programs) that are stored in read-only memory (ROM). These programs are placed in the computer by the manufacturer and cannot be altered by the analyst. However, ROM chips can be replaced and the system upgraded. This permits changes in instrument control and data manipulation without the necessity of extensive and costly hardware modifications.

In-line computer systems are used with chromatographs (gas and liquid) and spectrophotometers where precise control of instrument parameters and the ability to perform repetitive analyses are required. Instrument accuracy is ensured by automatic instrument calibration coupled with built-in diagnostic testing. Communication between the instrument and other computer systems provides capabilities for the storage of large amounts of data, analysis of intermediate results, and coordination between different instruments.

1.1.4 Intraline

Several microcomputers distributed within a single instrument constitute subsystems that have the capability to change the nature of the measurement system (Fig. 1.4). These subsystems replace both hard-wired circuits and more general minicomputers. This improves the cost-performance characteristics of the entire system because maintenance costs are decreased. The analysis is more complete and reliable because the analyst is prompted to enter and check all variables. The accuracy of results is improved because the instrument is calibrated periodically. Built-in diagnostic tests check the functions of the instrument components. The precision of the results is improved through digital signal processing.

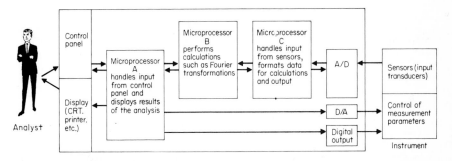

FIGURE 1.4 Intraline computer configuration. (*By permission from H. H. Willard, L. L. Merritt, Jr., J. A. Dean, and F. A. Settle, Jr., Instrumental Methods of Analysis, 7th ed., 1988. Courtesy of Wadsworth Publishing Company.*)

1.2 COMPUTER ARCHITECTURE—HARDWARE

The basic hardware configuration of every electronic digital computer comprises five standard hardware components which are connected by the internal signal pathways that make up the bus, as shown in Fig. 1.5. These units are the arithmetic-logic unit (ALU), the control unit, the input and output units (collectively denoted as I/O), and the memory. The heart of a computer is the central processing unit, which is made up of the control unit and the arithmetic-logic unit.

1.2.1 Arithmetic-Logic Unit (ALU)

The ALU performs the arithmetic and logic operations on data presented to it. Data are processed in the form of binary words, each word containing a specified number of binary bits. Arithmetic operations include addition and subtraction. Logic operations involve AND, OR, and shifting all the bits of a word to the left or right.

1.2.2 Control Unit

The control unit is responsible for coordinating the operation of the entire computer system. It generates and manages the operation of the computer by organizing and controlling the transfer of information between units and properly se-

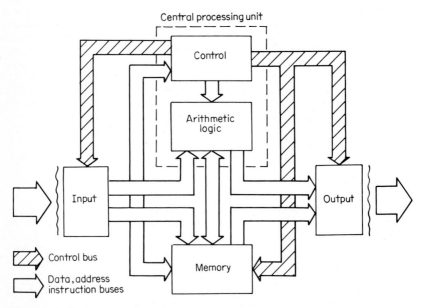

FIGURE 1.5 Basic computer organization. (*Taken by permission from H. H. Willard, L. L. Merritt, Jr., J. A. Dean, and F. A. Settle, Jr., Instrumental Methods of Analysis, 7th ed., 1988. Courtesy of Wadsworth Publishing Company.*)

quencing and executing the instruction or programs processed by the computer. The control unit manages the control signals necessary to synchronize the flow of data on all buses with the operation of the functional units. It also fetches, decodes, and executes successive instructions (a program) stored in the memory unit.

1.2.3 Central Processing Unit

The combination of the CU and ALU is known as the central processing unit (CPU). Critical parameters in evaluating CPU operation are the minimum time required to execute specific instructions and the number of bits in a computer word. These parameters determine the rates at which data can be acquired and processed.

1.2.4 Memory

All information required by the CPU is stored in memory. Memory stores two basic classes of information—data and instructions. Each instruction represents a specific operation to be performed by the computer. Instructions are retrieved individually from memory and placed into specific registers in the CPU where they are interpreted (decoded) and executed by the computer. A sequence of instructions which performs a specific task is called a *program*. Instructions also operate on data. When required, data are moved from memory into the ALU. Results from operations performed in that unit may again be stored in memory.

Performance considerations of memory are volatility, capacity, and access time or speed. Volatile memory loses its information when the power is removed; this includes most semiconductor or solid-state memories. Memory capacity is commonly expressed as the number of data storage locations within a computer. It is usually referred to in units of kilobytes (K), actually 1024 (2^{10}) bits. A bit is a binary digit and is the smallest unit of information that can be used by a computer; it can only have a value of 0 (off) or 1 (on). One byte equals one character of information. The number of bits in a given location matches the computer word length. Thus 1024 (1K) locations of 16-bit words contain twice as many bits as 1K of 8-bit words. Access to (primary) memory must be rapid because it determines the overall operating speed of the computer. It is usually measured in micro- (10^{-6}) or nanoseconds (10^{-9}) (abbreviated as μs and ns, respectively).

1.2.4.1 Primary Memory. Primary memory refers to that portion of memory that is directly accessed and addressed by the CPU. All instructions being executed as well as data being manipulated are stored in primary memory. The memory size will be limited for reasons of cost and addressability.

1.2.4.2 Storage (or Secondary Memory). Storage is accessed indirectly through a controller. Storage offers increased capacity but at the expense of access time. To store or retrieve data located in storage devices, the control unit passes addresses to the controller of the device. Examples of such devices are magnetic (hard) disks, floppy disks, punched cards, and magnetic tape.

1.2.4.3 Other Memory Designations. Random-access memory (RAM) is memory that is used for information storage and retrieval within the computer. RAM may be read or written by the control unit and its access time is independent of the position of the storage location in memory. ROM usually contains specific instructions for the machine it is in and is usually placed there by the manufacturer. The information in ROM can be transferred (read) only in one direction, namely from memory to the CPU; it cannot be altered by the user. A ROM that can be reprogrammed is called erasable programmable memory (EPROM). Special hardware is required to program ROM and EPROM.

1.2.4.4 Memory Technology. The variety of main memory technologies includes metal oxide semiconductor (MOS), complementary MOS (CMOS), ferrite core, bipolar, and charge-coupled devices.

1.2.5 Input/Output Units

Input and output (I/O) units provide the computer with the external links which enable information to enter and leave the computer system. I/O units interface directly with the arithmetic-logic unit or the main memory (Fig. 1.6). External sources of data include keyboards, remote terminals or memory devices, card readers, optical scanners, sensors, and instruments. Output units transfer data

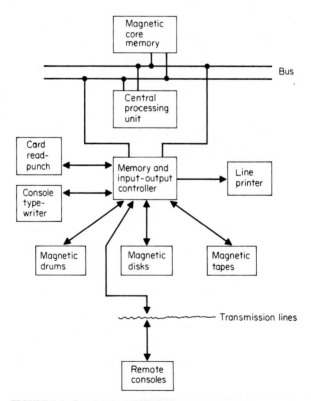

FIGURE 1.6   Input-output (I/O) devices.

from the ALU or internal memory to external devices such as printers, plotters, external memory devices, visual display devices such as light-emitting diodes and video terminals, and control devices such as relay switches and stepper motors.

The efficiency of the I/O units is largely determined by which of two techniques is used—programmed I/O or direct memory access. In programmed I/O the arithmetic-logic unit sequentially executes instructions whose function is to transfer data between an I/O unit and memory. During this time other execution is inhibited. With direct memory access the memory controller autonomously transfers data between the I/O unit(s) and memory while the arithmetic-logic unit continues to process other instructions.

1.2.6 Buses

The bus contains the paths that connect the components of the computer system. Address lines, data lines, and control lines link the memory to the CPU (Fig. 1.7). Address lines carry required binary information to specific parts of the memory. The location of a word in memory is called an *address*. Specific locations in memory are identified by binary data placed on the address lines by the control unit. Data lines carry information between memory locations and the CPU. Control lines direct the sequence of data transfers.

Internal buses link the components within the CPU whereas those joining the CPU to memory, peripheral I/O units, or other computers and external instruments are denoted as external buses. All buses transmit data in binary form.

Buses may be serial or parallel in nature. Parallel buses require less complicated interface hardware and allow high-speed communication. The size of the word in bits will determine the number of wires or leads in an internal parallel bus. For example, a 16-bit computer uses an internal memory bus made up of 16 lines. Separate address bus lines control the source and destination locations on the data bus. A 16-line address bus can identify up to 65 536 (2^{16}) locations. A control bus carries status, timing, and control information for the data bus.

The IEEE 488-1975 (IEEE is the acronym for Institute of Electrical and Electronics Engineers) system is a bit parallel I/O system which allows as many as 15 instruments to be connected to a single controller or computer. Connectors are stackable, which permits easy installation. This is a two-way system which both

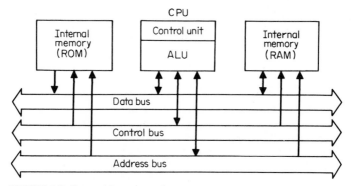

FIGURE 1.7 Internal bus connections.

"talks and listens." Of the 16 lines, 8 are used for data (information) and 8 are used for management and control. The cable length should not exceed 20 m (65 ft).

Serial buses require fewer lines, but they transfer data more slowly and require more complex interface hardware. Serial buses usually connect communication terminals and remote instruments to computers. Information can be sent across the laboratory or to distant points via telephone modems. Information can be further analyzed and formatted by using spread sheets and database software. When data are transferred serially, the data bits are validated sequentially. The advantage of sending data serially is that the connection between the two pieces of equipment requires only two wires. The RS232 interface, the serial interface most commonly found in personal computers, defines the electrical as well as the mechanical requirements of the system.

1.2.7 Universal Asynchronous Receiver-Transmitter

A single 40-pin chip provides a programmable digital communications center. The transmitter register accepts bit parallel data and produces bit serial output. The receiver register does just the opposite. These universal asynchronous receiver-transmitter chips are used to link instruments with output devices, controllers, or computers at remote locations.

1.3 INTERFACES

An interface system is used to connect a variety of I/O and memory devices to programmable and nonprogrammable instruments, computers, and peripherals as needed when building an instrumentation system (Fig. 1.8).

1.3.1 Digital-to-Analog (D/A) Converters

An analog output can be produced from a digital input through the use of a summing operational amplifier circuit. The operational amplifier sums the input currents and converts them to a scaled output voltage. Converters with 4 to 16 bits

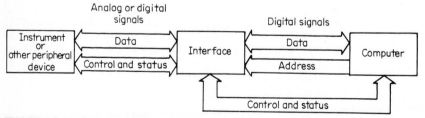

FIGURE 1.8 An instrument interface system.

are common. D/A converters are used to operate any device that requires an analog input signal and that is interfaced to a digital source such as a computer.

1.3.2 Analog-to-Digital (A/D) Converters

An analog signal must be digitized before it can be read and understood by the computer. This step is accomplished by an A/D converter whose sole job is to assign a set of binary digits that is indicative of the analog signal's intensity. At theoretical zero or baseline the registers are filled with 0s. When the signal intensity increases, the values in the registers increase accordingly.

The difference between an 8-bit and a 12-bit A/D converter is the number of registers which contain the associated number. At theoretical zero the 8-bit unit has 8 zeros whereas the 12-bit unit has 12.

Resolution is determined by the number of bits in the digital signal. A 12-bit converter can break up a 0- to 1-V signal into 4096 parts (2^{12}) and detect signals as small as 0.000 24 V. By contrast, an 8-bit converter can only break up a signal into 256 parts. Something like a pH meter does not require anything more than an 8-bit A/D converter, and a spectrophotometer can operate nicely with 12 bits. However, some gas chromatographs have greater requirements for resolution and dynamic range and need a 16-bit A/D converter.

Another characteristic of the A/D converter is its data acquisition rate. This is the number of times per second that the signal must be digitized in order to maintain integrity. This factor is usually controlled by the software program. The implication is that the computer should be able to generate a sufficient number of data points per second so that the raw signal or waveform can be reproduced sometime in the future. Data acquisition rates vary from one methodology to another. Nuclear magnetic resonance may require rates as high as 1000 to 2000 times per second, whereas chromatographic methods may only need rates of 1 to 40 times per second.

1.3.3 Voltage-to-Frequency Converters

Voltage-to-frequency converters transform an analog signal to a train of digital pulses at a rate that is proportional to the input voltage. The train is then counted over a fixed time interval to create a digital value. Voltage-to-frequency conversion offers excellent resolution and minimizes noise when data are transmitted over some distance but requires a conversion time of about 0.4 s. Speed is traded for the ability to represent accurately low-level input signals.

1.3.4 Integrating (Counter) Converters

Integrating converters average the analog input over a fixed period of time and count out the average as a digital value. The size of the count is proportional to the magnitude of the analog input signal. The conversion time is directly proportional to the magnitude of the input signal, a limiting factor in most applications involving computer interfacing.

1.3.5 Successive-Approximation Converter

A successive-approximation converter uses a series of logical guesses (approximations) to determine the digital equivalent of the analog input. It offers high resolution and fast conversion times which are independent of the magnitude of the analog input signal. The converter requires N approximations for an N-bit conversion.

1.4 SOFTWARE

Selection of hardware should be one of the last steps in the purchase of any computer system. The software provides instructions for the hardware, and without the software, the most impressive hardware specifications are meaningless. One should first choose the software, and then select the computer that is compatible with software.

Data acquisition and control applications require computers to respond to events in the real world. These external events intrinsically set the parameters of the response required of the computer and are independent of the computer's optimal mode of operation. These applications make some unique demands on a computer system, particularly when the system is operating near its maximum capacity.

Designers and software programmers must balance a number of contributing factors. The increasing availability of a wide range of integrated data acquisition hardware from a variety of sources has eased the task of systems hardware design. Unfortunately, the selection of available software lags behind. Sometimes a user must generate the software required for a particular application.

The user communicates with the digital computer through a sequence of instructions (a program) written in a computer language. Programs are referred to as *software* in contrast to the hardware or physical components of the computer. All computations done by the computer must be predefined by a series of logical steps (algorithms). The lowest level of programming is in binary code and is machine-oriented. Higher-level languages are machine-independent.

1.4.1 Machine-Language Programs

Computers execute binary-coded instructions stored in the primary memory. Machine-language programs are sequences of these binary instructions which take the form—operation/address. The operation is a function to be performed by the computer; the address is the location of the data to be manipulated. Writing machine-language programs is a tedious, time-consuming task which should be avoided whenever possible. Programs written in higher-level languages must eventually be converted into machine-language instructions (referred to as object programs).

1.4.2 Assembly-Language Programs

Next in the hierarchy of programming languages is assembly language. Each machine instruction is represented by a set of mnemonics. The instruction "add" in assembly language replaces the machine binary code of 10000110. Assembly-language programs are easier to write yet are comparable in execution and efficiency to machine-language programs.

A program known as an assembler translates symbolic assembly-language instructions into machine-language programs. When resident in the main memory, a program written in assembly language can be written, translated into object instruction, and executed on a single computer. Assemblers are machine-dependent so that instructions defined for one type of computer cannot be executed by another type of computer.

1.4.3 High-Level Languages

The high-level languages are convenient and user-oriented. They include Basic, Cobol, Fortran, Algol, PL/l, Pascal, and C. These languages are either algebraic or English in nature. Each line of the source program generates many lines of object code. High-level language programs are machine-independent so a source-code program can run on different computers provided a translator program (to translate the code into machine language) is available for each brand of computer.

Structured programming is a technique often used with high-level languages. Algorithms are systematically formulated utilizing a top-down approach in which individual tasks at each level of complexity are broken down into a series of commands called subroutines. These subroutines can be written and tested as separate units. Structured programs consist of a hierarchy of interacting subroutines. Commands are flexible and easily understood by the operator.

1.4.4 Translation Programs

Translation programs, which are resident in main memory, may be either interpreters or compilers. Both types of translators produce machine object code; however, they differ in their manner of translation. Translators also detect syntax errors and certain kinds of run-time errors. Debugging software augments this function.

Interpreters translate source programs line by line each time the program is run and execute each series of machine instructions before translating the next line of code. Thus, this type of program must be interpreted and executed by the same computer. Where speed of data manipulation is important, the use of an interpretive translator may be inappropriate. However, interpreters do allow maximum interaction between the user and the computer by using caution and error messages during program execution.

The compiler translates the entire high-level program into object code which is not executed until translation is complete. Compiled programs may be compiled on one computer and executed on another. When a source program is converted into object code by a compiler, it can be stored and recalled quickly for execution. Compiled programs are usually executed faster than interpreted ones.

1.5 SOFTWARE CONTROL OF THE COMPUTER-INSTRUMENT INTERFACE

Interactions between the central processor and peripheral devices are controlled in one of three ways: scanning (programmed I/O), interrupt, and direct memory access.

1.5.1 Scanning

Scanning is carried out by the central processor, which periodically surveys a device (or each of several devices) to ascertain whether it requires service. A status register is maintained by each instrument to indicate its need for service and the specific nature of the request. This is the register which is interrogated by the microprocessor during interface operations. A positive response causes the processor to prompt the instrument for the specific nature of the request, which the processor then satisfies. If no service is required, the processor either scans another device or returns to rescan the first device at a later time. Scanning techniques are simple to implement both in hardware and software. However, they place a moderate amount of overhead on the central processor and should be used only when processor time is available or when a large number of positive service requests are expected.

1.5.2 Interrupt

An interrupt mechanism is used when either the response time for the processor to service the peripheral device is critical or when too much processing time would be consumed by scanning. A special control lead is run over the control bus between the device(s) and the central processor unit. When this lead is activated, the CPU immediately interrupts its current task and begins to service the request. The CPU must first identify the interrupting peripheral. Then it determines the nature of the service request via the status register, and finally fulfills the request. This type of operation requires more complex hardware and software. Since the CPU is affected only when an interrupt occurs, this approach is more efficient when there are infrequent requests for service. Several tasks can be handled simultaneously (multitasking ability). Operations can be scheduled for definite time periods.

1.5.3 Direct Memory Access

With a direct memory access controller, a peripheral device can autonomously have read/write access to the microprocessor's main memory without having to interact with the central processor. A high degree of efficiency is achieved with respect to the simultaneous utilization of the instrument and the microprocessor. This type of controller is appropriate when a large amount of data must be exchanged between the CPU and the peripheral devices or a great deal of processing is expected from the microprocessor.

1.5.4 Operating Systems and ROM

An operating system (Fig. 1.9) is a type of software that directs such tasks as sequencing of jobs and controlling access to I/O devices. These systems control and direct the overall operation of the computer and are usually purchased directly from the vendor. Common operating systems include OS, DOS on large mainframe computers, and PC/DOS on microcomputers.

Read-only memory is used to control the computer-instrument interface and is not intended to be written or modified by a user. Applications include microcode implementation of the basic machine-language instruction sets in large systems or the entire program for dedicated or specialized microprocessor systems where

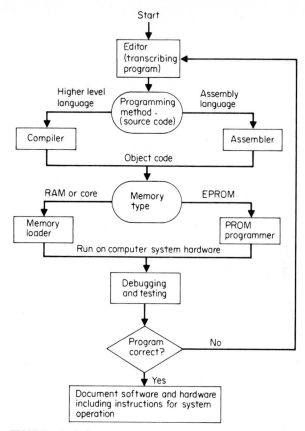

FIGURE 1.9   Software design.

user modification is not desired. Instrument vendors often place programs in ROM.

1.5.5   Debuggers and Editors

In program development, errors are detected and corrected during execution by debugging software. Debuggers can stop program execution at designated instruction points to display or change the contents of memory. These programs can also generate diagnostic error messages while the program is running.

An editor allocates a portion of memory as a scratch pad where program instructions can be written, deleted, or changed.

1.5.6   Analog-to-Digital Interface

In its simplest form, the A/D interface has two major functioning sections—a data conversion subsystem (A/D converter) and a computer-bus interface section.

These functional groups may be supported by local controllers, data buffer memory, or auxiliary communication channels.

1.5.7 Analog I/O Peripheral

In the form most familiar to system designers, the analog I/O peripheral is a self-contained system, which is both physically and electrically compatible with a particular microcomputer's bus structure. These systems (boards or cards) are generally housed in the microcomputer's card cage and are powered by the computer's power supply.

The increased use of analog I/O peripherals has been of particular benefit to systems designers. The broad selection of analog interface capabilities available for the most widely used microprocessors allows them to design multiple systems with differing requirements around a single microcomputer. This permits transfer of previously written application's software from one design to the next and results in large savings in development costs and time.

1.6 DATA REPRESENTATION

The natural mode of operation of all computers is base 2 or binary arithmetic. All digital computers store information in the form of binary digits or bits. Each bit can take the value 0 or 1. Individual bits are further grouped into octets called "bytes." These bytes are further aggregated into words which may consist of 1, 2, or 4 bytes each and correspond to 8-bit, 16-bit, or 32-bit machines. Each byte can be used to store a value ranging from 0 to 255.

Commonly a byte is subdivided into two 4-bit groups. Each can store a binary-coded decimal (BCD) digit. As shown in Table 1.1, the 4 bits have 16 potential values and can be used to store numerals from 0 to 15. Also shown in the table is the hexadecimal notation which includes the letters A through F. Typical digital instrumentation uses the BCD form.

Another common form of data representation is known as ASCII (American National Standard Code for Information Interchange). In this system 7 bits of the full byte, or 128 possible values, can be used to represent a complete alphanumeric set of characters including 52 capital and lowercase letters, 10 digits, and 22 special symbols. Computer manuals provide the user with a complete listing of the numerical codes for specific characters.

1.6.1 Data Acquisition Rate

Acquiring too much data is wasteful in terms of storage space, particularly if there is no gain in sensitivity. If the analytical requirement is 10 data points per second, acquiring 40 data points per second will not increase the sensitivity of the device. The latter is limited by the detection system. However, you are decreasing your available storage three-fourths. One approach to this problem is to find software which allows the operator to set the data acquisition rate. Alternatively one can acquire data at a higher rate and then discard points until the integrity threshold is discovered empirically for the particular technique.

The acquisition rate is given by the expression

TABLE 1.1   Binary Number Equivalents

Binary	Hexadecimal	Decimal
00000	00	0
00001	01	1
00010	02	2
00011	03	3
00100	04	4
00101	05	5
00110	06	6
00111	07	7
01000	08	8
01001	09	9
01010	0A	10
01011	0B	11
01100	0C	12
01101	0D	13
01110	0E	14
01111	0F	15
10000	10	16
10001	11	17
10010	12	18
...	...	...
01001	19	25
11010	1A	26
...	...	...
11111	1F	31

$$\text{Frequency (in Hz)} \times \text{time (in s)} = \text{RAM}$$

where RAM is the random-access memory. Assuming that an 8-bit microprocessor with 64K RAM and an 8-bit A/D converter is being used, the amount of RAM limits the amount of data that can be acquired. For example, for how long can data be acquired continuously at 10 Hz?

$$10t = 64\ 000$$

and
$$t = 6400 \text{ s} \qquad \text{(or 17.78 h)}$$

which is a fairly long run. Of course, driving your personal computer to the limit in this way is not required or advocated. Once data acquisition has been accomplished, the data can be "saved" (i.e., stored on disk), and RAM can then be emptied of its contents and used for a new data set. At this point the previously acquired data are in the form of coded messages on the disk, and RAM is empty so it can acquire data for the next operation.

1.6.2   Data Acquisition in the Laboratory

A typical data acquisition system is shown in Fig. 1.10. Many data acquisitions require the transmission of very low level signals, such as outputs from thermocouples, strain gauges, or instruments with a strip-chart recorder. A preamplifier

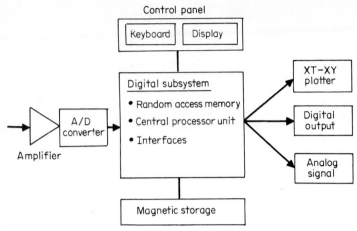

FIGURE 1.10   A data acquisition system.

prevents normal electrical noise from masking the information contained in the signal. After amplification, signals can be sent over a longer distance with less expensive cable.

In analog amplification the input signal is amplified and scaled from 0 to 10 V. The amplifier bandwidth is at least 50 kHz and, for special applications, may go as high as 1 MHz.

Continuous data acquisition and writing to disk is performed using a real-time operating system. This is a simple menu-driven program which acquires data at a rate specified by the operator and then continuously writes the data to the disk. In the interactive mode, such a program continuously displays the data in graphical form on the system console. In batch mode, the program proceeds with the acquisition of data and leaves the system console free for editing, data analysis, and other tasks.

1.6.3 Input Devices

A standard keyboard is the basic input device for most computers. The keyboard resembles that of most typewriters, with the addition of a keypad that provides an alternative means of inputting numbers and other data.

Most terminals and personal computers now include an additional set of keys known as function keys. They can be programmed, either by software or by the user, to perform a function or series of steps that would normally take many keystrokes.

Another input device is the "mouse," a small device which rolls on a flat desktop and is connected to the computer by a cable. The movement of the mouse controls the movement of the cursor on the screen. Buttons on the mouse work like function keys.

1.6.4 Recorders

High-performance strip chart recorders are designed to capture and display high-speed analog phenomena. The converted analog signal is transferred to the electronic data storage, which is continually being updated. On initiation of a predetermined trigger (selected by the operator), bits of data are retained in memory.

Electronic recorders accept and amplify analog electric signals, digitize them, store them temporarily or permanently, and plot them either on-line or after processing in signal-time (Y-T) or X-Y format. These recorders are used to record and process a wide variety of transducer and instrument signals resulting from chemical analysis, physical and mechanical testing, and materials characterization; they are also used in many areas of engineering.

The recorder is the front end (analog-to-digital) of the laboratory computer and the hard-copy plotter for data processed by the computer. All data are stored in digital form and can be transferred at any time. Recorder curves can be processed when desired. Scale expansion, normalization, inversion, and smoothing are standard operations. Curves can be overlaid. Any portion of a curve can be magnified or several curves can be compressed to fit on a single page.

1.6.5   Storage Disks and Tapes

The disk device is the primary device used in personal computers (PC) and mainframe computers for reading, writing, and storing programs and data. Disk drives store data magnetically. They are quickly accessed, and old data can be overwritten when no longer needed. The disk is read by an electromagnetic head.

In PC systems, hard disks have data capacities of 10 to 100 Mbytes. Floppy disks (or diskettes) are designed to be moved in and out of their drives but have less storage space (360K to 1.4 Mbytes). They are the primary means of moving information from one system to another, and most vendors supply their software in this form.

Magnetic tapes, either cassettes or larger reel-to-reel tapes, are used for long-term storage on many mainframe systems. Access time may be very long, and for this reason, fixed disks are preferred for day-to-day use and tapes are reserved for archiving purposes.

1.6.6   Printers

The final output of a computer is usually generated on paper by a printer or by a plotter. There are three main types of printers: the daisy-wheel printer, the dot-matrix printer, and the laser printer.

In the daisy-wheel printer a hammer hits the end of a particular spoke (which has a raised character on its end) onto a ribbon, which is between the paper and the spoke, thus imprinting a character. Dot-matrix printers use patterns of tiny dots in a matrix array to form each individual character. The quality of the printing depends largely on the number of pins in the print head array. The pins hit a ribbon, thus transferring the pattern to the paper. Laser printers work with a computer input to determine which characters are to be printed. The actual printing process is similar to that of a copy machine. High-quality graphics can be printed along with text.

1.7   AN AUTOMATED LABORATORY

A microprocessor is a single integrated circuit capable of providing centralized control and data manipulation for a number of attached devices. Its heart is a tiny computer. Add a set of preprogrammed instructions (a read-only memory) and some form of input and output for communication with the outside world, and a fully functional microcomputer is created. The few electronic components which make up the microcomputer may replace hundreds of discrete components which would be required to do the same control and computational tasks.

In an automated system, dedicated microprocessors and software operate interactively with interim data from a prescribed series of analytical measurements. Throughout these interactive operations the instrument automatically and iteratively adjusts key parameters as the series of measurements continues. The instrument learns from experimental results. All the analytical power of a totally automated system can be accessed through a few push buttons on a keyboard.

1.7.1 Interactive Dialogue

The key to the operational ease of a computer-aided laboratory system is interactive dialogue. Sophisticated programming operates unobtrusively. The operator makes only simple parameter-value decisions. Management of these system functions is activated through the keypad, while the current system status is displayed on the screen (CRT).

In interactive dialogue the operator is asked simple questions in the English language. The computer asks what the user wants to measure, and he or she replies by choosing an analytical program stored in memory. Operations and parameters are selected from displayed choices. Each step is logical and self-explanatory. In this way the user is guided through the setup of acquisition and analysis procedures. As each displayed question is answered, the next logical question appears on the screen until all necessary information has been stored in the system. The computer automatically establishes analytical conditions and selects appropriate algorithms to achieve the specific analytical result as the operator refines the data. Only minimal operator training is required. Even occasional users will find that this type of computer system leads them through procedures and reminds them of analysis details which would have to be memorized on most other systems.

In its interactive mode, the software package allows direct modification of all displayed intermediate results. The noninteractive mode automates all phases of data acquisition in batch mode and analysis for both qualitative and quantitative work. At any time, raw data, as well as intermediate and final results, can be accessed, and all data are retained for further use at a later date, perhaps with alternative mathematical techniques.

Such flexibility renders many manual techniques inappropriate. For example, peak location using the second-derivative method resolves overlaps missed by manual methods and provides the sensitivity needed to identify minor components. The highly accurate convolution method improves second-derivative estimates for even more precise peak location.

1.7.2 Advantages of Automation

The use of a microcomputer with an analytical instrument, such as a spectrophotometer, creates a versatile system which can be configured to meet the user's particular needs and resources. Also, automation of equipment allows more effective use of personnel time and capabilities, and it gives the laboratory the ability to handle larger numbers of samples without errors in data recording and calculation. The advantages of automating individual instruments become obvious as we look at the potential functions of the computer.

1. *Automatic instrument setup and control:* The software automatically programs the instrument with the parameters for the particular test being run. This ensures that the instrument parameters have been set correctly each time for the particular experiment. Operator prompting decreases the possibility of operator error and also decreases the learning time required for a new user. Once the instrument is programmed, the computer will control the instrument until the experiment is over or has been terminated.

2. *Easier, faster, and more accurate data collection:* Computers record data far more quickly and accurately than a person can. In the case of continuous data, operator judgment in setting up a chart recorder and supporting equipment is eliminated. Analog-to-digital converters provide greater resolution and accuracy. Automated collection of individual readings eliminates difficulties arising from manual data entry such as operator transcription error, time constraints, and the boredom of mundane redundant tasks.

3. *Real-time data presentation:* During data acquisition the data can be displayed to allow the user to view the data immediately.

4. *Faster analysis:* Once the data have been obtained, they can be rapidly manipulated. This is particularly important when there is a complex mathematical relationship between the raw data and the final answer. The power of a computer allows results that require extensive calculations to be generated in a few moments by means of algorithms.

5. *More uniform test procedures:* The use of a computer helps to ensure the reproducibility of data and results. System monitoring with problem diagnosis can decrease downtime and warn the user of erroneous data. Statistical evaluation of data can be part of the analytical results.

6. *New data reduction capabilities:* Analyses that were previously unreasonable because of time restraints and limited data collection capability now become feasible.

7. *Ability to archive raw data:* Floppy disks (and hard disks) allow raw data and results to be stored for long periods of time.

8. *Ability to reanalyze old data:* If results are questioned or additional calculations need to be performed, the raw data are always available.

9. *Rapid comparison of results:* Tables and graphs can be quickly and easily prepared. The format can be prepared for the final report.

1.7.3 Design Criteria for a Laboratory Data System

Software taps the potential of the computer by providing the instructions necessary for actual use. Software is useless unless the computer is reliable and powerful, has the appropriate peripherals required to interface with laboratory instruments, and can understand and act on the instructions provided. Computer software allows more complex systems to be analyzed. It also provides functions such as interelement and background correction, and data fitting, stripping, and superimposition. Often routine determinations can be accomplished by non-specialized personnel.

For example, the program instructs the computer to save and record parameter settings that tune the spectrophotometer to specific wavelengths, adjust the slit settings, and select fuel and oxidant flows in addition to accumulating data on blanks, standards, and samples. Finally, it will compute a calibration curve, determine the prescribed analyte content of serial samples, and compute the results along with statistical information in the appropriate format.

1.7.4   Decision-Making Processes in Acquisition of Computerized Laboratory Instrumentation

The first and most important questions are:

1. Does the system perform the desired task?
2. Is appropriate software available?
3. Are analytical requirements uncertain?
4. Are a number of different analytical techniques involved?
5. Is future expansion possible?
6. Is the computer to be linked to another station?
7. Are limited resources a constraint?

These questions imply that there are instrumentation firms which are interested in offering hardware, but the important thing is the software. Software should be "user friendly"; that is, it should be easy to use and yet powerful enough to handle the particular task and flexible enough to accommodate new techniques. Today, interactive dialogue is the easiest way to set up and perform tasks. The computer poses the queries as a series of "help" screens or menus, the operator responds, and the computer proceeds with the analysis. All responses are saved for possible future recall and modification.

A carefully chosen system can reduce sample costs and enhance productivity by allowing more samples to be processed in a given time. Two major considerations are (1) the cost of the computer system, which includes the hardware, software, and employee training, and (2) selection of the best system. These decisions must be made by qualified personnel who have expertise in the field. The initial cost of the computer system is often deceptive because there are usually hidden charges for new software, maintenance, and loss of employee productivity when the system is first put in operation.

There are companies that specialize in the evaluation, design, and implementation of computer-assisted laboratories. Their professional advice may not only save money, headaches, and associated problems but will provide exactly what is needed for the best price. A computer system is worthwhile only if laboratory personnel are willing to use it. Since most potential users are not interested in becoming computer experts, but in securing help with their problems, a computer system that is designed to be easy to use and that is adapted to the specific problems faced by the operator will be accepted most easily.

1.7.5   Modular Computer Systems

Modular computer systems provide a wide range of capabilities for people working in laboratory environments. They offer maximum flexibility for current requirements along with expandability for future needs. Modular design permits the mixing and matching of various features to meet specific requirements, that is, tailoring a system.

The laboratory workstation provides a flexible solution to the interface and data processing needs of the research laboratory. A workstation may be interfaced to instruments which can be operated via an interactive dialogue with the operator. Workstations may be used independently or connected to other workstations via a local area network. Interconnections greatly enhance the power and capabilities of the workstation by providing extensive resource sharing and communication capabilities. For example, the network allows a common storage device (fixed disk) to be shared, thereby eliminating the need for every workstation to have a disk.

1.7.6 Dedicated Instruments

In dedicated instruments the physical configuration of sample handling; data entry, display, and printing; or recording modules is integrally designed and optimized for the particular purpose of the equipment. "Dedicated" implies that the equipment will be used for one specific purpose. Consequently, the simplicity and efficiency of the equipment's operation can be maximized. However, such systems are limited to entering working parameters and procedural choices within the scope of the procedural method that they were designed to handle.

The instrumentation features a dedicated microprocessor and a variety of I/O and memory devices that are directly interfaced with the laboratory instrument(s) and possibly linked to an auxiliary computer system (Fig. 1.11). With the advent of small, powerful chips, microprocessors may now incorporate enormous computing and data-manipulating power within an individual analytical instrument. Consequently, the analytical community has moved toward using instrumentation that has powerful self-contained microprocessors and using highly customized and dedicated software. This frees the analytical chemist from the inflexibility experienced with shared computer systems. The user has push-button command of sophisticated analytical routines at low cost. This cannot but help experimentation. Increased accuracy of data and vastly reduced drudgery of routine operations have been achieved without sacrificing user control of, and interaction with, the experiment.

Replacement of obsolete instruments can be costly. However, when modular hardware and versatile software are properly designed, the chemist has the option to upgrade rather than replace instruments.

1.7.7 The True Test of a Laboratory Automation System

The true test of a laboratory automation system is a complete database management system which organizes storage and retrieval of data for all samples logged onto the system. Data retrieval may be based on diverse criteria. Once the data have been retrieved, they can be formatted and printed as a report or plotted in various ways.

The most frequently requested database retrievals and report formats can be stored as templates in the database itself. For immediate implementation on entering the procedural name, this execution can be initiated manually or can be programmed to occur automatically at a specific time. An automatic report feature is particularly useful when reporting loads are heavy and overnight turn-

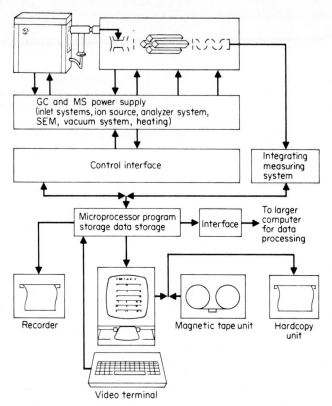

FIGURE 1.11 A dedicated instrument system for gas chromatography–mass spectrometry.

around is satisfactory. One use for this kind of system is the preparation of periodic status reports.

The purchase and implementation of a laboratory automation system requires a large investment, which should begin to show the promised returns as quickly as possible. The critical factors in a successful installation and rapid full-scale implementation are thorough planning and detailed specifications of what the system contains and how it is expected to perform. Although every situation is different, a project implementation team should specify the exact requirements in regard to the instruments to be interfaced, the initial positioning of equipment, environmental factors, cabling plans, system training, and system documentation. The initial software configuration is finalized after defining these items: (1) calculations required, (2) method and/or test review protocols, (3) security structure, (4) final report formats, (5) sample and test status reports, and (6) special reports required. This same team should be responsible for installation of software and on-site training of the staff.

1.7.8 Artificial Intelligence

From the user's viewpoint, intelligence, or lack thereof, is judged by analogy to human intelligence. Instrument intelligence is a combination of functionalization and its attributes, along with the user's perception thereof. A pH meter which cannot detect that it is malfunctioning is less smart than one which can. Thus, the number of functions an instrument can handle forms one basis for instrumental intelligence. A really smart pH meter not only measures pH and millivolts but also measures temperature, evaluates its own health and that of its electrodes, and performs self-calibration.

A solvent-mixing system which allows only a few predefined mixing programs is not as smart as one which permits the user to specify complex mixing programs. Similarly, a gas flow controller which can be dynamically programmed is smarter than one which can only control flow at some preset value.

The quality of the function has not yet been addressed. Let us consider data-handling devices which perform integration. An integrator which can only handle discrete components is less intelligent than one which can handle overlapped components. Similarly a system which can integrate a complex system with an error of less than one percent for the smallest component is smarter than one which can integrate the same system but produces an error of 10 percent relative to the smallest component.

An instrument which can only be programmed using a marked card system is adjudged less intelligent than one which permits keyboard entry. Instruments which permit user definition of command labels and can accept wordlike commands are smarter still. At the highest level of user-interaction, the term "intelligence" could be used to describe systems which allow conversational control and provide vocal feedback as an adjunct to their function.

Thus the intelligence of a given instrument should be judged by how many of the possible functional areas it handles and by how well it performs each function. Moreover, the intelligibility of the user interface is a major factor in assessing instrument intelligence. However, the quality of the results provided by the instrument is more important than its intelligence. During any evaluation of an instrument, the primary focus of the evaluator should be on result accuracy. Only then should intelligence factors be considered.

1.7.9 Personal Computers

A personal computer *add-on* can provide valuable assistance and speed projects because of its programmable microprocessor capabilities (Fig. 1.12). It can also function as an intelligent terminal. Fitted with inexpensive peripherals, it can acquire and process laboratory data quickly and efficiently. These combinations bring improved accuracy and convenience to data-gathering tasks that might normally be performed by multichannel chart recorders. An add-on I/O board can serve as an alternative to a mainframe control.

It is advantageous to locate a personal computer near a sensor, which eliminates the need to transmit sensor output to a central computer, which in turn reduces noise.

There are limitations to the use of personal computers in the laboratory. Si-

FIGURE 1.12   Chromatographic analysis using a personal computer.   (*Courtesy of Interactive Microware.*)

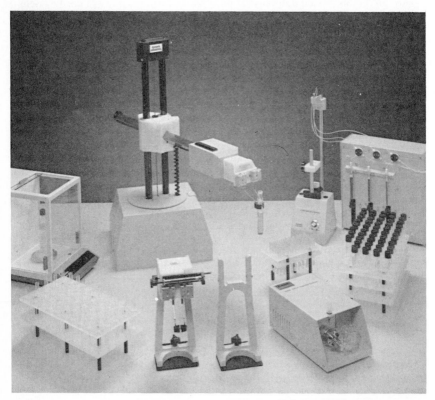

FIGURE 1.13   A computer-controlled laboratory robotic system.   (*Courtesy of Zymark Corporation.*)

multaneous data acquisition and writing to disk is outside their capabilities because personal computers do not have real-time multitasking capabilities.

1.7.10 Robots

A laboratory automation system uses robots to perform automatically existing manual laboratory procedures such as sampling. A sample preparation system might consist of a robotic arm and laboratory stations such as balances, dispensers, mixers, and centrifuges (Fig. 1.13). Samples are moved from one station to another according to user-defined procedures. Upon completion of the preliminary procedures, the robot places the sample directly into an analytical instrument or into a rack for subsequent serial analysis. Robots can be programmed for multiple procedures and, of course, programmed for new or revised procedures. This flexibility allows robots to be used in a variety of applications, including those in hazardous environments.

Laboratory procedures are broken down into common tasks; the sequence of operations is specified; and each task is defined. The robots emulate the manual procedure. The accompanying software uses common laboratory terms. The operator programs the system by teaching the robot the locations and laboratory stations using specific names. A program that combines these locations in the proper sequence is thus created. The result is increased productivity, faster sample turnaround, multimethod capability, and greater convenience and safety.

BIBLIOGRAPHY

Barker, P., *Computers in Analytical Chemistry,* Pergamon, Elmsford, N.Y., 1983.

Carr, J., *Elements of Microcomputer Interfacing,* Reston, Reston, Va., 1984.

Cofforn, J., and W. Long, *Practical Interfacing for Microcomputer Systems,* Prentice-Hall, Englewood Cliffs, N.J., 1983.

Dessy, R., ed., *The Electronic Laboratory: Tutorials and Case Histories,* American Chemical Society, Washington, D.C., 1985.

Ratzlaff, K., *Computer-Assisted Experimentation,* Wiley, New York, 1987.

CHAPTER 2

CHROMATOGRAPHY—GENERAL PRINCIPLES

Chromatography is a technique for separating a sample into its constituent components and then measuring or identifying the components in some way. The components to be separated are distributed between two mutually immiscible phases. The heart of any chromatograph is the stationary phase, which is sometimes a solid but is most commonly a liquid. The stationary phase is attached to a support, a solid inert material. The sample, often in vapor form or dissolved in a solvent, is moved across or through the stationary phase. It is pushed along by a liquid or a gas—the mobile phase. As the mobile phase moves through the stationary phase, the sample components undergo a large number of exchanges (partitions) between the two phases. The differences in the chemical and physical properties of the components in the sample are used to bring about the separation and govern the rate of movement (called migration) of the individual components. When a sample component emerges from the end of a chromatograph, it is said to have been eluted. Ideally, components emerge from the system as gaussian-shaped peaks and in the order of their increasing interaction with the stationary phase. Separation is obtained when one component is retarded sufficiently to prevent overlap with the peak of an adjacent neighbor.

2.1 CLASSIFICATION OF CHROMATOGRAPHIC METHODS

The mobile phase can be a gas or a liquid, whereas the stationary phase can only be a liquid or a solid. When the stationary phase is contained in a column, the term *column chromatography* applies. The stationary phase can also occupy a plane surface, such as filter paper. This is called *planar chromatography* and includes thin-layer and paper chromatography and electrophoresis.

Column chromatography can be subdivided into *gas chromatography* (GC) and *liquid chromatography* (LC) to reflect the physical state of the mobile phase (Fig. 2.1). If the sample passing through the chromatograph is in the form of a gas, the analytical technique is known as gas chromatography. Gas chromatography comprises gas-liquid chromatography (GLC) and gas-solid chromatography (GSC), names which denote the nature of the stationary phase.

Liquid column chromatography embraces several distinct types of interactions between the liquid mobile phase and the various stationary phases. When the

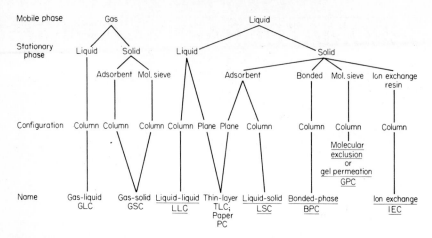

FIGURE 2.1   Classification of chromatographic systems. Underlined techniques are included in high performance liquid chromatography (HPLC).

separation involves predominantly a simple partition between two immiscible liquid phases, one stationary and one mobile, the process is called *liquid-liquid chromatography* (LLC). In liquid-solid chromatography, or *adsorption chromatography* (LSC), physical surface forces are mainly involved in the retentive ability of the stationary phase. Ionic or charged species are separated by selective exchange with counterions of the stationary phase; this may be accomplished through *ion-exchange chromatography* (IEC) or *ion-pair chromatography*. In columns filled with porous polymers, components may be separated by *exclusion chromatography* (EC) or *gel-permeation chromatography* (GPC). Separation is based largely on molecular size and geometry.

2.2 TERMS AND RELATIONSHIPS IN CHROMATOGRAPHY

We present the parameters and equations used in chromatography. The chromatographic behavior of a solute can be described by its retention volume V_R (or the corresponding retention time t_R) and the partition ratio (or capacity ratio) k'.

2.2.1 Partition Coefficient

The partition coefficient K is given by the ratio of the solute concentration in the stationary (liquid) phase to that in the mobile (gas or liquid) phase:

$$K = \frac{\text{concentration of solute in mobile phase}}{\text{concentration of solute in stationary phase}} \tag{2.1}$$

It is a thermodynamic quantity which depends on the temperature and on the change in the standard free energy of the solute when it goes from the mobile to the stationary phase.

2.2.2 Retention Time

The time required by the mobile phase to convey a solute from the point of injection onto the stationary phase, through the stationary phase, and to the detector (to the apex of the solute peak in Fig. 2.2) is defined as the retention time. The retention volume is the retention time multiplied by the volumetric flow rate F_c discussed later.

$$V_R = t_R F_c \qquad (2.2)$$

2.2.3 Nonretained Solute Retention Time

The quantity t_M or t_o (see Fig. 2.2) is the transit time of a nonretained solute through the column. It represents the time for the average mobile-phase molecule to pass through the stationary phase, that is, traverse the column or planar phase. When converted to volume V_M, it represents the void volume or holdup volume of the column. The column dead time is often recognized as the first disturbance in the baseline. In gas chromatography a solvent peak is common at t_M, but in HPLC separations there is often no baseline disturbance at t_M.

In practice, t_M is obtained by injecting a solute with $k' = 0$ (that is, all the solute remains in the mobile phase, and none partitions into the stationary phase). For a gas chromatograph that uses a thermal conductivity detector, air can be injected to obtain t_M. For other GC detectors, the peak of a solute whose boiling point is 90 or more degrees below the column temperature gives an estimate of t_M.

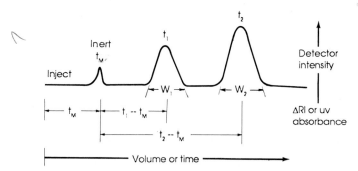

FIGURE 2.2 Chromatogram illustrating retention times and bandwidths W of a nonretained solute t_M and two retained materials 1 and 2.

2.2.4 Adjusted Retention Time

The adjusted retention time t'_R or volume V'_R is given by:

$$t'_R = t_R - t_M \quad \text{or} \quad V'_R = V_R - V_M \tag{2.3}$$

2.2.5 Volumetric Flow Rate

The volumetric flow rate, in terms of the column parameters, is as follows:

$$F_c = \frac{\pi d^2}{4} \epsilon \frac{L}{t_M} \tag{2.4}$$

where d = inner diameter of column
$\quad L$ = column length
$\quad \epsilon$ = total porosity of column packing

For solid packings the total porosity is 0.35 to 0.45, whereas for porous packings it is 0.70 to 0.90. In capillary columns the total porosity is unity.

From Eq. 2.4, the proper flow rate for columns of differing diameters can be approximated by assuming that the packing densities of the two columns are the same. If so, then

$$(F_c)_2 = (F_c)_1 \frac{d_2}{d_1} \tag{2.5}$$

For example, an analysis has been performed on a 4.6-mm-inside-diameter (i.d.) column at 2 mL $\cdot$ min^{-1}. If the same linear velocity is desired for a 9.4-mm-i.d. preparative column, the appropriate flow rate is $2 \times (9.4/4.6)^2$ or 8.4 mL $\cdot$ min^{-1}. Lengthening the column proportionally increases efficiency and analysis time, but does not affect the flow rate.

2.2.6 Velocity of the Mobile Phase

The average linear velocity u of the mobile phase

$$u = \frac{L}{t_M} \tag{2.6}$$

is measured by the transit time of a nonretained solute through the column.

EXAMPLE 2.1   The linear velocity was 43 cm $\cdot$ s^{-1} through a 15-m column. What is the value of t_M?

$$t_M = \frac{L}{u} = \frac{1500 \text{ cm}}{43 \text{ cm} \cdot \text{s}^{-1}} = 34.8 \text{ s}$$

2.2.7 Partition (Capacity) Ratio

The partition ratio (or capacity ratio) k' is a measure of the solute retention relative to that of a nonretained solute. It is the additional time a solute takes to be eluted (as compared with an unretained solute, for which $k' = 0$), divided by the elution time of an unretained solute:

$$k' = \frac{t_R - t_M}{t_M} = \frac{KV_S}{V_M} \tag{2.7}$$

where V_S and V_M are the volume of the stationary and the mobile phases, respectively.

For diagnostic purposes, k' values accurate to the nearest integer are satisfactory. Retention times are also related to k' by the equation:

$$t_R = t_M(1 + k') = \frac{L}{u}(1 + k') \tag{2.8}$$

It is immediately clear that k' and the solute retention time are functions of the amount of the stationary phase and of the temperature (via K). The solute retention time is also dependent on the mobile-phase velocity and the column length. Remember that although t_M changes when the flow rate is changed, k' remains constant.

EXAMPLE 2.2 If $k' = 2.25$ for the material in Example 2.1, what are the values of t_R and t'_R?

$$t_R = t_M(1 + k') = (34.8 \text{ s})(1 + 2.25) = 113 \text{ s (or 1.88 min)}$$

$$t'_R = t_R - t_M = 113 - 34.8 = 78 \text{ s (or 1.31 min)}$$

2.2.8 Relative Retention

The relative retention α, which is a selectivity term, is given by

$$\alpha = \frac{k'_2}{k'_1} = \frac{t'_{R,2}}{t'_{R,1}} \tag{2.9}$$

where solute 1 elutes before solute 2. The larger the relative retention, the better the resolution. For a given system, the relative retention is a function of temperature only.

2.2.9 Plate Height and Plate Number

Efficiency governs how narrow the peaks will be when elution occurs. It is usually measured in terms of the number of plates N that a column can deliver for a given peak. The plate height H given by L/N, where L is the column length, represents the distance a solute moves while undergoing one partition. The *effective plate number* N_{eff}, reflects the number of times the solute partitions between the stationary and mobile phases during its passage through the column.

$$N_{eff} = \frac{L}{H} = \left(\frac{t'_R}{\sigma}\right)^2 \tag{2.10}$$

where σ^2 is the band variance in time units.

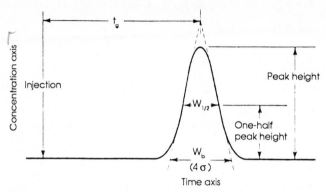

FIGURE 2.3   Evaluation of a chromatographic band for column effi- >
ciency.

Under ideal operating conditions the profile of a solute band closely approaches
that given by a gaussian distribution curve (Fig. 2.3). At the baseline the span of
4σ includes 95 percent of the solute band, while that of 6σ encompasses 99 percent with only 0.5 percent left on each wing.

The plate number can be measured for a test chromatogram and compared
with the column manufacturer's value under the same working conditions. The
width at the base of the peak W_b is ascertained experimentally from the intersec-
tions of the tangents to the inflection points with the baseline. It is equal to four
standard deviations (thus $\sigma = W_b/4$) and

$$N_{\text{eff}} = 16 \left(\frac{t'_R}{W_b}\right)^2 \tag{2.11}$$

Oftentimes it is preferable to measure the width at half the peak height $W_{1/2}$.
Then

$$N_{\text{eff}} = 5.54 \left(\frac{t'_R}{W_{1/2}}\right)^2 \tag{2.12}$$

The plate number is a good measure of the quality (efficiency) of a column.
The distance (or time) from injection to peak gives the apparent plate number; the
distance from the unretained solute peak to the sample peak gives the effective
plate number. Columns with large N values will produce narrow peaks and better
resolution than columns with lower N values. When measuring N, k' should be at
least 3 and preferably greater than 5.

EXAMPLE 2.3   What is the effective plate number of a solute whose $t_R = 9.55$ min
when a 1.5-m column is operated at 25°C? The value of $W_b = 0.71$ min. On the same
chromatogram the retention time of methane was 0.39 min.

To find t_M, use the retention time of methane (b.p. -182.5°C). Consequently,

$$t'_R = 9.55 - 0.39 = 9.16 \text{ min}$$

$$N_{\text{eff}} = 16 \left(\frac{9.16}{0.71}\right)^2 = 2660 \text{ plates}$$

and $$H = 1500 \text{ mm}/2660 = 0.56 \text{ mm}$$

2.2.10 Band Asymmetry

The peak asymmetry factor (AF) is defined as the ratio of the peak half-widths at a given peak height, usually at 10 percent of peak height. As shown in Fig. 2.4,

$$AF = \frac{b}{a} \tag{2.13}$$

When the asymmetry factor lies outside the range 0.95 to 1.15 for a peak of $k' = 2$, the apparent plate number for a column (as calculated by Eq. 2.11) is too high and should be calculated by the expression:

$$N_{\text{eff}} = \frac{41.7(t'_R/W_{0.1})^2}{(b/a) + 1.25} \tag{2.14}$$

2.2.11 Resolution

The degree of separation or resolution Rs of two adjacent bands is defined as the distance between their peaks (or centers) divided by the average bandwidth, as shown in Fig. 2.5. When measured in time units

$$Rs = \frac{t_{R,2} - t_{R,1}}{0.5(W_2 + W_1)} \tag{2.15}$$

The foregoing equation defines resolution in a given situation, but it does not re -

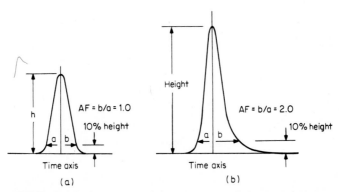

FIGURE 2.4 Peak asymmetry factor: (*a*) symmetrical band and (*b*) band tailing present.

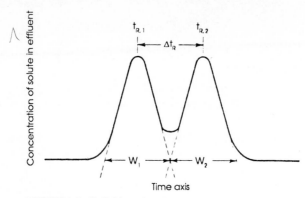

FIGURE 2.5 Definition of resolution.

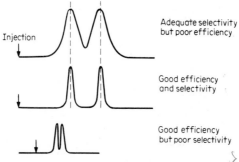

FIGURE 2.6 Selectivity, efficiency, and partition ratio for columns.

late resolution to the conditions of separation nor does it suggest how to improve resolution. For the latter purpose, the fundamental resolution equation is

$$Rs = \frac{1}{4}\left(\frac{\alpha - 1}{\alpha}\right)\left(\frac{k'}{1 + k'}\right)\left(\frac{L}{H}\right)^{1/2} \tag{2.16}$$

As shown in Fig. 2.6, a column may have adequate selectivity but exhibit poor efficiency (upper chromatogram, as compared with the middle chromatogram). The lower chromatogram exhibits good efficiency but poor selectivity.

EXAMPLE 2.4   What is the resolution between *trans*-2-butene (t_R = 3.12 min) and *cis*-2-butene (t_R = 3.43 min; W_b = 0.31 min)?

$$Rs = \frac{t_{R,cis} - t_{R,trans}}{W_{b,cis}} = \frac{3.43 \text{ min} - 3.12 \text{ min}}{0.31 \text{ min}} = 1.00$$

This is not baseline resolution, but because the resolution is proportional to the square root of the column length, the column length should be increased by

$$\frac{Rs_2}{Rs_1} = (\frac{1.5}{1.0})^2 = 2.25$$

Sections 3.9 and 4.4 deal with optimization of operating conditions for gas chromatography and HPLC, respectively.

2.2.12 Trennzahl (Separation) Number

The Trennzahl (TZ) or separation number has been used to express the separation efficiency of a column. It is useful for temperature-programmed runs where plate number or effective plate number would be meaningless. The TZ number is the number of completely separated peaks that, in theory, can be chromatographed between two other peaks. The two peaks are usually of a homologous series and commonly are two straight-chain alkanes differing by one methylene group.

$$TZ = \left(\frac{t_{R,b} - t_{R,a}}{W_{0.5,a} + W_{0.5,b}}\right) - 1 \tag{2.17}$$

where a and b are adjacent homologs.

2.2.13 The van Deemter Equation

Column efficiency, unlike selectivity, is a function of the average mobile-phase velocity, the column i.d. or average particle diameter, the type of carrier gas (in GC), as well as the type of solute and its retention, and the stationary-phase film thickness. The van Deemter equation relates the plate height to the mobile-phase velocity and other experimental variables.

$$H = A + \frac{B}{u} + C_{\text{stationary}}u + C_{\text{mobile}}u \tag{2.18}$$

The A term is defined as

$$A = \lambda d_p \tag{2.19}$$

where d_p is the particle diameter and λ is a function of the packing uniformity and column geometry. The A term arises from the inhomogeneity of flow velocities and path lengths around packing particles. In open tubular columns, the A term is zero.

The B term is defined as

$$B = 2\gamma D_M \tag{2.20}$$

where γ is an obstruction factor that recognizes that axial diffusion is hindered by the bed structure and D_M is the solute diffusion coefficient in the mobile phase. In open tubular columns γ is unity; in packed columns its value is about 0.6.

The $C_{\text{stationary}}$ term is proportional to d_f/D_S, where d_f is the thickness of the stationary phase film and D_S is the diffusion coefficient of the solute in the stationary phase.

The C_{mobile} term is proportional to d_f^2/D_p, where d_p is the particle diameter of the packing material and D_M is the diffusion coefficient of the solute in the mobile phase. From the above expressions, it is seen that the plate height is an explicit function of the support particle size, the mobile-phase velocity, the amount and thickness of the stationary phase film, the nature of the solute and stationary phase, the nature of the mobile phase, and the geometry and efficiency of the column packing. Implicitly, the plate height depends on the pressure drop from entrance to exit of the column and on the column temperature. See Fig. 3.2 for a plot of the van Deemter equation.

2.2.14 Time of Analysis and Resolution

The retention time is related to the plates required for a given resolution, the partition ratio, the plate height, and the linear velocity of the mobile phase.

$$t_R = N_{\text{req}} (1 + k') \left(\frac{H}{u}\right) \tag{2.21}$$

or

$$t_R = 16 \text{Rs}^2 \left(\frac{\alpha}{\alpha - 1}\right)^2 \frac{(1 + k')^3}{(k')^2} \left(\frac{H}{u}\right) \tag{2.22}$$

Each column has an optimal mobile-phase velocity. As mobile-phase velocities increase above optimum, efficiency is gradually lost. However, speed of analysis is increased (shorter retention times). Thus, efficiency can be sacrificed for speed when the column has excess resolving power (large gaps between the peaks).

2.3 QUANTITATIVE DETERMINATIONS

Chromatographic detectors that respond to the concentration of the solute yield a signal that is proportional to the solute concentration that passes through the detector. For these detectors the peak area is proportional to the mass of the component and inversely proportional to the flow rate of the mobile phase. Thus, the flow rate must be kept constant if quantitation is to be performed.

In differential detectors that respond to mass flow rate, the peak area exhibits no dependency on the flow rate of the mobile phase.

2.3.1 Peak-Area Integration

2.3.1.1 Peak Height. The distance from the peak maximum to the baseline is measured. Although inherently simple, peak heights are sensitive to small changes in operating conditions and sample injection. However, the method yields better precision than measuring the peak area, particularly of narrow peaks.

2.3.1.2 Height Times Width at Half-Height. This measurement is based on the assumed triangular shape of an ideal gaussian peak. The baseline of the peak is drawn, and the height of the peak is measured. The measuring scale is positioned parallel to the baseline at half the height, and the bandwidth is then measured at this position. The area is the product of the height times the width.

2.3.1.3 Disk Integrator. Good accuracy, independent of peak shape, is provided by a disk integrator. Figure 2.7 illustrates how to read the integrator trace. A full stroke of the "sawtooth" pattern (either up or down) represents 100 counts. Every horizontal division crossed by the trace has a value of 10. Values that are less than 10 are estimated. On some models the space between the "blips" that project slightly above the uppermost horizontal line is equivalent to 600 counts, making it easier to count the number of full strokes.

2.3.2 Computing Integrator

Computing integrators automatically determine the peak area bound by the point at which the chromatographic trace leaves the baseline and the point when it returns to the baseline. In the case of overlapping peaks, special algorithms allot areas to each component. During isothermal runs the software can automatically alter the slope sensitivity with time. This allows both sharp narrow peaks and low flat peaks to be measured with equal precision.

2.3.3 Evaluation Methods

Once the peak height or peak areas have been measured, there are four principal evaluation methods which can be used to translate these numbers to amounts of solute.

2.3.3.1 Calibration by Standards. Calibration curves for each component are prepared from pure standards, using identical injection volumes and operating

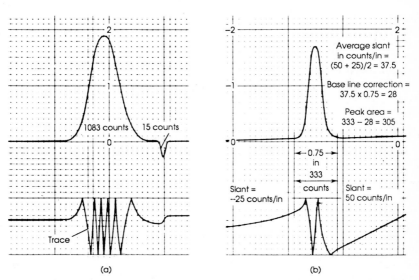

FIGURE 2.7 (a) Estimation of peak areas with ball-and-disk integrator. (b) Method for handling baseline correction.

conditions for standards and samples. The concentration of solute is read from its calibration curve or, if the curve is linear,

$$X = K(\text{area})_x \tag{2.23}$$

where X is the concentration of solute and K is the proportionality constant (slope of the calibration curve). In this evaluation method only the area of the peaks of interest need to be measured. However, the method is very operator-dependent and requires good laboratory technique.

Relative response factors must be considered when converting area to volume and when the response of a given detector differs for each molecular type of compound.

> **EXAMPLE 2.5**   The relative response factors for o-xylene and toluene (relative to the value for benzene, which is assigned unity) were found to be 0.570 and 0.793, respectively. An unknown mixture of these three solutes gave these peak heights (in millimeters): benzene, 98; o-xylene, 87; and toluene, 86.
>
> $$\text{Total adjusted response} = \frac{H_{bz}}{1.00} + \frac{H_{xyl}}{0.570} + \frac{H_{tol}}{0.793}$$
>
> $$= \frac{98}{1.00} + \frac{87}{0.570} + \frac{86}{0.793} = 98 + 153 + 108 = 359$$
>
> For benzene: $\left(\dfrac{98}{359}\right)(100) = 27.3\%$
>
> For o-xylene: $\left(\dfrac{153}{359}\right)(100) = 42.6\%$
>
> For toluene: $\left(\dfrac{108}{359}\right)(100) = 30.1\%$

2.3.3.2 Area Normalization. For this method to be applicable, the entire sample must have eluted, all components must be separated, and each peak must be completely resolved. The area under each peak is measured and corrected, if necessary, by a response factor as described. All the peak areas are added together. The percentage of individual components is obtained by multiplying each individual calculated area by 100 and then dividing by the total calculated area. Results would be invalidated if a sample component were not able to be chromatographed on the column or failed to give a signal with the detector.

2.3.3.3 Internal Standard. In this technique a known quantity of the internal standard is chromatographed, and area versus concentration is ascertained. Then a known quantity of the internal standard is added to the "raw" sample prior to any sample pretreatment or separator operations. The peak area of the standard in the sample run is compared with the peak area when the standard is run separately. This ratio serves as a correction factor for variation in sample size, losses in any preliminary pretreatment operations, or incomplete elution of the sample. The material selected for the internal standard must be completely resolved from adjacent sample components, must not interfere with the sample components, and must never be present in samples.

> **EXAMPLE 2.6**   Assume that 50.0 mg of internal standard is added to 0.500 g of the sample. The resulting chromatogram shows five components with areas (in arbitrary

units) as follows: $A_1 = 30$, $A_2 = 18$, $A_{std} = 75$, $A_3 = 80$, $A_4 = 45$, and the area sum equals 248. The amount of component 3 in the sample is

$$W_3 = W_{std}\left(\frac{A_3}{A_{std}}\right) = 0.0500\left(\frac{80}{75}\right) = 0.0533 \text{ g}$$

Percentage component 3:

$$\frac{0.0533 \text{ g}}{0.0500 \text{ g}} \times 100 = 10.66\%$$

Although component 3 appears to be a major component, it represents only about 10 percent of the total sample. A large part of the sample does not appear on the chromatogram, as would be the case if the organic mixture included some inorganic salts.

Usually it will be necessary to ascertain the ratio of response factors (such as K_{std}/K_3). When this is so,

$$\frac{K_{std}}{K_3} = \frac{W_3 A_{std}}{W_{std} A_3}$$

2.3.3.4 Standard Addition. If only a few samples are to be chromatographed, it is possible to employ the method of standard addition(s). The chromatogram of the unknown is recorded. Then a known amount of the analyte(s) is added, and the chromatogram is repeated using the same reagents, instrument parameters, and procedures. From the increase in the peak area (or peak height), the original concentration can be computed by interpolation. The detector response must be a linear function of analyte concentration and yield no signal (other than background) at zero concentration of the analyte. Sufficient time must elapse between addition of the standard and actual analysis to allow equilibrium of added standard with any matrix interferant.

If an instrumental reading (area or height) R_x is obtained from a sample of unknown concentration x and a reading R_1 is obtained from the sample to which a known concentration a of analyte has been added, then x can be calculated from the relation:

$$\frac{x}{x + a} = \frac{R_x}{R_1} \tag{2.24}$$

A correction for dilution must be made if the amount of standard added changes the total sample volume significantly. It is always advisable to check the result by adding at least one other standard. Additions of analyte equal to twice and to one-half the amount of analyte added to the original sample are optimum statistically.

2.4 SAMPLE CHARACTERIZATION

2.4.1 Use of Retention Data

The retention time under fixed operating conditions is a constant for a particular solute and, can, therefore, be used to identify that solute. This is accomplished by comparing the retention times of the sample components with the retention times of pure standards.

For isocratic (HPLC) or isothermal (GC) elution, retention times usually vary in a regular and predictable fashion with repeated substitution of some group i into the sample molecule as, for example, the $-CH_2-$ group in a homologous series.

$$\log t_{R,i} = mN_i + \text{constant} \tag{2.25}$$

where m is a constant and N_i is the number of repeating groups (or the number of carbon atoms) in the homologous series (Fig. 2.8). Partition ratio k' values are actually superior to retention data because k' values are not influenced by mobile-phase flow rate or column geometry.

A suspected solute can be verified by "spiking" the sample with a added amount of the pure solute. Only the peak height should vary if the two compounds are the same. Of course, pure standards must be available if spiking is to be used.

2.4.2 Chromatographic Cross Check

The ability to identify a sample component by means of retention times (or k' values) is significantly enhanced by the use of different stationary phases (in gas chromatography). The use of a polar liquid phase in one column and a nonpolar liquid phase in a second column will provide much information. If the retention times for the two stationary phases are plotted against each other, lines that radiate from the origin are obtained (one for each homologous series). If the logarithms of the retentions are plotted against each other, a corresponding series of

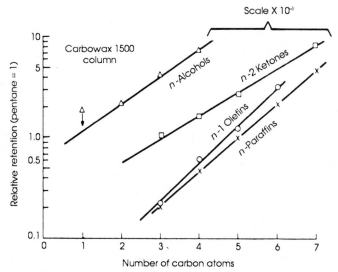

FIGURE 2.8 Plot of retention time (log scale) versus number of carbon atoms for several homologous series of compounds.

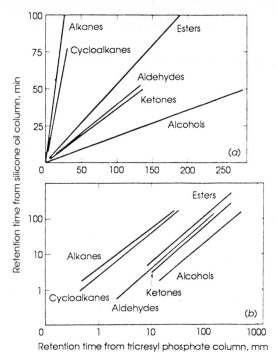

FIGURE 2.9 Two-column plots: (*a*) linear and (*b*) logarithmic.

parallel lines is obtained with points spaced linearly according to the number of repeating groups (or carbon number) for each homologous family (Fig. 2.9).

Retention indexing systems in gas chromatography, such as Rohrschneider or McReynolds constants are also useful for qualitative analysis. They are discussed in Chap. 3.

2.4.3 Identification by Ancillary Techniques

None of the foregoing techniques based on retention times are definitive because many compounds have similar retention times. However, structural information can be independently obtained from several spectroscopic techniques. This has led to hyphenated techniques, such as gas chromatography–mass spectroscopy and gas chromatography–infrared spectroscopy. If spectroscopic reference spectra are available, conformation of solute structure is likely.

2.5 PREPARATIVE CHROMATOGRAPHY

Certain procedures require that the component(s) of a sample be collected after they have passed through the detector, so that they can be isolated and studied further or collected for industrial use. The object of preparative chromatography is to separate the compound(s) of interest from a mixture in the desired purity and amount within a reasonable time and at a reasonable cost. The tradeoffs here are no different than those in analytical chromatography.

In order to determine the proper conditions for the preparative chromatography, use an analytical column of 2 to 4 mm i.d. Be sure that the type of packing used in the analytical column is also available in the preparative column. A match between the surface chemistry of the analytical and the preparative columns is required in order to allow a direct transition from one column to the other. In HPLC the overall cost of column packing and solvents (including the problem and cost of disposal) should be considered. Once the mode and packing are selected, the scale-up process is relatively straightforward. The procedure is as follows:

1. Determine the loading capacity of the stationary phase. This is dependent on the degree of purity required for the final product, which in turn influences the resolution needed. Gradually increase the amount of sample per injection just until resolution begins to suffer.

2. Use the throughput desired per injection to scale up the analytical column to the preparatory column by the ratio of the injection amounts. For example, if the overload conditions for a normal HPLC analytical column (4.6 mm i.d. $\times$ 25 cm) is 15 mg, and 300 mg per injection is needed for the final product, the preparative column volume should be 20 times that of the analytical column. Since the volume ratio is

$$\frac{V_{prep}}{V_{anal}} = \frac{(\pi r^2 L)_{prep}}{(\pi r^2 L)_{anal}} \tag{2.26}$$

$$20 = \frac{r^2 L}{(0.23 \text{ cm})^2 (25 \text{ cm})}$$

$$r^2 L = 26.4 \text{ cm}^3$$

Thus any combination of column radius squared times column length that equals 26.4 cm^3 would be suitable. Since semipreparative columns are generally 7 mm i.d., a 53-cm length would work. The actual column dimension will be dictated by what is available commercially.

To maintain a given separation time after scaleup, you should use the same mobile-phase linear velocity for the analytical and the preparative columns. Thus, the mobile-phase flow rate should be increased in direct proportion to the square of the ratio of the preparative-to-analytical column diameters. For optimum resolution, the flow rate used on the preparative column should be lower than that calculated for the analytical column.

If resolution is less than $R = 1.2$, the column probably cannot be usefully overloaded and, therefore, sample capacity will be limited.

For commercially available materials, average particle diameters can start at 3 μm for fast LC applications, increase to 5 μm for conventional separations, and extend to 15 or 20 μm for preparative applications. If the separation on a small-particle column is inadequate, it is unlikely to improve on a large-particle wide-bore column. Of course, if excellent resolution can be achieved on the analytical column, then the use of a larger-particle wider-bore column is warranted.

When scaling up in preparative chromatography, sample capacity is often the factor of major importance. The system may actually be run in an overloaded or less-than-optimum condition to increase throughput. With particularly valuable samples, for which purity is of the highest concern and the best resolution is demanded, speed and capacity can be sacrificed.

BIBLIOGRAPHY

Giddings, J. C., "Principles and Theory," in *Dynamics of Chromatography*, Part 1, Dekker, New York, 1965.

Miller, J. M., *Chromatography: Concepts and Contrasts*, Wiley, New York, 1988.

CHAPTER 3
GAS CHROMATOGRAPHY

Since its introduction in 1952, the use of gas chromatography (GC) has grown spectacularly. The major limitation of GC is that samples, or derivatives thereof, must be volatile. Any substance, organic or inorganic, which exhibits a vapor pressure of at least 60 torr (the column temperature may be raised to 350°C) can be eluted from a GC column.

The basic components of a gas chromatograph, shown in Fig. 3.1, are as follows:

1. A supply of carrier gas with attendant pressure regulator and flow meter
2. An injection port, possibly followed by a splitter
3. A separation column
4. A detector
5. A thermostatically controlled oven that can also be programmed for various heating rates
6. A recorder or other readout device

At the first functional level, the gas chromatograph accepts an input (the sample) and produces an output (the chromatogram). The second functional level is composed of the primary chromatographic components: injection port, column, and detector.

Each of the principal modules is composed of several electronic and pneumatic subsystems. The inlet, which is heated, conducts carrier gas to the column. The detector contains excitation and amplification circuitry. The column module is made up of the gas chromatography column itself and the column oven. The parts replaced most often in these subsystems are injector liners, septa, columns, ferrules, detector parts, heat assemblies, and amplifier boards.

Modern gas chromatographs span a wide range of complexity, capability, and options. More than 90 percent of systems available today are microprocessor-controlled and have temperature-programmable column ovens. The vast majority have dual-channel detection capability. Approximately half are sold with a capillary inlet system, and many are sold with companion data-handling systems and autosamplers.

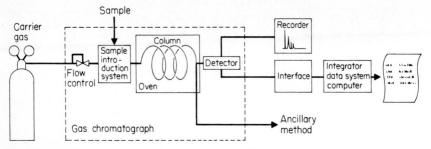

FIGURE 3.1   Basic components of a gas-chromatographic system.

3.1 CARRIER GAS

The carrier gas is usually helium or nitrogen. It carries the sample through the system but does not interact with the sample components. The gas is obtained from a high-pressure gas cylinder and should be free from oxygen and moisture. Gas flow is monitored by a flow meter, and the gas pressure is controlled by a pressure regulator.

A light carrier gas such as helium or hydrogen always permits faster analysis than a denser carrier gas such as nitrogen or argon. This is so because the van Deemter curve (Fig. 3.2) for the same sample component obtained on the same column will have its minimum at higher velocities with the light carrier gases, and at the same time the slope of the ascending part of the curve will be smaller with the light carrier gas.

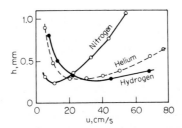

FIGURE 3.2   Van Deemter curves for three different carrier gases.

Most chromatographers hesitate to convert to hydrogen as a carrier gas because of safety considerations. However, with adequate equipment, hydrogen poses no greater hazard in the laboratory than any other compressed gas. The most reliable source of pure hydrogen carrier gas is a laboratory-size generator that produces ultrapure, ultradry gas from the electrolysis of water; the hydrogen is separated from other electrolysis products by permeation through a palladium membrane. Gaseous hydrogen can be stored safely and conveniently by using solid metal hydrides; the hydrogen is stored at the pressure of a comparable cylinder.

3.1.1 Gas Purifiers

Moisture and oxygen must be removed from the carrier gases by using appropriate scrubbers. Even small amounts of oxygen or water can damage GC columns and detectors. The thin films of the stationary phase in capillary columns are especially vulnerable to oxidation or hydrolysis. This is good practice for any phase but is essential for very polar phases. A gas purifier specifically designed to ensure maximum gas purity should be inserted in the carrier-gas line ahead of the injection port. One commercial product will remove all oxygen and water from 60 tanks of heavily contaminated gas before requiring replacement. Oxygen and moisture removal should be just as efficient at concentrations up to 2000 ppm as at levels below 100 ppm and should handle gas flow rates of up to 1100 mL · min^{-1}. The solid chemical compounds should be placed in an all-metal tube with metal Swagelok fittings. The purifier tube is placed in an oven so that oxygen and water are chemically reacted with the material in the tube. Once trapped, these contaminants cannot be returned to the gas stream. It is also desirable to install an indicating tube downstream from the high-capacity purifier.

Do not use plastic supply lines for carrier-gas lines. Plastic and O-rings allow both permeation and leaks.

3.2 SAMPLE INTRODUCTION SYSTEM

The functions of the injection port are (1) to provide an entry for the syringe, and thus the sample, into the carrier-gas stream and (2) to provide sufficient heat to vaporize the sample. For optimum performance the sample must be deposited on the column in the narrowest bandwidth possible. Usually liquid samples are injected onto a heated block (Fig. 3.3) whose function is to instantly convert the liquid sample into the gas phase (flash vaporization) without decomposition or fractionation. The flash-vaporization chamber of the injection port should be as small as possible to preserve efficiency. Sufficient volume is required, however, to accommodate the sudden vaporization and expansion of the sample after injection (1 μL of methanol produces 0.31 mL of vapor at 200°C and 30 lb . in^{-2}). Sometimes the sample is deposited directly on top of the stationary phase (on-column injection); it is then evaporated by heating the column at a programmed rate.

Gas samples are introduced into the carrier-gas stream by special gas syringes or by a rotary valve with sample loop. For a capillary column, injection volumes are on the order of 0.1 nL (nanoliter), which requires the carrier-gas stream to be split so that only a small fraction of the sample actually enters the column. Requirements for a sample splitter are stringent. Every component in the sample must be split in exactly the same ratio.

Split sampling involves dividing the mixture of the sample vapor and the carrier gas into two highly unequal parts, the smaller one being conducted into the

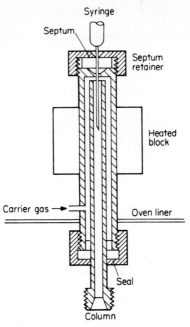

FIGURE 3.3   Schematic representation of a typical flash vaporizer injection port.

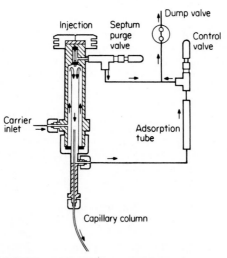

FIGURE   3.4 Splitter   system   for   gas chromatography. (*From   H.   H.   Willard, L.L. Merritt, Jr., J. A. Dean, and F. A. Settle, Jr., In- strumental Methods of Analysis, 7th ed. Copyright 1988, Wadsworth, Inc. Used by permission.*)

column. The concentric tube design dominates (Fig. 3.4). There are two basic requirements for linear splitting:

1. No sample loss should be encountered during evaporation.
2. Sample vapor and carrier gas must be homogeneously mixed prior to splitting.

An important point to remember when using splitless injection is that the initial column temperature should be approximately 20°C below the boiling point of the solvent in order to realize the solvent effect, which reduces band broadening.

It is desirable to purge the septum with a flow of vented carrier gas. This prevents sample absorption onto the septum and eliminates tailing and ghost peaks caused by septum bleed.

Valves may be installed in the column flow lines for such applications as backflushing and column selection.

3.2.1 Injecting the Sample with a Syringe

The syringe-septum system is the most popular of GC sample introduction systems. The sample size can be quickly selected and reproducibility adjusted (± 5 percent variation routinely and ±1 percent variation with skilled users). In gas chromatography, the results are only as good as the reproducibility of the sample and the sample injection. This is particularly true in quantitative work. When injecting the sample with a microliter syringe, follow these recommendations.

1. Develop a rhythm in your motion that is used each time an injection is made; that is, do the same things in the same manner at the same time.
2. Hold the syringe (Fig. 3.5) as close to the flange (in the face or unmarked area) as possible. This will prevent the heat transfer that occurs when the needle or barrel is held with the fingers. A syringe guide prevents heat transfer from the finger and the guide makes septum penetration easier.
3. Handle the plunger by the button, not the plunger shaft. This reduces the possibility of damage or contamination.
4. Develop a smooth rhythm that allows you to inject the sample as quickly as possible but with accuracy. The syringe should be left in the injection port for about two seconds after depressing the plunger.
5. Use the syringe at less than maximum capacity for greatest accuracy. When samples containing components with a wide range of boiling points are injected, fractionation may occur unless the syringe is prepared with a solvent slug (separated by an air bubble) that follows the sample into the injection port.
6. Wet the interior surfaces (barrel and plunger) of the syringe with the sample

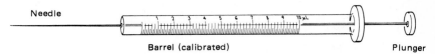

FIGURE 3.5 A microliter syringe. (*From H. H. Willard, L.L. Merritt, Jr., J. A. Dean, and F. A. Settle, Jr., Instrumental Methods of Analysis, 7th ed. Copyright 1988, Wadsworth, Inc. Used by permission.*)

by pumping the plunger before filling the syringe in order to ensure an accurate measurement.

7. Overfill the syringe in the sample bottle, withdraw from the bottle, move the plunger to the desired calibration line, and discharge the excess sample. Do not leave any drop hanging from the syringe tip.

8. Wipe the needle clean with a lint-free tissue before injecting; use a quick motion and take care neither to wipe sample out of the needle nor to transfer body heat from your fingers to the needle.

9. Check the syringe visually for bubbles or foreign matter in the sample.

10. Take extra care when filling syringes with a detachable needle because there is dead volume in the needle. Pressurize your sample bottle by using a gastight syringe filled with inert gas. Repeat as needed to build pressure in the bottle.

11. Use a larger bore needle when handling viscous samples.

12. Do not use dull or damaged syringe needles or overtighten the septum retainer because septum life will be shortened. Since high temperature also reduces septum life, the use of finned septum retainers, which stay cooler, is recommended. Since all septa start to leak eventually, it is good practice to replace the septum each day, or even more often during heavy use. Some septa may last up to 150 injections, but they should be routinely replaced well before this.

3.2.2 Care and Repair of Syringes

A syringe is a delicate instrument and should be treated carefully. Carefully clean the syringe before and after use. Cleaning the small bore of the needle, the glass barrel, and the closely fitting plunger is not a simple matter, but it is essential for a long syringe life. Moreover, cleaning is more effective and easier to accomplish if it is done immediately after the syringe has been used. The syringe must never be used when the needle is blocked or the barrel is dirty.

A single cleaning procedure consists of pumping a solution of a surface-active cleaning agent through the syringe and then rinsing both the syringe and plunger through the needle with distilled water followed by an organic solvent such as acetone or other ketone. Never touch the plunger surface with your fingers. Fingerprints, perspiration, or soil from the fingers can cause the plunger to freeze in the glass barrel.

Another recommended cleaning procedure is: (1) pump room-temperature chromic acid solution through the syringe with the plunger, (2) rinse both with distilled water, (3) blow the syringe dry with oil-free compressed gas, and (4) carefully wipe the plunger with lint-free tissue. For more persistent contamination, dismantle the syringe and soak the parts in a cleaning solution. A small ultrasonic bath will speed the cleaning process.

Never rapidly cool or heat assembled syringes, and never heat them over 50°C because the different expansion coefficients of the metal needles and plungers and the glass barrel may cause the barrel to fracture.

3.2.3 Pyrolysis Gas Chromatography[1]

Pyrolysis gas chromatography is a technique that has long been used in a variety of investigative fields because it produces volatile compounds from macromolecules that are themselves neither volatile nor soluble. Examples are polymers, rubbers, paint films, resins, bacterial components, soils and rocks, coals, textiles, and organometallics. Volatile fragments are formed and introduced into the chromatographic column for analysis.

Let us use a polymer as an example to illustrate the use of this technique, which consists of two steps. The injection port is heated to perhaps 270°C; when the sample is injected, the volatile ingredients that are driven off provide a fingerprint of the polymer formulation. Then the pyrolysis step develops the fingerprint of the nonvolatile ingredients. Known monomers of suspected polymers can be injected with a microsyringe for identification of the peaks of an unknown pyrogram. Specific identification of the peaks appearing in pyrograms is most effectively carried out by directly coupled gas chromatography–mass spectrometry together with the retention data of the reference samples. Gas chromatography–Fourier transform infrared also can provide effective and complementary information.

Pyrolyzers can be classified into three groups: (1) resistively heated electric filament type, (2) high-frequency induction (Curie-point) type, and (3) furnace type.

The filament type uses either a metal foil or a coil as the sample holder. Heat energy is supplied to the sample holder in pulses by an electric current. This permits stepwise pyrolysis at either fixed or varied pyrolysis temperatures. This feature sometimes permits a discriminative analysis of volatile formulations and the high polymers in a given compound without any preliminary sample treatment.

The Curie-point type (Fig. 3.6) uses the Curie points of ferromagnetic sample holders to achieve precisely controlled temperatures when the holder containing the sample is subjected to high-frequency induction heating. The Curie point is the temperature at which the material loses its magnetic property and ceases to absorb radio-frequency energy. Foils of various ferromagnetic materials enable an operator to select pyrolysis temperatures from 150 to 1040°C.

In the furnace (continuous) pyrolyzer type, the sample is introduced into the center of a tubular furnace held at a fixed temperature. Temperature is controlled by a proportioning controller that utilizes a thermocouple feedback loop.

3.2.4 Purge-and-Trap Technique

The analysis of volatile organic compounds in samples is commonly performed using the technique of purge-and-trap gas chromatography. It is the required technique for a number of the Environmental Protection Agency (EPA) methods for drinking water, source and wastewater, soils, and hazardous waste. Other applicable sample types include perfumes, food products, blood, urine, and human breath.

In the purge-and-trap method, samples contained in a gastight glass vessel are purged with an inert gas, causing volatile compounds to be swept out of the sample and into the vapor phase. Organic compounds are then trapped on an adsorbent (often Tenax-GC, a porous polymer based on 2,6-diphenyl-p-phenylene oxide) that allows the purge gas and any water vapor present to pass through. In this manner volatiles can be efficiently collected from a relatively large sample, producing a concentration factor that is typically 500- to 1000-fold greater than the original. After collection, the adsorbent is heated (thermally or in a microwave oven) to release the sample and then backflushed using the GC carrier gas. This sweeps the sample directly onto the GC column for separation and detection

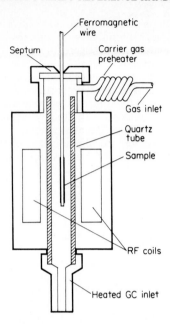

FIGURE 3.6   Diagram of a typical Curie-point pyrolyzer. [*By permission from C. J. Wolf, M. A. Grayson, and D. L. Fanter, Anal. Chem., 52:349A (1980). Copyright 1980 American Chemical Society.*]

by normal GC procedures. Trapped samples can easily be stored or shipped to another site for analysis.

3.2.5   Headspace Sampling[2,3]

When only the vapor above the sample is of interest and the partition coefficient allows a sufficient amount of analyte into the gaseous phase, headspace sampling can be done. Samples may be solid or liquid. A measured amount of sample, and often an internal standard, are placed in a vial, and then the septum and cap are crimped in place. The vials, contained in a carousel, are immersed in a silicone oil bath operating from ambient 15°C to 150°C. A heated flexible tube which terminates in a needle samples each vial in turn. A gas sampling valve provides a standard 1.0-mL vapor sample for transfer into the GC injection port.

3.3   GAS CHROMATOGRAPHIC COLUMNS

Separation of the sample components takes place in packed or open tubular columns through which the carrier gas flows continuously. The separation column is placed immediately after the injection port and any attendant sample splitter. The separation column contains the stationary phase which can be either (1) an adsorbent (GSC) or a liquid distributed over the surface of small diameter (capillary) tubing or (2) a granular substrate (support particles).

GC columns can be divided into three broad categories: (1) packed, (2) wide-bore capillary, and (3) high-resolution capillary. Two factors, selectivity and efficiency, should be considered when comparing different types of columns.

3.3.1 Packed Columns

In packed columns the stationary phase covers the surface of a substrate, which is usually a porous diatomaceous earth whose internal pore diameters range from 2 to 9 μm. Mesh sizes of the substrate are usually 125 to 150 μm.

Compared to the capillary column, which is described later, the packed column has a variety of mobile-phase flow path lengths and the discontinuous film of its stationary phase is much less uniform. Both of these differences contribute to the increased standard deviation of the packed-column peak. Surface mineral impurities of the substrate, which can serve as adsorption sites, must be removed by acid washing. Surface silanol groups (—Si—OH) tend to adsorb polar solutes and must be converted to silyl ethers (—Si—O—Si—) by treating the column packing with dimethyldichlorosilane (unused silylating agent is removed with methanol).

The poor heat transfer properties that characterize almost all column packings increase the standard deviation of the peaks. There is a temperature gradient across any transverse section of a packed column, so identical molecules of each given solute, ranged transversely across a short segment of column length, are at any given moment exposed to different temperatures. As a consequence, identical molecules of each solute also exhibit a range of volatilities which, in turn, leads to a larger standard deviation (wider peak) for the molecules that make up that peak.

3.3.2 Capillary Columns

The flexible, mechanically durable, and chemically inert fused-silica capillaries offer many advantages over packed columns and are becoming the dominant type of column design. There are six good reasons for changing from packed columns to capillary columns.

1. Shorter retention times
2. Greater inertness
3. Longer life
4. Lower bleed
5. Higher efficiency
6. Greater reproducibility

A highly efficient capillary column does not require as much selectivity toward sample components as a less efficient packed column to achieve the same resolution. Peaks are sharper. Sharper peaks provide better separation, and sharper peaks also deliver the solutes to the detector at higher concentrations per unit time, thus enhancing sensitivity.

The major advantage of capillaries regardless of the separation is an increase in the speed of analysis. When packed-column flow rates are used, the capillary column produces separation efficiencies that are equal to those of packed columns but at roughly three times the speed. When the flow rate is optimized for the tubing diameter, the capillary column produces far superior efficiencies with analysis times approximately equal to those for packed columns.

Although efficiency (plates per meter) is the same for both packed and capillary columns, capillary columns are open tubes. Therefore, they are more permeable and can be made much longer before the inlet pressure requirement becomes too large. Capillary columns with more than 100 000 plates are relatively common. Because of the greater permeability of capillary columns, the van Deemter (h/v) curves are also flatter, which means that the column can be operated at 2 to 3 times the optimum flow rate without losing much efficiency. Consequently, the time of analysis is shorter. Better peak shape and increased column stability are other advantages. Nothing is superior to a capillary column for resolving mixtures with many components. It has obvious advantages even in relatively simple applications that are usually solved with packed columns.

Present-day porous layer open tubular (PLOT) columns have diameters from 0.05 to 0.35 mm, film thicknesses from about 0.2 μm up to 5 to 6 μm, and lengths from 10 m or shorter up to 100 m. The availability of such a wide range readily permits the selection of the optimum parameters for a given application. For high resolution, long (30 to 50 m), narrow (0.1 to 0.25 mm i.d.), thin-film (0.2 μm) columns are used to generate 250 000 plates for 0.5- to 1.5-h separations of complex materials. These columns are operated at low velocity near the optimum. High resolution is possible because of the fast mass transfer of sample within the thin stationary phase films.

Other PLOT columns are coated with specially prepared gums, aluminum oxide, and molecular sieves. With aluminum oxide excellent separations of C_1 to C_{10} hydrocarbons are obtained. With molecular sieve 5A the quantification of permanent gases is made very easy. The 13X type molecular-sieve column easily separates aliphatic and naphthenic hydrocarbons.

3.3.3 Wide-Bore Capillary Columns[4]

With the use of special coating techniques in combination with suspension technology, a 10- to 30-μm layer of a porous polymer can be coated on the inner wall of a fused-silica capillary column. Fused-silica tubing is flexible, durable, and inert. Porous polymers are cross-linked polymers that are produced by copolymerizing styrene and divinylbenzene. The pore size and surface area can be varied by changing the amount of divinylbenzene added to the polymer. The introduction of functional groups such as acrylonitrile or vinyl pyrrolidone controls the selectivity of the polymers.

So-called megabore columns, with their wider bore (0.5 to 0.75 mm i.d.), thick films, and direct on-column injection, serve as a good compromise between capillary columns and packed columns. They can often be substituted directly for packed columns. Wide-bore columns give better resolution (because of their longer lengths) and more symmetrical peaks (because of their decreased adsorption) than packed columns, as shown in Fig. 3.7. They also combine the high capacity and ease of use of packed columns with the high efficiency, rapid analyses, and greater versatility of narrow-bore columns. Sample capacity and flow rates are similar to those for packed columns, so experimental parameters need not be altered much.

The inner surface of a wide-bore capillary column (0.53 mm i.d.) is coated

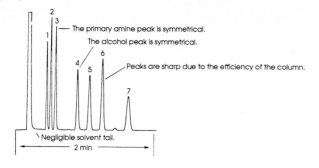

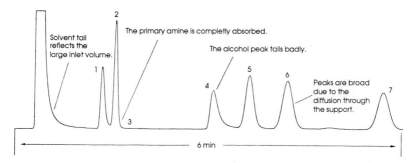

FIGURE 3.7 Comparison of chromatograms obtained on a wide-bore column (upper) and a packed column (lower). Peaks: (1) 4-chlorophenol, (2) dodecane, (3) 1-decylamine, (4) 1-undecanol, (5) tetradecane, (6) acenaphthene, and (7) pentadecane. (*Courtesy of J & W Scientific, Inc.*)

with the stationary polymer phase (a polysiloxane) to a thickness of 1 to 2 μm. The coating is both surface-bonded and cross-linked so it cannot be disturbed by repeated injections of a polar solvent or by prolonged heating. Thicker films permit the analysis of volatile materials without the use of subambient column temperatures. Thinner films are valuable for analyzing high-molecular-weight compounds. Columns of 10 m in length closely approximate the capacity and separation of the standard analytical packed column (20 m × 2 mm i.d.) with a 3 to 5 percent loading. A variety of functional groups can be blended into the polysiloxane chain to provide stationary phases of different polarity or selectivity.

Virtually all commonly used solvents can be injected onto these wide-bore bonded-phase capillary columns, but water does cause column deterioration (hydrolysis of the stationary phase).

3.3.4 Support-Coated Open-Tubular Columns

Support-coated open-tubular (SCOT) columns have a porous layer of stabilizing inert support built up or deposited on the original tubing wall. This layer is coated with the liquid phase. The SCOT column finds use in combination with wide-bore columns for multidimensional separations.

3.4   STATIONARY PHASES

For any resolution to be achieved, the components of the sample must be retained by the stationary phase. The longer and more selective the retention, the
better the resolution will be. In gas chromatography the inert carrier gas plays no
active role in solute selectivity, although it does affect resolution. Selectivity can
be varied only by changing the polarity of the stationary phase or by changing the
column temperature.

3.4.1   Liquid Phases

In the early days of GC, the stationary liquid phase was coated on a substrate
such as a diatomaceous earth. Problems arose because the liquids would pool and
the surface coverage was incomplete. The former led to longer but irregular residence times in the stationary phase (leading to decreased resolution), and the
latter gave rise to adsorption onto the substrate and thereby mixed partition effects. Bonded (cross-linked or chemically bonded) liquid phases are much superior. They can be cleaned by rinsing with strong solvents and baking at high temperatures. Column life can be extended, dirtier samples can be tolerated, and
sample cleanup can be reduced.

Knowing which type of stationary phase to select for a particular chromatographic analysis is a crucial skill. Compounds will be retained longer by a phase
to which they have a higher affinity; therefore, for best results, select a stationary
phase with a polarity similar to the sample. The old adage is applicable: namely,
"like dissolves like." Polar phases interact more strongly with polar or polariz-
able components such as alcohols, esters, and aromatics, whereas nonpolar
phases differentiate solutes on the basis of their vapor pressures (boiling points).
Adsorbent supports rely on differential adsorption/desorption to separate various
gases and low-molecular-weight components. Molecular sieves separate by molecular size.

When selecting new columns that are appropriate for a particular analysis, one
must define the goals of the analysis and the characteristics of the potential stationary phases, and then match the two to the sample at hand. Solutes can be
classified according to the functional group which is predominant in the molecule; that is, the classification is based upon the solute's polarity.

Using a similar classification scheme, the stationary phases in liquid-phase gas
chromatography can be grouped by their polarity. Table 3.1 provides a list of the
common stationary phases, their temperature range, and their McReynolds constant, a useful polarity indicator to be discussed later. These phases range from
nonpolar polydimethylsiloxanes to moderately polar phenyl- or cyanopropyl-
substituted polysiloxanes. Polyethylene glycols are the most frequently used polar phases and are sometimes chemically bonded with polysiloxanes to increase
the range of available polarities. Some very polar phases, such as
tris(cyanoethoxypropane), cannot easily be deposited as a stable film on the inner
wall of capillary columns without a layer of stabilizing inert support.

3.4.2   Kovats Retention Index System[5]

The Kovats retention indices (RI) indicate where compounds contained in the
sample will appear on a chromatogram with respect to unbranched alkanes. By

TABLE 3.1 Stationary Phases in Gas Chromatography

Liquid phase (chemical type; similar phases)	Minimum/maximum temperature, °C	McReynolds constants					
		x'	y'	z'	u'	s'	Σ
For boiling point separation of broad molecular weight range of compounds:							
Squalane (2,6,10,15,19,23-hexamethyltetracosane)	20/150	0	0	0	0	0	0
Polydimethylsiloxane (OV-101, SF 96, SP-2100, SE-30, CP SIL-5, Apiezon L, DC-200, UCW-982)	50/350	17	57	45	67	43	229
Polydiphenylvinyldimethylsiloxane, 5%/1%/94%* (SE-54)	50/300	33	72	66	99	67	337
Polydiphenyldimethylsiloxane, 10%/90% (OV-3%)	0/350	44	86	81	124	88	423
Dexsil 300 (polycarboranemethylsiloxane)	50/500	47	80	103	148	96	474
Dexsil 400 (polycarboranemethylphenylsilicone)	50/500	72	108	118	166	123	587
Polydiphenyldimethylsiloxane, 20%/80% (DC 550, OV-7)	20/350	69	113	111	171	128	592
For unsaturated hydrocarbons and other semipolar compounds:							
Polydiphenyldimethylsiloxane, 50%/50% (OV-17, SP-2250. DC-210, Hallcomide)	0/325	119	158	162	243	202	884
Dinonyl phthalate	20/150	83	183	147	231	159	803
Dexsil 410 (polycarboranemethylcyanoethylsilicone)	50/500	72	286	174	249	171	952
For moderately polar compounds:							
Polycyanopropylphenylmethylsiloxane (OV-225, XE 60, SP-2300, Silar 5CP,UCON HB 5100, AN-600, CP SIL-84)	0/265	228	369	338	492	386	1813
For polar compounds:							
Polyethylene glycol (Carbowax 20M, FFAP, SP-2300)	25/275	316	495	446	637	530	2424
Tetracyanoethylated pentaerythritol (Silar-10C, SP-2340)	25/275	523	757	659	942	801	3682
Polydicyanoallylsilicone (OV-275)	25/250	781	1006	885	1177	1089	4938

3.13

TABLE 3.1 Stationary Phases in Gas Chromatography (*Continued*)

Liquid phase (chemical type; similar phases)	Minimum/maximum temperature, °C	McReynolds constants					
		x'	y'	z'	u'	s'	Σ
Specifically retards compounds with keto groups; for halogen compounds:							
Polytrifluoropropylmethylsiloxane, 50%/50% (OV-215)	0/275	149	240	363	478	315	1545
Polytrifluoropropylsiloxane (QF-1, SP-2401, OV-210, UCON HB 280X)	0/275	146	238	358	468	310	1520
For nitrogen compounds:							
Polyamide (Poly-A 103)	70/275	115	331	149	263	214	1072
Polycyanoethylmethylsilicone, 50%/50% (XF-1150)	20/200	308	520	470	669	528	2495
For fatty acid methyl esters:							
Neopentyl glycol succinate (HI-EFF-3BP)	50/230	272	469	366	539	474	2120
Diethylene glycol adipate (SP-2330, HI-EFF-1AP, LAC-1-R-296)	25/275	378	603	460	665	658	2764
Diethylene glycol succinate (HI-EFF-1BP, LAC-3-R-728)	20/200	499	751	593	840	860	3543
Absolute index values on squalane for reference compounds		653	590	627	652	699	

*5%/1%/94% = 5% phenyl/1% vinyl/94% methyl.

definition, the RI for these hydrocarbons are assigned a number that is 100 times the number of carbon atoms in the molecule. Thus, the retention indices of butane, pentane, hexane, and octane are 400, 500, 600, and 800, respectively, regardless of the column used or the operating conditions. However, the exact conditions and column must be specified, such as liquid loading, particular support used, and any pretreatment. For example, suppose that on a 20% squalane column at 100°C, the retention times for hexane, benzene, and octane are found to be 1.5, 1.6, and 2.5 min, respectively. On a graph of ln t'_R (naperian logarithm of the adjusted retention time) of the alkanes versus their retention indices, an RI of 653 for benzene is read off the graph. The number 653 for benzene (see the last line of Table 3.1 in the column headed x') means that it can be eluted halfway between hexane and heptane on a logarithmic time scale. If the experiment is repeated with a dinonyl phthalate column, the RI for benzene is found to be 736 (lying between heptane and octane), which implies that dinonyl phthalate will retard benzene slightly more than squalane will; that is, dinonyl phthalate is slightly more polar than squalane by $I = 83$ units (the entry in Table 3.1 for dinonyl phthalate in the column headed x'). The difference gives a measure of solute-solvent interaction due to all intermolecular forces other than London dispersion forces. The latter are the principal solute-solvent effects of squalane.

3.4.3 McReynolds Constants[6,7]

To systematically categorize the multitude of liquid phases that have been suggested or that are commercially available, Rohrschneider and later McReynolds introduced a series of coefficients derived from standards. The overall effects due to hydrogen bonding, dipole moment, acid-base properties, and molecular configuration can be expressed for the several standards as

$$\sum \Delta I = ax' + by' + cz' + du' + es'$$

where $x' = \Delta I$ for benzene (column headed x' in Table 3.1; represents intermolecular forces typical of aromatics and olefins)

$\quad\ y' = \Delta I$ for 1-butanol (column headed y' in Table 3.1; represents electron attraction typical of alcohols, nitriles, acids, and nitro and alkyl monochlorides, dichlorides, and trichlorides)

$\quad\ z' = \Delta I$ for 2-pentanone (column headed z' in Table 3.1; represents electron repulsion typical of ketones, ethers, aldehydes, esters, epoxides, and dimethylamino derivatives)

$\quad\ u' = \Delta I$ for 1-nitropropane (column headed u' in Table 3.1; typical of nitro and nitrile derivatives)

$\quad\ s' = \Delta I$ for pyridine or 1,4-dioxane (column headed s' in Table 3.1)

The last entry in Table 3.1 shows the absolute retention indices for the individual reference compounds on squalane. Also tabulated in Table 3.1 is the minimum temperature at which normal gas-liquid chromatography behavior is expected. Below that temperature, the phase will be a solid or an extremely viscous gum. The maximum temperature is that above which the bleed rate will be excessive as a result of solvent vaporization.

With the present state of column technology, a few facts hold true: Silicones are the first choice for stationary phases, cross-linking of the phase is preferred, and capillary columns provide the quickest separations. Only a few stationary phases are statistically different in their chromatographic behavior. For general analytical uses, methyl silicone (OV-1, SE-30, and OV-101), polyethylene glycol (PEG, CW-20M, and Superox), and trifluoropropyl silicone (OV-210, QF-1, and SP-2401) span the range of selectivity and polarity of stationary phases.

Silicones are the most versatile, reproducibly made, stable, and popular liquid stationary phases. They are thermally and oxidatively more stable than most other stationary phases and can be coated on virtually any surface, yielding excellent efficiencies. Silicones are easily cross-linked and have a wide usable operating temperature range.

Methyl silicone is the least polar of the silicones; compounds are eluted primarily in order of their boiling points. Changes in selectivity are achieved by replacing the methyl groups with polar groups such as phenyl, cyanopropyl, or trifluoropropyl. These subtle changes are shown in Table 3.1. A phase such as SE-54 has 5 percent phenyl groups, 1 percent vinyl groups, and 94 percent methyl groups on the silicone backbone. As the percent phenyl substitution increases, the selectivity and polarity (the $\Sigma \Delta I$ in the right-hand column of Table 3.1) of the phase also increases. However, strong dipole and proton donor molecules are not resolved from each other. Both OV-225 (polytrifluoropropyl + methylsiloxane) and SP-2300 (polyethylene glycol) are more polar and exhibit more selective retention for polar solutes. For most new analyses they should be tried second and third, respectively, after methyl silicones. They have a greater chance of showing differences in retention for polar solutes.

Another general rule applies to polar phases. As the polarity of the phase increases, overall retention decreases. The total energy of interaction, dispersive (nonpolar) interactions plus hydrogen bonding and dipole interactions (the two most significant polar interactions), is less than the total interactions of the nonpolar phases. Analysis times decrease. However, one can compensate by decreasing column temperature. For the separation of high-boiling compounds, a polar phase should be used so they can be eluted at a lower temperature.

A further general rule is that as the polarity of the phase increases, the phase becomes less stable. This results in a lower column lifetime.

High-temperature GC stationary phases are available with fused-silica capillary columns. Cross-linked methyl- and methyl(phenyl)silicones are polymerized inside fused-silica tubing. Capable of operating above 400°C, they are useful for separating components over C_{100}.

If a particular stationary phase is not in Table 3.1, then the McReynolds constants should be found from the vendor's literature (or determined in your laboratory as outlined earlier in this section). If a 3 to 5 percent loading of a standard packed column stationary phase currently separates your components, then the wide-bore column equivalent will separate them under the same conditions of temperature and flow. If a 10 percent loading is being used, then the wide-bore phase will usually perform the same separation at an oven temperature that is approximately 15 to 20 degrees lower.

One of the best ways to identify potential columns and stationary phases is to review the large number of example applications provided by column manufacturers and suppliers. Their catalogs contain many sample chromatograms obtained with a variety of columns.

3.4.4 Enantiomeric Separations

A novel stationary phase separates chiral molecules by providing chiral selectivity in the stationary phase. Thermostable alkylpolysiloxanes have been modified by chiral substitution. These types of phases have been used for the separation of many types of chiral molecules, including amino acids, some carbohydrates, and pharmaceuticals. Most of the phases can only be used up to 220°C.

3.4.5 Multidimensional Separations

In multidimensional gas chromatography, the components of a sample are separated by using series-connected columns of different capacity or selectivity. Two common multicolumn configurations are packed column–capillary column and two capillary columns in series. Two independently controlled ovens may be needed, and such a configuration is available commercially. In addition to decreased analysis time, this arrangement provides an effective way of handling samples containing components that vary widely in concentration, volatility, and polarity. Used in conjunction with techniques such as heart-cutting, backflushing, and peak switching, useful chromatographic data have been obtained for a variety of complex mixtures.

3.5 DETECTORS

After separation in the column, the sample components enter a detector. The detector should have the following characteristics:

1. High sensitivity
2. Low noise level (background level)
3. Linear response over a wide dynamic range
4. Good response for all organic component classes
5. Insensitivity to flow variations and temperature changes
6. Stability and ruggedness
7. Simplicity of operation
8. Positive compound identification

In Fig. 3.8 gas-chromatographic detectors are compared with respect to sensitivity and their linear dynamic ranges. What is desired is either a universal detector that is sufficiently sensitive or a dedicated detector that is very sensitive and specific to particular classes of molecules.

3.5.1 Thermal Conductivity Detector

The thermal conductivity detector (TCD) is the most common universal detector used in GC. It is rugged, versatile, and relatively linear over a wide range. In operation it measures the difference in the thermal conductivity between the pure carrier gas and the carrier gas plus components in the gas stream (effluent) from the separation column. Its construction is shown in Fig. 3.9. The detector uses a heated filament (often rhenium-tungsten) placed in the emerging gas stream. The amount of heat lost from the filament by conduction to the detector walls depends on the thermal conductivity of the gas. When substances are mixed with the carrier gas, its thermal conductivity goes down (except for hydrogen in he-

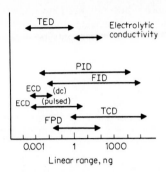

FIGURE 3.8 Linear dynamic ranges of gas chromatography detectors: TED, thermionic emission; PID, photoionization; FID, flame ionization; ECD, electron capture; TCD, thermal conductivity; and FPD, flame photometric.

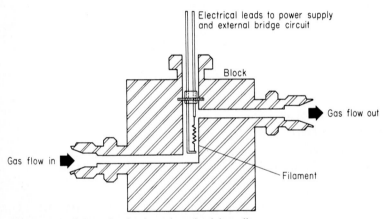

FIGURE 3.9 Schematic of a thermal conductivity cell.

lium); thus, the filament retains more heat, its temperature rises, and its electrical resistance goes up. Monitoring the resistance of the filament with a Wheatstone bridge circuit (Fig. 3.10) provides a means of detecting the presence of the sample components. The signals, which are fed to a chart recorder, appear as peaks on the chart, which provides a visual representation of the process.

Of all the detectors, only the thermal conductivity detector responds to anything mixed with the carrier gas. Being nondestructive, the effluent may be passed through a thermal conductivity detector and then into a second detector. The former serves as a general survey detector since it responds to all types of compounds. This detector is particularly suitable for fraction collection and preparative gas chromatography.

The cavity volume of the detector should be matched with the separatory col-

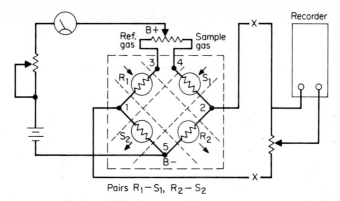

FIGURE 3.10 Circuitry for thermal conductivity (four-filament) cells.

umn used, 2.5 mL for detectors coupled to packed columns down to 30 µL in a detector designed for use with capillary columns.

The linearity of the detector is good at the lower concentration range but not in the high percent range. In the high percent range a multipoint calibration is the only way to ensure accurate measurements. At low parts-per-million concentrations, the trace impurities present in the carrier gas could be a limiting factor.

Oxygen is the most detrimental carrier-gas impurity. Even gold- and nickel-coated tungsten-rhenium hot wires are susceptible to oxidation, which may unbalance the bridge to the point where it cannot be rezeroed. Also, the oxide formation on the hot wire surface will minimize the detector's ability to sense changes in thermal conductivity and thus decrease its sensitivity.

3.5.2 Flame Ionization Detector

The flame ionization detector (FID) is the most popular detector because of its high sensitivity (0.02 C per gram of hydrocarbon), wide linear dynamic range, low dead volume, and responsiveness to almost all organic compounds. This detector (Fig. 3.11) adds hydrogen to the column effluent and passes the mixture through a jet where it is mixed with entrained air and burned. The ionized gas (charged particles and electrons produced during combustion) passes through a cylindrical electrode. A voltage applied across the jet and the cylindrical electrode sets up a current in the ionized particles. An electrometer monitors this current to derive a measure of the component concentration. An ignitor coil and flame-out sensor are placed above the jet to reignite the flame should it become extinguished. The entire assembly is enclosed within a chimney so that it is unaffected by drafts and can be heated sufficiently to avoid condensation of water droplets resulting from the combustion process.

The response of the flame ionization detector is proportionate to the number of $-CH_2-$ groups that enter the flame. The response to carbons attached to hydroxyl groups and amine groups is lower. There is no response to fully oxidized carbons such as carbonyl or carboxyl groups (and thio analogs) and to ether groups. Gaseous substances giving little or no response are listed in Table 3.2. The detector's insensitivity to moisture and the permanent gases is advantageous in the analysis of moist organic samples and in air-pollution studies. Column op-

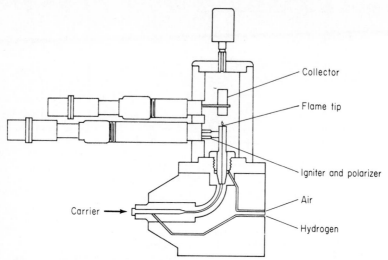

FIGURE 3.11   Flame ionization detector.

TABLE 3.2   Gaseous Substances Giving Little or No Response in the Flame Ionization Detector

He	CS_2	NH_3
Ar	COS	CO
Kr	H_2S	CO_2
Ne	SO_2	H_2O
Xe	NO	$SiCl_4$
O_2	N_2O	$SiHCl_3$
N_2	NO_2	SiF_4

erating temperatures can vary from 100 to 420°C, an obvious advantage in programmed temperature applications.

The quality of both the carrier gas and the hydrogen support gas is more critical than that of the flame support air. The presence of organics in any of these gases will increase the detector noise and the minimum detectable limit and, therefore, decrease the dynamic range.

3.5.2.1   Cleaning the Detector.   Problems such as excessive noise or random spikes on the chromatogram or a general lack of detector sensitivity may indicate a dirty detector. In-place cleaning is done with a halocarbon liquid, Freon 113. Inject 5 μL once or twice a day into the chromatographic column while the flame is lit. The combustion products of the cleaner remove silica deposits from the detector electrodes.

For light cleaning, turn off the power to the detector and remove the cap or top. Use a reamer to clean the jet opening, then brush the jet with a wire brush.

If more thorough cleaning is required, disassemble the detector carefully. Use a brush to remove all deposits (stainless steel for hard deposits and brass for softer ones). Clean all parts including the insulators, especially where Teflon or ceramic insulators contact metal. If necessary, the detector parts can be immersed in a surfactant cleaning solution, preferably in an ultrasonic bath. Commercial cleaning kits are available.

3.5.3 Thermionic Emission Detector

The thermionic emission detector (TED) (Fig. 3.12) responds only to compounds that contain nitrogen or phosphorus. Its fabrication is similar to that of a flame ionization detector except that a bead of rubidium silicate is centered 1.25 cm above the flame tip. The physical arrangement of the other component parts resembles that of a flame ionization detector. The ceramic bead is electrically heated and can be adjusted between 600 and 800°C. A fuel-poor hydrogen flame is used to suppress the normal flame ionization response of compounds that do not contain nitrogen or phosphorus. With a very small hydrogen flow, the detector responds to both nitrogen and phosphorus compounds. Enlarging the flame size and changing the polarity between the jet and collector limits the response to only phosphorus compounds.

Compared with the flame ionization detector, the thermionic emission detector is about 50 times more sensitive for nitrogen and about 500 times more sensitive for phosphorus.

3.5.4 Electron Capture Detector

The electron capture detector (ECD) responds only to electrophilic species, such as nitrogenated, oxygenated, and halogenated compounds. It consists of two electrodes (Fig. 3.13). On the surface of one electrode is a radioisotope (usually nickel-63, although tritium has been used) that emits high-energy electrons as it decays. Argon mixed with 5 to 10% methane is added to the column effluent. The high-energy electrons bombard the carrier gas (which must be nitrogen when this detector is used) to produce a plasma of positive ions, radicals, and thermal electrons. A potential difference applied between the two electrodes allows the col-

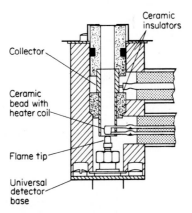

FIGURE 3.12 Thermionic emission detector.

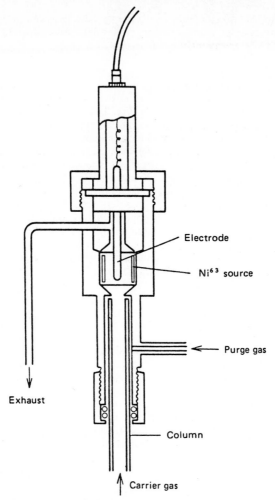

Electrode

Ni^{63} source

Purge gas

Exhaust

Column

Carrier gas

FIGURE 3.13 Electron capture detector.

lection of the thermal electrons. The current that results when only carrier gas is flowing through the detector is the baseline signal. When an electron-absorbing compound is swept through the detector, there will be a decrease in the detector current, that is, a negative excursion of the current relative to the baseline as the effluent peak is traced. The potential is applied as a sequence of narrow pulses with a duration and amplitude sufficient to collect the very mobile electrons but not the heavier, slower negative ions.

Residual oxygen and water must be rigorously removed from the carrier gas and makeup gases.

Next to the TCD and FID, the ECD has the greatest usefulness in the GC field. Unlike the FID, the ECD has neither ease of operation nor dynamic range. What it does have is detectability on the order of 1×10^{-13} g and good specificity.

Steroids, biologic amines, amino acids, and various drug metabolites can be converted to perfluoro derivatives which will give a signal with this detector.

3.5.5 Flame Photometric Detector

The flame photometric detector (FPD) is shown in Fig. 3.14. Phosphorus and sulfur can be detected simultaneously by attaching an interference filter for sulfur, an interference filter for phosphorus, an electrometer, and two photomultiplier tubes to one FPD cell.

The column effluent passes into a hydrogen-enriched low-temperature flame contained within a shield. Both air and hydrogen are supplied as makeup gases to the carrier gas. In this particular cell two flames are used to separate the region of sample decomposition from the region of emission. Flame blowout is no problem because the lower flame quickly reignites the upper flame. Phosphorus compounds emit green band emissions at 510 and 526 nm that are due to HPO species. Sulfur compounds emit a series of bands; the most intense is centered around 394 nm, but other bands overlap the phosphorus spectrum. The detector response to phosphorus is linear, whereas the response to sulfur depends on the square of its concentration.

Organic and carbon dioxide impurities in the makeup and carrier gases must be less than 10 ppm. The quenching effect of carbon dioxide is very significant.

The FPD has found application in the determination of pesticides and pesticide residues containing sulfur and phosphorus. It has also been used to detect gaseous sulfur compounds and to monitor air for traces of nerve gases (phosphorus compounds).

3.5.6 Photoionization Detector

The photoionization detector (PID), shown in Fig. 3.15, passes ultraviolet (uv) radiation through the column effluent from one of several lamps with energies

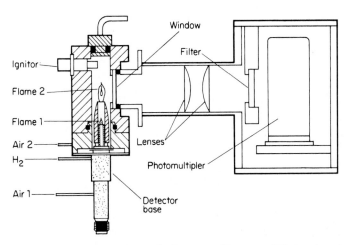

FIGURE 3.14 Flame photometric detector. (*Courtesy of Varian Associates.*)

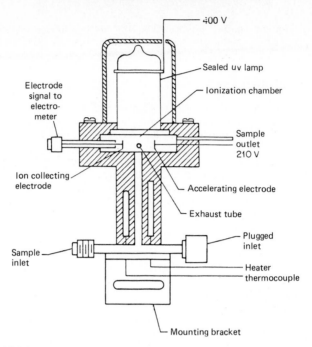

FIGURE 3.15   Photoionization detector.

ranging from 9.5 to 11.7 eV. Photons in this energy range are energetic enough to ionize most organic species but not the permanent gases. The ions formed are collected at a positively charged electrode, and the current is measured. Detector sensitivity can be varied by changing the uv lamp.

In comparison with other GC detectors, the PID has a wide linear range that extends into the parts-per-billion range.

The discharge ionization detector (DID) uses far-ultraviolet photons to ionize and detect sample components. Helium gas is passed through a chamber where high-voltage electrodes generate a glow discharge and cause it to emit a high-energy emission line at 58.84 nm (having an energy of 21.2 eV). This energy passes through an aperture to a second chamber where it ionizes all gas or vapor species present in the sample stream that have an ionization potential less than 21.2 eV (which embraces practically all compounds, including hydrogen, argon, oxygen, nitrogen, methane, carbon monoxide, nitrous oxide, ammonia, water, and carbon dioxide). A polarizing electrode directs the resulting electrons to a collector where they are quantitated with a standard electrometer.

3.5.7   Electrolytic Conductivity Detector

Organic compounds in the effluent are burned in a miniature furnace. Simple molecular species are formed that readily ionize and contribute to the conductivity of deionized water. Changes in electrolytic conductivity are measured. Ionic material is removed from the system by water that is continuously circulated through an ion-exchange column.

The combustion products may be mixed with hydrogen gas and hydrogenated over a nickel catalyst in a quartz-tube furnace. Ammonia is formed from organic nitrogen, HCl from organic chlorides, and H_2S from sulfur compounds.

This detector finds use in analysis of pesticides, herbicides, and alkaloids and certain pharmaceuticals.

3.5.8 Chemiluminescence-Redox Detector

This detector is based on specific redox reactions coupled with chemiluminescence measurement. An attractive feature of this detector is that it responds to compounds such as ammonia, hydrogen sulfide, carbon disulfide, sulfur dioxide, hydrogen peroxide, hydrogen, carbon monoxide, sulfides, and thiols that are not sensitively detected by flame-ionization detection. Moreover, compounds that typically constitute a large portion of the matrix of many environmental and industrial samples are *not* detected, thus simplifying matrix effects and sample cleanup procedures for some applications.

3.6 DATA PRESENTATION

The detector signal is normally displayed on a strip-chart recorder, producing the familiar chromatogram consisting of individual peaks. However, data handling and interpretation may be carried out with the help of more or less sophisticated data systems which are discussed in Chap. 1.

3.7 TEMPERATURE CONTROL

The temperature should be monitored, adjusted, and regulated at the injection port, in the oven surrounding the column, and at the detector. The temperature of the injection port must be sufficiently high to vaporize (virtually instantly) the sample, yet not so high that thermal decomposition or molecular rearrangements occur.

The temperature of the detector housing should be sufficiently high so that no condensation of the effluent occurs, yet not so high that the detector malfunctions.

The separation column is placed in a temperature-controlled oven. The temperature of the column should be maintained at a sufficiently high but constant level (isothermal operation), or it should increase with time in a manner which is reproducible from run to run (programmed temperature operation), so that the components of the sample are separated from one another and the retention time for the final peak is not too great.

3.7.1 Isothermal Operation

Selecting the column temperature for isothermal operation is a complex problem, and a compromise is usually the answer. Among members of a homologous series, there will be a decrease in retention time as the column temperature is increased. However, samples whose components have a wide range of boiling points cannot be satisfactorily chromatographed in a single isothermal run. A run at a moderate column temperature provides good resolution of the lower-boiling compounds but requires a lengthy period for the elution of high-boiling material. One solution is to raise the column temperature to a higher value at some point during the chromatogram so that the higher-boiling components will be eluted more rapidly and with narrower peaks. The better solution to this problem is to change the band migration rates during the course of separation by using temperature programming.

3.7.2 Temperature Programming

In temperature programming, the sample is injected into the chromatographic system when the column temperature is below that of the lowest-boiling component of the sample, preferably 90°C below. Then the column temperature is raised at some preselected heating rate. Typical programmed-temperature chromatograms are shown in Figs. 3.16 and 3.17. As a general rule, the retention time is halved for a 20 to 30°C increase in temperature. The final column temperature should be near the boiling point of the final solute but should not exceed the upper temperature limit of the stationary phase. Heating rates of 3 to 5°C $\cdot$ min^{-1} should be used initially and then fine-tuned to achieve optimum separation.

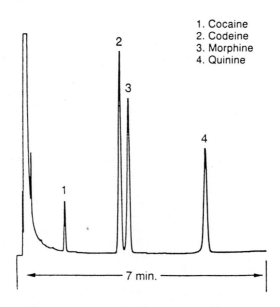

1. Cocaine
2. Codeine
3. Morphine
4. Quinine

7 min.

FIGURE 3.16   Programmed temperature separation of alkaloids on a polydiphenyldimethylsiloxane column (15 m × 0.53 mm i.d.), film thickness 1.5 μm; temperature programmed from 200 to 270°C at 10°/min; helium flow at 34 mL/min.

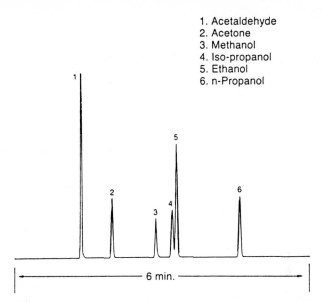

FIGURE 3.17   Separation of blood alcohols on a column coated with a 1.0-μm-thick film of polyethyleneglycol. Column was 30 m × 0.53 mm i.d. Temperature was programmed from 40°C (2-min hold) to 80°C at 10°/min. Helium flow was 7 mL/min.

EXAMPLE 3.1   Find the temperature increase that typically halves the k' value of one or several sample components. From sets (k', T) of data and realizing that, approximately,

$$\ln k' = \frac{B}{T} + C$$

where T is the temperature in kelvins and B and C are constants, determine the values of the constants. Next, estimate k' at both higher and lower temperatures. If k' equals 30 at 80°C and 15 at 100°C, then a 20°C increase in temperature will halve k'. After a temperature rise of 100°C to a column temperature of 180°C, the value of k' will be approximately one and the solute will have been eluted. Finally, try a heating rate of 5°C · min^{-1}. Increase the heating rate if the peaks are well separated but there is excessive space between the peaks. Consult catalogs of suppliers of chromatographic columns for sample chromatograms run with various temperature programs.

3.8   GAS-SOLID CHROMATOGRAPHY

In gas-solid chromatography (GSC) the column packing material is a molecular sieve or a porous polymer that consists of interconnected cavities with uniform openings. Only molecules smaller than these openings are able to enter the cavities.

3.8.1   Column Packings

Several types of column packings are available. These are as follows:

Type 3A has a pore diameter of 0.3 nm (or 3 Å) and will absorb molecules such as water and ammonia.

Type 4A has a pore diameter of 0.4 nm and absorbs all molecules with critical diameters up to 0.4 nm; these molecules include carbon dioxide, sulfur dioxide, hydrogen disulfide, ethane, ethylene, propylene, and ethanol.

Type 5A has a pore diameter of 0.5 nm. It separates straight-chain hydrocarbons (through C-22) from branched-chain and cyclic hydrocarbons.

Type 13X has a pore diameter of 1.0 nm.

Type B is a carbon molecular sieve that has a pore diameter of 1 to 3 nm. Water is eluted quickly and is well separated from methanol, ethanol, and formaldehyde.

Silica gel has a high retention for carbon dioxide.

Porous polymers are analogous to the packings used in exclusion chromatography; they are discussed in Chap. 4.

3.8.2   Operating Techniques

Many gas mixtures contain some components that cannot be separated on a particular column or cannot pass through a column in a reasonable time. There are two remedies.

1. *Backflushing:* In one scheme the flow of carrier gas is reversed after the last peak of interest has been eluted from the column. The higher hydrocarbons are backflushed from the upper end of the column packing and passed through the detector. In a second method the column is backflushed before any component has been eluted; after this the forward flow of carrier gas is resumed so that the lower-molecular-weight hydrocarbons can be eluted.

2. *Series-bypass method:* Certain sample components are temporarily stored in a second column while separations of the remainder are made on the first column. Then the carrier gas is switched to pass through the storage column and elute those components. For example, in a programmed time interval after sample injection, oxygen, nitrogen, methane, and carbon monoxide will have passed through the porous polymer column and entered the molecular sieve 5A column for temporary storage. Next, the carrier is directed only through the porous polymer column to elute hydrogen, carbon dioxide, ethylene, ethane, acetylene, and hydrogen sulfide. Finally, the molecular sieve column is switched back into the carrier-gas stream to elute the stored components in the order listed.

3.9   OPTIMIZATION IN GAS CHROMATOGRAPHY

In this section we show how separations can be optimized by varying some experimental condition. The process involves two sets of closely interrelated parameters: column characteristics and GC operational parameters. The question of what column should be used can be answered by selecting such variables as stationary phase, film thickness, column diameter and/or particle diameter, and column length. The operational parameters include carrier gas type, linear gas velocity, and oven temperature profile.

Efficiency and selectivity cannot be considered as separate entities. They combine to produce the observed resolution for a pair of solutes. When resolution is insufficient, one or both must be increased. If greater selectivity is not possible, efficiency must be increased. Cases of prime interest are those where resolution Rs is less than unity. The working equation is given by Eq. 2.16.

3.9.1 Carrier-Gas Velocity

From Eq. 2.18 and Fig. 3.2 one can see that there is a carrier-gas velocity which minimizes the plate height. It is u_{min} in a graph of H vs u. The only parameter which is velocity dependent in the resolution expression is H, hence Rs will have a maximum at u_{min}. It is good experimental practice to determine u_{min}. This can be best accomplished by obtaining H at several flow rates for some solute. Since

$$H = \frac{L}{N} \quad \text{and} \quad N = 5.54 \left(\frac{t'_R}{W_{1/2}} \right)^2$$

a chromatogram of some solute at several carrier-gas velocities should be obtained. Ascertain the plate number from the adjusted retention time and the peak width at half the maximum height. The carrier-gas velocity is measured from the expression

$$u = \frac{L}{t_M}$$

where L is the column length and t_M is the retention time of a nonretained solute.

3.9.2 Column Length

The resolution is proportional to the square root of the column length, so doubling the column length or halving the capillary column diameter will increase the resolution only by about 2. To a good approximation, the plate height is length-independent, as are k' and α. (Only at high carrier-gas velocities is H somewhat dependent on the column length.) Thus, to double the resolution, the column length must be increased by a factor of 4. At a constant inlet pressure any increase in length will result in an increase in the retention time. However, if the plot of H versus u is fairly flat, then the column length and the carrier velocity can be increased. This action will improve the resolution and help keep the analysis time reasonable. When the increase in plate height is large as the carrier-gas velocity increases, then only the column length should be changed to improve the resolution. The limiting factors here are the pressure drops needed across the column or the long analysis times.

3.9.3 Amount of Stationary Phase

Both H and k' are functions of the amount of stationary phase. Both quantities will decrease with decreasing volume of the liquid phase. Frequently u_{min} increases with decreasing volume of the liquid phase, which means shorter analysis time. However, at very low phase loading, not all the support surface (or the capillary walls) will be coated. Adsorption of the solute on the bare walls may occur with a deterioration in the efficiency. From the resolution equation we see that a decrease in k' is detrimental, especially if k' becomes fractional, while a decrease in plate height improves the resolution.

In general optimum loading and typical behavior of resolution are functions of percent stationary phase. As a rough guide when using Chromosorb W or P (or related supports), the optimum loading is around 15 percent. The optimum is lower for Chromosorb G. With untreated glass beads the optimum is probably around 0.1 percent. With open tubular (capillary) columns the optimum film thickness is roughly between 0.4 and 1 μm.

3.9.4 Column Diameter or Size of the Support Particle

The plate height is an explicit function of the diameter of the support particle. In addition, H is related (at least in packed columns) to the geometry of the packed bed, which, in turn, is support-size dependent. In general, the smaller the support particle, the better the column can be packed. Under normal conditions, a column packed with 100–120 mesh (149 to 125 μm) particles is more efficient than one packed with 60–80 mesh (250 to 177 μm) particles.

As column diameters or particle sizes decrease, increased pressures are required to maintain optimum flow rates. Moreover, carrier-gas compressibility acts to further reduce gains in column efficiency as higher inlet pressures are utilized. A practical maximum inlet pressure of approximately 100 lb · in² (700 kPa) limits the useful capillary column inside diameter to about 100 μm at a length of about 50 m.

When columns with reduced inside diameters are used, expensive high-performance inlets, column ovens, and detectors become necessary. Also, increasingly complex instruments are more difficult and expensive to maintain and service properly.

3.9.5 Temperature Effects

Since α, k', D_S, D_M, and u are all functions of temperature, resolution is a strong function of temperature. With rising temperature, the diffusion coefficients increase, while k' and the retention time decrease. Most frequently, the relative volatility, or selectivity, decreases as the temperature increases. The resolution can be improved by programming the temperature during a chromatographic run.

3.9.6 Treatment of Support and Tubing

Frequently resolution is lost because of peak tailing, which is a result of adsorption on active sites on the solid support or the tubing wall. It is good practice to treat the support with an acid (or base) wash to remove metallic impurities and to follow up by reaction with some silanizing reagent, such as dimethyl-dichlorosilane, to deactivate the SiOH groups on the support surface.

3.9.7 Conclusions

When trying to optimize the separation, the following steps should be taken.

1. *Reduce the column temperature:* Do not hesitate to work at temperatures far below the boiling points of the solutes since, once evaporated in the injector, condensation is a function of the dew point. Working at low temperatures and small loadings of the stationary phase is especially effective. Lower temperatures are conducive to longer contact periods between the solute and stationary phase, which results in increased retention and analysis times. Of course, a decrease in temperature can be coupled with a decrease in column length or in film thickness of the liquid phase. Higher temperatures lower the partition ratio and tend to reduce the effect of phase selectivity.

Several different temperature program rates may be needed through the run to tailor the profile for the particular analysis. A suitable initial trial rate is $2°C \cdot min^{-1}$.

2. *Carrier gas and linear velocity:* Once a suitable isocratic temperature or temperature program is found, change the carrier-gas velocity to operate around u_{min} or slightly higher if the van Deemter (h/u) plot is rather flat. Hydrogen gas is the best choice of carrier gas if analysis time is important (and adequate ventilation is available). Hydrogen is not recommended for use with a nitrogen-phosphorus detector or while temperature programming with a flame detector since a change in hydrogen glow may affect the flame detector response. Nitrogen is not recommended as a carrier gas for capillary chromatography because its use results in long analysis times compared to helium or hydrogen.

3. *Stationary phase:* Select a stationary phase with a polarity similar to the sample. Use the McReynolds constants as a guide. If the sample is a mixture of various polarities, choose an intermediate polarity phase such as 5 or 50% phenylmethyl silicone. A 5% phenylmethyl silicone capillary column should handle two-thirds of all applications.

4. *Column length:* To improve resolution, the column length can be increased, keeping in mind that the inlet pressure must be increased and that the retention times will increase. Remember that resolution is proportional to the square root of column length.

5. *Film thickness and column diameter:* The distribution of a solute between the stationary phase and the gas phase is directly proportional to the film thickness and inversely proportional to the column diameter. To improve resolution, especially of low-molecular-weight materials, a thick film may be advisable. This will result in a longer analysis time. Common film thicknesses are 0.11, 0.17, 0.33, 0.50, and 1.0 µm. A good starting point is 0.33 µm.

A smaller column diameter increases the partition ratio. For increased resolution, a smaller diameter is beneficial. However, this results in longer analysis times since flows are lower. To speed up an analysis, use a wider bore column or increase the oven temperature. Diameters of available capillary columns are 0.2, 0.32, and 0.53 mm i.d. A 0.2-mm column is a good starting point.

3.9.8 Influence of the Gas Chromatographic System

The observed variance of the chromatographic peak σ^2 is composed of the variance due to the partition process in the column ($\sigma_c{}^2$) and the variance due to band spreading outside the column ($\sigma_i{}^2$). For example, if the the ratio of the two standard deviations is 0.5, the actual number of theoretical plates will be 20 percent lower than could be obtained if no band spreading occurred outside the column.

The bandwidth outside the column increases in several ways. This spreading includes the dilution factor caused by (1) the injector, (2) any connecting tubing, and (3) the detector volume. Some problems are as follows:

1. The volume of the sample is too large, which causes the original bandwidth of the sample "plug" to be excessive.

2. The sample injection is not carried out rapidly enough, thus the width of the sample vapor plug is increased.

3. The (void) volume between the point of injection and the entrance to the stationary phase is too large, which causes excessive band spreading as a result of diffusion of the sample vapor into the carrier gas.

4. The tubing connecting the column outlet and the detector must be kept to an absolute minimum because diffusion in the carrier gas results in the remixing of the separated components. Addition of makeup gas to the column effluent will speed up the gas velocity and sweep the column effluent through this "dead" or void volume.

5. The volume of the detector is too large compared to the actual gas volumes corresponding to the peak width at half height. The detector volume should be roughly 0.1 the peak width.

6. All couplings should possess zero dead volume.

3.10 SYSTEMATIC TROUBLESHOOTING

Visualize your instrument from the diagram shown in Fig. 3.1. To locate a failure or a fault, systematically work downward through the functional levels and work from detector to column to inlet, following the signal path back to its origin.

Before any testing is done, a set of expected performance criteria must be established. The instrument manufacturer often will supply standard performance benchmarks along with the conditions under which they were measured. Usually a test column and a test mixture are involved. Detector sensitivity, noise, minimum detectable quantity, peak resolution, and other general measurements can be specified. These criteria should be determined by the user when the instrument is installed and periodically thereafter even though the test may not be directly related to the intended application.

Instrument manuals usually include a troubleshooting section. This section should be read both as a reference and to learn how to operate the instrument properly.

Establish performance standards for the applications for which the instrument was purchased. Then check these performance standards regularly. Perform frequent and regular calibration of detector response factors along with visual assessment of the calibration chromatogram. Control samples interspersed with the application samples will enable the user to detect decreasing detector response, gradually broadening peaks, or drifting retention times.

Anticipate trouble. Keep on hand a good supply of nuts, ferrules, and fittings as well as inlet and detector spares, such as septa, inlet liners, and flame jets. Extra syringes and autosampler parts are always useful. A set of tools purchased specifically for your instrument is a good investment; store them safely in a drawer. Perform the recommended periodic maintenance procedures on schedule.

An instrument log is an essential tool for problem diagnosis. This includes everything done with the instrument and auxiliary units. Keep notes on peculiarities that occur—they may telegraph upcoming trouble.

If the problem appears to be chromatographic in origin, check the various inlet, detector, and column temperatures, pressures, and flows and compare them to their normal values. If possible, measure some basic chromatographic parameters such as unretained peak times, capacity factors, or theoretical plate numbers. Check for loose column or tubing connections. If the problem appears to be electronic in nature, carefully inspect the instrument (with the power cord unplugged). Are the cables plugged in? Are boards seated firmly in their connectors?

If the column is determined to be at fault, consider the following problems and suggested remedies:

1. Loss of resolution due to band-broadening effects may have been caused by large solvent injections displacing the stationary phase. Although bonded-phase columns are relatively immune to this problem, backflushing such columns with solvent is recommended for restoration. (Do not backflush nonbonded columns.)

2. If solvent flushing is inappropriate or fails, remove the first two or three column turns of a capillary or the upper centimeter or so of a packed column.

3. Excessive column heating may cause the stationary film to coalesce into uneven "hills and valleys" on the inner column wall. Usually a replacement column is needed.

4. Microcontaminants may gradually deposit at the beginning of the column and interact with solutes, leading to adsorption and possibly to catalytic decomposition or rearrangement. Removing the beginning column section is an effective remedy. A guard column in front of the regular column is another solution.

5. Oxygen contamination in the carrier gas can cause excessive column bleed. The remedy is to install a fresh, high-quality oxygen trap.

6. Septum bleed or residues, or high-boiling, strongly retentive solutes from a previous run, may raise the baseline. Try several bake-out temperature cycles with no sample injection. If large peaks from previous injections are eluted as broad peaks during a subsequent run, a thinner stationary phase film should be used if higher column temperatures are not possible.

If the problem persists and is electronic, it is time to call the manufacturer's service technician.

REFERENCES

1. C. J. Wolf, M. A. Grayson, and D. L. Fanter, "Pyrolysis Gas Chromatography," *Anal. Chem.*, **52**:348A (1980).

2. B. V. Joffe and A. G. Vitenberg, *Headspace Analysis and Related Methods in Gas Chromatography,* Wiley, New York, 1984.

3. M. E. McNally and R. L. Grob, "Static and Dynamic Headspace Analysis," *Am. Lab.,* **17:**20 (January 1985); **17:**106 (February 1985).

4. R. T. Wiedemer, S. L. McKinley, and T. W. Rendl, "Advantages of Wide-Bore Capillary Columns," *Am. Lab.,* **18:**110 (January 1986).

5. L. S. Ettre, "The Kovats Retention Index System," *Anal. Chem.,* **36:**31A (1964).

6. L. Rohrschneider, *J. Chromatogr.,* **22:**6 (1966).

7. W. O. McReynolds, *J. Chromatogr. Sci.,* **8:**685 (1970).

BIBLIOGRAPHY

Berezkin, V. G., *Chemical Methods in Gas Chromatography,* Elsevier, New York, 1983.

Cowper, C. J., and A. J. DeRose, *The Analysis of Gases by Gas Chromatography,* Pergamon, New York, 1983.

Dressler, M., *Selective Gas Chromatographic Detectors,* Elsevier, New York, 1986.

Grob, R. L., and M. Kaiser, *Environmental Problem Solving Using Gas and Liquid Chromatography,* Elsevier, New York, 1982.

Lee, M. L., F. J. Yang, and K. D. Bartie, *Open Tubular Column Gas Chromatography: Theory and Practice,* Wiley, New York, 1984.

CHAPTER 4

HIGH PERFORMANCE LIQUID CHROMATOGRAPHY

High performance liquid chromatography (HPLC) uses two immiscible phases in contact with one another. The operations of sample introduction, chromatographic separation, and detection are performed in exactly the same way in HPLC as in gas chromatography except for those modifications necessary to accommodate a liquid rather than a gas as the mobile phase. That is, the sampling is sequential, the separated components are eluted with the mobile phase, and detection is a dynamic time-dependent process. Chromatographic separation in HPLC is the result of specific interactions of the sample molecules with both the stationary and mobile phases. With an interactive liquid mobile phase, another parameter is available for selectivity in addition to an active stationary phase. HPLC can be used whenever the sample can be dissolved in a liquid.

HPLC comprises a number of methods. A general guide for the selection of a method is given in Fig. 4.1. The division of sample components based on molecular weight is arbitrary but useful. Consider these factors when selecting an HPLC method.

1. All sample components are less than 2000 daltons (atomic mass units).
   a. The sample is water insoluble and possesses an aliphatic or aromatic character.
      (1) Adsorption (liquid-solid) chromatography (LSC) is suggested for class separations or for the separation of isomeric compounds.
      (2) Liquid-liquid chromatography (LLC) is suggested for the separation of homologs. Separation is achieved by matching the polarities of the sample and stationary phase and by using a mobile phase that has a markedly different polarity.
   b. The sample is ionic or possesses ionizable groups.
      (1) Ion-exchange chromatography (IEC) is suggested; use anion exchangers for acidic compounds and cation exchangers for basic compounds.
      (2) Ion-pairing chromatography (IPC) is another possibility.
2. Some or all of the sample components exceed 2000 daltons.
   a. Exclusion chromatography (EC) is suggested. This method is based on the ability of controlled-porosity substrates to sort and separate sample mixtures according to the size and shape of the sample molecules.
   b. If the sample is water soluble, an aqueous mobile phase is used.
   c. If water insoluble, a nonaqueous mobile phase is used.
3. Affinity chromatography (not shown in Fig. 4.1) uses immobilized bio-

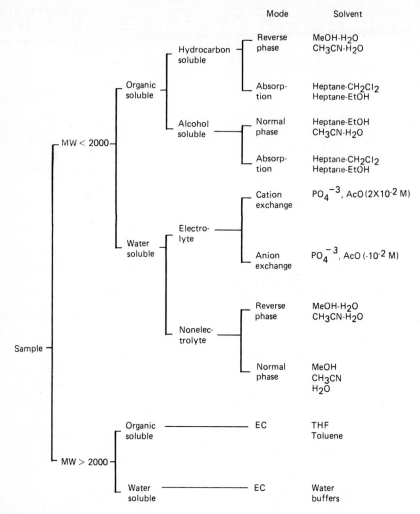

FIGURE 4.1 Guide to selecting HPLC methods.

chemicals as the stationary phase to achieve separations via the "lock and key" binding that is prevalent in biologic systems.[1]

4.1 COMPONENTS OF A LIQUID CHROMATOGRAPH

The essential components for HPLC instrumentation are shown in Fig. 4.2. These components include:

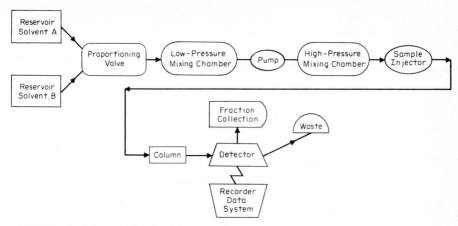

FIGURE 4.2 Schematic showing the components of an HPLC system.

1. A pump or pumps to force the mobile phase through the system. Suitable pressure gauges and flow meters are placed in the system.
2. Sampling valves and loops to inject the sample into the mobile phase just at the head of the separation column.
3. A separation column where the sample components are separated into individual peaks before elution.
4. A detector and readout device to detect the presence of solutes in the mobile phase and record the resulting chromatogram. To collect, store, and analyze the chromatographic data, computers, integrators, and other data processing equipment are being used more frequently in conjunction with the strip-chart recorder.

4.2 MOBILE-PHASE DELIVERY SYSTEMS

Pumps are used to deliver the mobile phase to the column. The pump, its seals, and all connections in the chromatographic system must be constructed of materials that are chemically resistant to the mobile phase. A degassing unit is needed to remove dissolved gases from the solvent.

Less expensive LC systems require a pump that is capable of operating to at least 1500 lb · in^2. A more desirable upper pressure limit is 6000 lb · in^2. The pump's holdup volume should be small. Flow rates for many analytical columns fall within 0.5 to 2 mL · min^{-1}. For microbore columns the flow rates need only be a few microliters per minute.

4.2.1 Reciprocating Piston Pumps

The reciprocating piston pump (Fig. 4.3) is the most popular type because it is relatively inexpensive and permits a wide range of flow rates. A hydraulic fluid transmits the pumping action to the solvent via a flexible diaphragm which min-

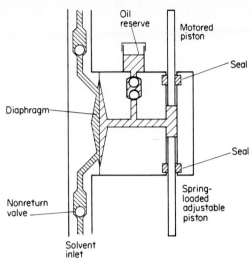

FIGURE 4.3   Reciprocating pump.

imizes solvent contamination and corrosion problems. Some of the characteristics of this pump are:

1. Flow rates can be varied either by altering the stroke volume during each cycle of the pump or the stroke frequency.
2. Solvent delivery is continuous.
3. No restriction exists on the reservoir size or operating time.
4. Solvent changes are rapid and accurate, an advantage when doing gradient elution or solvent scouting.
5. Some type of pulse-dampening system must be used.
6. Dual-head and triple-head pumps having identical piston-chamber units operated 180° or 120° out of phase smooth out all but small pulsations from the solvent delivery.

4.2.2 Syringe-Type Pumps

Syringe-type pumps operate through positive solvent displacement and use a piston mechanically driven at a constant rate (Fig. 4.4). The characteristics of this pump are:

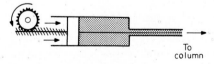

FIGURE 4.4   Syringe-type pump.

1. Solvent delivery is controlled by changing the voltage on the digital stepping motor.
2. Solvent chamber capacity is finite (250 to 500 mL) but is adequate with small-bore columns.
3. Pulseless flow is obtained.
4. High-pressure capability (200 to 475 atm, 3000 to 7000 lb $\cdot$ in^2) can be achieved.
5. Solvent gradients are made feasible by the tandem operation of two or more pumps.

4.2.3 Constant-Pressure Pumps

The mobile phase is driven by the pressure from a gas cylinder, which is delivered through a large piston (Fig. 4.5). With this type of pump,

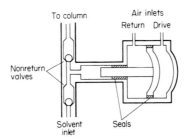

FIGURE 4.5   Constant pressure pump.

1. A low-pressure gas source (1 to 10 atm) can generate high liquid pressures (up to 400 atm, 6000 lb $\cdot$ in^2) because of the large difference in area of the two pistons.
2. Pulseless flow is provided.
3. High flow rates are available for preparative applications; a valving arrangement permits the rapid refill of the solvent chamber.
4. Gradient elution is inconvenient.

4.2.4 Pulse Dampers

Some detectors are sensitive to variations in flow; all detectors will benefit from the decreased noise resulting from pulseless operation. The following types of pulse dampers are available:

1. Flexible bellows or a compressible gas in the capped upright portion of a tee tube. These types have large fluid volumes but require pressures in excess of 1000 lb $\cdot$ in^2 for effective operation.
2. A compressible fluid separated from the mobile phase by a flexible diaphragm.

This system offers easy mobile-phase changeover and minimal dead volume, and is effective at low system pressures.

3. Electronic pulse dampers are useful with reciprocating pumps. They provide a small, rapid forward stroke of the piston following the pump's rapid refill stroke.

4.2.5 Connecting Tubing

Polymeric tubing is limited in its use to supply lines that connect the mobile-phase reservoirs to the pump, to the outlet side of the detector, and to lines such as the injector waste. Stainless-steel tubing must be used in regions subject to high pressure, that is, between the pump outlet and the detector outlet. LC-grade tubing should be used. The tubing is clean and ready to use, and the vendor takes care to ensure concentricity (an important feature in small-diameter tubing) so that the tubing bore will align properly with the hole in connecting fittings.

If tubing is cut in your laboratory, be certain to remove any burrs or filings and to clean the tubing thoroughly. Before using the tubing, flush it with dichloromethane or tetrahydrofuran and then with a detergent solution to remove grease and oil residues.

4.2.6 Sampling Valves and Loops

A calibrated sample loop (volume usually 10 or 20 μL) is filled by means of an ordinary syringe which thoroughly flushes the sample solution through the loop. Manual rotation of the valve rotor places the sample-filled loop into the flowing stream of the mobile phase, as shown in Fig. 4.6. A sliding valve uses an internal sample cavity consisting of an annular groove on a sliding rod that is thrust into the flowing stream. With either injecting system, samples are dissolved, if possible, in the mobile phase to avoid an unnecessary solvent peak in the final chromatogram.

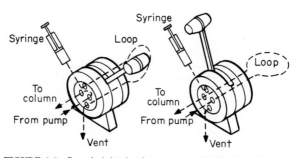

FIGURE 4.6 Sample injection loop.

4.3 GRADIENT ELUTION (SOLVENT PROGRAMMING)

4.3.1 Equipment

Either low- or high-pressure systems may be used in gradient elution. In the low-pressure system, two or more solvents are blended in the desired proportions by a precise valving system which often is controlled by a microprocessor. The mixing vessel is packed with an inert fiber which creates sufficient turbulence to mix the solvents while simultaneously causing the release of any generated gas bubbles. A low-pressure system requires only one pressurization pump—a distinct advantage.

In a system where the gradient is formed after the HPLC pump, the output from two or more high-pressure pumps is programmed into a low-volume mixing chamber before flowing into the column. One pump is required for each solvent involved.

4.3.2 Methodology

Gradient elution is normally used for separating samples that vary widely in polarity (the k' ratio exceeds 20), as well as for those in which the isocratic retention bunches peaks together at the beginning of the chromatogram and spreads them out (with broadening) at the end of the chromatogram. The gradient separation provides a more even spacing of bands and better resolution, as well as narrower bands at the end of the chromatogram. Gradient elution works by changing k' during the separation (k' programming) because mobile-phase strength changes. Weakly retained compounds leave the column first in a weak mobile phase. Strongly retained bands leave last in a strong mobile phase.

With commercial equipment a variety of gradients are available. The linear gradient is best for scouting initially. If nothing elutes at the beginning of the run, start the gradient at an organic content that is 5% lower than that needed to elute the first peak; this will eliminate wasted solvent at the beginning of the run. Stopping the gradient as soon as the last band has been eluted will eliminate waste at the end of the run. Thus the gradient range will have been changed without adversely affecting the separation.

The starting mobile-phase composition should be sufficiently weak to give good separation of the early bands. The final percentage of strong solvent in the gradient should be adjusted so that the last band leaves the column at about the time the gradient is completed. If early bands are poorly resolved, the starting percentage of strong solvent is too large and/or the gradient shape is too convex. The remedy is to decrease the starting percentage of strong solvent. If the early bands are well resolved and sharp and there is no excessive space between bands, the gradient should be started only after they have been eluted or else a convex-shape gradient should be used.

Resolution can be improved in gradient separations by optimizing k', N, and/or α. The k' value in gradient elution is given by

$$k' \quad \frac{(\text{constant})t_G F_c}{(\Delta \%B)(V_M)} \qquad (4.1)$$

where t_G = gradient retention time
$\quad\quad F_c$ = flow rate
$\quad\quad V_M$ = column volume of mobile phase
$\quad\quad \Delta\%B$ = difference between starting and final percentage of strong solvent in gradient

Here, the term denoted by "(constant)" is approximately 20 for reverse-phase HPLC. An increase in the flow rate for a gradient separation reduces the resolution, the bandwidths, and the retention times and causes k' to be higher. In order to change the flow rate without changing k', other parameters in Eq. 4.1 must be changed. One possibility is to double the flow rate while halving the gradient time. This speeds the separation without altering the selectivity.

It is possible to increase resolution by increasing the plate number. For example, if the column length is doubled (which doubles V_M), either the flow rate or the gradient time must also be doubled (or some suitable variation in each quantity to yield a factor of 2 for their product).

Selectivity is achieved by altering α and is accomplished by changing mobile-phase composition, column type, or temperature. Experience is the main guide.

When running high-molecular-weight samples or very shallow gradients, special problems may occur. Small changes in mobile-phase composition affect the retention times as if they were varying within a run and spurious peaks appear.

4.4 COLUMNS[2,3]

Separation columns are available in various lengths and diameters. To withstand the high pressures involved, columns are constructed of heavy-wall glass-lined metal tubing or stainless-steel tubing. Connectors and end fittings must be designed with zero void volume. Column packing is retained by frits inserted in the ends of the column.

4.4.1 Standard Columns

Columns with an internal diameter of 4 to 5 mm are standard in HPLC. Packings should be uniformly sized and mechanically stable. Particle diameters lie in the range from 2 to 5 μm. Column lengths range from 10 to 30 cm.

When analytical speed is of prime consideration, as in quality-control work, a short (3 to 6 cm) column is useful.

4.4.2 Narrow-Bore Columns

With narrow-bore (2-mm) columns, the mobile-phase velocity remains the same and the analysis time remains unchanged in comparison with standard columns. Narrow-bore columns offer several advantages over the standard columns:

1. The detector signal of a sample component is increased by a factor of 4 when the internal diameter of the column is decreased by about a factor of 2 (Fig. 4.7).

2. Solvent consumption is 4 times less, which is a considerable savings in solvent purchase and waste disposal costs.

3. Unconventional or high-purity solvents can be used because solvent consumption is lower.

4. Packing density is more homogeneous.

5. Better column permeability allows smaller particles to be packed without exceeding the conventional pressures used in HPLC.

6. Frictional heat is dissipated better, so the temperature gradients across the column are smaller.

7. A number of short columns can be joined together to increase the total column length without loss of efficiency (plate count).

Extracolumn effects must also be considered. Generally, this calls for use of a detector cell with a smaller volume and for very short lengths of 1.3- to 1.8-mm-i.d. tubing if short columns with 3-μm particle packings are used.

4.4.3 Guard Columns and Filters

Particulate matter from the sample or solvent(s) will plug the frit (0.5- to 2-μm pore diameter) at the column entrance or, in the absence of a frit, the upper portion of the column packing. The cause is inadequate sample cleanup or contaminated solvents. The guard column acts as a chemical filter to remove strongly retained materials that might otherwise foul the analytical column and thus shorten its lifetime. Guard columns, 1 to 5 cm in length, are available in dispos-

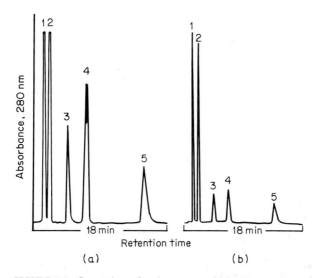

FIGURE 4.7 Comparison of peak response of (*a*) a 30-cm × 2-mm column and (*b*) a 30-cm × 4-mm column. Both columns were packed with the same stationary phase. Components: 1, ascorbic acid; 2, niacinamide; 3, pyridoxine HCl; 4, riboflavin; and 5, thiamine HCl. Linear velocity is same for both columns.

able cartridge designs or as pack-it-yourself kits. They contain a stationary phase similar to that in the analytical column and fit between the injector and the analytical column and result in a low dead volume.

The mobile phase and sample should usually be filtered through a 0.45-μm membrane filter, especially if they contain buffers, salts, or any other dissolved solid chemicals. Only HPLC-grade solvents need not be filtered.

Plugged frits can be cleaned by backflushing or by cleaning the frit in an ultrasonic bath. Frits are easily replaced. Upper column packing that has become contaminated can be removed and replaced by fresh packing material.

4.5 STATIONARY PHASES

The stationary phase may be either a totally porous particle or macroporous polymer, a superficially porous support (porous-layer beads), or a thin film covering a solid core (pellicular supports). Each type may have a polymer bonded to the support surface (bonded-phase supports).

4.5.1 Totally Porous Particles

Represented by silicas and aluminas, totally porous particles have a large surface area and pores throughout the structure. Particles can be packed in columns to give 800 plates per centimeter if a 5-μm particle size is used. Wide-pore silicas are the basis of many families of specialty packings. A pore size of 30 nm is popular for proteins and is suitable for the separation of most biopolymers. For biopolymers not fully resolved on the smaller-size materials, 50-, 100-, and 400-nm pore sizes are available.

4.5.2 Macroporous Polymers

Macroporous polymers have large channels as well as micropores leading off the channels and surface. Beads of these polymers do not swell or shrink appreciably with changes in the ionic strength of the mobile phase or deform at high flow velocities. They are well suited to separations conducted in nonaqueous media and for gel permeation chromatography. When functional ion-exchange groups are incorporated in the polymer structure, these packings are used in ion-exchange chromatography.

4.5.3 Porous-Layer Beads

Porous-layer beads are solid, spherical glass beads covered by a thin, porous outer shell or film. Mass transfer is improved in a thin film or layer. Several coating thicknesses are available; thicker coatings give rise to slower mass transfer but have increased sample capacity.

4.5.4 Extracolumn Effects

Most commercial LC systems are designed so that extracolumn effects are insignificant if tubing runs are kept short and if standard columns (15 cm × 4.6 mm, 5 μm particles) are used. If smaller-dimension or smaller-particle columns are used, however, extracolumn effects can be important. These extracolumn effects are the added volume contribution from the injector sample loop, the detector cell and time constant, and the tubing connecting the column to the injector and detector.

4.5.5 Void-Volume Markers

Suitable void-volume markers are few in number. The best is a D_2O-enriched mobile phase. Uracil gives the most uniform results with changes in either solute concentration or mobile-phase composition, and it is easily detected at 254 nm. The use of ionic species is ill-advised; depending upon the partial charge present on the residual silanol groups, negatively charged ionic species can be excluded from the pores of the packing material and from the surface in general. This exclusion process leads to a significant decrease in the apparent void volume of the column.

4.6 DETECTORS

Before considering individual types of detectors, general properties are discussed.

1. *Response time:* Response time should be at least 10 times less than the peak width of a solute in time units to avoid distortion of the peak area.
2. *Detection limit:* Typically the concentration of the solute peak in the detector is $\frac{1}{5}$ to $\frac{1}{350}$ of the initial sample concentration at injection because of the dilution factor. For precise quantitation, a tenfold greater concentration than this estimate is needed. Keep in mind that narrow-bore columns dilute small samples less than do large-bore columns.
3. *Linearity and dynamic range:* For quantitative work the signal output of the detector should be linear with concentration (concentration-sensitive detector) or mass (mass-sensitive detector). A wide linear dynamic range, perhaps 5 orders of magnitude, is desirable in order to handle major and trace components in a single analysis.

Microprocessors are often used to control the operations for HPLC. The chromatographer sets a threshold detector level at a value desired to trigger the control subroutine. The threshold is the level to which the detector signal must rise from the baseline signal for the controller to recognize that a peak is entering the sensitive volume of the detector. Several tasks can be performed:

1. Ascertaining the completion of an analysis.
2. Ascertaining and controlling the proper times for stop-flow analysis methods.
3. Developing methods.
4. Collecting individual peaks with a fraction collector.
5. Monitoring of column performance.

Table 4.1   Guide to HPLC Solvent Properties

Solvent	Viscosity,* cP	UV cutoff†	Refractive index at 20°C	B.P.
Acetic acid	1.31^{15}	255	1.372	117.9
Acetone	0.30^{25}	330	1.359	56.3
Acetonitrile	0.34^{25}	190	1.344	81.6
Benzene	0.65	280	1.501	80.1
Carbon tetrachloride	0.97	265	1.460	76.8
Chloroform	0.58	245	1.446	61.2
o-Dichlorobenzene	1.32^{25}	295	1.551	180.5
Ethyl acetate	0.46	255	1.372	77.1
Diethyl ether	0.24	220	1.352	34.6
Heptane	0.42	197	1.388	98.4
Hexane	0.31	192	1.375	68.7
Isobutyl alcohol	4.70^{15}	220	1.396	107.7
Methanol	0.55	206	1.328	64.7
Methylene chloride	0.45^{15}	233	1.424	39.8
Methyl ethyl ketone	0.42^{15}	330	1.379	79.6
Pentane	0.24	200	1.357	36.1
2-Propanol	2.86^{15}	210	1.377	82.3
Tetrahydrofuran	0.55	212	1.407	66
Toluene	0.59	285	1.497	110.6
2,2,4-Trimethylpentane (iso-octane)	0.50	205	1.391	99.2
Water	1.00	190	1.333	100

*Viscosity in centipoise (cP) at 20°C unless otherwise noted.
†The ultraviolet cutoff is the wavelength in nanometers at which the absorbance (1-cm path vs. water) reaches 1.0 scanning from longer wavelengths. Maximum allowable value listed.

4.6.1   Ultraviolet-Visible Detectors

Optical detectors based on ultraviolet-visible absorption constitute over 70 percent of HPLC detection systems. It is not necessary that the wavelength selected for HPLC coincide with the wavelength where maximum absorption occurs. True, sensitivity will suffer, but any wavelength within the absorption envelope will be usable.

A list of common HPLC solvents and their transmittance cutoffs in the ultraviolet region is given in Table 4.1.

Detectors are available that span the full range from single-wavelength to simultaneous multiwavelength detection capabilities. Diagrams of the several optical systems are found in Chap. 6. Two key advantages of this mode of detection are as follows:

1. Wavelength selectivity
2. Excellent analyte sensitivity

Restrictive operating conditions are as follows:

1. Thermostatic control to 0.01°C is required if the detector is to approach the shot (random) noise limitation of 10^{-6} absorbance unit.
2. Detector cell volume should not exceed 8 µL per centimeter of optical path

length for the standard separation columns, and 2 μL for narrow-bore columns.

3. The time constant should be 0.04 s.

The mobile-phase solvent should absorb only weakly or not at all. Water, methanol, acetonitrile, and hexane all permit operation in the far ultraviolet to at least 210 nm.

The concentration detection limit in units of $mol \cdot cm^{-1} \cdot L^{-1}$ is given by

$$\frac{2\,(noise)}{b\epsilon}$$

where b is the path length of the optical cell and ϵ is the molar absorptivity.

The detection limit in terms of the sample weight is given by

$$\frac{(2)\,(noise)\,(dilution\ factor)\,(sample\ size,\ L)\,(mol.\ wt.)}{b\epsilon}$$

For conventional HPLC conditions, typical detectability with commercial absorbance detectors is 1 ng of injected analyte or an injected concentration of $5 \times 10^{-7}\ M$.

The basic types of ultraviolet-visible detectors are described in order of increasing complexity (and cost) along with their advantages and limitations.

4.6.1.1 Fixed-Wavelength Detector.   This type of detector uses a light source that emits maximum light intensity at one or several discrete wavelengths that are isolated by appropriate optical filters. Its *advantages* are (1) low cost, (2) a minimum of noise, usually less than 0.0001 absorbance unit, and (3) sensitivity at the nanogram level for compounds that absorb at an available fixed wavelength. Its *disadvantage* is that there is no free choice of wavelength in many situations.

4.6.1.2 Variable-Wavelength Detector.   Usually a wide-bandpass ultraviolet-visible spectrophotometer is coupled to the chromatographic system. Its *advantage* is the wide selection of wavelengths from 190 to 600 nm, which permits the user to choose the wavelength at which solute absorbance is maximum. *Disadvantages* are (1) increased cost and (2) increased detector noise level that is usually a factor of five to ten greater than that of filter photometers.

4.6.1.3 Scanning-Wavelength Detector.   Although the scanning-wavelength detector is the most costly of the three types, it has many *advantages*:

1. It offers a real-time spectrum from 190 to 600 nm that can be obtained in as little as 0.01 s for each solute as it elutes.

2. Spectra are available for visual presentation as three-dimensional chromatograms of time vs. wavelength vs. absorbance.

3. When diode arrays are used as the detector, it is possible later to extract data at other wavelengths from the memory.

4. Some instruments can be configured to monitor simultaneously a number of wavelength intervals with bandwidths ranging between 4 and 400 nm. This permits the integration of all signals between two preset wavelengths.

5. With simultaneous absorbance detection at several wavelengths, the presence

of coeluting peaks is diagnosed from the shifting of retention times as a function of detection wavelength.

6. Peak purity can be determined by taking a spectrum at the front and at the rear of a peak.

4.6.2 Fluorometric Detector

Fluorescence detection in HPLC provides excellent selectivity and sensitivity. Typical instrument diagrams are discussed in Chap. 7. The choice as to which type of unit to purchase depends primarily on the type of work that will be done. If the detector will be used exclusively for a specific task and the chromatography is fairly clean, a filter fluorometer is quite appropriate. On the other hand, if the detector will be used for many different purposes, a spectrofluorometer will be best. An instrument that employs a monochromator for the excitation wavelength and a filter system for isolating the emission provides selectivity for excitation and sensitivity for emission.

Design criteria for flow cells involve a compromise between the cell volume and the excitation-emission collection efficiency. One commercial unit (Kratos) employs a 5-µL flow cell with a narrow depth (1.07 mm) and large surface area for excitation-emission collection. The emitted radiation is collected by a concave mirror placed around the flow cell, and the rear of the cell is reflective. This configuration captures greater than 75 percent of the fluorescence while minimizing inner filtering. With flow cells, scattered radiation from the excitation source is removed with cutoff filters placed before the photomultiplier tube.

The mobile-phase composition should be considered as some compounds can quench fluorescence. Avoid chlorinated hydrocarbons and salts which include heavy atoms (such as Br, I, Cl, and heavy metals). Organic solvents such as methanol, acetonitrile, or hexane, and aqueous buffers do not present a problem.

In practice, fluorescence detection is limited by the presence of background light, which includes various types of light scattering, luminescence from the flow cell walls, and emission from impurities in the solvent. All of these increase with excitation intensity to produce no net gain.

By comparison with an absorption detector, the fluorescence detector measures a signal against a blank of zero (assuming the solvent does not fluoresce). This leads to considerably less noise and thus significantly better sensitivity. Selectivity is also improved because the analyst sets the desired excitation wavelength and the desired emission wavelength. Three-dimensional (excitation, emission, and retention time) chromatograms often distinguish species that are not chromatographically resolved. All species that absorb light do not necessarily fluoresce. Two or more solutes may absorb at the same wavelength but emit at different wavelengths (or some may not emit). A wide variety of derivative-forming reagents have been developed to extend the realm of fluorescent detection to nonfluorescing compounds.

4.6.3 Electrochemical Detectors[4]

Electrochemical detection depends on the voltammetric characteristics of solute molecules. Sensitive detection is possible for species exhibiting a reversible electron transfer for a particular functional group. Detectabilities are quite favorable for aromatic amines and phenols, ranging from about 10 pg to 1 ng injected material. This type of detector has found its greatest application when polar mobile phases are used. The detector offers considerable selectivity, since relatively few components in a complex mixture are likely to be electroactive.

The flow cell (5 μL) is a channel in a thin polyfluorocarbon gasket sandwiched between two blocks, one plastic and the other stainless steel, which serves as the auxiliary electrode (Fig. 4.8). A working electrode is positioned along one side of the channel. Farther downstream a reference electrode is connected to the working region by a short length of tubing..

4.6.4 Differential Refractometers

A differential refractometer monitors the difference in refractive index between the mobile phase and the column eluant. In LC it comes closest to being a universal detector. Unless the analyte happens to have exactly the same refractive index as the mobile phase, a signal will be observed. Unfortunately, the detection sensitivity, which typically is in micrograms, is poor compared with other detectors, and the detector is extremely sensitive to temperature and flow changes. Temperature must be controlled within 0.001°C. Solvent delivery systems must be pulse free to avoid noise. Use of a gradient is restricted to a few solvent pairs that have virtually identical refractive indices. These detectors find use mostly

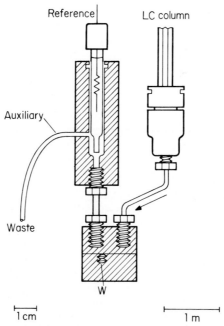

FIGURE   4.8 Thin-layer   amperometric detector. (*Courtesy of Bioanalytical Systems.*)

for the initial survey of samples and in exclusion (gel permeation) chromatography.

The optical diagram of the reflection-type refractometer is shown in Fig. 4.9. In the optical path two collimated beams from the light source (with masks and lens) illuminate the reference (mobile phase only) and sample (eluant) cells. The cell's volume (3 μL) is a depression formed by a Teflon gasket clamped between the prism and a reflecting backplate (finely ground to diffuse the light). The diffuse reflected light passes through the flowing liquid film and is imaged onto dual photodetectors. Since the percentage of reflected light at the glass-liquid interface changes as the refractive index of the liquid changes, a signal arises when a solute emerges. The detector is adjusted to zero with mobile phase in both cells.

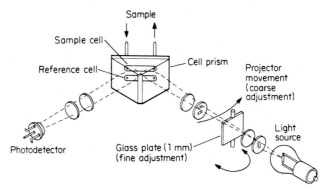

FIGURE 4.9 Optical diagram of a reflection-type (Fresnel) refractometer. (*Courtesy of Laboratory Data Control Division of Milton Roy.*)

4.7 OPTIMIZATION OR METHOD DEVELOPMENT IN HPLC[5–8]

Method development for an HPLC separation usually begins with retention optimization. A column type is selected bearing in mind the information in Fig. 4.1. Any arbitrary column geometry is selected, although a standard-length (15 or 25 cm) column with an internal diameter of 4.6 mm and packed with 5- or 10-μm particles would be a good first choice. The mobile-phase composition is varied as described later under the individual HPLC methods to achieve a uniform spacing of sample bands within a range $1 < k' < 10$.

The next step is the selection of the best column and packing configuration, that is, (1) length, (2) internal diameter, and (3) particle size. The separation will require a certain plate number to achieve adequate resolution between the most poorly resolved band pair. The plate number can be varied without changing retention by changing (1) the column dimension, (2) the particle size of the packing, or (3) the mobile-phase flow rate. Although the final choice can be made by trial-and-error, this is not recommended.

The general theory presented in Chap. 2 is the basis for predicting the plate number. Values of N can be related by Eq. 2.16 to the required resolution of the most poorly separated band pair in the chromatogram. Here k' refers to the average capacity-factor value for the band pair.

4.7.1 Generation of Required Plate Number in the Minimum Separation Time

The pressure drop ΔP across a bed packed with spherical particles of diameter d_p is

$$\Delta P = \frac{N^2 h^2 \phi \eta}{t_M} \tag{4.2}$$

Inserting the desired plate count (assumed to be $N = 10\ 000$) and $t_M = 60$ s, and using an optimum value of the reduced plate height, $h = 3$, $\phi = 1000$ (for fully porous packings; use 500 for pellicular packings), and η (viscosity of the mobile phase) $= 10^{-3}$ N $\cdot$ s $\cdot$ m^{-1}, Eq. 4.2 gives

$$\Delta P = \frac{(10\ 000)^2 (3)^2 (1000)(10^{-3}\ \text{N} \cdot \text{s} \cdot \text{m}^{-2})}{60\ \text{s}}$$

$$= 1.5 \times 10^7\ \text{N} \cdot \text{m}^{-2} \qquad (\text{or } 1500\ \text{lb} \cdot \text{in}^{-2})$$

The required particle size is given by Eq. 4.3:

$$d_p = \frac{L}{Nh} = \frac{L}{(10\ 000)(3)} = \frac{L}{30\ 000} \tag{4.3}$$

For columns 10 and 25 cm long, the corresponding particle diameters needed are 3.3 and 8.3 μm, respectively. The combinations of column length and particle size, plus operating pressures for different plate counts and retention times, are given in Table 4.2. Of course, in practice most laboratories must choose among available commercial columns, and they are limited to a small number of configurations.

A rapid estimate of column plate number can be calculated from the standard equation:

$$N = \frac{L}{h d_p} \tag{4.4}$$

where L = column length, cm
$\quad d_p$ = particle diameter, μm
$\quad h$ = reduced plate height (= 3.0 to 3.5 for real samples)

Taking 3.3 as an average value for h, a 5-cm long column packed with 3-μm particles should give about 5000 plates. To minimize measurement errors, remember that for measurement of the plate number the k' value should be at least 3 and preferably greater than 5 (but it should not exceed 20).

TABLE 4.2   Typical Performances for Various Experimental Conditions*

Performances		Column parameters			Peak bandwidth (4σ), μL
N	t_M, s	L, cm	d_p, μm	ΔP, atm (lb · in^{-2})	
2,500	30	2.3	3	18.4 (270)	23
2,500	30	3.7	5	18.4 (270)	37
2,500	30	7.5	10	18.4 (270)	75
5,000	30	4.5	3	74 (1088)	41
5,000	30	7.5	5	74 (1088)	68
5,000	30	15.0	10	74 (1088)	136
10,000	30	9.0	3	300 (4410)	82
10,000	30	15.0	5	300 (4410)	136
10,000	30	30.0	10	300 (4410)	272
10,000	30	9.0	3	300 (4410)	82
10,000	60	9.0	3	150 (2200)	82
10,000	90	9.0	3	100 (1470)	82
15,000	90	2.3	3	223 (3275)	23
15,000	120	2.3	3	167 (2459)	23
11,100	30	10.0	3	369 (5420)	91
11,100	37	10.0	3	300 (4410)	91
11,100	101	10.0	3	100 (1470)	91
27,800	231	25.0	3	300 (4410)	75

*Assumed reduced parameters: $h = 3$, $v = 4.5$.

4.8   ADSORPTION CHROMATOGRAPHY

In adsorption chromatography the solute and solvent molecules are in competition for adsorption sites on the surface of the column packing, usually silica gel (alumina and carbon in special situations). The adsorption sites are the slightly acidic silanol (Si—OH) groups which extend out from the surface of the porous particles and from the internal channels of the pore structure. These hydroxyl groups interact with polar or unsaturated solutes by hydrogen bonding or dipole interaction. The competition between the solute molecules and the solvent molecules for an active site provides the driving force and selectivity in separations.

Variations in solute retention are achieved by changes in the composition of the mobile phase. It is solvent strength which controls the k' value of solute peaks. Table 4.3 lists the common solvents used in adsorption chromatography in order of increasing solvent strength (and roughly in the order of increasing polarity). This listing also ranks the adsorption strength of the various functional groups (acting singly) of solute molecules; log k' varies linearly with $\epsilon°$, the solvent strength parameter. Not included in Table 4.3 are sulfides (less polar than ethers), nitro compounds (more polar than ethers), aldehydes (similar to esters and ketones), amines (similar to alcohols), and sulfones, sulfoxides, amides, and carboxylic acids (between the alcohols and water, with their polarities increasing in the order listed).

In practice, a solvent is chosen to match the most polar functional group in the sample. If the k' values are too small (sample elutes too rapidly), a weaker (less polar) solvent is substituted. A stronger (more polar) solvent is used if the sample

TABLE 4.3   Solvent Strength Parameter of Selected Solvents

Solvent	Solvent strength parameter $\epsilon°$		Boiling point, °C
	On silica	On alumina	
Fluoroalkanes		−0.25	
Pentane	0.00	0.00	36
Hexane		0.01	69
Cyclohexane	−0.05	0.04	81
Carbon disulfide	0.14	0.15	46
Carbon tetrachloride	0.14	0.18	77
Diisopropyl ether		0.28	68
Benzene	0.25	0.32	80
Diethyl ether	0.38	0.38	35
Chloroform	0.26	0.40	62
Dichloromethane		0.42	40
Methyl isobutyl ketone		0.43	118
Tetrahydrofuran		0.45	66
Acetone	0.47	0.56	56
1,4-Dioxane	0.49	0.56	107
Ethyl acetate	0.38	0.58	77
Acetonitrile	0.50	0.65	82
2-Propanol		0.82	82
1-Propanol		0.82	97
Ethanol		0.88	78
Methanol		0.95	64
Acetic acid		Large	118

does not elute in a reasonable time because of high k' values. Two solvents may be blended together in various proportions to provide continuous variation in solvent strength between that of each pure solvent. An increase of 0.05 unit in the value of the solvent strength parameter usually decreases all k' values by a factor of 3 to 4.

Binary solvent mixtures offer additional selectivity through the adjustment of the dipole, proton acceptor, and proton donor forces. For example, both the hydroxyl group and the aromatic ring of phenol have adsorption capabilities. The hydroxyl group is more strongly adsorbed because of its greater polarity and its capability to engage in hydrogen bonding. A binary solvent mixture containing an ether group (in contrast to dichloromethane) would provide donor sites for hydrogen bonding with either the silica gel or the phenol molecules.

Adsorption chromatography is influenced more by specific functional groups and less by molecular weight differences than liquid-liquid chromatography. This makes possible the separation of complex mixtures into classes of compounds with similar chemical functionality. For example, polynuclear aromatics are easily separated from aliphatic hydrocarbons, and the triglycerides are easily separated from a lipid extract. On the other hand, separation among members of a homologous series is usually poor.

Adsorption chromatography excels in the separation of positional isomers which differ in the geometric arrangement of functional groups with respect to the adsorption sites.

4.9 BONDED-PHASE CHROMATOGRAPHY

Bonded-phase supports are made from microparticulate silica (usually with a 5- or 10-μm average particle diameter) by chemical bonding an organic moiety to the surface through a siloxane (Si—O—Si—C) bond. Unreacted silanols and hydrolyzed end groups from bonded phases are removed with trimethylchlorosilane, a process known as endcapping. The bonded phase acts like an immobilized stationary liquid phase without the problems that plagued the older columns where the stationary phase simply coated the walls of the column or the substrate. Nonpolar hydrocarbonaceous moieties of low polarity are ethyl, octyl, and octadecyl linear alkanes. The octadecyl packing is useful when maximum retention is required. By contrast, the ethyl group is useful in applications that involve very strongly retained solutes. Octyl packings are a good compromise for the separation of samples with wide-ranging polarities.

Bonded phases of medium polarity will have functional groups, such as cyanoethyl ($—CH_2CH_2CN$) groups, bonded to the silica surface. These packings are useful in separations involving ethers, esters, nitro compounds, double-bond isomers, and ring compounds that differ in double-bond content. These types of compounds are found in the middle of Table 4.3.

Several other bonded phases have unique selectivity. The phenylsiloxane group is used for separations that rely on the interaction of aromatic components with the packing. The aminoalkyl group provides a highly polar surface. It may function as either a Brönsted acid or base, or it may interact with solutes through hydrogen bonding.

Among the more selective of the specialty columns are the chiral phases developed for the separation of enantiomers. The packing consists of plasma protein α_1-acid glycoprotein immobilized on silica microparticles and used as a bonded phase.

Silica-based bonded phases are not recommended for use in aqueous solutions that have pH values above about pH 7. For headily loaded, polymeric, or endcapped phases (especially those with long alkyl-chain lengths), pH limits may be somewhat higher. Hydrolysis of the silica matrix occurs at any pH value, and bonded phases ultimately will be degraded by aqueous mobile phases, the rate being enhanced at high salt concentrations and in the presence of some ion-pairing reagents.

When the octadecyl group is covalently bound to a polystyrene-divinylbenzene (PS-DVB) matrix, the packing material is stable from pH 0 to 14. These polymers are rigid, macroporous structures with pore sizes of approximately 8 nm. When a tertiary nitrogen is bonded into the polymer, the packing displays ion-exchange character at low pH values; it shows weak ionic character at intermediate pH; and it functions as a reverse-phase column at high pH.

4.10 REVERSE-PHASE CHROMATOGRAPHY

The distinction between reverse-phase and normal-phase chromatography is based on the nature of the stationary and mobile phases. Reverse-phase chromatography is by far the most popular HPLC method in use today. A hydrophobic (nonpolar) packing, usually with an octadecyl or octyl functional group is used in

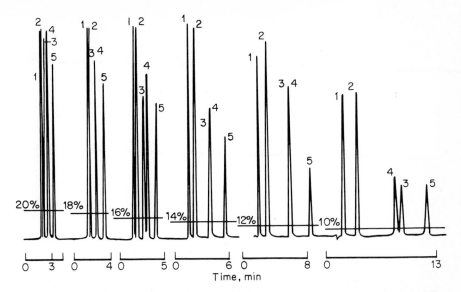

FIGURE 4.10 Isocratic separation of methyl xanthines on a bonded (C-8) phase column. Mobile phase is 20-mM phosphate buffer (adjusted to pH 2.6) and the percent acetonitrile marked on the graph. Peaks: (1) theobromine, (2) theophylline, (3) hydroxypropyl theophylline, (4) caffeine, and (5) 8-chlorotheophylline.

conjunction with a polar mobile phase, often a partially or fully aqueous mobile phase.

In designing laboratory methods, keep in mind these general principles:

1. The elution order often is predictable because retention time usually increases as the hydrophobic character of the solute increases. For the functional groups listed in Table 4.3, the elution order is reversed, thus the name reverse-phase chromatography. The nonpolar solutes are retained more strongly.

2. The eluant strength of the mobile phase follows the reverse order given in Table 4.3 and illustrated in Fig. 4.10. The predominant mobile phase, water, is the weakest eluant. Methanol and acetonitrile are popular modifiers; they are commercially available with excellent chromatographic purity and have low viscosity.

When scouting for the optimum pairing of column packing and mobile-phase composition (Fig. 4.11), try these steps:

1. If the sample components are of low to moderate polarity (that is, soluble in aliphatic hydrocarbons), use an octadecyl bonded-phase column and a methanol-water mixture as eluant.

2. For solutes of moderate polarity (soluble in methyl ethyl ketone), use an octyl bonded-phase packing in conjunction with an acetonitrile-water mobile phase.

3. High-polarity solutes (soluble in the lower alcohols) are best handled with a bonded ethyl packing and 1,4-dioxane–water mobile phase.

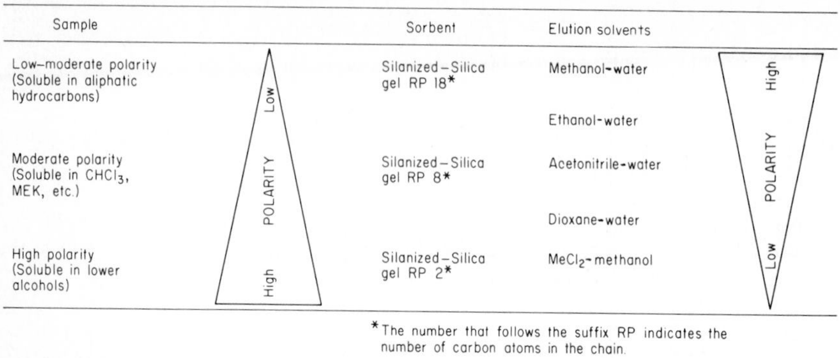

Sample		Sorbent	Elution solvents	
Low–moderate polarity (Soluble in aliphatic hydrocarbons)	*POLARITY* (Low → High)	Silanized–Silica gel RP 18*	Methanol–water	*POLARITY* (High → Low)
			Ethanol-water	
Moderate polarity (Soluble in CHCl₃, MEK, etc.)		Silanized–Silica gel RP 8*	Acetonitrile-water	
			Dioxane-water	
High polarity (Soluble in lower alcohols)		Silanized–Silica gel RP 2*	MeCl₂–methanol	

*The number that follows the suffix RP indicates the number of carbon atoms in the chain.

FIGURE 4.11   Adsorbents and eluants for reverse-phase chromatography. (MEK is methyl ethyl ketone, and MeCl$_2$ is methylene hydrochloride.)

If the sample components elute too rapidly with a 1:1 mixture of organic and water, a lower concentration of the stronger eluant (methanol, acetonitrile, or 1,4-dioxane) should be tried. Of course, the reverse should be tried if the solutes elute too slowly. How much stronger or weaker? The 10 percent rule is a handy guide: k' changes two- to threefold for a 10 percent change in mobile-phase organic. For example, if the last band elutes with a k' of approximately 20 in a 60:40 acetonitrile-water mobile phase, k' would be expected to be between about 6 and 10 for a 70:30 mobile phase. When establishing binary solvent gradients, the polarity of the eluant is continuously decreased during the chromatographic run—for example, by gradually increasing the organic solvent in methanol-water, acetonitrile-water, or 1,4-dioxane–water mixtures.

Even if k' is in the proper region, the separation still may not be adequate. In that case, try another organic solvent in the mobile phase—for example, change from acetonitrile to methanol or tetrahydrofuran. The nomogram in Fig. 4.12 can be used to convert from one solvent mixture to another. Locate the present solvent composition on the appropriate line, and then draw a vertical line through the other solvent lines. The intersection of these lines indicates the equivalent other mobile-phase compositions. Thus, 40:60 acetonitrile-water is equivalent to 50:50 methanol-water or about 25:75 tetrahydrofuran-water. Changing to ternary mixtures, such as methanol-acetonitrile-water, or binary mixtures, such as acetonitrile–2-propanol or 1,4-dioxane–methanol, can often improve selectivity.

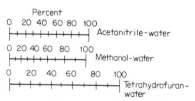

FIGURE 4.12 Nomogram for conversion from one solvent composition to another. (Percent scale is from 0 to 100% of the organic component.)

4.11   NORMAL-PHASE CHROMATOGRAPHY

Normal-phase chromatography is just the reverse of reverse-phase chromatography. A polar stationary phase is used in conjunction with a less polar mobile phase. For example, aromatic compounds have been separated on a nitrile-bonded phase using a 25% tetrahydrofuran–75% cyclohexane mixture as the mobile phase.

For a normal-phase system, a polar modifier, such as 2-propanol or water, often is added to the mobile phase to minimize secondary retention by interaction with the silica surface. That is, the column is deactivated.

4.12   ION-PAIR CHROMATOGRAPHY

Ion-pair chromatography overcomes difficulties with ionized or ionizable species that are very polar, multiply ionized, and/or strongly basic. This technique allows for unique separations not otherwise obtainable by either reverse-phase or ion-exchange HPLC. Unlike conventional ion exchange, ion-pair chromatography can separate nonionic and ionic compounds in the same sample.

An ion-pair reagent (a large organic counterion which is ionized) is added at low concentration (usually 0.005 M) to the mobile phase. The stationary phase is a monolayer octadecyl or octyl bonded-phase packing. Two popular ion-pair reagents are triethylalkyl quaternary amines (deliberately unsymmetrical to provide a better association of the long alkyl chain with the paraffinic surface of the octadecyl stationary phase) and alkyl sulfonates. The former bind solute anions; the latter bind organic cations to form ion pairs, as follows:

$$RNH_3^+ + C_7H_{15}SO_3^- \rightleftharpoons [RNH_3^+, {}^-O_3C_7H_{15}]° \qquad (4.5)$$

Although the exact mechanism has not been established, either the solute molecule forms a reversible ion-pair complex (a coulombic association species of zero charge formed between two ions of opposite electric charge) which partitions into the nonpolar stationary phase or the counterion is loaded onto the packing via the alkyl moiety with its ionic group oriented at the surface and able to participate in the formation of an ion-pair complex.

Water-methanol is the common mobile phase. Any buffer required should be 0.001 to 0.005 M and should have good solubility but poor ion-pair properties.

Three factors that influence retention are:

1. *Control of pH:* Maintaining a pH around 2.0 ensures that both strong and weak bases are in their protonated form and that any weak acids present are primarily in their nonionic forms. At a pH around 7.5 both strong and weak acids are in their ionic form and weak bases are in their nonionic form.

2. *Alkyl chain length:* The longer the alkyl chain on the counterion, the greater the retention of given ions.

3. *Counterion concentration:* Increasing the concentration of the counterion increases retention up to the limit set by the solubility of the counterion in the mobile phase.

If one is separating nonionic and ionic compounds in the same sample, follow these steps:

1. Optimize the separation of the nonionic solutes as described under "Reverse Phase Chromatography."
2. Select and add the counterion to the mobile phase.
3. Fine-tune the separation by changing the chain length of the counterion (or use mixtures of two counterions) and perhaps the counterion concentration.

4.13 ION-EXCHANGE CHROMATOGRAPHY

Column packings for ion-exchange chromatography (IEC) have charge-bearing functional groups attached to a polymer matrix (Fig. 4.13). A macroreticular resin bead has channels with exchange groups protruding into the channels and from the surface of the bead. The exchanger may also be bonded to silica microparticles or polymerized into the pores of a porous gel.

4.13.1 Functional Groups

Typical column packings are outlined. They are based on a stationary phase that is hydrophobic (styrene-divinylbenzene matrix) in the absence of ionic functional groups (Fig. 4.14).

4.13.1.1 Sulfonate Exchangers. The $—R—SO_3^-$ groups are strongly acidic and completely dissociated whether they are in the H form or the cation form. These exchangers are used for cation exchange.

4.13.1.2 Carboxylate Exchangers. The $—R—COOH$ groups have weak acidic properties and will only function as cation exchangers when the pH is sufficiently high to permit dissociation of the $—COOH$ site.

4.13.1.3 Quaternary Ammonium Exchangers. The $—R_4^+N$ groups are strongly basic and completely dissociated in the H form and the anion form.

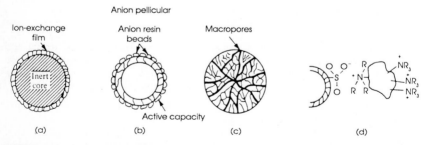

FIGURE 4.13 Structural types of ion-exchange resins: (*a*) pellicular with ion-exchange film, (*b*) superficially porous resin coated with exchanger beads, (*c*) macroreticular resin bead, and (*d*) surface sulfonated and bonded electrostatically with anion exchanger.

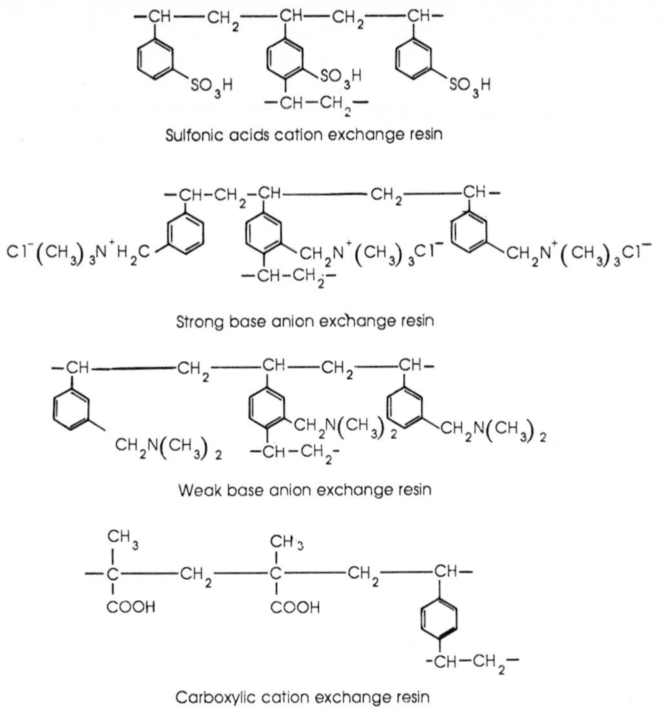

Sulfonic acids cation exchange resin

Strong base anion exchange resin

Weak base anion exchange resin

Carboxylic cation exchange resin

FIGURE 4.14 Functional groups of ion-exchange resins.

4.13.1.4 Tertiary Amine Exchangers. The $-R_3NH_2$ groups have exchanging properties only in an acidic medium when a proton is bound to the nitrogen atom.

4.13.1.5 Aminodiacetate Exchanger. The $-N(CH_2COOH)_2$ group has affinity for heavy metal cations and, to a lesser extent, for alkaline earth cations. This exchanger is the column packing often used for ligand exchange.

4.13.2 Hydrophilic Polymers

Hydrophilic polymers are based on a polyamide matrix. The separation of proteins, nucleic acids, and other large ionic molecules becomes possible without side effects from denaturation or irreversible absorption. Each monomer has three hydroxymethyl groups that impart hydrophilic character.

4.13.3 Exchange Equilibrium

Retention differences among cations with an anion exchanger, or among anions with a cation exchanger, are governed by the physical properties of the solvated ions. The stationary phase will show these preferences:

1. The ion of higher charge.
2. The ion with the smaller solvated radius. Energy is needed to strip away the solvation shell surrounding ions with large hydrated radii even though their crystallographic ionic radii may be smaller than the average pore opening in the resin matrix.
3. The ion that has the greater polarizability (which determines the van der Waals attraction).

To accomplish any separation of two cations (or two anions) of the same net charge, the stationary phase must show a preference for one more than the other. No variation in the eluant concentration will improve the separation. However, if the exchange involves ions of different net charges, the separation factor does depend on the eluant concentration. The more dilute the counterion concentration in the eluant, the more selective the exchange becomes for polyvalent ions.

4.13.4 Applications

4.13.4.1 Acids

1. Use a strong anion exchanger and elute with a strong acid; the acids emerge in order of increasing strength with the weakest acid first.
2. In gradient elution use buffers of decreasing pH.
3. Group separation of weak acids from strong acids can be accomplished on weakly basic anion exchangers. Weak acids will be retained to only a slight extent or not at all.

4.13.4.2 Amino Acids.   Use a strong cation exchanger and gradient elution with buffers of increasing pH; the most acidic components emerge first.

4.13.4.3 Cations

1. Use a strong cation exchanger and elute with buffered solutions that contain a complexing agent that will convert some cations into neutral or negatively charged ionic species. The separation will be a function of pH and small differences in metal formation constants.
2. Use a strong anion exchange in conjunction with a masking system that converts selected metal ions to a negatively charged complex ion. Control of the ligand concentration (either directly or indirectly by control of pH) in the eluant provides the separation with the weakest metal complex emerging first.

4.13.4.4 Organic Compounds

1. Polyhydroxy compounds (e.g., sugars) and borate ions form a series of complexes with varying stabilities that dissociate as weak acids. Use a borate buffer as eluant and a pH gradient that rises from 7 to 10 in conjunction with a strong anion exchanger.
2. Carbonyl groups form addition compounds with the hydrogen sulfite ion to give $>C(OH)(SO_3^-)$ which is strongly adsorbed by anion exchange resins. Since the addition product of ketones is less stable than that of aldehydes at

elevated temperature, a group separation is possible. Ketones are desorbed by hot water, and the aldehydes are eluted with a NaCl solution.

4.14 ION CHROMATOGRAPHY

One difficulty encountered with ion-exchange chromatography is the lack of suitable detectors. A second is the adaptability of the resin packings for HPLC equipment.

In a technique called ion chromatography (IC), low-capacity ion-exchange resins are used to separate analyte ions. A dilute aqueous solution is used as eluant. Eluant conductivity is suppressed by using what is now called a "suppressor column." Ionic analytes are then detected by a conductivity meter or by indirect uv (vacancy) detection.

4.14.1 Dual-Column Method

Dual-column IC requires the use of two columns: the first, a separator, is a high-capacity ion-exchange column which separates the anions (or cations); the second, a suppressor column, is used to convert the highly conductive eluant into a less conductive form, thereby enhancing detection. Special ion-exchange column packings are used. For an anion exchanger, the packing consists of a rigid inert, polymer core (about 10 μm in diameter), an intermediate layer of sulfonate groups, and a thin layer of anion-exchange beads bonded electrostatically to the sulfonate groups. All the active sites are close to the eluant-resin interface, which provides favorable mass transfer characteristics. For a cation exchanger, there is an intermediate layer of aminated groups on the polymer core covered by a layer of sulfonated resin beads.

4.14.1.1 Anion Analysis. Elute the sample from an anion-exchange column with dilute sodium salicylate. Then pass the effluent through a cation-exchange column in the H form. The effluent contains salicylic acid (only slightly ionized) and the acids of the sample's anions. Thus a little postcolumn chemistry removes the eluant ions that cause a large baseline signal in the conductivity detector. A microprocessor controls the detector-scale setting, which can be changed in the middle of an analysis so that peaks caused by trace ions can be detected in the presence of a major ion. The microprocessor can also reset the detector response to zero before each sample injection.

4.14.1.2 Cation Analysis. Pass the sample through a cation-exchange column with HCl as the eluant. Then pass the effluent through an anion-exchange packing in the OH form. The conductivity is exactly equal to that of the cation chlorides as the hydrogen ion of the HCl is removed by the hydroxyl ion to form innocuous water.

4.14.2 Hollow-Fiber Column

The ion-exchange sites in a conventional suppressor column gradually become saturated so that the column requires regeneration. Recently, hollow-fiber suppressors have been developed that can be continuously regenerated. Separator-column effluent flows in one direction and regenerating solution flows in the opposite direction.

4.14.3 Single-Column Method

The single-column technique uses a low-capacity ion-exchange column as the separator. The low-capacity resin allows the anions to be separated using low conductance eluants such as potassium hydrogen phthalate, sodium benzoate, and p-hydroxybenzoic acid. Anions of weak acids, such as cyanide, silica, and borate, cannot be separated unless poly(styrene-divinylbenzene) packings are used. These packings make it possible to operate the column to pH 11.7 (whereas with the former type of packing one is restricted to pH 2 to 8.5). Separation possibilities are shown in Fig. 4.15.

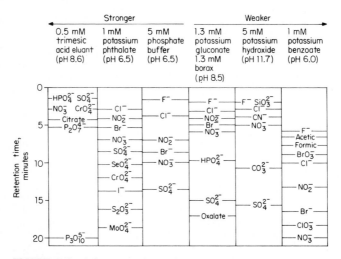

FIGURE 4.15   Anion separations using various eluants are outlined for ion chromatography. Trimesic acid is 1,3,5-benzenetricarboxylic acid.

4.15   HYDROPHOBIC-INTERACTION CHROMATOGRAPHY

Hydrophobic-interaction chromatography is a technique that uses reverse salt gradients to elute proteins without organic solvents. Thus, the protein can be recovered without denaturation. The technique uses packed column (4 to 10 mm i.d.) with stationary phases of low hydrophobic character. The packing is composed of 6.5-μm macroporous silica covalently bonded with a polyamide coating which is derivatized with a hydrophobic ligand such as methyl, hydroxypropyl, butyl, benzyl, and pentyl. The separation of proteins is usually accomplished using a descending salt gradient that preserves enzymatic activity and tertiary structure.

The least hydrophobic ligands are best for the separation and purification of very hydrophobic proteins.

4.16 EXCLUSION (GEL-PERMEATION) CHROMATOGRAPHY

Compounds can be separated according to their molecular size by exclusion chromatography, which is based on the diffusion or permeation of solute molecules into the inner pores of the column packing. The size and shape of the molecules to be separated govern their ability to enter the pore. The smaller molecules enter the pores without hindrance and are the last to be eluted. Molecules that are too large to enter the pores are completely excluded and must travel with the solvent front. Between these two extremes, intermediate-size molecules can penetrate some passages but not others and, consequently, are retarded in their progress down the column and exit at intermediate times. The selection of column packings, each with its corresponding exclusion limit, enables separations to be achieved. This technique is well suited for the separation of polymers, proteins, natural resins and polymers, cellular components, viruses, steroids, and dispersed high-molecular-weight compounds. Useful detectors include the differential refractometer and the spectrophotometric detectors that operate in the ultraviolet and infrared spectral regions.

4.16.1 Column Packings

Column packings are either cross-linked macromolecular polymers or controlled-pore-size glasses or silicas. Semirigid materials will swell slightly; these materials are limited to a maximum pressure of 300 lb $\cdot$ in^2 because of the bed's compressibility. Packings prepared from methacrylate polymers can withstand pressures up to 3000 lb $\cdot$ in^2. Hydrophilic packings are usable with aqueous system and polar organic solvents. Bead diameters are usually 5 μm.

Inorganic packings have advantages over organic packings. After calibration, columns can be used routinely and indefinitely. The bed volume remains constant at high flow rates and high pressures.

4.16.2 Retention Behavior

Totally excluded molecules elute in one column-void volume, and so the distribution coefficient $K = 0$. For small molecules that can enter all the pores of the packing, $K = 1$. Intermediate-size molecules elute between these two limits. An elution graph is shown in Fig. 4.16. The upper portion is the graph of the logarithm of molecular weight versus retention volume. There is a linear range of effective permeation between the limiting values that correspond to exclusion and to total permeation. The maximum elution volume is often only twice the column-void volume.

The various pore sizes available in commercial packings provide selective permeation ranges that permit separation of small molecules with molecular weights less than 100 to compounds with molecular weights of up to 500 million.

If nothing is known regarding the probable molecular dimensions or weights of

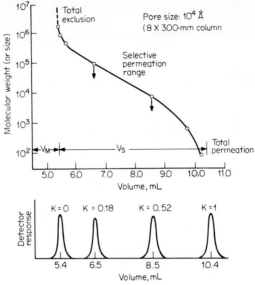

FIGURE 4.16   Retention behavior in exclusion chromatography.

the sample components, a preliminary run can be made using a column packed with 100-nm material. Use a flow rate of 3 mL/min. If most of the sample elutes near the exclusion limit, a column packed with larger-pore-size material should be tried. Elution of the majority of the sample halfway between the exclusion and total permeation volume suggests a column packing of 50-nm pore size. Near-total permeation indicates a smaller pore-size packing, perhaps 10 nm. If sample components elute over a wide range, a series of column cartridges, each with a specific packing, should be used.

4.16.3   Column Calibration

Exclusion columns are calibrated by eluting calibration standards and monitoring the elution volume. Narrowly dispersed standards of polystyrene, poly-tetrahydrofuran, and polyisopropene are available for use in organic solvents. Samples of dextrans, polyethylene glycols, polystyrene sulfonates, and proteins are available for use in hydrophilic solvents.

4.16.4   Ion-Exclusion Chromatography

Separations are carried out on high-capacity polystyrene-based exchange resins using dilute mineral acid eluants. Work is usually done at elevated temperatures (but below any boiling points). Acetonitrile may be used as an organic modifier to decrease the retention time of relatively nonpolar compounds. Wine, beer, fruit juice, and many dairy products are quickly analyzed by this technique.

4.17 TROUBLESHOOTING

Most problems result from bubbles, dirt, or normal wear. Preventive maintenance is the key to eliminating them.

4.17.1 General Precautions

For reliable HPLC operation, take these precautions:

1. Use only clean, degassed (by helium sparging) HPLC-grade solvents. Filter all salt- or buffer-containing solvents through a 0.45-μm filter before use. Use an inlet filter on the reservoir end of the solvent inlet lines to prevent dust from entering the system.
2. Check the septum daily for leaks and change frequently.
3. Use a guard column to act as a superfilter to trap particulate matter and chemical contaminants before they reach the analytical column.
4. Check the flow rate regularly at a specified pressure to detect buildup of pressure (or decrease of flow). Pressure buildup can be caused by small pieces of septum which become deposited at the head of the column (if no guard column is used) after many injections. To correct this situation, remove a few millimeters of packing from the top of the column and repack with new material.
5. Dissolve samples in the mobile phase or in a less polar solvent than the mobile phase, if possible. This technique tends to concentrate the injection on top of the column and yields better resolution. Many times it is possible to inject very large samples when more sensitivity is needed with no deleterious effects apparent in the separation.
6. Flush the system after each day's use with about 10 column volumes of the strong component of the mobile phase. If buffers are being used, flush with unbuffered mobile phase before changing to pure organic. This precaution prevents salt-deposit buildup, precipitation of crystals, microbial growth, and corrosion of LC hardware.
7. Replace pump seals regularly (about every 3 months).
8. Check detector (lamp life 0.5 to 1 year). Replace deuterium lamps every 6 months and mercury lamps once a year.
9. Beware of the effects of preservatives in solvents when using normal-phase chromatography. In the case of chloroform, common preservatives are ethanol and 2-methyl-2-butene at levels of about 0.5 to 1.0% and 0.01 to 0.02%, respectively. Ethanol in amounts of 0.5% will significantly modify the retention characteristics of a chloroform-hexane mobile phase with a silica column. If possible, avoid solvents that tend to form peroxides upon storage, such as ethers, chloroform, or cyclohexane.
10. Check for stabilizers (antioxidant) in solvents used for reverse-phase chro-

matography because the stabilizers absorb in the ultraviolet region below 320 nm, which causes problems when ultraviolet-visible detectors are used. Butylated hydroxytoluene is often added to tetrahydrofuran (THF) to prevent peroxide formation; only the unstabilized THF should be used in conjunction with uv detectors.

11. In adsorption chromatography the presence of trace levels of transition-metal ions in the adsorbent can catalyze oxidation of easily oxidized samples. Acid washing (1 M HCl) removes transition-metal impurities from the adsorbent.

4.17.2 Precautions with Bonded-Phase Columns

Bonded-phase columns, as with any silica-based column, may show secondary retention by interaction with the silica surface. In reverse-phase systems, the silica is often endcapped and amine modifiers are added to the mobile phase to minimize the extent of silanol interactions. For a normal-phase system, a polar modifier, such as 2-propanol or water, often is added to the mobile phase to deactivate the column. The primary aminopropyl bonded-phase column is not inert and precludes concomitant use of strongly oxidizing samples and samples that can condense with a primary amine group such as aldehydes, ketones, and ketosteroids (including reducing carbohydrates, such as aldoses and ketoses). The condensation is accelerated by polar alcohol solvents (so use acetonitrile rather than methanol). Peroxide and hydroperoxide samples will suffer loss on this type of column and will slowly transform the amine groups into nitro groups. An amine salt column is more resistant to oxidation; conversion to the ammonium phosphate form is recommended. Also derivatizing agents such as dansyl chloride should be avoided.

4.17.3 Extracolumn Effects

Extracolumn effects should be suspected under these laboratory observations:

1. Early bands broaden more than later ones (smaller plate numbers).
2. Coupled columns give higher plate numbers than the sum of the individual columns; also the asymmetry factor values are closer to 1.0.
3. Shorter and/or narrower columns give more band-broadening problems than normal columns.
4. Conversion from gradient to isocratic operation gives poor plate numbers and/or band shape.
5. Tailing bands are present under ideal column-test conditions.

The idea that extracolumn effects cause equal broadening for all bands is a common misconception. For isocratic separations, the later the bands come out, the broader they become. In other words, band broadening within the column is retention related, whereas extracolumn effects result mostly from the plumbing and the detector cell volume which are constant for the set of one particular system.

Earlier peaks may broaden more than later ones if the value of k' for the early peak is less than 2 and particularly if it is less than 1. Now the problem lies with the k' value, and one should suspect that the wrong mobile phase (too strong an eluant) is being used. It is the k' value, not the retention time, that is the important factor. The k' value is independent of the column volume and flow rate, whereas retention is not.

4.17.4 Change in Retention Time

A gradual change in retention time of sample components can be caused by changes in either (1) the column or (2) the LC pumping system. Column changes that will affect retention time are as follows:

1. Coverage of active sites by extraneous sample material that is irreversibly adsorbed. An increase in operating pressure accompanies this problem. The remedy is to improve sample cleanup.

2. Gradual cleavage of the bonded phase from the silica support as a result of hydrolysis. The remedy is to increase the organic content of the mobile phase, lower the pH, decrease the ionic strength of the mobile phase, or switch to a neutral-bead polystyrene-divinylbenzene packing material.

3. Reaction of the column packing with a sample component is confined largely to primary alkylamine bonded phases when sample components contain aldehydes and reactive (unhindered) ketones. The remedy is to avoid this type of packing when working with carbonyl-containing compounds.

4.18 SAFETY

The three major areas for which proper safety precautions are imperative are (1) toxicity of liquids and fumes, (2) flammability of solvents, and (3) high fluid pressures.

4.18.1 Toxicity and Fumes

Toxicity of solvents and samples can be avoided by taking sensible safety precautions. All solvents should be stored in vented fireproof cabinets. A well-ventilated laboratory and the use of an exhaust hood will minimize exposure to these compounds. Careful handling of solvents, including the proper use of funnels and safety glasses, will protect the chromatographer.

Waste collection vessels should be larger than the total volume of the solvent(s) to be pumped through the system; they should be kept covered to minimize the evaporation of waste solvent. Spills should be wiped up as quickly as possible; acids and bases should be neutralized prior to wiping.

4.18.2 Solvent Flammability

Another possible source of danger in HPLC exists from the use of highly flammable solvents. In this respect, leaks are of very great concern since ignition may occur without warning. Fire extinguishers should be kept in close proximity to the HPLC system. Smoking and open flames should be prohibited without exception in the laboratory, as should the unnecessary operation of any electronic instrumentation or other potential sources of electrical discharges.

4.18.3 Mechanical

When using highly flammable solvents with a pumping system which has a low-limit, shut-down feature, set that limit approximately 300 lb · in^2 below the minimum operating pressure necessary for the particular method. This shut-down feature automatically stops the pump when a leak occurs or when the system runs out of solvent, since the pressure drops below the limit setting as a result of fluid loss.

The compressibility of liquids is very small, so the dangers related to the high-pressure system are minimal. Generally a high-pressure system component will leak before the pressure involved causes the plumbing or components to fracture. However, it is good safety practice always to wear eye protection in the laboratory. A small fracture in tubing under high pressure can produce a stream of solvent that is capable of puncturing tissue. High pressure must not be used with glass columns since shattering of the column may easily occur.

REFERENCES

1. R. R. Walters, "Affinity Chromatography," *Anal. Chem.*, **57**:1099A (1985).
2. E. Katz, K. Ogan, and R. P. W. Scott, "LC Column Design," *J. Chromatogr.*, **289**: 65–83 (1984).
3. B. L. Karger, M. Martin, and G. Guiochon, "Role of Column Parameters and Injection Volume on Detection Limits in Liquid Chromatography," *Anal. Chem.*, **46**:1640 (1974).
4. D. A. Roston, R. E. Shoup, and P. T. Kissinger, "Liquid Chromatography/ Electrochemistry: Thin-Layer Multiple Electrode Detection," *Anal. Chem.*, **54**:1417A (1982).
5. J. L. Glajck and J. J. Kirkland, "Optimization of Selectivity in Liquid Chromatography," *Anal. Chem.*, **55**:319A (1983).
6. J. H. Knox, "Practical Aspects of LC Theory," *J. Chromatogr. Sci.*, **15**:352 (1977).
7. M. Martin, G. Blu, C. Eon, and G. Guiochon, "Optimization of Column Design and Operating Parameters in High Speed Liquid Chromatography," *J. Chromatogr. Sci.*, **12**: 438 (1975).
8. M. Martin, C. Eon, and G. Guiochon, "Trends in Liquid Chromatography," *Res/Dev.*, p. 24 (April 1975).

BIBLIOGRAPHY

Gjerde, D. T., and J. S. Fritz, *Ion Chromatography*, 2d ed., Huthig, Heidelberg, 1987.

Horvath, C., ed., *High-Performance Liquid Chromatography*, Vol. 1, 1980; Vol. 2, 1980; Vol. 3, 1983; Academic, Orlando, Fla.

Scott, R. P. W., ed., *Small Bore Liquid Chromatographic Columns: Their Properties and Uses,* Wiley, New York, 1984.

Simpson, C. F., ed., *Techniques in Liquid Chromatography,* Wiley-Heyden, New York, 1982.

Snyder, L. R., and J. J. Kirkland, *Introduction to Modern Liquid Chromatography,* 2d ed., Wiley-Interscience, New York, 1979.

Yau, W. W., J. J.Kirkland, and D. D. Bly, *Modern Size Exclusion Liquid Chromatography,* Wiley-Interscience, New York, 1979.

Yeung, E. S., ed., *Detectors for Liquid Chromatography,* Wiley-Interscience, New York, 1986.

CHAPTER 5
PLANAR CHROMATOGRAPHY

Planar chromatography includes thin-layer and paper chromatography and electrophoresis. In *thin-layer chromatography,* a stationary phase is coated on an inert plate of glass, plastic, or metal. The samples are spotted or placed as streaks on the plate. Development of the chromatogram takes place as the mobile phase percolates through the stationary phase and the spot locations. The sample travels across the plate in the mobile phase, propelled by capillary action. Separation of components occurs through adsorption, partition, exclusion, or ion-exchange processes, or a combination of these. In *paper chromatography,* the chromatograph is simply a piece of porous paper. *Electrophoresis* involves the migration of charged species through the stationary phase under the influence of an electric field. Each charged species moves at a rate which is a function of its charge, size, and shape. In planar chromatography the position of the resultant bands or zones after development is observed or detected by appropriate methods. Because of its convenience and simplicity, sharpness of separations, high sensitivity, speed of separation, and ease of recovery of the sample components, planar chromatography finds many applications.

5.1 THIN-LAYER CHROMATOGRAPHY

Resolution of the sample in thin-layer chromatography is accomplished by passage of the mobile solvent mixture through the initial spot and beyond until the solvent front reaches within 2.5 cm of the opposite edge (top of plate in vertical development). Interaction of individual sample components with the stationary phase and the mobile phase, if each is properly selected, will cause the components to separate into individual spots. In thin-layer chromatography (TLC) the variety of coating materials and solvent systems, as discussed in HPLC (Chap. 4), are also suitable for this technique.

The differences between HPLC and TLC is that HPLC operates with a closed system whereas TLC functions in an open thin-layer plate whose surface is exposed to the atmosphere. In TLC the separation is by development (the components remain in the chromatographic bed) and the detection is static (independent of time). HPLC is carried out in a steady-state mobile-phase environment, whereas in TLC the driving force is solely capillary action; the latter may cause an uneven flow velocity. TLC is a more economical separation technique than HPLC because multiple samples (including standards) may be introduced and run simultaneously. With a high-performance TLC plate, it is often sufficient to carry the development no more than 3 or 4 cm in order to achieve satisfactory separation, and this can be accomplished in 5 min, more or less, depending on the development solvent. Two-dimensional runs in TLC enable more complex samples to be analyzed.

A major advantage of TLC is the speed of analysis on a per sample basis. This is due to the short development distance and the resultant short development time. Spotting samples along with standards on the same plate allows them to be processed under identical conditions in contrast to sequential analysis on a column. By far the greatest number of chromatographic analyses are concerned with the measurement of only one or two components of a sample mixture. Therefore, development conditions are optimized for resolution of only the components of interest, while the remainder of the sample material is left at the origin or is moved away from the region of maximum resolution.

The choice of mobile-phase components is not restricted by concerns about deterioration of the coating material since layers are not reused or by compatibility with a detector; thus solvents that are highly absorbing in the ultraviolet region can be used.

A complete view of a complex sample can be provided on a thin-layer chromatogram by use of a sequence of detection reagents. A variety of techniques can be used to optimize sensitivity and selectivity of detection. Often the entire plate can be examined to make sure that all of the sample has moved, whereas in HPLC one is never certain if every peak has been eluted at any given time.

TLC has the ability to separate nanogram to picogram quantities of analytes by rapid development (usually less than 15 min over short migration distances, i.e., 3 to 15 cm). Because there is less band broadening, compact zones are produced, leading to high detection sensitivity, resolution, and efficiency. Smaller samples are used so that a larger number of samples per plate can be applied.

Unlike GC or HPLC, the operations that make up the TLC system are not online. This lends considerable flexibility in the scheduling of the separate steps of sample application, development, and detection.

5.1.1 Retardation Factor

In systems such as TLC, where the cross section and partition coefficient may not be constant along the length of the development path, the ratio of distances traveled by the solute and the mobile phase is called the retardation factor R_f:

$$R_f = \frac{\text{distance traveled by the solute}}{\text{distance traveled by the mobile phase}} \tag{5.1}$$

TLC can be used as a pilot technique for HPLC. The simple equation

$$k' = \frac{1}{R_f} - 1 \tag{5.2}$$

relates k', the capacity factor from an HPLC column, and the R_f value measured from the TLC plate. With this equation, isocratic experiments can sometimes be directly extrapolated from one technique to the other. R_f values of 0.1 to 0.5 correspond to the optimum 1 to 10 range for k' values in HPLC.

If the results of isocratic TLC are to be transferred to a gradient column

HPLC, the gradient should be started about 20 percent weaker than the best mobile phase determined for TLC and run to a point about 20 percent stronger than the TLC solvent.

5.1.2 Solvent Selection

The selection of a solvent system to accomplish the required separation may involve a number of trials, but the choice of solvent is quite unrestricted by considerations of interference with detector response or of possible deterioration of the stationary phase. The solvent systems are usually a two-component mixture of water and a polar organic solvent miscible with water. Assuming that methanol was chosen with water as the strength-adjusting solvent, development is performed with different proportions of methanol and water to find the mixture that produces k' values for all compounds, or at least the ones of interest, within the optimum 1 to 10 range. Once this solvent mixture is experimentally determined, the total solvent strength (P') is calculated from the relationship:

$$P' = F_a P_a + F_b P_b + \cdots \tag{5.3}$$

where F is the volume fraction of the pure solvents (a, b, and so on) and P is their strength (Table 5.1).

Using similar equations, binary mixtures of acetonitrile-water and tetrahydrofuran-water, and even ternary and quaternary mixtures of solvents are formulated so that they have the same overall strength (P' value) but different selectivities. Plates are then developed with these mixtures to determine the best overall mobile phase. Remember that the highest P values represent the strongest solute for adsorption (silica gel) TLC but the weakest for reverse-phase TLC.

TABLE 5.1 Solvent Strength Parameter (P) Values

Solvent	P
Hexane	0.0
Diisopropyl ether	2.4
1-Propanol	3.9
Tetrahydrofuran	4.2
Ethyl acetate	4.3
2-Propanol	4.3
Chloroform	4.4
Ethyl methyl ketone	4.7
1,4-Dioxane	4.8
Ethanol	5.2
Pyridine	5.3
Acetone	5.4
Acetic acid	6.2
Acetonitrile	6.2
Methanol	6.6
Water	9.0

5.1.3   Efficiency

The number of theoretical plates in TLC is given by

$$N = 16 \left(\frac{\text{distance of migration of spot center}}{\text{width of spot}} \right)^2 \qquad (5.4)$$

Typical efficiencies range from 1000 to 3000 plates.

The increased efficiency realized by the smaller particle size bed in HPTLC plates described later results in less band broadening and, hence, improved resolution and greater sensitivity of detection of the separated fraction.

5.1.4   TLC Plates

As a general rule, any of the stationary phases used in HPLC can be used in TLC provided that they are available in a uniformly fine particle size or can be bonded to the substrate. Conventional TLC plates are 20 × 20 cm, 5 × 20 cm, and 1 × 3 cm (microscope slide) sizes. Analytical plates are usually 0.250 mm thick. The 20 × 20 cm plates can be purchased with 19 channels and/or a 3 cm distance of inert preadsorbent. Preparative TLC plates are 0.5 to 2.0 mm thick.

There are also plates prepared with 5-μm packings. Known as high-performance TLC (HPTLC) plates, they are smaller (10 × 10 cm) and faster, and require smaller samples. In addition, the average size of the silica gel particles is smaller and the size distribution is tighter than found in standard plates. The layer itself is usually somewhat thinner and the surface more uniform. Although the small particle size reduces the velocity of mobile-phase flow, the length of the chromatographic bed required is markedly less than that encountered in conventional TLC plates.

5.1.5   Silica Gel

Silica gel is used more often than any other coating material. The silica gel layer consists of an extremely dense packing of small particles of very uniform size with a smooth, homogeneous surface. The material is acid washed, water washed to neutrality, and dried to coalesce geminal hydroxyls into siloxane bonds. The silica gel is held together with various proprietary binders.

Silica gel is used for the resolution of acidic and neutral substances, and mixtures with relatively low water solubility. Separation of organic acids is facilitated by the addition of a weak acid (usually acetic acid) to the mobile phase in order to lower the pH below the pK_a of the acid solutes. Adding specific material to the base silica adsorbent can aid in the resolution of certain analytes. The additive can be present in the slurry when the plate is prepared, or it can be impregnated into the stationary phase of the plate. Examples of such additives are as follows:

1. Boric acid is used for the separation of sugar isomers since the borate ion forms complexes with the sugars to varying degrees.

2. Chelate-forming reagents are useful in the separation of inorganic cations and of phenol carboxylic acids.

3. Silver nitrate (argentation) is useful for compounds containing carbon-carbon double bonds.

4. Dilute sodium hydroxide allows separation of alkaloids.

5. Potassium oxalate is used for the separation of polyphosphoinositides and acidic magnesium acetate for phospholipids.

5.1.6 Cellulosic Sorbents

TLC plates coated with microcrystalline cellulose resembles paper chromatography in terms of separation properties, but the plates are generally more rugged, have shorter separation times, and give better resolution and sensitivity. Cellulose plates are suited for water-soluble (polar) materials, in particular, carboxylic acids and carbohydrates. Modified cellulose plates find special uses:

1. Acetylated cellulose can be used for reverse-phase chromatography. Caution: The degree of acetylation must be matched to the application.

2. Cellulose with diethylaminoethyl (DEAE) groups attached carries a positive charge on the amino ion-exchange group and can be used in the anion-exchange mode for separation of proteins and nucleic acids.

3. Polyethyleneimine (PEI) impregnated into the cellulose is used for the separation of nucleic acid components.

5.1.7 Alumina

Plates coated with alumina, or alumina mixed with silica gel, are used for the resolution of basic mixtures.

5.1.8 Chemically Bonded Reverse-Phase Plates

Reverse-phase TLC plates are available with C-2, C-8, C-18, and diphenyl bonded phases. A high molecular weight, aliphatic cross-linked polymer is added to the formulation to form a durable coating. Since these coatings match the HPLC phases, TLC can be easily used for screening or methods development. Advantages of chemically bonded plates compared to classical impregnated plates include the absence of need to saturate mobile phases with the stationary phase; no contamination with stationary liquid or solutes recovered from the layer by elution; and uniform and reproducible R_f values. Unlike plates coated with silica gel or other adsorbent, reverse-phase plates do not normally require activation or other preparation prior to use. Reverse-phase plates do suffer from poor wettability with pure water or solvents with high water content, especially those containing shorter hydrocarbon chains.

Placed on a polarity scale, the chemically bonded phases can be arranged as follows:

$$Si > NH_2 > CN = diol > C_2 > C_8 > C_{18}$$

Polar solvents for reverse-phase TLC typically consist of solvent mixtures such as water-methanol or water-acetonitrile. Substances migrate in a general order of decreasing polarity (the most polar solute moves the fastest), and mobile-phase strength increases with decreasing polarity (for example, acetonitrile is a stronger solvent than water).

For ion-pair chromatographic separations, lipophilic counterions are added to the mobile phase. The pH of the mobile phase is maintained at a value such that the solutes are ionized and can bond with the counterion.

Chiral compounds can be resolved with plates modified with Cu(II) and a proline derivative for ligand exchange separation.

5.1.9 Normal-Phase Bonded Plates

For normal-phase TLC, bonded phases available include the amino, cyano, and diol functional groups. The amino bonded phase displays weak-base ion-exchange properties, and can be an alternative to PEI cellulose plates. The diol plates have some of the same characteristics of unmodified silica gel but differ in that the diol group resembles an alcohol as compared to a silanol.

5.1.10 Preparation of the Bed

Precoated plates are commercially available. Those on plastic or aluminum sheets can be cut easily with scissors. However, some workers prefer to prepare their own plates. Slurries of silica gel (or alumina) are applied to glass plates or plastic sheets in a thin, uniform layer with the aid of a commercial applicator. The applicator is pulled across a series of plates laid out on a mounting board. Up to five 20 × 20 cm plates or twenty 20 × 5 cm plates can be coated in one operation. The height of the exit gate controls the thickness of the layer.

After the plate is coated with adsorbent slurry, it is dried for 30 min in air and then heated to 110 to 120°C to give active (water-free) adsorbent. After activation, the plates must be stored in a special desiccator or storage cabinet with controlled humidity.

5.1.11 Sample Application and Development

Since approximately 80 to 90 percent of all TLC analyses are done on reverse-phase plates, directions will be given for this method. Standards and samples should be dissolved in the weakest (most polar) possible solvent to minimize excessive band spreading. Because of stability and wettability limitations, sample solutions should be made up in totally or predominantly organic solvent rather than water. In most instances, methanol, which is quickly evaporated and wets the layer properly, is convenient for substances capable of hydrogen bonding. For less polar compounds, dichloromethane and acetone are convenient solvents for sample applications.

Standard TLC procedure involves these steps:

1. For manual spotting, a microcap micropipet of appropriate size or a capillary tube is used to apply samples and standards in concentrated spots on the plate, paper sheet, or paper strip. The spots should be no closer than 2.5 cm to the bottom edge (in the direction of development) and 2.5 cm to the edge which is parallel to the development. Spots are located 1 cm apart (Figs. 5.1 and 5.2). If

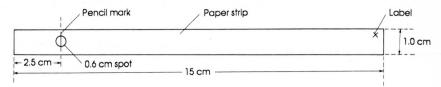

FIGURE 5.1 Dimensions of a paper strip for chromatography.

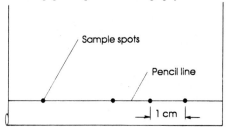

FIGURE 5.2 Spotting the paper or TLC plate.

desired, a plate can be divided into individual columns by scoring the layer with a scriber.

2. The initial zones are dried completely before development using a heat gun (air dryer) or infrared lamp. Care must be taken if the solute can decompose with heating, in which case a stream of room-temperature air or nitrogen can be used.

3. The plate (or sheet) is positioned in the developing tank (Fig. 5.3). The solvent mixture should cover the bottom of the plate (but at least 1 cm below the sample spots); the top can lean against the side of the tank.

4. Figure 5.4 illustrates how a sheet is assembled in the shape of a cylinder with the adjoining edges held together with plastic clips.

5. A simple arrangement for the development of strips is shown in Fig. 5.5.

6. The atmosphere of the tank need not necessarily be saturated with the solvent vapors. In adsorbent TLC, solvent equilibration improves the reproducibility of separation but has an adverse effect on sample resolution.

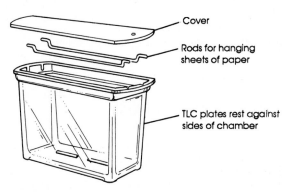

FIGURE 5.3 Developing tank for thin-layer chromatography.

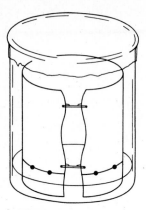

FIGURE 5.4 Developing a chromatogram on a paper cylinder.

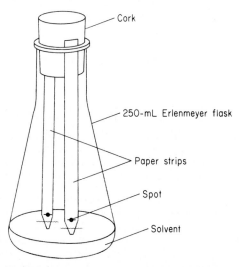

FIGURE 5.5 Simple chamber for developing paper strips.

7. When the solvent front has reached the desired position, usually 2 cm from the top of the plate, the plate [sheet or strip(s)] is removed and dried in the same manner as were the initial spotted zones.

Sample volumes of 1 to 5 μL are used in ordinary TLC, whereas volumes of 100 nL are used in HPTLC (high-performance TLC) when using fairly polar spotting solvents (200 nL for less polar solvents). A convenient means for manipulating such small volumes is a micropipet constructed from platinum-iridium capillary tubing fused into the end of a length of glass tubing.

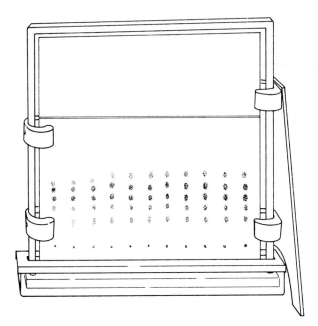

FIGURE 5.6 A sandwich chamber with a developing chromatogram.

In HPTLC spot diameters are less than 1 mm at the origin and about 2 to 3 mm diameter after development. The spot diameters in TLC will be 3 to 6 mm at the origin and increase to 6 to 15 mm after development. Small spots yield better separation and development. Multiple spotting of a sample, allowing the solvent to evaporate in between applications, keeps the initial spot small. A spotting template aids in reproducible placement of samples when not using plates with an inert preadsorbent area or concentrating zone.

The optimum sample size should not exceed 10 ng. The number of theoretical plates (efficiency) drops sharply with an increase in sample size up to 100 ng. A streaking pipet is used for applying larger amounts of sample as a streak.

In the sandwich technique (Fig. 5.6) a blank plate is placed over the spotted plate, the assembly clamped together, and placed in the developing chamber.

High performance plates contain a 2-cm wide inert spotting strip (concentrating zone) composed of purified diatomaceous earth along the lower edge of the plate. Use of these plates allows TLC to be performed without the need for special spotting apparatus or laborious spotting techniques.

5.1.12 Development Methods

Development may be carried out in an ascending manner (Fig. 5.7) or with the plate in a horizontal position when any of the following techniques are used.

5.1.12.1 Continuous Development. A continuous flow of the mobile phase passes through the stationary phase. As the mobile phase reaches the opposite

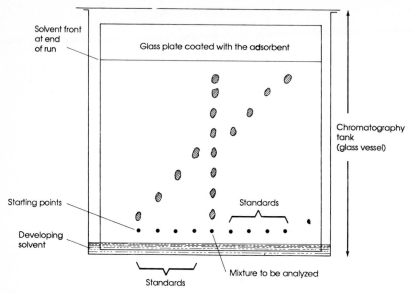

FIGURE 5.7   Separation by thin-layer chromatography.

end of the plate, it is removed by wicks or simply allowed to drip off the edge where it is sucked up by an adsorbent pad. This technique is useful when the R_f values of the more rapidly moving spots are small, but the spots show signs of separation. It allows further resolution among the components until the most rapidly moving spot reaches the opposite end of the plate. Mobile phases of low solvent strength are used to optimize selectivity. Continuous development occurs over a short distance (2 to 7 cm) with a constant, relatively high velocity solvent flow. Since spots do not travel very far, they will be small and compact, thus enhancing sensitivity.

5.1.12.2 Stepwise Development.   The sample is developed with a succession of solvents of different eluting strengths. This technique is useful for separating compounds in a mixture that contains a group of substances with low R_f values and other materials with relatively large R_f values; that is, when the components have significantly different polarities. If initially the developer is the more polar solvent, compounds with the higher R_f values travel with the front. If development is stopped about midway and the plate is dried, development with a less-polar second solvent serves to separate the compounds with high R_f values farther up the plate. The compounds originally separated in the lower half of the plate are not appreciably moved by the second solvent.

5.1.12.3 Two-Dimensional Development.   The sample is spotted in the lower left-hand corner of the plate 2.5 cm from each edge. Standards are placed in a similar position in the lower right-hand corner and also in the upper left-hand corner of the plate. One edge of the plate is immersed in the first solvent and separation is conducted in one direction until the solvent front approaches the opposite edge of the plate but doesn't overrun the spot containing the standards. After thoroughly drying the plate to remove the first solvent, the other edge adjacent to the original spot is placed in the second solvent and the chromatogram is developed at right angles to the preceding development.

Two-dimensional development with two different solvent phases is recommended when additional resolving power is required for complex mixtures. This technique brings about more effective development of adjacent components which overlap after a one-dimensional separation using either solvent mixture alone (Fig. 5.8).

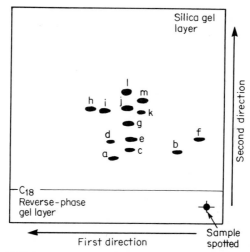

FIGURE 5.8 Two-dimensional development of 13 sulfonamides. First direction is on a C-18 reverse-phase gel layer with methanol/0.5 M NaCl (60:40). Second direction is mostly on a silica gel layer with ethyl acetate–methanol–aqueous ammonia (85:14.4:0.6). Solutes are: a, sulfisoxazole; b, sulfathiazole; c, sulfadiazine; d, sulfaquinoxaline; e, sulfachloropyridazine; f, sulfaguanidine; g, sulfamerazine; h, sulfabromomethazine; i, sulfadimethoxine; j, sulfamethazine; k, sulfaethoxypyridazine; l, sulfanilamide; and m, sulfapyridine.

This method also provides separation of standards in both directions, so the migrations of the two reference mixtures form the coordinates of a graph which permits the unknowns to be located in two dimensions.

5.1.12.4 Radial (or Circular) Development. In the radial method the sample is applied to the center of the plate. The plate is held in a closed developing chamber. The solvent is fed at a constant, controlled rate through a small-bore pipet to the central point via a syringe controlled by a stepping motor. With radial development the component zones are continuously expanding as the mobile phase flows outward from the center. This compresses the ensuing zones into progressively narrower concentric rings (Fig. 5.9). Furthermore, solvent feed is always faster at the trailing edge of a component zone which tends to compress it. Sep-

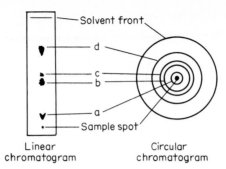

Linear
chromatogram

Circular
chromatogram

FIGURE 5.9 Radial (or circular) development
compared with linear development.

aration is very fast, and resolution is high, especially in the low R_f region. Samples can also be applied to the plate after the layer has been impregnated with mobile phase; in this case the TLC development more closely resembles HPLC.

5.1.13   Visualization of Solutes

Detection in TLC is completely separate from the chromatographic operation and may be considered a static process. After chromatography and evaporation of the mobile phase, chromophoric substances can be located visually; colorless substances require other means. Any given chromatographic fraction can be examined for as long as necessary to extract the maximum amount of information. This freedom from time constraints is potentially the most important aspect of TLC because it permits utilization of a variety of techniques to enhance the sensitivity of detection and wavelength selection for optimum response of each component to be measured. In addition, in situ spectra can be obtained for component identification.

Spots that are neither naturally colored nor fluorescent can be visualized after reaction with a chromogenic or fluorogenic detection reagent or by fluorescence quenching. For fluorescence quenching, plates are coated with fluorescent materials, such as zinc silicate; the spot will obscure this fluorescence when the plate is irradiated. Chromogenic reagents (Table 5.2), selected to react with a particular functional group, can be applied by either dipping or aerosol spraying to reveal the spot location. After application of reagents, many detection reactions require uniform heating for a specified time at some controlled temperature or irradiation by ultraviolet light. Charring, after spraying with dilute sulfuric acid, or better, ammonium sulfate, is a general method that reveals organic material.

The spray should be very fine and applied with a horizontal motion, beginning in the upper left of the plate, proceeding to the right, down and across the plate to the left, in an alternating motion to right and then left, and finally moving up and down the plate until the entire area has been covered twice.

After development, spots or zones can be removed from the plate by scraping. A miniature suction device collects scrapings directly in an extraction thimble held in a vacuum flask. Compounds are eluted from the scrapings and examined by any suitable method, often spectrophotometrically.

A permanent record of the developed chromatogram can be prepared by pho-

TABLE 5.2 Spray and Staining Reagents

Reagent	Application
Aniline phthalate	Reducing sugars
Antimony(III) chloride (25%) in chloroform; may yield fluorescence when heated	Carotenoids, steroid glycosides
Ammonium sulfate, 1 M, and heating to 170°C	Natural substances
Antimony(V) chloride (20%) in chloroform	Resins, terpenes, and oils
Bromocresol green, 0.05%	Acids and bases
2′, 7′-Dichlorofluorescein, 0.1%	Lipids
4-Dimethylaminobenzaldehyde, 0.5%	Free amino groups
Diphenylcarbazone-mercury(II) spray with DPC, followed by mercury sulfate in methanol	Barbiturates
Fluorescamine	Primary amines (including primary amines and sulfonamides)
Iodine vapor (10 min)	General for organics
Iodine, 0.5% in ethanol	Organic nitrogenous compounds
Iodoplatinate in methanol (2:1 v/v)	Morphine
Iodosulfuric acid [0.1 M iodine, 16% sulfuric acid (1:1)]	Organic nitrogenous
Methanol–sulfuric acid (9:1 v/v); heat at 110–170°C for 2–5 min	Fluorescent spots for sterols, bile acids, steroids, and cholesteryl esters
Ninhydrin, 0.1%	Amino acids and amines
Phosphomolybdic acid, 3.5%	Reducing compounds
Rhodamine B, 0.25%	Higher fatty acids and lipids
Rhodamine B, 0.025%	Insecticides
Sulfuric acid; heated to 120°C	Natural substances
o-Toluidine, 1% in acetone	Fluoresces blue to green

tography with a Polaroid camera or by spraying a plastic dispersion on the plate. After setting, the layer is peeled from the plate and stored as a flexible film.

5.2 QUANTITATION

Quantitation is performed directly on the developed plate, sheet, or strip chromatograms by transmittance, reflectance, fluorescence, or fluorescence quenching modes. The plate is mounted on a movable stage, usually motor-driven in the direction perpendicular to the slit length, and either manually operated or motor driven in the orthogonal direction. A chromatogram analysis system is shown in Fig. 5.10.

5.2.1 Absorption Densitometry

In absorption densitometry the spots on the TLC plate are scanned by a beam of monochromatic light formed into a slit image with the length of the slit selected according to the diameter of the largest spot.

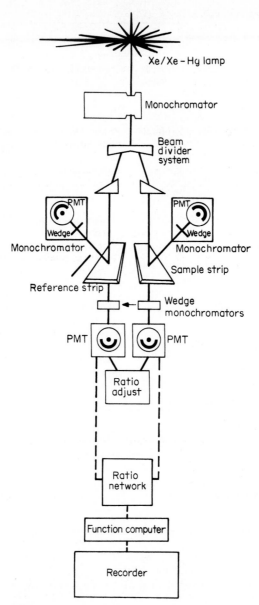

FIGURE 5.10   A chromatogram analysis system. *(Courtesy of Schoeffel, Inc.)*

Because the response of reflectance-absorbance scans is nonlinear with concentration, calibration standards are included with each sample run. Consequently, all samples, both standards and unknowns, are subjected to exactly the same chromatographic conditions, and systematic errors remain very much at a minimum. Typical minimum detection levels for measurement of visible or ultraviolet absorption range from 100 pg to 100 ng per spot.

These are some tips on densitometric scanning of TLC plates.

1. *Single-beam scanners* often give excellent quantitative results. Baseline drift may be troublesome due to extraneous adsorbed material in the thin layer which can move during chromatographic development or small irregularities in the plate surface.

2. *Double-beam scanners* (Fig. 5.10) have a reference beam scanning the intervening space between sample lanes. The difference signal eliminates the contribution of general plate background.

3. *Dual-wavelength scanners* have two monochromators that alternately furnish to the same sample lane a reference wavelength with minimal sample absorbance and a sample wavelength chosen for maximum absorbance by the analyte. This helps cancel out the general background and correct for plate irregularities.

5.2.2 Reflectance Mode

In the reflectance mode, the diffusely reflected (or scattered) light is detected with a photomultiplier. Those areas of the plate that are free of light-absorbing materials will yield a maximum signal, and the separated spots will cause a diminution of reflected light that is concentration dependent.

5.2.3 Fluorescence Measurement

When applicable, fluorescence measurement is preferred because of its greater selectivity and sensitivity and the high degree of linearity irrespective of the form of the zone. Often fluorimetry involves measuring a positive signal against a dark or zero background. Consequently, the signal can be highly amplified. A single-beam mode of operation of the fluorimeter is usually quite sufficient. Typical sensitivities for fluorescent compounds are in the low picogram range.

5.2.4 Peak Area Measurements

Quantitation is based on peak area measurements (see Sec. 2.3). A calibration curve is prepared for each compound using standards. The amount of compound in the unknown sample is obtained from the curve, and the concentration is calculated based on the amount of sample spotted. The quality of results in the fluorescence quenching mode depends upon the homogeneity of the phosphor distribution in the layer, and calibration curves tend to be nonlinear.

5.2.5 Standard Addition Method

The standard addition assay method compares each unknown sample with three standards. Four bands of the sample solution are applied first. After intermediate drying the first band is oversprayed with 10 μL of a standard solution containing known concentrations of the impurities. The second band is oversprayed with 4 μL and the third with 2 μL. The fourth band is not oversprayed; it remains unchanged.

5.3 PAPER CHROMATOGRAPHY

In paper chromatography the substrate is a piece of porous paper with water adsorbed on it. The sample is placed on the paper as a spot or streak and then irrigated by the solvent system which percolates within the porous structure of the paper. Usually development of the chromatogram is stopped before the mobile phase reaches the farther edge of the paper, so the solute zones are distributed in space instead of time. The major limitations of paper chromatography are the relatively long development times and less sharply defined zones as compared to thin-layer techniques. Thin-layer chromatography with powdered cellulose as stationary phase has displaced paper chromatography from many of its previous applications.

5.3.1 Solvent Systems

Chromatography on paper is essentially a liquid-liquid partition in which the paper serves as carrier for the solvent system. Aqueous systems are used for strongly polar or ionic solutes. Water is held stationary on the paper as a "water-cellulose complex" or puddles of water, organized and dense near the amorphous regions of the cellulose chains. The stationary phase is attained by exposing the suspended paper to an atmosphere saturated with water vapor in a closed chamber. If an aqueous buffer or salt solution is to be used as the stationary phase, the paper is drawn through the solution, allowed to dry, and then exposed to the atmosphere saturated with water vapor.

The mobile phase might be butanol for a neutral system, butanol–acetic acid–water (40:10:50) for an acidic system, or butanol-ammonia-water (75:8:17) for a basic system. The latter two systems are prepared by shaking all components in a separatory funnel; the less-polar phase serves for development. Even water itself can serve as developer.

5.3.2 Equipment

The only essential piece of equipment is the developing chamber, often simply an enclosed container with an airtight lid. For one-dimensional paper chromatography, either ascending or descending development can be carried out in simple units shown in Figs. 5.3 to 5.5. Descending development is more often used because it is faster and more suitable for long paper sheets which give higher efficiencies.

Radial development is carried out by cutting a tab from a circular piece of paper, spotting the sample at the upper end of the tab or streaked in a circle a short distance from the center of the paper, and then placing the tab in the solvent reservoir. Small chambers are made from two petri dishes, one inverted over the other.

5.3.3 Development

Each of the development techniques described for thin-layer chromatography is equally applicable for paper chromatography. One difference should be noted. The paper is equilibrated with both mobile and stationary phases for 1 to 3 h before development. The two phases are placed in separate reservoirs in the bottom of the chamber. The mobile phase should also be equilibrated with the stationary phase. Close temperature control is required for reproducible R_f values. With volatile solvents, careful equilibration is required to avoid band tailing.

5.3.4 Procedure

1. Cut filter paper to size desired.
2. Draw a *light* pencil line about 2.5 cm from the edge and make marks 2.5 cm apart, leaving 2.5 cm from each vertical edge.
3. Place a spot of solution at each mark (Figs. 5.1 and 5.2).
4. If a sheet, roll paper into a cylinder, holding ends together with a plastic clip (Fig. 5.4).
5. Dry the spots thoroughly.
6. Prepare the solvent mixture. Transfer the solvent mixture to the bottom of the developing tank or chamber, being careful that none touches the sides of the chamber.
7. Place the dried, spotted paper in the developing chamber. The initial zones of sample should be 1 cm above the solvent level in the chamber. Cover the chamber with a glass plate or plastic wrap.
8. Remove the paper chromatogram after the solvent front has traveled the desired distance, but no closer than 2 cm of the farther edge of the paper. Mark the final solvent front with a pencil.
9. Dry the paper. Develop the chromatographic zones if they are not already visible. Measure the distance from the origin to the center of each spot. Divide this distance by the distance between the origin and the solvent front; it is the R_f value.

5.3.5 Visualization and Evaluation of Paper Chromatograms

All the discussion on the subject in TLC is applicable for paper chromatography with the exception of fluorescent quenching using TLC plates impregnated with a phosphor. Special attachments for commercial transmission spectrophotometers provide a means for drawing the paper chromatogram across a window in front of a photodetector. The reflection mode can also be used. An assessment of the area under the photometric curve completes the measurement.

5.4 ZONE ELECTROPHORESIS

Zone electrophoresis involves the migration of charged species in a supporting medium (bed) under the influence of an electric field. Each charged species moves along the field gradient at a rate which is a function of its charge, size, and shape. The supporting medium is soaked with an electrolyte (usually a buffer), and the sample is spotted near the center of the bed. Upon application of a controlled dc source of potential to the ends of the supporting medium, each component of the sample begins to move according to its own mobility. Ideally, each sample component will eventually separate from its neighbors, forming a discrete zone. The chromatographic analog of zone electrophoresis is elution chromatography.

Electrophoresis is the premier method of separation and analysis of proteins and polynucleotides. For example, the primary classification of protein components is based on their relative positions in a developed electrophoretic pattern. Components are sometimes identified by their electrophoretic mobilities.

5.4.1 Equipment

The essential components of equipment for zone electrophoresis are shown in Fig. 5.11. Electrodes are affixed at opposite ends of the supporting medium and are separated from the supporting medium (or bed) by diffusion barriers whose function is to prevent diffusion and convection processes from carrying electrolytic decomposition products into the bed. The entire bed and electrodes are enclosed in an airtight chamber to prevent excessive evaporation of solvent.

Under the influence of the applied emf, usually between 5 to 10 V $\cdot$ cm^{-1}, the charged species move with a velocity (ionic mobility) which depends upon the field strength. The power supply must develop between 50 to 150 V. With paper or cellulose acetate as supporting media the current drain is less than 1 mA per strip.

To compare results on separate runs, the temperature must be closely controlled because the mobility of species increases with temperature. Thermostatting the bed minimizes the effect of electrical heating that accompanies electrophoresis.

5.4.2 Supporting Media

Early work involved zone electrophoresis on filter paper which developed parallel to paper chromatography. Paper is now largely replaced by cellulose acetate strips or polyacrylamide gels which provide superior resolution.

The electrophoresis bed may be a trough or a vertical column of gel. The latter method is called *disk electrophoresis* because of the shape assumed by the fractions as they migrate in the cylindrical bed. Some 20 components in about 3 μL of serum have been separated in only 20 min.

5.4.2.1 *Cellulose Acetate Beds.* Cellulose polyacetate forms strong and flexible membranes which possess a very uniform, foamlike structure. The pore size is closely controlled in the 1-μm range. Uniform porosity results in sharper boundaries and permits better resolution and shorter migration times as compared with filter paper. The inertness of the membrane virtually eliminates adsorptive effects and trailing boundaries. Running times are 20 to 90 min. After electrophoresis the membrane can be cleared to glasslike transparency for densitometry, thus permitting accurate quantitation. The developed strip is dipped into a mixture of acetic acid–ethanol or in Whitemor oil 120.

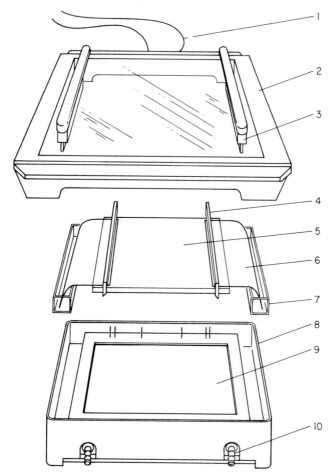

FIGURE 5.11 Electrophoresis equipment component parts: 1, electric power lead; 2, cell lid with safety microswitch; 3, electrode unit; 4, clamping rod; 5, thin-layer plate; 6, conductor wicks; 7, buffer trough; 8, housing; 9, insulated cooling block; and 10, cooling water connections.

For a simple analysis of proteins in serum or other body fluids, zone electrophoresis on cellulose acetate membranes is sufficient. A variety of staining procedures is available.

5.4.2.2 Polyacrylamide Gels. Gels, such as polyacrylamide, are formed directly in the electrophoretic bed. By controlling the relative proportion of the cross-linking agent bisacrylamide, gels can be formed with well-defined molecular sieving properties. The sieving effect, unique to gels, provides extremely high resolving power. Techniques for gel sieving in polyacrylamide are the heart of the powerful and efficient methods developed for sequencing polynucleotides.

Polyacrylamide can also be used in slabs or rods and prepared so that the concentration of acrylamide increases in a continuous manner over the length of the gel. This is known as a *gradient gel*. In gradient electrophoresis the development occurs through a gradient of decreasing pore size, similar to separations by exclusion (gel permeation) chromatography. Proteins are separated in the order of their molecular sizes. The shape of the gradient determines the resolution and is chosen to give a linear relationship between the migration distance of a globular protein and the logarithm of its molecular weight.

5.4.2.3 Agarose Gels. For molecules of molecular weight in excess of 1 million, even the lowest concentration of polyacrylamide that forms a physically stable gel yields pores too small to permit migration. An alternative medium is agarose derived from agar. Whole chromosomes with molecular weights up to one billion are separated.

5.4.3 Gel Sieving

The most effective application of gel sieving to proteins involves the use of sodium dodecyl sulfate, which binds to a wide variety of proteins in a constant ratio, producing complexes with constant charge per unit mass. Protein aggregates (dimers, tetramers, etc.) are broken up (denatured or unfolded). Electrophoretic mobilities are a direct function of the protein subunit molecular weight. The rate of migration of a protein is compared to the rate of migration of standard proteins, and fairly accurate estimates of protein molecular weight are obtained.

5.4.4 Electrophoretic Mobilities

The positively charged species will move toward the cathode, and the negatively charged species will move toward the anode. The rate of movement of both species is determined by the product of the net charge of the species and the field strength X defined as

$$X = \frac{E}{s} \tag{5.5}$$

where E is the potential difference in volts between test probes inserted into the supporting medium and spaced s centimeters apart.

The electrophoretic mobility of an ion μ is given by

$$\mu = \frac{d}{t}\left(\frac{E}{s}\right) \tag{5.6}$$

where d is the linear distance (in centimeters) traveled by the migrating species relative to the bed in time t (in seconds) (that is, its velocity) in a field of unit potential gradient (in volts per centimeter). Electrophoretic migration is directly proportional to the field strength. Increasing the field strength will speed fractionation; it will also enhance the resolution because the decreased development time minimizes the opportunity for diffusion of the fractions. Permissible field strength is predicated by the removal of heat from the supporting medium by the cooling system.

The net charge carried by most species is pH-dependent. Buffer pH and ionic strength are important. The net mobility of a partially ionized solute in solution is given by

$$\mu' = \frac{\mu K_a}{[H^+] + K_a} \tag{5.7}$$

where K_a is the ionization constant. The extent to which two incompletely ionized acids may be separated depends upon the difference in their apparent mobilities.

For cations, electrophoresis is generally run in the intermediate pH range using complexing agents such as lactic, tartaric, or citric acids to ensure solubility. Most procedures for separating anions or weakly acidic substances, such as carboxylic acids and phenols, require alkaline solutions to avoid any possible existence of two ionization states in equilibrium. Conversely, weakly ionized bases, such as amines and alkaloids, should be separated at low pH values.

The ionic strength of a buffer is adjusted to 0.05 to 0.1 This range minimizes the production of heat and the quantity of electrode products formed.

5.4.5 Electro-osmosis

During the migration to the appropriate electrodes, hydrated ions carry with them their associated water molecules. This assemblage consists of water molecules held directly in the solvation sheath plus additional layers of water molecules attracted and held by the inner layers. Thus, it is possible for hundreds of solvent molecules to be dragged along by each migrating ion. Inasmuch as cations usually carry more water than anions, the net flow of liquid is almost always toward the cathode. This results in an apparent movement of neutral molecules toward the cathode. Because this movement is approximately directly proportional to the voltage gradient, it serves as a mobility correction that is added for anions and subtracted for cations. The value of this correction is determined by observing the migration of an uncharged substance.

5.4.6 Detection of Separated Zones

Common approaches to detection involve the use of stains, autoradiography of radiolabeled analytes, or immunoreaction with specially prepared antisera.

5.4.6.1 Staining. Staining procedures are quite lengthy and usually involve a fixative, such as trichloroacetic acid, which precipitates proteins and prevents their diffusion out of the gel during subsequent soaking with the staining reagent(s).

5.4.6.2 Autoradiography. In autoradiography the sample is run on a slab gel. Following development the gel is covered with a thin sheet of plastic and placed on top of X-ray film. Regions of the gel where radioactivity is localized will create corresponding exposed regions on the film. For isotopes producing low-energy beta particles, the gel is soaked first in a scintillation fluid, dried, and clamped on top of X-ray film between glass plates, and maintained at approximately −70°C for 1 day to 1 week.

5.4.6.3   Crossed Immunoelectrophoresis.   In crossed immunoelectrophoresis a polyspecific antibody-containing gel is poured next to the developed electrophoresis gel. After this gel has solidified, the sample bands are electrophoretically migrated perpendicular to their original direction of migration and into the antibody-containing gel. Sample bands will continue to migrate through this gel until they have encountered sufficient antibody to result in precipitation of the antigen-antibody complex. Precipitated bands are then stained. Peak heights and areas are indicative of the quantity of protein in each band. The resolution of overlapping bands is remarkable.

BIBLIOGRAPHY

Braithwaite, A., and F. J. Smith, *Chromatographic Methods,* Chapman and Hall, London, 1985.

Fennimore, D. C., and C. M. Davis, "High Performance Thin-Layer Chromatography," *Anal. Chem.,* **53**:252A (1981).

Jorgenson, J. W., "Electrophoresis," *Anal. Chem.,* **58**:743A (1986).

CHAPTER 6
ULTRAVIOLET-VISIBLE SPECTROPHOTOMETRY

6.1 INTRODUCTION

Energy can be transmitted by electromagnetic waves. They are characterized by their frequency v, the number of waves passing a fixed point per second, and their wavelength λ, the distance between the peaks of any two consecutive waves (Fig. 6.1). As the frequency increases, the wavelength decreases and, conversely, as the frequency decreases, the wavelength increases.

In any electromagnetic wave, wavelength and frequency are related to the energy of a photon E by Planck's constant h, 6.62×10^{-34} J $\cdot$ s, and c, 2.998×10^{10} cm $\cdot$ s, the velocity of radiant energy in a vacuum:

$$E = hv = \frac{hc}{\lambda} \tag{6.1}$$

6.1.1 Types of Electromagnetic Radiation

The values of the wavelength and frequency are what differentiate one kind of radiation from another within the electromagnetic spectrum, the name given to the broad range of radiations that extend from cosmic rays with wavelengths as short as 10^{-9} nm all the way up to radio waves longer than 1000 km. The various regions in the electromagnetic spectrum are displayed in Fig. 6.2 along with the nature of the changes brought about by the interaction of matter and radiation.

6.1.2 Conversion of Units

Wavelengths are expressed in different units throughout the electromagnetic spectrum. This may lead to confusion and error when attempting to determine equivalent values or when reading older literature that used values different from the International Union of Pure and Applied Chemistry (IUPAC) recommendations. Table 6.1 shows how to interconvert wavelengths.

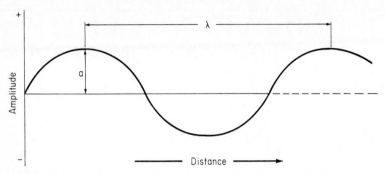

FIGURE 6.1 An electromagnetic wave. λ = wavelength; a = amplitude, a measure of intensity.

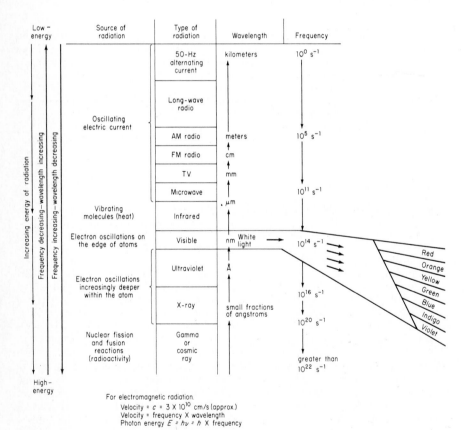

FIGURE 6.2 Electromagnetic radiation spectrum.

TABLE 6.1   Wavelength Conversion Table

To obtain number of	Angstroms Å	Nanometers nm	Micrometers μm	Millimeters mm
			Multiply number of	
			By	
Angstroms (Å)	1	10	10^4	10^7
Nanometers (nm)*	10^{-1}	1	10^3	10^6
Micrometers (μm)†	10^{-4}	10^{-3}	1	10^{-3}
Millimeters (mm)	10^{-7}	10^{-6}	10^{-3}	1
Centimeters (cm)	10^{-8}	10^{-7}	10^{-4}	10^{-1}
Meters (m)	10^{-10}	10^{-9}	10^{-6}	10^{-3}

*Often called millimicrons in older literature.
†Often called microns in older literature.

6.2 DEFINITIONS AND SYMBOLS

6.2.1 Radiant Power

Radiant power (P) is the amount of energy transmitted in the form of electromagnetic radiation per unit time, expressed in watts. This is what a photodetector measures. It should not be confused with intensity (I) which is the radiant energy emitted within a time period per unit solid angle, measured in watts per steradian.

6.2.2 Transmittance

Transmittance (T) is the ratio of the radiant power (P) in a beam of radiation after it has passed through a sample to the power of the incident beam (P_o) and is defined as

$$T = \frac{P}{P_o}$$ (6.2)

What is of more interest to the analyst is the internal transmittance of the cuvette contents, that is, the ratio of the radiant power of the unabsorbed (transmitted) radiation that emerges from the absorbing medium to that transmitted through a blank (usually a solvent blank):

$$T = \frac{P_{\text{sample}}}{P_{\text{blank}}}$$ (6.3)

when using the same or matched cuvettes. Percent transmittance is the transmittance multiplied by 100.

6.2.3 Absorbance

Absorbance (A) is defined as

$$A = \log \frac{P_{\text{blank}}}{P_{\text{sample}}} = \log \frac{1}{T} \tag{6.4}$$

which is the logarithm (base 10) of the reciprocal of the transmittance.

6.2.4 Specific Absorptivity

When absorbance is defined as

$$A = abC \tag{6.5}$$

the proportionality constant a is known as the absorptivity and is given in units $L \cdot g^{-1} \cdot cm^{-1}$.

6.2.5 Molar Absorptivity

Molar absorptivity (ϵ), also known as the molar extinction coefficient, is the absorbance of a solution divided by the product of the optical path b in centimeters and the molar concentration C of the absorbing species:

$$\epsilon = \frac{A}{bC} \tag{6.6}$$

where ϵ is in units of $L \cdot mol^{-1} \cdot cm^{-1}$.

6.3 FUNDAMENTAL LAWS OF SPECTROPHOTOMETRY

6.3.1 Lambert-Beer Law

Lambert's law states that, for parallel, monochromatic radiation that passes through an absorber of constant concentration, the radiant power decreases logarithmically as the path length b increases arithmetically.

Beer's law states that the transmittance of a stable solution is an exponential function of the concentration of the absorbing solute.

If both concentration and thickness are variable, the combined Lambert-Beer law (often known simply as Beer's law) becomes

$$A = \epsilon bC \tag{6.7}$$

A plot of absorbance versus concentration should be a straight line passing through the origin, as shown in Fig. 6.3. Readout scales are often calibrated to

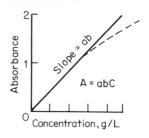

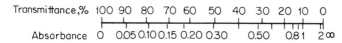

FIGURE 6.3 Representation of Beer's law and comparison between scales in absorbance and transmittance.

read absorbance as well as transmittance. In Fig. 6.3 note also the comparison between scales in absorbance and transmittance.

6.3.2 Linearity and Beer's Law

Deviations from Beer's law fall into three categories: real, instrumental, and chemical. *Real deviations* arise from changes in the refractive index of the analytical system; these changes will be significant only in high-absorbance differential measurements.

6.3.2.1 Instrumental Deviations. Instrumental deviations arise primarily from the bandpass (bandwidth) of filters or monochromators. Beer's law assumes monochromatic radiation. Deviations from Beer's law are most serious for wide bandpasses and narrow absorption bands. What is important is the ratio between the spectral slit width and the bandwidth of the absorption band.

6.3.2.2 Chemical Deviations. Chemical deviations are caused by shifts in the position of a chemical or physical equilibrium involving the absorbing species. Generally the pH and ionic strength of the system must be kept constant. For weakly ionized acids the pH should be adjusted to at least three units more or less than the pK_a value of the monoprotic acid. When the absorbing species is a complex ion, the concentration of the free (excess) ligand must be constant and preferably in 100-fold excess.

6.3.3 Calibration Curves

Calibration curves, if known to be linear over the concentration range to be used, can be prepared in several ways.

1. Use a reagent blank to set zero absorbance and one standard (preferably close in concentration to the sample).

2. Use a reagent blank to set zero absorbance and known concentration factor (slope of a previously determined calibration curve of absorbance versus concentration).

3. Use two standards whose concentrations bracket that of the sample(s).

4. Use one standard and a known concentration (slope) factor.

When establishing a linear calibration curve from a number of absorbance-concentration data points, a certain degree of data scatter usually occurs. It is necessary to perform a linear regression analysis on the standards to determine from the best-fit calibration line the degree of correlation among the data. For this work a computer is invaluable.

For *nonlinear calibration curves,* a stored program is needed. Values of absorbance versus concentration are stored. When point-to-point interpolation between entered standards is needed, the program generates a series of small straight lines that approximate the nonlinear curve.

6.4 RADIATION SOURCES

Radiation sources must provide sufficient radiant energy over the wavelength region where absorption is to be measured, and they should maintain a constant intensity over the time interval during which measurements are made.

6.4.1 Hydrogen or Deuterium Discharge Lamps

Deuterium or hydrogen discharge lamps are used in the ultraviolet regions of the spectrum. A current-regulated power supply is required. In a monochromator the lamp arc is positioned at one of the foci of an elliptical reflector which provides more than 60 percent collection efficiency. At less than 360 nm these discharge lamps provide a strong continuum; weak line emissions are superimposed on the continuum at longer wavelengths. With fused silica envelopes, work to wavelengths as short as 160 nm is feasible. The use of deuterium in place of hydrogen enhances the lamp output 3 to 5 times.

Warning Ultraviolet radiations are harmful. Never look directly into the source while it is in operation. Protect the eyes with ultraviolet-absorbing goggles. Protective clothing should be worn to protect the skin. Persons on medications which produce photosensitivity should observe special precautions and avoid any exposure.

6.4.2 Incandescent Filament Lamps

Incandescent filament lamps give a continuous spectrum from 350 nm (where the glass envelope absorbs strongly) and into the near infrared to 2.5 μm. A tungsten filament is heated to incandescence by an electric current. The filament is enclosed in a sealed bulb of glass filled with an inert gas or a vacuum. The units are rugged and inexpensive.

The tungsten incandescent lamp has its maximum emissivity at about 1000 nm, and drops off very rapidly in the ultraviolet region to 1/100 of that value at about 300 nm. Only about 15 percent of the output radiant energy falls within the visible region of the spectrum. Stray light can become a serious problem. Often a heat-absorbing filter or a multilayer interference beam splitter (cold dichroic mirror) is inserted between the lamp and sample holder to remove the infrared radiation.

In tungsten-iodine lamps, iodine is added to the normal filling gases. These lamps maintain more than 90 percent of their initial brightness throughout their life.

6.5 WAVELENGTH SELECTION

For spectrophotometric work, discrete bands of radiation are desired. Furthermore, Beer's law is based on the assumption of monochromatic radiation.

6.5.1 Filters

6.5.1.1 Absorption Filters. Absorption filters are available as two series of sharp cutoff filters. The red and yellow series pass long wavelengths whereas the blue and green series cut off the long wavelengths. They are widely used as blocking filters to suppress unwanted spectral orders from interference filters and diffraction gratings. Composite filters are constructed from two sharp cutoff filters, one from each series, as shown in Fig. 6.4. The bandwidth ranges from 20 to 70 nm. Peak transmission is 5 to 25 percent and decreases as the bandwidth narrows.

In Fig. 6.4 note the meaning of these terms: nominal wavelength, peak transmittance, spectral bandwidth or full width at half maximum (FWHM; that is, at ½ peak transmittance), and spectral bandpass (total wavelength region passed by the filter).

6.5.1.2 Interference Filters. Interference filters consist of a dielectric spacer film sandwiched between two parallel, partially reflecting metal films of silver. When the spacer thickness is half the wavelength of the radiation to be transmitted (in the refractive index of the dielectric spacer), transmitted and reflected beams are in phase and interfere constructively. Other wavelengths are destructively interfered. Since partial reinforcement occurs for wavelengths centered around the nominal wavelength, the bandwidth is 10 to 15 nm. The maximum transmission is about 40 percent. Unwanted transmission bands from other orders are eliminated by using appropriate sharp cutoff absorption filters as the protective glass covers.

6.5.1.3 Wedge Filter. A wedge filter is a wedge-shaped slab of dielectric deposited between the semireflecting metallic layers. At each point along the length of the filter a different wavelength is transmitted. Different wavelengths are isolated by moving the wedge past a slit, or the opposite.

6.5.2 Monochromators

A monochromator provides a beam of radiation of finite wavelength interval to be passed through a sample; it also can adjust the energy throughput. As shown in Fig. 6.5, a monochromator will have these components:

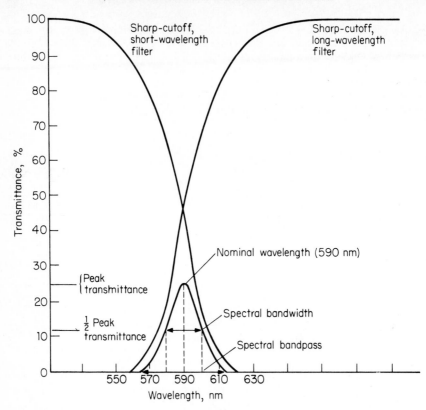

FIGURE 6.4 Spectral transmittance characteristics of a composite glass absorption filter and its components.

1. An entrance slit that provides a narrow optical image of the radiation source
2. A collimating mirror that renders the radiation entering the entrance slit parallel
3. A grating or prism for dispersing the incident radiation into discrete wavelengths
4. A collimating mirror to re-form images of the entrance slit on the exit slit
5. An exit slit to isolate the desired spectral bandwidth.

A concave holographic grating eliminates the need for items 2 and 4. The fewer reflecting optical surfaces in a monochromator, the higher the light throughput and the fewer stray light problems.

6.5.3 Plane Diffraction Gratings

A diffraction grating consists of parallel, equally spaced grooves. The radiation incident on each groove is diffracted (spread out) over a range of angles and in several overlapping orders, as shown in Fig. 6.6. Modern gratings have an inclined groove profile (blazed) which enables them to concentrate most of the

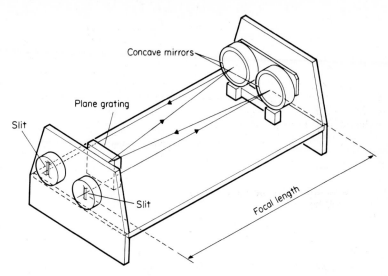

FIGURE 6.5   The Czerny-Turner mounting.

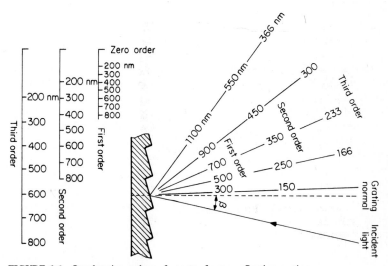

FIGURE 6.6   Overlapping orders of spectra from a reflection grating.

incident energy into a single order. Unwanted orders are eliminated by a bandpass filter or by cross dispersion with another grating or prism. The range in the first order extends from two-thirds the blaze wavelength to twice the blaze wavelength. (The *blaze wavelength* is the wavelength at which the angle of diffraction equals the angle of specular reflection from the face of the grating groove. *Specular reflection* occurs when the grating acts like a mirror.)

A grating monochromator has nearly constant dispersion throughout the spectrum and, consequently, a linear scale for wavelength. This is an important advantage of gratings over prisms.

6.5.4 Holographic Gratings

Holographic gratings are produced by photographic means using lasers. These gratings have no ghosts and a much lower stray light level than ordinary gratings. Holographic gratings can be produced with as many as 6000 grooves per millimeter as compared with a usual 1200 grooves per millimeter ruled grating. A concave holographic grating, plus entrance and exit slits, is in itself a monochromator (see Fig. 6.14).

6.5.5 Grating Monochromator Systems.

The Czerny-Turner mount (Fig. 6.5) is a popular monochromator configuration. Entrance and exit slits are on either side of the grating. Twin concave mirrors are used as collimating and focusing mirrors. The bandpass is controlled by the number of grooves per millimeter, the focal length, and the entrance and exit slit widths. As an example, a monochromator possessing a grating with 1200 grooves per millimeter, both slits 1.0 mm in width, and a focal length of 0.25 m, offers a spectral bandpass of about 4 nm. A holographic plane grating with 2400 grooves per millimeter improves the bandpass to 2 nm.

6.5.6 Resolving Power

The resolving power of a monochromator is its ability to distinguish as separate entities adjacent spectral features whether absorption bands or emission lines. For most samples in solution, typical natural bandwidths are 5 to 50 nm. Sharper peaks than this occur in some rare earth salts and gases. When the spectral bandwidth is one-tenth of the natural bandwidth of the absorption or emission band, deviation from the true peak height is less than 0.5 percent.

6.5.7 Wavelength Precision

Wavelength precision is more important than absolute accuracy, especially if a measurement is being made on the slope of a peak or on the top of a very sharp peak. With the use of microprocessor-controlled stepping motors, it is possible to achieve a wavelength precision of 0.02 nm. For wavelength calibration, didymium glass has sharp, multiple absorption bands in the visible region. The emission lines of a mercury arc are useful in the region 200 to 1100 nm.

6.5.8 Photometric Precision

All spectrophotometers measure transmittance, which then must be converted to absorbance. The sample, dark, and reference signals are digitalized immediately after the photomultiplier output. All comparisons, correction factors, and calculations of absorbance are made on this digital signal. This allows full digital signal processing over the entire absorbance scale, whereas in the older instruments the analog amplifiers were adjusted to be correct only at certain fixed points, such as 0, 1, 2 absorbance units.

6.5.9 Stray Radiation

Stray radiation arises from scattering of radiant energy by dust or smudges on the optical surfaces, optical imperfections of mirrors and cuvette walls, and reflections from interior surfaces of the monochromator. It affects photometric accuracy by causing apparent deviations from Beer's law. Negative deviations occur if stray radiation is not absorbed by the sample, which usually is the case. For example, if an instrument has stray radiation of 1 percent at a particular wavelength, the instrument can never read more than two absorbance units regardless of the sample concentration. Stray radiation becomes very serious in spectral regions where (1) the photodetector sensitivity is low or (2) the radiant energy that reaches the detector is weak.

6.6 CELLS AND SAMPLING DEVICES

Square rectangular cells are always used in precision work; cylindrical cells are used in inexpensive instruments. The latter must be carefully positioned each time if reproducible results are to be expected. Cells are usually 1 cm in path length, but cells are available from 0.1 to 10 cm or more. Silicate glasses can be used from about 350 to 3000 nm. Fused silica or quartz is transparent down to about 190 nm. Inexpensive plastic cells are available for use in the visible region. Cells should be matched both for internal width and cell wall widths. Walls in the radiation beam must be perfectly parallel to each other and perpendicular to the beam.

6.6.1 Care of Cuvettes

1. Cells should be scrupulously cleaned before and after each use. To clean cuvettes, lens paper soaked in spectrograde methanol, which is held by a hemostat or a similar device, should be used. Very dirty cuvettes can be cleaned with nitric acid or aqua regia (*never* with dichromate solution).
2. After cleaning with methanol, allow the methanol to evaporate. After an acid cleaning, thoroughly rinse with distilled water and then with spectrograde methanol. Dry at room temperature (*never* in an oven as path length may change).
3. *Never* touch the cell faces when handling the cells because fingerprints, grease, and lint may alter the transmittance.

6.6.2 Fiber Optics

Fiber-optic bundles are composed of numerous strands of glass or plastic fused at the ends. Radiation is transmitted along the individual fibers by total internal reflection. Fiber optics are useful when standard cells are not suitable, as when sampling a flowing stream. One fiber optic bundle brings the incident beam to the solution and another bundle returns the transmitted beam to the spectrophotometer.

6.7 DETECTORS

A detector converts electromagnetic radiation into an electron flow. Important parameters are spectral sensitivity, wavelength response, gain, and response time.

6.7.1 Photoemissive Tubes

A single-stage vacuum phototube contains a radiation-sensitive cathode in the form of a half cylinder and an anode wire located along the axis of the cylinder or a rectangular wire that frames the cathode (Fig. 6.7). The spectral sensitivity for several types of materials coating the photocathode surface is shown in Fig. 6.8. Cathode quantum efficiency (QE) is the average number of photoelectrons emitted from the photocathode divided by the number of incident photons.

The Sb-Cs type is popular for the visible region from 390 to 600 nm; fitted with a fused silica window it is usable down to 200 nm. Red response is provided by the Ag-O-Cs type which is often used from 600 to 1100 nm.

When a photoemissive tube is operated in darkness, a current still flows. This "dark" current limits the sensitivity as well as the difficulty of amplifying low

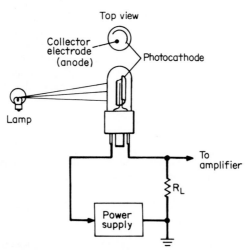

FIGURE 6.7 Photoemissive tube and detection circuit.

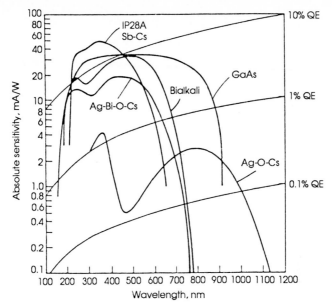

FIGURE 6.8 Spectral response curve of selected photoemissive surfaces. (*Courtesy of Burle Industries, Incorporated.*)

radiation levels that are smaller than the ohmic leakage across the tube envelope. The time constant is about 150 ps.

6.7.2 Photomultiplier Tubes

The photomultiplier tube (PMT) combines the photoemissive cathode with an internal electron-multiplying chain of dynodes. The circular cage design is shown in Fig. 6.9. Response time is as low as 0.5 ns. Dark current can be minimized by cooling the PMT. The gain varies from 3 to 50 per dynode stage; this figure is raised to the power of the number of stages to estimate the overall tube gain. Photomultiplier tubes are excellent detectors for low levels of radiant energy.

6.7.3 Photodiodes

Construction of a planar-diffused *p-n* junction diode is shown in Fig. 6.10. Present devices detect only visible and near-infrared radiation. Sensitivity is at least an order of magnitude more than vacuum photoemissive tubes but many orders of magnitude less than photomultiplier tubes. Rise times vary from less than 1 ns to more than 10 μs.

Many individual photodiodes can be assembled in a linear or two-dimensional array in which each diode senses simultaneously the signal from a unique portion of the spectrum. Because the detector array is read out very rapidly using high-speed electronic techniques, an entire spectrum can be acquired in milliseconds. No scanning is involved; therefore, wavelength calibration is precise and

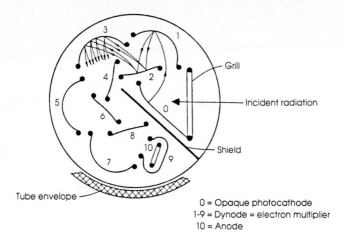

0 = Opaque photocathode
1-9 = Dynode = electron multiplier
10 = Anode

FIGURE 6.9   Photomultiplier design. The circular-cage multiplier structure in a "side-on" tube.   (*Courtesy of Burle Industries, Incorporated.*)

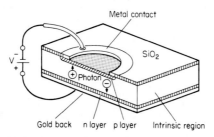

FIGURE 6.10 Construction of a planar-diffused *p-n* junction photodiode.

nonvarying with time. Commercial units are available with 128 to as many as 4096 elements per array. Each array is spaced 15 μm apart.

6.8 MEASUREMENT OF ABSORPTION SPECTRA

Instrument modules for measuring absorption of radiation are shown in Fig. 6.11. The modules are: (a) a radiation source, (b) focusing optics such as entrance and exit slits and collimating mirrors, (c) sample holder(s), (d) a wavelength isolation device, such as filters or gratings, and (e) some type of detector with amplifier and readout system. In terms of construction there are single-beam and double-beam radiation paths, and the photometer module may be direct reading, or it can use a balancing circuit.

6.8.1 Filter Photometers

A relatively inexpensive absorption photometer may be designed around a set of filters, a tungsten bulb (flashlight type), and a barrier-layer photocell or a

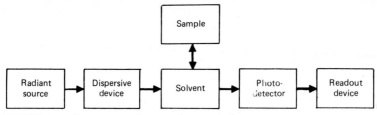

FIGURE 6.11   Components of spectrophotometric instruments.

photoemissive tube (Fig. 6.12). Higher signal-to-noise ratios can usually be obtained than with a monochromator, so less sensitive detectors can be coupled with a filter photometer.

6.8.2   Spectrophotometers

Instruments are classified as spectrophotometers when they incorporate a monochromator. They offer the advantage of continuous wavelength selectability, but they are limited by the radiant energy transmitted. A photomultiplier tube is generally required.

6.8.3   Single-Beam Instruments

In a single-beam instrument the reference and the sample are measured sequentially to determine the amount of radiant energy absorbed at a given wavelength

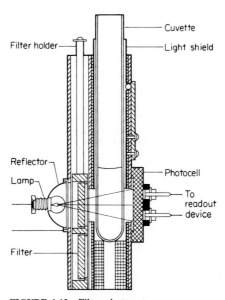

FIGURE 6.12   Filter photometer.

by the sample. Source intensity must remain constant for both measurements. The sequence of operations is as follows:

1. Set the wavelength to the selected value.
2. Fill the cuvette with the solvent blank and position the cuvette in the radiation path.
3. Adjust the instrument to read 0 percent transmittance when a shutter or some occluder blocks all radiation from the detector.
4. With the shutter removed adjust the instrument to read 100 percent transmittance.
5. Replace the solvent blank with the sample. Read the transmittance.

In some instruments there may be two readout scales, one in transmittance and the other in absorbance. Remember that zero transmittance is infinity on the absorbance scale, and 100 percent transmittance is 2.00 on the absorbance scale.

6.8.4 Double-Beam Instruments

Only those spectrophotometers using a completely symmetrical double-beam system can compensate for instabilities in the light source, detector, and system electronics. In double-beam instruments, a beam of radiant energy is split (or chopped) into two components of equal radiant power. One beam passes through the sample cuvette and the other through a reference solution or blank (which may simply be an air path). The output from the reference beam is kept constant by an automatic gain control (which involves only electronic circuitry). Now the transmittance of the sample can be recorded directly as the output of the sample beam.

The optical diagram of a popular direct-reading double-beam spectrophotometer is shown in Fig. 6.13. Whenever the sample container is removed from the instrument, an occluder falls into the light beam so that the phototube is dark. In this condition the amplifier control (dark current) can be adjusted to bring the meter needle to zero on the transmittance scale. With a solvent blank in the light beam, the instrument is balanced at 100 percent transmittance by means of a variable V-shaped slit (light control) in the dispersed light beam. The range of the instrument is from 340 to 650 nm with a blue-sensitive phototube and can be extended to 950 nm by the substitution of a red-sensitive phototube. A second phototube, located to the rear of the light source, serves as a reference to monitor fluctuations in the light source.

In double-beam operation, radiation source instability and amplifier drift affect both beams similarly. Automatic gain control provides a constant slit width and thus constant resolving power during the scan when a grating monochromator is used. Solvent absorption is automatically subtracted by placing solvent in the reference beam. Also, two samples may be placed in the instrument and the absorbance of one subtracted from the other or their absorbance ratios may be measured.

6.8.5 Scanning Spectrophotometers

Instruments of this type automatically compare the transmittance of sample and reference beams while scanning continuously through the wavelength region. The operator has a choice of several scan speeds, often ranging from 0.01 to 10 nm · s^{-1}, and several spectral bandwidths from 0.07 to 3.6 nm in the ultraviolet-visible region

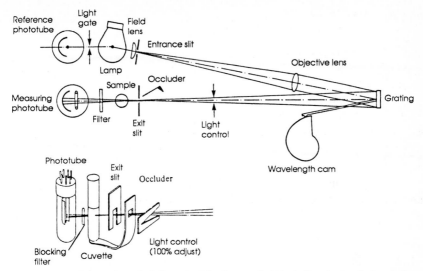

FIGURE 6.13 Schematic optical diagram of the Spectronic 20 including the detail of occluder, cuvette, and phototube. (*Courtesy of Milton Roy, Inc.*)

and from 0.16 to 14.4 nm in the near infrared region. Automatic operation eliminates many time-consuming manual adjustments, especially for qualitative analysis. A microprocessor controller provides automatic wavelength calibration, serial and parallel wavelength scans, repetitive scans, time scans, multiple wavelength scans, automatic zero, automatic wavelength selection, data storage and manipulation, derivative capability, peak integration, and signal averaging.

6.8.6 Reversed-Optics Spectrophotometer

An unusual spectrophotometer is shown in Fig. 6.14. Two radiation sources are used: a quartz-halogen lamp for the visible range and a deuterium lamp for the ultraviolet. A blazed holographic grating is the dispersing medium; all wavelengths pass through the sample. The dispersed radiation is detected by two photodiode arrays, one for the visible region (400 to 800 nm) and the other for the ultraviolet region (200 to 400 nm). Each diode array consists of 200 diodes; thus each diode in the uv range covers 1 nm and in the visible region, 2 nm. One built-in microprocessor controls the main instrument functions and another controls the monochromator. The microcomputer provides for operations such as absorbance ratios or differences, statistical calculations, and multiple-wavelength (up to 12) measurements on a sample.

6.9 VISIBLE AND ULTRAVIOLET SPECTROSCOPY

Visible and ultraviolet spectroscopy can be used to measure materials qualitatively, providing information regarding the structure, formula, and stability of the

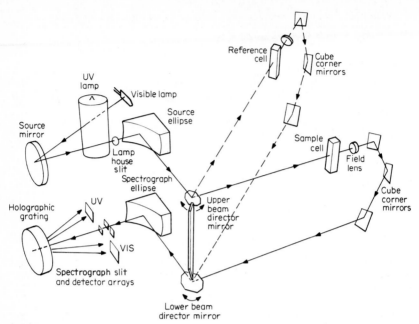

FIGURE   6.14 Diagram   of   the   Hewlett-Packard   8450A   UV-VIS   spectrophotometer.   (*Courtesy of Hewlett-Packard.*)

materials. Both solid and liquid samples can be analyzed, and the procedures are
sensitive in concentrations on the order of parts per million.

Absorption spectra are produced when ions or molecules absorb electromagnetic radiation in the ultraviolet or visible regions. The absorption of energy is a
result of displacing an outer electron in the molecule. The spectrum is a function
of the whole structure of a substance. No unique electronic spectrum is found;
rather the information obtained from this region should be used in conjunction
with other evidence to confirm the identity of a compound. However, molar
absorptivity values frequently exceed 10,000. Thus, dilute solutions are easily
measured.

The intensity of the absorption is proportional to the number of that type of
chromophore present in the molecule. Representative chromophores are given in
Table 6.2. The solvent chosen must dissolve the sample, yet be relatively transparent in the spectral region of interest. In order to avoid poor resolution and
difficulties in spectrum interpretation, a solvent should not be employed for measurements that are near the wavelength at which absorbance for the solvent alone
approaches one absorbance unit (10 percent transmittance). Ultraviolet cutoffs
for solvents commonly used are given in Table 4.1.

Equally important in spectral interpretation is the "negative" information that
can be deduced regarding molecular structures. If the compound is highly transparent throughout the region from 220 to 800 nm, it contains no conjugated unsaturated or benzenoid system, no aldehyde or keto group, no nitro group, and
no bromine or iodine. If the sample exhibits chromophores, the wavelength(s)
of maximum absorbance is ascertained, tables are searched for known

TABLE 6.2 Electronic Absorption Bands for Representative Chromophores

Chromophore	System	λ_{max}	ϵ_{max}
Acetylide	$-C^{.}C-$	175–180	6 000
Aldehyde	$-CHO$	210	Strong
		280–300	11–18
Amine	$-NH_2$	195	2 800
Azido	$=C=N-$	190	5 000
Azo	$-N=N-$	285–400	3–25
Bromide	$-Br$	208	300
Carbonyl	$=C=O$	195	1 000
		270–285	18–30
Carboxyl	$-COOH$	200–210	50–70
Disulfide	$-S-S-$	194	5 500
		255	400
Ester	$-COOR$	205	50
Ether	$-O-$	185	1 000
Ethylene	$-C=C-$	190	8 000
Iodide	$-I$	260	400
Nitrate	$-ONO_2$	270 (shoulder)	12
Nitrile	$-C^{.}N$	160	—
Nitrite	$-ONO$	220–230	1 000–2 000
		300–400	10
Nitro	$-NO_2$	210	Strong
Nitroso	$-NO$	302	100
Oxime	$-NOH$	190	5 000
Sulfone	$-SO_2-$	180	—
Sulfoxide	$=S=O$	210	1 500
Thiocarbonyl	$=C=S$	205	Strong
Thioether	$-S-$	194	4 600
		215	1 600
Thiol	$-SH$	195	1 400
	$-(C=C)_2-$ (acyclic)	210–230	21 000
	$-(C=C)_3-$	260	35 000
	$-(C=C)_4-$	300	52 000
	$-(C=C)_5-$	330	118 000
	$-(C=C)_2-$ (alicyclic)	230–260	3 000–8 000
	$C=C-C^{.}C$	219	6 500
	$C=C-C=N$	220	23 000
	$C=C-C=O$	210–250	10 000–20 000
		300–350	Weak
	$C=C-NO_2$	229	9 500
Benzene		184	46 700
		204	6 900
		255	170
Diphenyl		246	20 000
Naphthalene		222	112 000
		275	5 600
		312	175
Anthracene		252	199 000
		375	7 900

TABLE 6.2   Electronic Absorption Bands for Representative Chromophores (*Continued*)

Chromophore	System	λ_{max}	ϵ_{max}
Phenanthrene		251	66 000
		292	14 000
Naphthacene		272	180 000
		473	12 500
Pentacene		310	300 000
		585	12 000
Pyridine		174	80 000
		195	6 000
		257	1 700
Quinoline		227	37 000
		270	3 600
		314	2 750
Isoquinoline		218	80 000
		266	4 000
		317	3 500

Source: J. A. Dean, ed., *Handbook of Organic Chemistry*, McGraw-Hill, New York, 1986.

chromophores, and the absorption spectrum is compared with the spectra in standard compilations of ultraviolet-visible spectra.

6.10   COMPUTER PROGRAMS

Various capabilities are available when a microcomputer is interfaced with the spectrophotometer. When a program has been loaded into the computer memory, programmable function keys on the keyboard are assigned specific commands. Execution of a particular command only requires depressing the appropriate key.

Although programs vary slightly among different vendors, the commands fall into these categories: instrument control, library, graphics, data extractive, and data manipulative.

6.10.1   Instrument Control

Instrument controls facilitate communication between the spectrophotometer and the computer. These commands include starting and ending wavelengths, data interval, slit width, scan speed, and response time. During data collection, the information can be monitored on the video display.

6.10.2   Spectrum Storage

The spectrum can be stored on a disk for future viewing or further manipulation. Library commands are used for retrieving, saving, or obtaining a listing of spectral files stored on diskettes.

6.10.3 Graphics and *XY* Axes Scaling

A segment of the spectrum can be viewed or expanded (or compressed) to highlight detail. The ability to change scale expansions eliminates one of the most tedious problems of conventional spectroscopy, namely the need to rescan samples because the proper scale expansion was not selected initially. Pertinent experimental information can be written on the screen and printed with the hard copy.

6.10.4 Data-Extractive Commands

Commands such as peak, area, and intensity involve extracting information from the spectrum that includes peak area, peak tables, and absorbance values at specific wavelengths. A movable cursor is placed at appropriate points on the screen.

6.10.5 Derivative Analysis of Spectra

A variety of high-level software routines allows the user to display first-, second-, and fourth-order derivative spectra for fast, easy, and accurate resolution of spectrum features. This feature is useful if the absorbance band is a shoulder of a broad background absorbance or if there are several overlapping peaks which prevent accurate measurement of an individual peak. Derivative analysis can resolve shoulder peaks or overlapping peaks into discrete peaks for measurement.

6.10.6 Spectra Subtraction

One spectrum can be subtracted from another to give precise difference spectra within seconds. One use is to ascertain the spectrum of an impurity. The removal of an interfering component in the mixture allows the measurement of the remaining components. This technique is known as spectral stripping and can be done without physical or chemical extraction.

Two spectra can be normalized, if their concentrations are unequal, before subtraction of the two spectra. The software automatically calculates the normalization factor by using two analytical wavelengths selected by the operator.

6.10.7 Spectrum Overlay

The user can compare two or more spectra plotted under the same experimental conditions. This feature is useful when comparing standard spectra with those of samples. Any changes in the spectrum of production products over a period of time can be found. Similarly, kinetic experiments can be performed.

6.10.8 Smooth Command

A smoothing function fits a moving point polynomial to the data and rounds the rapid transients characteristic of noise. This feature is useful for improving spectra acquired at very high or at low absorbance levels and for reducing noise in multigeneration-processed spectra.

6.11   TURBIDIMETRY AND NEPHELOMETRY

Turbidity is an expression of the optical property of a sample that causes radiation to be scattered and absorbed rather than transmitted in straight lines through the sample. A turbidimeter measures the amount of radiation that passes through a suspension in the forward direction; that is radiation not absorbed. A nephelometer measures the radiation which is scattered from a suspension of finely divided material. Usually the scattered radiation is measured at an angle of 45° or 90° to the direction of the beam of radiation through the sample. Analytical determinations are empirical. Even differences in the physical design of an instrument cause differences in the measured values for turbidity.

6.11.1   Instrumentation

The arrangement shown in Fig. 6.15 is used for the measurement of very small amounts of suspended material (sample transmittance greater than 90 percent), typical of water and waste-water analysis. A 90° detection angle, while not the most sensitive to concentration, is the least sensitive to variations in particle size and affords a simple optical system that is relatively free from stray radiation. If

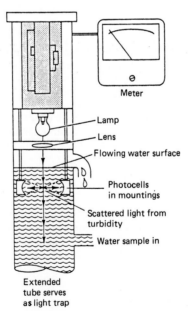

Meter

Lamp
Lens
Flowing water surface
Photocells in mountings
Scattered light from turbidity
Water sample in

Extended
tube serves
as light trap

FIGURE 6.15 Low-range turbidimeter. (*Courtesy of Hach Chemical Co.*)

no suspended matter is present in the sample, no scattered radiation reaches the photodetectors and the meter reading is zero. A linear response to suspended matter extends only to a certain turbidity, after which the response begins to level off.

Excessive turbidity causes the instrument to go blind. For larger amounts of suspended matter, the path length of the scattered radiation can be decreased. However, with short path lengths, any scratches or other imperfections in the cell windows, dirt, or films on the walls scatter radiation and give a positive error to the turbidity measurement.

6.11.2 Standards

Accurately weigh and dissolve 5 g of hydrazinium (2+) sulfate ($N_2H_4 \cdot H_2SO_4$) and 50 g of hexamethylenetetramine in 1 L of distilled water. Let stand for 48 h. The white turbidity that develops is defined as 4000 nephelometric turbidity units. The mixture can be diluted to prepare weaker standards.

BIBLIOGRAPHY

Burgess, C., and A. Knowles, eds., *Techniques in Visible and Ultraviolet Spectroscopy,* Vol. 1, Chapman and Hall, New York, 1981.

Silverstein, R. M., G. C. Bassler, and T. C. Morrill, *Spectrometric Identification of Organic Compounds,* 4th ed., Wiley, New York, 1981.

CHAPTER 7

FLUORESCENCE AND PHOSPHORESCENCE SPECTROPHOTOMETRY

7.1 INTRODUCTION

Luminescence, the term applied to the reemission of previously absorbed radiation, includes fluorescence and phosphorescence. If a molecule absorbs sufficient energy, its electrons may be promoted to higher energy electronic states. Subsequent return of the electrons from an excited state to the ground (unexcited) electronic state is often accompanied by the emission of light. When the source of energy used to produce luminescence is light radiation, the process is termed photoluminescence. If the source is a chemical reaction, the term is chemiluminescence. The wavelength of light emitted during photoluminescence is the same or longer than that of incident radiation. By contrast with phosphorescence, in fluorescence the energy transitions do not involve a change in electron spin.

7.1.1 Fluorescence

When absorbing radiation, a molecule first interacts with the radiation in a time interval of approximately 10^{-15} s, the period of oscillation of the electromagnetic wave. The molecule is raised to an excited electronic state, usually from the lowest vibrational level of the ground electronic state and proceeding to the $v = 2$ vibrational level of the excited electronic state. Other transitions occur but with lower probabilities. In returning to the ground electronic state, the favored route involves first dropping to the lowest vibrational level of the lowest excited singlet state within 10^{-12} s by means of radiationless processes, usually collisions with other molecules. Upon reaching the lowest vibrational level of the lowest excited singlet state, fluorescence radiation can occur when the electron returns to any of the vibrational levels of the ground electronic state. Each transition involves radiation of a specific wavelength within 10^{-8} s and that wavelength is usually longer than that of the exciting wavelength. The excitation and fluorescence emission spectra of anthracene are shown in Fig. 7.1.

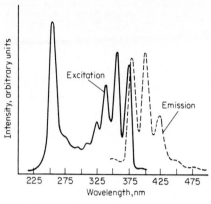

FIGURE 7.1 Excitation and fluorescence emission spectra of 300 ng · mL^{-1} anthracene in methanol.

7.1.2 Phosphorescence

Phosphorescence is the term for the emission of light that is delayed by more than 10^{-8} s following excitation by absorption of electromagnetic radiation. When the potential energy curve of the excited singlet state crosses that of the triplet state, some excited molecules may pass over to the lowest triplet state via an intersystem crossing. In a triplet state all the electrons in the molecule are paired except two. A spin reversal must take place when the triplet state forms, and another spin reversal must occur upon phosphorescence emission. Consequently, the decay time of phosphorescence is relatively long, approximately 10^{-4} to 10 s. The long lifetime of the triplet state greatly increases the probability of collisional transfer of energy with solvent molecules. Because of this solvent interaction, phosphorescence is usually observed only in viscous solutions or in the solid state (such as a solvent that freezes to form a rigid glass at the temperature of liquid nitrogen). Phosphorescence wavelengths are longer than fluorescence wavelengths if both exist for a molecule.

7.2 PHOTOLUMINESCENCE RELATED TO CONCENTRATION

Whether fluorescence or phosphorescence, the luminescent power P_L is proportional to the number of molecules in excited states. In turn, this is proportional to the radiant power absorbed by the sample. Thus

$$P_L = \Phi_L(P_o - P) \qquad (7.1)$$

where Φ_L = luminescence efficiency or quantum yield of luminescence
P_o = radiant power incident on sample
P = radiant power emerging from sample

Applying Beer's law to Eq. 7.1, one obtains

$$P_L = \Phi_L P_o (1 - e^{-\epsilon bc}) \tag{7.2}$$

When expanded in a power series, this equation yields

$$P_L = \Phi_L P_o \epsilon bc \left[1 - \frac{\epsilon bc}{2!} + \frac{(\epsilon bc)^2}{3!} - \cdots \right] \tag{7.3}$$

If ϵbc is 0.05 or less, only the first term in the series is significant and Eq. 7.3 can be written as

$$P_L = \Phi_L P_o \epsilon bc \tag{7.4}$$

Thus, when the concentrations are very dilute, there is a linear relationship between luminescent power and concentration.

Of particular interest in Eq. 7.4 is the linear dependence of luminescence on the excitation power. This means sensitivity can be increased by working with high excitation powers. The b term is not the path length of the cell but the solid volume of the beam defined by the excitation and emission slit widths together with the beam geometry. Therefore, slit widths are the critical factor and not the cell dimensions.

7.2.1 Problems with Photoluminescence

7.2.1.1 Self-Quenching. Self-quenching results when luminescing molecules collide and lose their excitation energy by radiationless transfer. Serious offenders are impurities, dissolved oxygen, and heavy atoms or paramagnetic species (aromatic substances are prime offenders). Always use "Spec-pure" solvents.

7.2.1.2 Absorption of Radiant Energy. Absorption either of exciting or the luminescent radiation reduces the luminescent signal. Remedies involve (a) diluting the sample, (b) viewing the luminescence near the front surface of the cell, and (c) using the method of standard additions for evaluating samples.

7.2.1.3 Self-Absorption. Attenuation of the exciting radiation as it passes through the cell can be caused by too concentrated an analyte. The remedy is to dilute the sample and note whether the luminescence increases or decreases. If the luminescence increases upon sample dilution, one is working on the high concentration side of the luminescence maximum. This region should be avoided.

7.2.1.4 Excimer Formation. Formation of a complex between the excited-state molecule and another molecule in the ground state, called an excimer, causes a problem when it dissociates with emission of luminescent radiation at longer wavelengths than the normal luminescence. Dilution helps lessen this effect.

7.3 STRUCTURAL FACTORS AFFECTING PHOTOLUMINESCENCE

Fluorescence is expected under these conditions.

1. In molecules that are aromatic or contain multiple-conjugated double bonds with a high degree of resonance stability.
2. Polycyclic aromatic systems are very fluorescent.
3. Substituents, such as $-NH_2$, $-OH$, $-F$, $-OCH_3$, $-NHCH_3$, and $-N(CH_3)_2$ groups, often enhance fluorescence.
4. These groups decrease or quench fluorescence completely: $-Cl$, $-Br$, $-I$, $-NHCOCH_3$, $-NO_2$, and $-COOH$.
5. Molecular rigidity enhances fluorescence. Substances fluoresce more brightly in a glassy state or in viscous solution. Formation of chelates with metal ions also promotes fluorescence. However, the introduction of paramagnetic metal ions gives rise to phosphorescence but not fluorescence in metal complexes.
6. Changes in the system pH, if they affect the charge status of the chromophore, may influence fluorescence.

7.4 INSTRUMENTATION FOR FLUORESCENCE MEASUREMENT

The basic components of a luminescent instrument is shown in Fig. 7.2. The primary filter or excitation monochromator selects specific wavelengths of radiation from the source and directs them through the sample. The resultant luminescence, usually observed at 90° to excitation radiation, is isolated by the secondary filter or emission monochromator and directed to the photodetector. Front

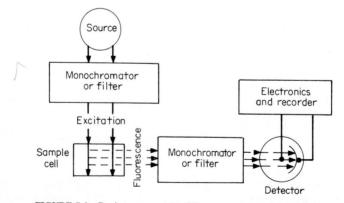

FIGURE 7.2 Basic components of fluorescence instrumentation.

surface or small-angle viewing (37°) is used for high-absorbance samples or solids.

7.4.1 Radiation Sources

As seen from Eq. 7.4, it is desirable to use a source that is as powerful as possible. High-pressure xenon arc lamps are used in nearly all commercial spectrofluorometers. The xenon lamp emits an intense and relatively stable continuum of radiation that extends from 300 to 1300 nm.

Low-pressure mercury vapor lamps are most frequently used in filter fluorometers. With a clear bulb of ultraviolet-transmitting material, individual mercury emission lines (Fig. 7.3) occur at 253.7, 302, 313, 365, 404.7, 435.8, 546, 577.0, and 579.1 nm. Interference filters are used to select the desired mercury line. When the lamp bulb is coated with a phosphor, a more nearly continuous spectrum is emitted.

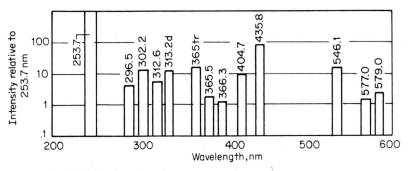

FIGURE 7.3 Emission lines from the mercury arc lamp.

7.4.2 Filter Fluorometers

Filter photometers offer a low-cost instrument (Fig. 7.4). It uses a mercury lamp as an excitation source. A primary filter selects the desired mercury emission line for passage through the sample cuvette. A secondary filter, placed between the sample and the photodetector, has two functions: (1) to transmit the desired fluorescence radiation and (2) to absorb scattered excitation radiation. Remember that the excitation radiation will be much more intense than the fluorescent radiation.

Generally filter fluorometers use a single-beam arrangement with some arrangement for monitoring and compensating for fluctuations and drift in source intensity and detector response. The ratio system shown in Fig. 7.4 is quite effective. A portion of the source radiation passes through the excitation optical system and is attenuated by means of a reference aperture disk before reaching the reference photodetector. The signals from the sample and reference detectors

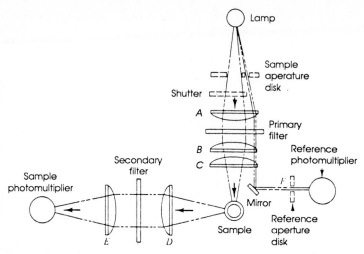

FIGURE 7.4 Optical diagram for a ratio fluorometer. (*Courtesy of Farrand Optical Co., Inc.*)

are fed into a solid-state electronic divider, which computes the ratio of sample-to-reference signals.

7.4.3 Spectrofluorometers

Spectrofluorometers replace the filters of a filter fluorometer with grating monochromators of the Czerny-Turner type with a 0.25-m focal length. Filters are used to block out higher-order diffracted radiation from the gratings. The emission spectrometer should be able to resolve two lines 1.0 nm apart. When the excitation wavelength can be selected, the fluorescence intensities of different molecules may vary dramatically, as shown in Fig. 7.5.

7.4.4 Fluorescence Measurements

Fluorescent measurements are usually made by reference to some arbitrarily chosen standard. The standard is placed in the instrument and the circuit is balanced with the reading scale at some chosen setting. Without readjusting any circuit components, the standard is replaced by known concentrations of the analyte and the fluorescence of each is recorded. Finally, the fluorescence of the solvent and cuvette alone is measured to establish the true zero concentration readings if the instrument is not equipped with a zero-adjust circuit. Fluorescence standards may be (1) rhodamine B in ethylene glycol, (2) quinine hydrogen sulfate in 0.1 N H_2SO_4, (3) tryptophan in water, and (4) anthracene in cyclohexane or ethanol. With separate emission and excitation monochromators, the emission and excitation spectra can be ascertained. If no knowledge of the spectra is available, place a solution of the analyte into the cuvette. Select an emission wavelength that produces a fluorescent signal either visible to the eye or from the detector signal. While scanning through the spectrum with the excitation monochromator,

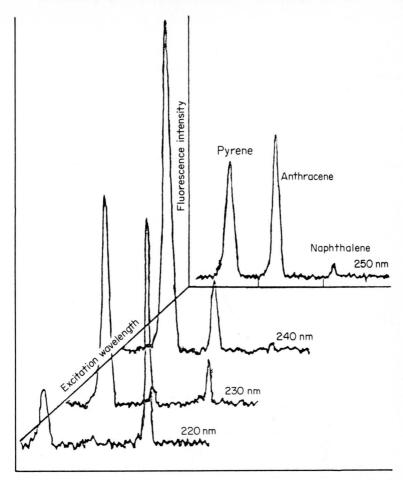

FIGURE 7.5   Effect of changing the excitation wavelength upon fluorescence emission intensities.   (*Courtesy of S.I./McPherson.*)

plot the strength of the fluorescence signal. This gives an uncorrected excitation spectrum which should resemble the normal absorption spectrum obtained in the ultraviolet-visible region. Next, select a suitably strong excitation wavelength and scan the fluorescence spectrum with the emission monochromator.

7.5 INSTRUMENTATION FOR PHOSPHORESCENCE MEASUREMENTS

The only change in instrumentation for phosphorescence measurements is the addition of a radiation interrupter (both excitation and emission) and provision for immersion of the sample in a Dewar flask for liquid nitrogen temperatures. Sche-

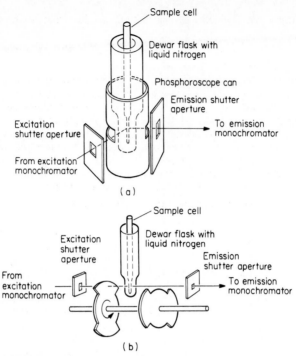

FIGURE 7.6 Schematic diagram of the interruption of excitation radiation and phosphorescence emission: (*a*) rotating can device and (*b*) rotating shutter. [*Reprinted with permission from T. C. O'Haver and J. D. Winefordner, Anal. Chem., 38: 602 (1966). Copyright 1966 American Chemical Society.*]

matic diagrams of two types of interrupters are shown in Fig. 7.6. Either the excitation radiation is admitted to the sample via a fixed slit system and a rotating can chopper or via a set of slotted disks with equally spaced ports. The phosphorescent signal is measured without interference from scattered radiation and short-lived fluorescence. The solvent frequently used is a mixture of diethyl ether, isopentane, and ethanol in a volume ratio of 5:5:2. When cooled to liquid nitrogen temperature, it gives a clear transparent glass.

Decay curves can be recorded if the detector circuit is equipped with an oscillograph. The resolution time is a function of the motor speed of the interrupter, the size and spacing of the openings, and the relative radial positions of the ports to one another.

7.6   COMPARISON OF LUMINESCENCE AND ULTRAVIOLET-VISIBLE ABSORPTION METHODS

Luminescence, if applicable, is usually the method of choice for quantitative analytical purposes, especially trace analysis. The significant advantages of luminescence are enumerated below:

1. Fewer luminescing species than absorbing species exist in the ultraviolet-visible region.
2. Luminescence is more selective. A pair of wavelengths, excitation and emission, characterize the process instead of one.
3. Luminescence is more sensitive. This is because (a) a luminescence signal is measured directly against a very small background and (b) the signal is proportional to the intensity of the incident radiation. Greater sensitivity for weakly emitting compounds can be obtained by using more intense sources. Luminescence analyses can determine less than parts-per-billion concentrations of many substances. By contrast, in absorption spectrophotometry concentration is proportional to absorbance, which is the logarithm of the ratio between incident and transmitted radiant power. This corresponds to the measurement of a small difference between two large signals.
4. The lifetime of the luminescence offers another way to discriminate among compounds.

BIBLIOGRAPHY

Lakowicz, J. R., *Principles of Fluorescence Spectroscopy,* Plenum, New York, 1983.

Rendell, D., *Fluorescence and Phosphorescence,* Wiley, New York, 1987.

Schulman, S. G., ed., *Molecular Luminescence Spectroscopy,* Vol. 1, Wiley, New York, 1985; Vol. 2, 1988.

Wehry, E. L., ed., *Modern Fluorescence Spectroscopy,* Plenum, New York, 1981.

CHAPTER 8
INFRARED SPECTROMETRY

8.1 INTRODUCTION

The infrared region of the electromagnetic spectrum includes radiation at wavelengths between 0.7 and 500 μm or, in wave numbers, between 14,000 and 20 cm^{-1}. (The relationship between the frequency and wavelength scales in the infrared region is shown in Fig. 8.1.) Molecules have specific frequencies which are directly associated with their rotational and vibrational motions. Infrared absorptions result from changes in the vibrational and rotational state of a molecular bond. Coupling with electromagnetic radiation occurs if the vibrating molecule produces an oscillating dipole moment that can interact with the electric field of the radiation. Homonuclear diatomic molecules like hydrogen, oxygen, or nitrogen, which have a zero dipole moment for any bond length, fail to interact. These changes are subtly affected by interaction with neighboring atoms or groups, as well as resonating structures, hydrogen bonds, and ring strain. This imposes a stamp of individuality on each molecule's infrared absorption spectrum as portions of the incident radiation are absorbed at specific wavelengths. The multiplicity of vibrations occurring simultaneously produces a highly complex absorption spectrum that is uniquely characteristic of the functional groups that make up the molecule and of the overall configuration of the molecule as well. It is therefore possible to identify substances from their infrared absorption spectrum.

For qualitative analysis, one of the best features of an infrared spectrum is that the absorption or the *lack of absorption* in specific frequency regions can be correlated with specific stretching and bending motions and, in some cases, with the relationship of these groups to the rest of the molecule. Thus, when interpreting the spectrum, it is possible to state that certain functional groups are present in the material and certain others are absent. The relationship of infrared spectra to molecular structure is treated in a later section.

8.2 PREPARATION OF SAMPLES

8.2.1 Liquids and Solutions

Pure liquids can be run directly as liquid samples, provided a cell of suitable thickness is available. This represents a very thin layer, about 0.001 to 0.05 mm thick.

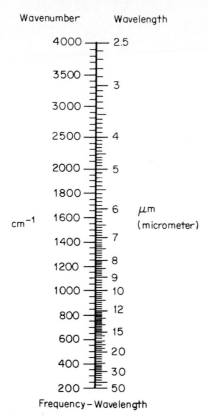

FIGURE 8.1 Frequency-wavelength conversion scale.

⌐ For solutions, concentrations of 10% and cell lengths of 0.1 mm are most practical. All solvents absorb in the infrared region. Transparent regions of selected solvents are shown in Fig. 8.2. When possible, the spectrum is obtained in a 10% solution in CCl_4 in a 0.1-mm cell in the region 4000 to 1333 cm^{-1} (2.50 to 7.50 μm) and in a 10% solution of CS_2 in the region 1333 to 650 cm^{-1} (7.50 to 15.38 μm). If the sample is insoluble in these solvents, chloroform, dichloromethane, acetonitrile, and acetone are useful solvents. For any solvent, a reference cell of the same path length as the sample cell is filled with pure solvent and placed in the reference beam.⌐

Liquid-sample cells are very fragile and must be handled very carefully. Cells are usually filled by capillary action, and the solution or pure sample is introduced with a syringe. Each cell is labeled with its precise path length as measured by interference fringes. Permanent solution cells are constructed with two window pieces sealed and separated by thin gaskets of copper or lead that have been wetted with mercury. The whole assembly is clamped together and mounted in a holder. A demountable cell uses Teflon as gasket material and the cell is slipped into a mount and knurled nuts are turned until finger tight.

⌐The minicell consists of a threaded, two-piece plastic body and two silver

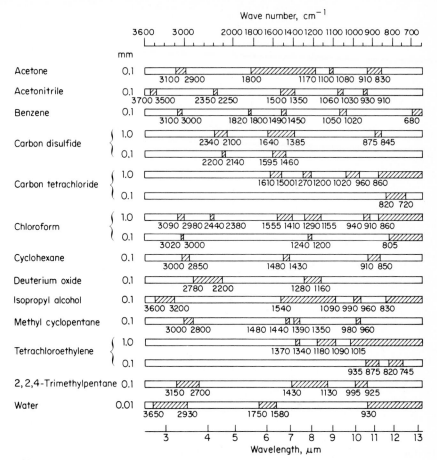

FIGURE 8.2 Transmission characteristics of selected solvents. The material is considered transparent if the transmittance is 75 percent or greater. Solvent thickness is given in millimeters.

chloride cell window disks with either a 0.025 or a 0.100-mm circular depression. The windows fit into one portion of the cell; the second portion is then screwed in to form the seal (AgCl flows slightly under pressure). The windows can be (1) placed back-to-back for films or mulls, (2) arranged with one back to the circular depression, or (3) positioned with facing circular depressions.

8.2.2 Cell Thickness

In the interference fringe method for measuring the internal path length, the empty cell is placed in the spectrometer on the sample side and no cell in the reference beam. Cell windows must have a high polish. Operate the spectrometer as nearly as possible to the 100 percent line. Run sufficient spectrum to produce

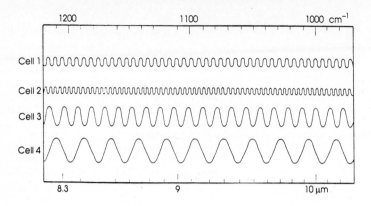

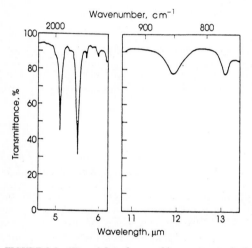

FIGURE 8.3   (*Upper*) Interference fringe patterns obtained with four different cells. (*Lower*) Transmittance curve from a cell filled with benzene.

20 to 50 fringes, as shown in Fig. 8.3. The cell internal thickness b in centimeters, is calculated from:

$$b = \frac{1}{2\eta}\left(\frac{n}{\bar{\nu}_1 - \bar{\nu}_2}\right) \tag{8.1}$$

where n is the number of fringes (peaks or troughs) between two wave numbers, $\bar{\nu}_1$ and $\bar{\nu}_2$, and η is the refractive index of the sample material. If measurements are made in wavelength, the expression is

$$b = \frac{1}{2\eta}\left(\frac{n\lambda_1\lambda_2}{\lambda_2 - \lambda_1}\right) \tag{8.2}$$

where λ_2 is the starting wavelength and λ_1 is the final wavelength between which the fringes are counted. Film thickness can also be measured by the interference fringe method.

The *standard absorber method* may be used with a cell in any condition of polish, including cavity or minicells. Fill the cell with pure benzene. Use the 1960-cm^{-1} (5.10-μm) band of benzene for path lengths 0.1 mm or less, where benzene has an absorbance of 0.10 for every 0.01 mm of thickness. For longer path lengths, use the benzene band at 845 cm^{-1} (11.83 μm), where the benzene absorbance is 0.24 for every 0.1 mm of thickness.

⌐ **EXAMPLE 8.1**   From Fig. 8.3, calculate the path length for cell 1. In the interval from 1200 to 1000 cm^{-1}, there are 34 peaks. Don't forget to count the first peak as 0. From Eq. 8.1 and taking the refractive index of air as 1,

$$b = \frac{1}{2} \left(\frac{34}{1200 - 1000} \right) = 0.085 \text{ cm}$$

Cell 2 is 0.13 cm, cell 3 is 0.044 cm, and cell 4 is 0.022 cm.

Also from Fig. 8.3, calculate the path length for a cell filled with benzene. Using the baseline technique, the absorbance at 1960 cm^{-1} (the logarithm of the ratio between 93 and 45.5 percent T) is found to be 0.310. Because benzene has an absorbance of 0.10 for every 0.01 mm of thickness at this frequency, the cell path length is 0.031 mm.⌐

8.2.3 Film Technique

⌐A large drop of the neat liquid is placed between two infrared-transmitting windows, which are then squeezed together and placed in a mount.⌐

For polymers and noncrystalline solids, dissolve the sample in a volatile solvent and pour the solution onto the window material. The solvent is evaporated⌐ by gentle heating to leave a thin film which then can be mounted in a holder. ⌐Sometimes polymers can be hot pressed onto window material or cut into a film of suitable thickness with a microtome.⌐

8.2.4 Mull Technique

For qualitative analysis the mull technique is rapid and convenient, but quantitative data are difficult to obtain. The sample is finely ground in a clean mortar or a "wiggle bug." After grinding, the mulling agent (often mineral oil or Nujol, but may be perfluorokerosine or chlorofluorocarbon greases) is introduced in a small quantity just sufficient to convert the powder into the consistency of toothpaste.⌐ The cell is opened and ⌐a few drops of the pasty mull are placed on one plate, which is then covered with the other plate. The thickness of the cell is governed by squeezing the plates when the screws are tightened. Sample thickness should be adjusted so that the strongest bands display about 20 percent transmittance.⌐

Always disassemble the cell by sliding the plates apart.⌐ *Do not attempt to pull the plates apart.*

Be aware of changes in the sample that may occur during grinding.

8.2.5 Pellet Technique

Mix a few milligrams of finely ground sample with about 1 g of spectrophotometric grade KBr. Grinding and mixing are done in a vibrating ball mill or wiggle bug. Place the mixture in an evacuable die at 60 000 to 100 000 lb · in^{-2}. The pressure can be applied by using either a hydraulic press or a lever-screw press. Remove the pressed disk from the mold for insertion in the spectrometer.

KBr wafers can be formed, without evacuation, in a Mini-Press. When two highly polished bolts are turned against each other in a screw-mold housing with a wrench for about 1 min, a clear wafer is produced. The housing with wafer inside is inserted into the cell compartment of the spectrometer. Use 75 to 100 mg of powder. No moisture may be present.

CsI or CsBr are used for measurements in the far-infrared region.

Caution Never apply pressure unless the powdered sample is in the mold. If no sample is present, the faces of the bolts or pistons will be scored.

8.3 INFRARED WINDOW MATERIALS

In selecting a window material for an infrared cell, these factors must be considered.

1. The wavelength range over which the spectrum must be recorded and thus the transmission of the window material in this wavelength range.

2. The solubility of window material in the sample or solvent, and any reactivity of window with sample. Cell windows constructed from the alkali halides are easily fogged by exposure to moisture and require frequent repolishing.

3. The refractive index of the window material, particularly in internal reflectance methods. Materials of high refractive index produce strong, persistent interference fringes and suffer large reflectivity losses at air-crystal interfaces.

4. The mechanical characteristics of the window material. For example, silver chloride is soft, easily deformed, and also darkens upon exposure to visible light. Germanium is brittle as is zinc selenide (Irtran IV) which also releases H_2Se in acid solutions.

There is no rugged window material for cuvettes or internal reflectance methods that is transparent and also inert over the entire infrared region. Properties of infrared-transmitting materials are compiled in Table 8.1.

8.3.1 Mid-Infrared Region

The alkali halides are widely used; all are hygroscopic and should be stored in a desiccator when not in use.

When working with wet or aqueous samples, windows of fused silica, calcium fluoride, or barium fluoride may be useful, though limited by their long wavelength transmission. These transmission limitations can be overcome to some extent by using ZnS, ZnSe, or CdTe windows.

Although soft, silver chloride is useful for moist samples or aqueous solutions. Combined with its comparatively low cost and wide transmission range, silver

TABLE 8.1 Infrared Transmitting Materials

Material	Wavelength range, μm	Wave number range, cm^{-1}	Refractive index at 2 μm
NaCl, rock salt	0.25–17	40 000–590	1.52
KBr, potassium bromide	0.25–25	40 000–400	1.53
KCl, potassium chloride	0.30–20	33 000–500	1.5
AgCl, silver chloride*	0.40–23	25 000–435	2.0
AgBr, silver bromide*	0.50–35	20 000–286	2.2
CaF$_2$, calcium fluoride (Irtran-3)	0.15–9	66 700–1 100	1.40
BaF$_2$, barium fluoride	0.20–11.5	50 000–870	1.46
MgO, magnesium oxide (Irtran-5)	0.39–9.4	25 600–1 060	1.71
CsBr, cesium bromide	1–37	10 000–270	1.67
CsI, cesium iodide	1–50	10 000–200	1.74
TlBr-TlI, thallium bromide-iodide (KRS-5)*	0.50–35	20 000–286	2.37
ZnS, zinc sulfide (Irtran-2)	0.57–14.7	17 500–680	2.26
ZnSe, zinc selenide* (Irtran-4)	1–18	10 000–556	2.45
CdTe, cadmium telluride (Irtran-6)	2–28	5 000–360	2.67
Al$_2$O$_3$, sapphire*	0.20–6.5	50 000–1 538	1.76
SiO$_2$, fused quartz	0.16–3.7	62 500–2 700	
Ge, germanium*	0.50–16.7	20 000–600	4.0
Si, silicon*	0.20–6.2	50 000–1 613	3.5
Polyethylene	16–300	625–33	1.54

*Useful for internal reflection work.

bromide is attractive for use with aqueous solutions and wet samples. It is much less sensitive to visible light compared to silver chloride.

Teflon has only C—C and C—F absorption bands.

8.3.2 Near-Infrared Region

The near-infrared region, which meets the visible region at about 12 500 cm^{-1} (0.80 μm) and extends to about 4000 cm^{-1} (2.50 μm), is accessible with fused quartz optics.

8.3.3 Far-Infrared Region

Window materials for the far-infrared region include high density polyethylene, silicon, and crystal quartz (cut with the optic axis parallel to the face of the window). Polyethylene has one weak, broad absorption band at approximately 70 cm^{-1}; its principal disadvantage is its lack of rigidity. High-resistivity silicon is rigid but its high index of refraction leads to large reflectivity losses.

8.4 RADIATION SOURCES

8.4.1 Nernst Glower

A popular radiation source is the Nernst glower, which is constructed from a fused mixture of oxides of zirconium, yttrium, and thorium molded in the form of hollow rods 1 to 3 mm in diameter and 2 to 5 mm long. It is fragile and must be preheated to be conductive; a ballast system is needed to prevent overheating (above 1500°C). A glower must be protected from drafts, but adequate ventilation is needed to remove surplus heat and evaporated oxides and binder.

8.4.2 Globar

The Globar is a rod of silicon carbide with an operating temperature near 1300°C. It finds use in the far-infrared region.

8.4.3 Nichrome Coil

A less expensive and very rugged radiation source is a coil of Nichrome wire (a film of black oxide forms on the coil) raised to its operating temperature (1100°C) by resistive heating. It requires no cooling and little maintenance. This source is recommended where reliability is essential, such as in process analyzers and filter photometers. The Nichrome coil provides less intense radiation than other infrared sources.

8.5 INFRARED SPECTROMETERS

Infrared instrumentation is divided into dispersive and nondispersive types. The dispersive instruments are similar to ultraviolet-visible spectrometers except that different sources and detectors are required for the infrared region.

8.5.1 Dispersive Spectrometers

Most dispersive spectrometers are double-beam instruments. Two equivalent beams of radiant energy from the source are passed alternately through the reference and sample paths. In the optical-null system, the detector responds only when the intensity of the two beams is unequal. An optical wedge or comb shutter coupled to the recording pen moves in or out of the reference beam to restore balance. The electrical beam-ratioing method is the other measuring technique.

To cover the wide wavelength range, several gratings with different ruling densities and associated higher-order filters are necessary. Two gratings are mounted back-to-back; each is used in the first order. The gratings are changed at 2000 cm^{-1} (5.00 μm) in mid-infrared instruments. Undesired overlapping grating orders are eliminated with a fore-prism or by suitable filters. The optical schematic for a filter-grating double-beam grating spectrometer is shown in Fig. 8.4.

The use of microprocessors has alleviated much of the tedious work required to obtain usable data. The operator selects a single recording parameter (scan time, slit setting, or pen response) and the microprocessor automatically optimizes these and other conditions.

8.5.2 Fourier Transform Infrared (FT-IR) Spectrometer

The FT-IR spectrometer provides speed and sensitivity. A Michelson interferometer, a basic component, consists of two mirrors and a beam splitter

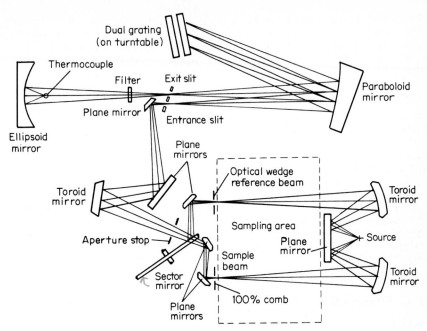

FIGURE 8.4 Optical schematic of a filter-grating double-beam infrared spectro-photometer. (*Courtesy of Perkin-Elmer Corp.*)

(Fig. 8.5). The beam splitter transmits half of all incident radiation from a source to a moving mirror and reflects half to a stationary mirror. Each component reflected by the two mirrors returns to the beam splitter, where the amplitudes of the waves are combined either destructively or constructively to form an interferogram as seen by the detector. By means of algorithms, the interferogram is Fourier-transformed into the frequency spectrum.

This technique has several distinct advantages over conventional dispersive techniques.

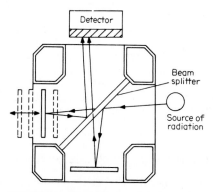

FIGURE 8.5 The Michelson interferometer.

1. The FT-IR spectrometer scans the infrared spectrum in fractions of a second at moderate resolution, and this is constant throughout its optical range. It is especially useful in situations that require fast, repetitive scanning (for example, in gas or high performance liquid chromatography).

2. The spectrometer measures all wavelengths simultaneously. Scans are added. The signal is N times stronger and the noise is $N^{1/2}$ as great, so the signal-to-noise advantage is $N^{1/2}$.

3. An interferometer has no slits or grating; its energy throughput is high and this means more energy at the detector where it is most needed.

8.6 DETECTORS

Below 1.2 μm, the detection methods are the same as those used for ultraviolet-visible radiation. At longer wavelengths the detectors can be classified into two groups: (1) thermal detectors and (2) photon or quantum detectors. The wavelength response of some infrared detectors is shown in Fig. 8.6.

8.6.1 Thermal Detectors

With thermal detectors the infrared radiation produces a heating effect that alters some physical property of the detector. The active element is blackened and thermally insulated from its substrate. Thermal detectors operate at room temperature. When radiation ceases, the element returns to the temperature of the substrate within a finite time interval, usually milliseconds.

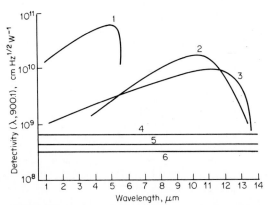

FIGURE 8.6 Wavelength response of some infrared detectors: (1) InSb at 77 K, (2) PbSnTe at 77 K, (3) PbSnTe at 4.2 K, (4) pyroelectric at 300 K, (5) thermistor at 300 K, and (6) thermopile at 300 K. Detectivity is obtained by irradiating the detector with monochromatic power at the same wavelength as that where the detector produced its peak output, chopping frequency is 900 Hz, and noise is measured in a 1-Hz bandwidth.

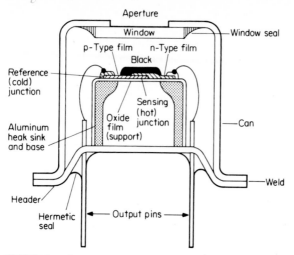

FIGURE 8.7 Schematic construction of a thermocouple (or thermopile) in cross section.

8.6.1.1 Thermocouple.   A thermocouple is fabricated from two dissimilar metals; a *thermopile* is several thermocouples used in series. When incident radiation strikes the detector, a voltage proportional to the temperature of the junction is produced. Response time is about 30 ms. The schematic construction of a thermocouple is shown in Fig. 8.7.

8.6.1.2 Thermistor.   A thermistor is made up of sintered oxides of manganese, cobalt, and nickel and has a high-temperature coefficient of resistance. It functions by changing resistance when heated. Response time is the order of a few milliseconds. When connected in a bridge circuit, a shielded thermistor compensates for ambient temperature changes.

8.6.1.3 Pyroelectric Detector.   A pyroelectric detector depends on the rate of change of the detector temperature rather than on the temperature itself. Response time is much faster than for the preceding types of detectors. It is the detector of choice for Fourier transform spectrometers. But it also means that the pyroelectric detector responds only to changing radiation that is modulated (chopped or pulsed); it ignores steady background radiation.

8.6.1.4 Golay Pneumatic Detector.   The detector, shown in Fig. 8.8, uses the expansion of xenon gas within an enclosed chamber to expand and deform a flexible diaphragm (mirror). To amplify distortions of the mirror surface, light from a lamp inside the detector housing is focused on the mirror, which reflects the light beam onto a phototube after passing through a moire grid. Response time is about 20 ms. It is a superior detector for the far-infrared region.

8.6.2 Photon Detectors

In a photon detector the incident photons interact with a semiconductor. The result produces electrons and holes—the internal photoelectric effect. Response times are less than 1 μs. These detectors require cryogenic cooling.

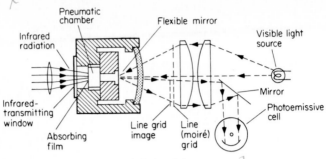

FIGURE 8.8   Golay pneumatic infrared detector.

8.6.2.1 Photoconductive Detector.   In a photoconductive detector the resistance changes with radiation level. A bias current or voltage registers this change.

8.6.2.2 Photovoltaic Detectors.   These detectors generate a small voltage at a diffused *p-n* junction when exposed to radiation. One type of lead tin telluride detector cooled by liquid nitrogen has optimum sensitivity throughout the 5- to 13-μm region. The second type is cooled by liquid helium and has optimum performance in the 6.6- to 18-μm region.

8.7   INTERNAL REFLECTANCE

When a beam of radiation enters a plate surrounded by or immersed in a sample, it is reflected internally if the angle of incidence at the interface between sample and plate is greater than the critical angle (which is a function of refractive index). Although all the energy is reflected, the beam appears to penetrate slightly beyond the reflecting surface and then return. By varying the angle of incidence, the depth of penetration into the sample may be changed. At steep angles (near 30°) the depth of penetration is considerably greater by about an order of magnitude as compared with grazing angles (60°). This is significant in the study of surfaces as, for example, a film or plastic in which chemical additives are suspected of migrating to the surface or where a surface coating has been exposed to weathering.

When a sample is placed in contact with the reflecting surface, the incident beam loses energy at those wavelengths where the material absorbs as a result of an interaction with the penetrating beam. This attenuated radiation is an absorption spectrum that is similar to an infrared spectrum obtained in the normal transmission mode. Internal reflectance enables one to obtain a qualitative infrared absorption spectra from most solid materials or samples available only on a nontransparent support. This eliminates the need for grinding or dissolving or making a mull. Aqueous solutions can be handled without compensating for very strong solvent absorption.

An internal reflectance attachment, shown in Fig. 8.9, is inserted into the sampling space of an infrared spectrometer. One version has three standard positions of 30°, 45°, and 60°. Another version allows a range of angles to be selected by a scissor-jack assembly linking the four mirrors and sample platforms in a panto-

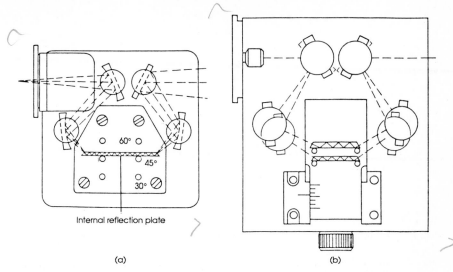

(a)   (b)

FIGURE 8.9 Internal reflection attachment: (*a*) three-position pin plate for 30°, 45°, and 60° plates and (*b*) variable-angle model. (*Courtesy of Foxboro/Wilks, Inc.*)

graph system. Twenty-five internal reflections are standard for a 2-mm thick plate. A reflector plate with a relatively high index of refraction should be used. Thallium(I) bromide iodide (KRS-5) is satisfactory for most liquid and solid samples. The plate must not be brittle as some pressure is required to bring some samples in contact with the plate. AgCl is suitable for aqueous samples.

In the single-pass plate, radiation enters through a bevel (effective aperture) at one end of the plate and, after propagation via multiple internal reflections down the length of the plate (1 to 10 cm in length), leaves through an exit bevel either parallel or perpendicular to the entrance bevel (Fig. 8.9*a*). In the double-pass technique, the radiation propagates down the length of the plate, is totally reflected at the opposite end by a surface perpendicular to the plate length, and returns to leave the plate at the entrance end. The end of the plate where total reflection occurs can be dipped into liquids or powders and placed in closed systems.

8.8 CORRELATION OF INFRARED SPECTRA WITH MOLECULAR STRUCTURE

Only a very general overview will be given. For a detailed tabulation of infrared absorption frequencies consult the *Handbook of Organic Chemistry*[1] and other references.[2,3]

8.8.1 Near-Infrared Region

The absorptivity of near-infrared bands is from 10 to 1000 times less than that of mid-infrared bands. Thicker sample layers (0.5 to 10 mm) must be used to compensate. On the other hand, minor impurities in a sample are less troublesome.

Significant spectral features are:

1. O—H stretching vibration near 7140 cm^{-1} (1.40 μm).
2. N—H stretching vibration near 6667 cm^{-1} (1.50 μm).
3. C—H stretching and deformation vibrations of alkyl groups at 4548 cm^{-1} (2.20 μm) and 3850 cm^{-1} (2.60 μm).
4. Absorption bands due to water at 2.76, 1.90, and 1.40 μm (3623, 5263, and 7143 cm^{-1}).
5. Aromatic amines: (a) Primary amines have bands near 1.97 and 1.49 μm (5076 and 6711 cm^{-1}). (b) Secondary amines show only the band at 1.49 μm. (c) Tertiary amines exhibit no appreciable absorption at either wavelength.

8.8.2 Mid-Infrared Region

Useful correlations in the mid-infrared region are shown in Fig. 8.10. It is best to divide this region into two parts—the group-frequency region (4000 to 1300 cm^{-1}; 2.50 to 7.69 μm) and the fingerprint region (1300 to 650 cm^{-1}; 7.69 to 15.38 μm). First, note the absorption bands in the group frequency region; these bands are more or less dependent on only the functional group that gives the absorption and not on the complete molecular structure although structural influences do reveal themselves as shifts about the fundamental frequency. Last, observe the fingerprint region, for here the major factors are single-bond stretching frequencies and skeletal frequencies of polyatomic systems that involve motions of bonds linking a substituent group to the remainder of the molecule. Collectively these absorption bands aid in identifying the material.

Proceed systematically through the group frequency region. Hydrogen stretching frequencies with elements of mass 19 or less appear from 4000 to 2500 cm^{-1} (2.50 to 4.00 μm).

8.8.2.1 C—H frequencies. The C—H stretching frequencies of alkyl groups are less than 3000 cm^{-1}, whereas in alkenes and aromatics they are greater than 3000 cm^{-1}. The CH$_3$ group gives rise to an asymmetric stretching mode at 2960 cm^{-1} and a symmetric mode at 2870 cm^{-1}. For —CH$_2$— these bands occur at 2930 cm^{-1} and 2850 cm^{-1}.

C≡C—H occurs around 3300 cm^{-1} (3.03 μm), and aromatic and unsaturated compounds around 3000 to 3100 cm^{-1} (3.33 to 3.23 μm).

For alkanes, bands at 1460 cm^{-1} (6.85 μm) and 1380 cm^{-1} (7.25 μm) are indicative of a terminal methyl group. If the latter band is split into a doublet at about 1397 and 1370 cm^{-1} (7.16 and 7.30 μm), geminal methyls are indicated. The latter band is shifted to lower frequencies when the methyl group is adjacent to =C=O (1360 to 1350 cm^{-1}), —S— (1325 cm^{-1}), and silicon (1250 cm^{-1}). A band at 1470 cm^{-1} indicates the presence of —CH$_2$—. Four or more methylene groups in a linear arrangement give rise to a weak band at about 720 cm^{-1}, and in the solid state, there will be a series of sharp bands around 1200 cm^{-1}, one band for every two methylene groups. Figure 8.11 illustrates some of the features mentioned for the alkane portion of palmitic acid.

The substitution pattern of an aromatic ring can often be deduced from a series of weak bands in the region 2000 to 1670 cm^{-1} coupled with the position of strong bands between 900 and 650 cm^{-1}.[4]

8.8.2.2 Alkenes (C=C). Double bond frequencies fall in the region from 2000 to 1540 cm^{-1}. An unsaturated C=C group introduces a band at 1650 cm^{-1}; it may be weak or nonexistent if symmetrically located in the molecule. Conjugation with a second C=C or C=O shifts the band 40 to 60 cm^{-1} to a lower frequency with a substantial increase in intensity.

Bands from the bending vibrations of hydrogen on a C=C bond are very valuable. A vinyl group gives rise to two bands at about 990 and 910 cm^{-1}. The =CH$_2$ band appears near 895 cm^{-1}. Cis- and trans-disubstituted olefins absorb in the region 685 to 730 cm^{-1} and 965 cm^{-1}, respectively. The single hydrogen in a trisubstituted olefin appears near 820 cm^{-1}.

8.8.2.3 Alkynes. Triple bonds, and little else, appear from 2500 to 2000 cm^{-1}. The absorption band for —C≡C— is located around 2100 to 2140 cm^{-1} when terminal, but from 2260 to 2190 cm^{-1} if nonterminal. The intensity of nonterminal alkynes decreases as the symmetry of the molecule increases and may not appear. Conjugation with a carbonyl group increases the strength of the band markedly. The ethynyl hydrogen appears as a very sharp and intense band at 3300 cm^{-1}.

8.8.2.4 Carbonyl Group. The carbonyl group is often the strongest band in the spectrum. It will lie between 1825 to 1575 cm^{-1}; its exact position is dependent upon its immediate substituents (Table 8.2). Anhydrides usually show a double absorption band and, in addition, will exhibit a C—H stretching frequency of the CHO group at about 2720 cm^{-1}. The carboxyl group has bands at 2700, 1300, and 943 cm^{-1} which are associated with the carboxyl OH; these bands disappear when the carboxylate ion is formed. When a dimer exists, the band at 2700 cm^{-1} disappears.

8.8.2.5 Ethers. For ethers, one important and quite strong band appears near 1100 cm^{-1}.

8.8.2.6 Alcohols. The absorption due to the stretching of the O—H bond is most useful. In the unassociated state, it appears as a weak but sharp band at about 3600 cm^{-1}. Hydrogen bonding increases the band intensity and moves it to lower frequencies. If the hydrogen bonding is quite strong, the band becomes very broad.

The differentiation between the several types of alcohols is often possible on the basis of the C—O stretching bands. Saturated tertiary alcohols have a band in the region 1200 to 1125 cm^{-1}. For saturated secondary alcohols, the band lies between 1125 to 1085 cm^{-1}. Saturated primary alcohols show a band between 1085 to 1050 cm^{-1}.

8.8.2.7 Amines. Very useful are the N—H stretching frequencies at about 3500 and 3400 cm^{-1} for a primary amine (or amide), the N—H bending at 1610 cm^{-1}, and the —NH$_2$ bending at about 830 cm^{-1} (which is broad for primary amines). A secondary amine exhibits a single band at about 3350 cm^{-1}.

8.8.3 Compound Identification

The total structure of an unknown may not be readily identified from the infrared spectrum but perhaps the type or class of compound can be deduced. Once the

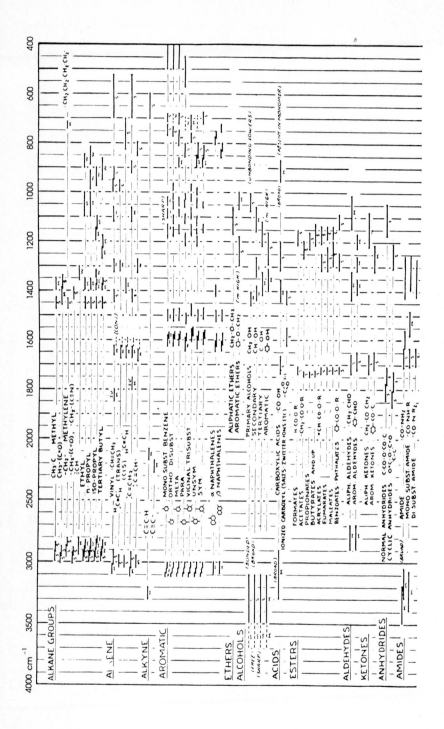

8.16

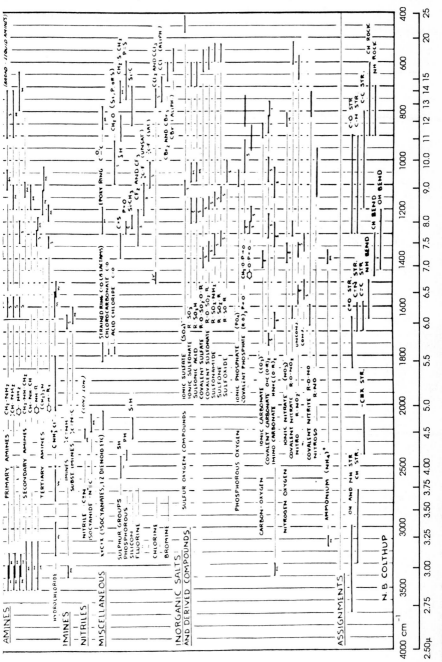

FIGURE 8.10 Some characteristic absorption bands. Band positions are given for dilute solution in nonpolar solvents. Intensities are expressed as strong (s), medium (m), weak (w), and variable (v). *(Courtesy of American Cyanamid Company.)*

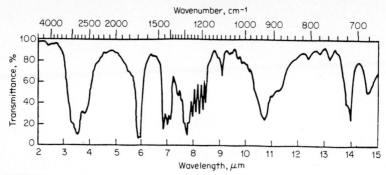

FIGURE 8.11 Infrared spectrum of stearic acid taken on a KBr pellet.

TABLE 8.2 Carbonyl Infrared Absorption Bands

Typical of	Wave number, cm^{-1}	Wavelength, μm
Anhydrides, saturated	1850–1800	5.41–5.55
	1790–1740	5.59–5.75
aryl and α,β-unsaturated	1830–1780	5.46–5.62
	1740–1710	5.75–5.85
Acid chlorides, saturated	1815–1790	5.51–5.59
aryl and α,β-unsaturated	1790–1750	5.59–5.71
Acid peroxide, saturated	1820–1810	5.49–5.52
	1800–1780	5.41–5.62
aryl and α,β-unsaturated	1805–1780	5.40–5.62
	1785–1755	5.60–5.70
Esters and lactones, saturated	1750–1735	5.71–5.76
aryl and α,β-unsaturated	1730–1715	5.78–5.83
aryl and vinyl	1800–1750	5.55–5.71
six-ring and larger lactones	1750–1735	5.71–5.76
five-ring lactones	1780–1760	5.62–5.68
Aldehydes, saturated	1740–1720	5.75–5.81
aryl	1715–1695	5.83–5.90
Ketones, saturated	1725–1705	5.80–5.87
aryl and unsaturated	1700–1660	5.88–6.02
Carboxylic acids, saturated	1725–1700	5.80–5.88
aryl and unsaturated	1715–1680	5.83–5.95
Carboxylate ions	1610–1550	6.21–6.45
Amides, primary	1690–1650	5.92–6.06
	1640–1600	6.10–6.25
secondary	1700–1630	5.80–6.13
	1570–1515	6.37–6.60
tertiary	1670–1630	5.99–6.13
Lactams, six-ring and larger	~ 1670	~ 5.99
five-ring	~ 1700	~ 5.80
Imides, cyclic six-ring	~ 1710	~ 5.85
	~ 1700	~ 5.80
cyclic five-ring	~ 1770	~ 5.65
	~ 1700	~ 5.80
Ureas, aliphatic	~ 1660	~ 6.02
six-ring	~ 1640	~ 6.10
five-ring	~ 1720	~ 5.81
Urethanes	1740–1690	5.75–5.92
Thioesters, alkyl	~ 1690	~ 5.92
aryl	~ 1710	~ 5.85

Source: J. A. Dean, ed., *Handbook of Organic Chemistry,* McGraw-Hill, New York, 1986. Contains much more detail and remarks.

key functional groups have been established as present (or, equally important, as absent), the unknown spectrum is compared with spectra of known compounds. Several collections of spectra are available.[2,3]

8.9 QUANTITATIVE ANALYSIS

The baseline method for quantitative analysis involves the selection of an absorption band that is separated from the bands of other matrix components (Fig. 8.12). Draw a straight line tangent to the absorption band. From the illustration observe how the value of P_o and P are obtained. The value of the absorbance, $\log (P_o/P)$, is then plotted against concentration for a series of standard solutions, and the unknown concentration is determined from this calibration curve. The use of such ratios eliminates many possible errors, such as changes in instrument sensitivity, source intensity, and adjustment of the optical system.

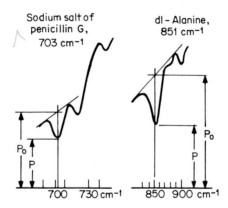

FIGURE 8.12 Baseline method for calculation of the transmittance ratio in quantitative analysis.

The KBr-pellet technique, when combined with the internal standard method of evaluation, can be used. Potassium thiocyanate makes an excellent internal standard. After grinding and redrying, KCNS is reground with dry KBr to make a concentration of about 0.2% by weight of KCNS. A standard calibration curve is constructed by mixing known weights of the test substance (usually about 10 percent of the total weight) with a known weight of the KBr-KCNS mixture and then preparing the pellet or thin wafer. The ratio of the thiocyanate absorption at 2125 cm^{-1} to a chosen absorption band of the test substance is plotted against the concentration of the test substance.

REFERENCES

1. J. A. Dean, ed., *Handbook of Organic Chemistry*, McGraw-Hill, New York, 1986.
2. C. J. Pouchert, ed., *The Aldrich Library of Infrared Spectra*, Aldrich, Milwaukee, Wisc. A series of dispersive and FT-IR spectra compilations.

3. Sadtler Research Laboratories, *Catalog of Infrared Spectrograms,* Philadelphia (a continuously updated series).
4. C. W. Young, R. B. Duvall, and N. Wright, *Anal. Chem.,* **23:**709 (1951).

BIBLIOGRAPHY

Brame, E. G., and J. G. Grasselli, eds., *Infrared and Raman Spectroscopy,* Vol. 1, Parts A, B, and C, *Practical Spectroscopy Series,* Dekker, New York, 1977.

Colthup, N. B., L. H. Daly, and S. E. Wiberley, *Introduction to Infrared and Raman Spectroscopy,* 2d ed., Academic, New York, 1975.

Griffiths, P. R., and J. A. deHaseth, *Chemical Infrared Fourier Transform Spectroscopy,* Wiley, New York, 1986.

Harrick, N. J., *Internal Reflection Spectroscopy,* Wiley, New York, 1967.

Silverstein, R. M., G. C. Bassler, and T. C. Morrill, *Spectrometric Identification of Organic Compounds,* 4th ed., Wiley, New York, 1981.

Williams, D. H., and J. Fleming, *Spectroscopic Methods in Organic Chemistry,* 4th ed., McGraw-Hill, New York.

CHAPTER 9
RAMAN SPECTROSCOPY

Raman spectroscopy is used to determine molecular structures and compositions of organic and inorganic materials. Materials in the solid and liquid are easily examined; even gas samples can be handled under special conditions. Normally the minimum sample requirements are on the order of tenths of a gram.

Raman spectroscopy embraces the entire vibrational spectrum with one instrument. It can be used to study materials in aqueous solutions. Sample preparation for Raman spectroscopy is generally much simpler than for the infrared.

9.1 PRINCIPLES

When an intense beam of monochromatic light impinges on a material, light scattering can occur in all directions; that is, most collisions are elastic with the frequency of the scattered light (v) being the same as that of the original light (v_o). This effect is known as *Rayleigh scattering*. Another type of scattering that can occur simultaneously with the Rayleigh scattering is known as the *Raman effect*. The Raman effect arises when a beam of intense monochromatic radiation passes through a sample that contains molecules that undergo a change in molecular polarizability as they vibrate. In other words, the electron cloud of the molecule must be more readily deformed in one extreme of the vibration than in the other extreme. By contrast, in the infrared region the vibration must cause a change in the permanent dipole moment of the molecule.

Only a small proportion of the excited molecules (10^{-6} or less) may undergo changes in polarizability during a normal vibrational mode. The incident radiation does not raise the molecule to any particular quantized level; rather, the molecule is considered to be in a virtual or quasi-excited state whose height above the initial energy level equals the energy of the exciting radiation. In fact, the wavelength of the incident radiation does not have to be one that is absorbed by the molecule. Through the induced oscillating dipole(s) that it stimulates, the radiation leads to the transfer of energy with the vibrational modes of the sample molecules. As the electromagnetic wave passes, the polarized molecule ceases to oscillate and this quasi-excited state then returns to its original ground level by radiating energy in all directions except along the direction of the incident radiation. A vibrational quantum of energy usually remains with the scattering radiation so there is a decrease in the frequency ($v_o - v_v$) of the emitted radiation (*Stokes lines*). However, if the scattering molecule is already in an excited vibrational level of the ground electronic state, a vibrational quantum of energy may be abstracted from the scatterer, leaving the molecule in a lower vibrational level and thus increasing the frequency of the scattered radiation (*anti-Stokes lines*). The latter condition is less likely to prevail and, consequently, the anti-Stokes lines are less intense than the Stokes lines. For either case, the shift in frequency of the scattered Raman radiation is proportional to one of the vibrational energy levels involved in the transition. Thus the spectrum of the scattered radiation consists of a relatively strong component with frequency unshifted (Rayleigh scattering) and the Stokes and anti-Stokes lines. The Stokes (or anti-Stokes) lines in a Raman spectrum will have the general appearance of the corresponding infrared spectrum.

In the usual Raman method the excitation frequency of the Raman source is selected to lie below any singlet-singlet electronic transitions and above the most fundamental vibrational frequencies.

9.2 INSTRUMENTATION

The laser Raman spectrometer, shown in Fig. 9.1, consists of the laser excitation unit and the spectrometer unit using gratings with 1200 grooves $\cdot$ mm^{-1}. The laser beam enters from the rear of the spectrometer into the depolarization autorecording unit and, after passing through this unit, it illuminates the sample. The Raman scattering, collected at 90° to the exciting laser beam, is focused on the entrance slit of a grating double monochromator. Immediately ahead of the spectrometer is a polarization scrambler which overcomes the grating bias caused

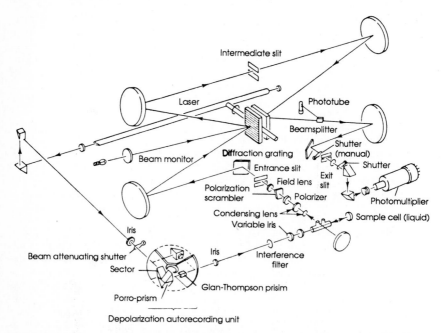

FIGURE 9.1   Optical schematic of a laser Raman spectrophotometer.   (*Courtesy of Jeol, Ltd.*)

by polarized radiation. When the polarization of the Raman spectrum is measured, a polarization analyzer is placed between the condenser lens and the polarization scrambler. The scattered radiation is detected by a photomultiplier tube that is placed in a thermoelectric cooler ($-30°$) to lower the dark current and reduce noise. Often the signals from the detector are amplified and counted with a photon-counting system—the most effective means of recovering low-level Raman signals. Inexpensive dc amplifiers are excellent for strong signals.

9.2.1 Detectors

The choice of phototube depends on which laser line is used. Laser selection and detector choice are interwoven, as shown in Fig. 9.2. Raman shifts of 3700 cm^{-1} require response to 827 nm for the He-Ne laser (excitation line at 632.8 nm) and slightly further for the Ar-Kr laser (Kr* excitation line at 647.1 nm). For these excitation lines, the extended red-sensitive multialkali cathode and the gallium arsenide photocathode have the needed quantum efficiency in the far-red portion of the electromagnetic spectrum. Blue-sensitive photomultiplier tubes are near their peak sensitivity at 488.0 nm, one of the other emission lines of the Ar-Kr laser.

9.2.2 Laser Sources

The He-Ne laser line at 632.8 nm is favorably located in the spectrum where the least amount of fluorescent problems appears in routine analyses. For some ex-

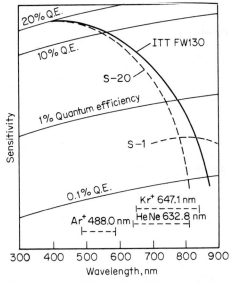

FIGURE 9.2 Sensitivity of several photomultiplier tubes. The dashed horizontal lines represent the range of 3500 cm^{-1}, which normally includes the Raman shift from the designated laser exciting lines.

periments the Ar-Kr laser is ideal; it has two intense argon lines at 488.0 and 514.5 nm and two major krypton lines at 568.2 and 647.1 nm.

9.3 SAMPLE HANDLING

Raman spectroscopy can be performed on specimens in any state: liquid, solution, transparent or translucent solid, powder, pellet, or gas. Neat liquids are examined with a single pass of the laser beam either axially or transversely. Multiple passes offer considerable gain in Raman intensity and permit work with samples in the microliter range or even down to about 8 nL. Photo- or heat-labile materials are studied in spinning cells.

Water is a weak scatterer and is therefore an excellent solvent for Raman work. Other solvents and their obscuration ranges are shown in Fig. 9.3.

Powders are tamped into an open-ended cavity for front-surface illumination or into a transparent glass capillary tube for transverse excitation. Sample illumination at 180° provides better signal-to-noise ratios whereas right-angle viewing improves the ratio of Raman to Rayleigh scattering.

For a translucent solid, the laser beam is focused into a cavity on the face of the sample, which is either a cast piece or a pellet formed by compression of the powder.

Gas samples are handled with powerful laser sources and efficient multiple passes or interlaser-cavity techniques. Gases are difficult to study because of their low scattering.

9.3.1 Sample Fluorescence

If fluorescence arises from impurities in the sample, one can clean up the sample by techniques such as chromatographic fractionation, recrystallization, or distil-

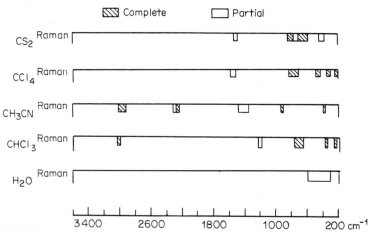

FIGURE 9.3   Obscuration ranges of the most useful solvents for Raman spectrometry in solution.

lation. However, if the fluorescence arises from the sample itself, one must select a different excitation line—a line that excites the Raman spectrum but not the fluorescent spectrum.

9.4 DIAGNOSTIC STRUCTURAL ANALYSIS

The Raman spectrum contains a number of distinct spectral features. Raman frequencies are tabulated in the *Handbook of Organic Chemistry*.[1] By comparison with infrared spectra, in Raman spectroscopy the most intense vibrations are those that originate in relatively nonpolar bonds with symmetrical charge distributions. Vibrations from —C=C—, —C≡C—, —C≡N, —C=S—, —C—S—, —S—S—, —N=N—, and —S—H bonds are readily observed (Table 9.1). Raman spectroscopy has a distinct advantage in the detection of low-frequency vibrations. In most cases, information can be taken to within 20 to 50 cm^{-1} of the exciting line. This corresponds to the far-infrared region where the important vibrations in metal bonding or inorganic and organometallic compounds reside.

Skeletal vibrations of finite chains and rings of saturated and unsaturated hydrocarbons are prominent in the region 800 to 1500 cm^{-1} and highly useful for cyclic and aromatic rings, steroids, and long chains of methylenes. All aromatic compounds have a strong band at 1600 ± 30 cm^{-1}. Monosubstituted compounds have an intense band at about 1000 cm^{-1}, a strong band at about 1025 cm^{-1}, and a weak band at 615 cm^{-1}. Meta- and 1,3,5-trisubstituted compounds have only the line at 1000 cm^{-1}. Ortho-substituted compounds have a line at 1037 cm^{-1}, and para-substituted compounds have a weak line at 640 cm^{-1}.

The band near 500 cm^{-1} is characteristic of the —S—S— linkage; the —C—S— group has a band near 650 cm^{-1}, and the intense band near 2500 cm^{-1} indicates the —S—H stretch.

Raman spectra are helpful whenever the NH and CH stretching frequencies are obscured by intense OH absorption. In Raman spectroscopy the OH band is weak, whereas the NH and CH stretching frequencies exhibit moderate intensity.

Figure 9.4 shows the infrared (*top*) and Raman (*bottom*) spectrum of a com-

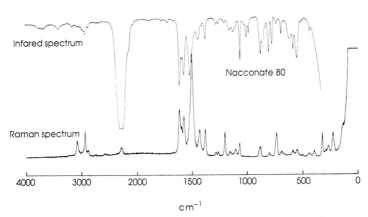

FIGURE 9.4 Complementary nature of infrared and Raman spectroscopy for a blend of 2,4-toluene diisocyanate and 2,6-toluene diisocyanate (80:20).

TABLE 9.1   Raman Frequencies of Multiple Bonds*

Group	Band, cm^{-1}	Remarks
R—C≡CH	2160–2100 (vs)	Monoalkyl substituted
	650–600 (m)	
	356–335 (s)	
R$_1$—C≡C—R$_2$	2300–2190 (vs)	Disubstituted alkyls; sometimes two bands
—C≡C—C≡C—	2264–2252 (vs)	
—C≡N	2260–2240 (vs)	Unsaturated nonaryl substituents lower the frequency
	2234–2200 (vs)	Lowered ~30 cm^{-1} with aryl and conjugated aliphatics
	840–800 (s-vs)	
	385–350 (m-s)	Aliphatic nitriles
Aldehydes	1740–1720 (s-vs)	
Ketones, aryl saturated	1700–1650 (m)	
	1725–1700 (vs)	
Carboxylic acids, mono-	1686–1625 (s)	
Carboxylate ions	1440–1340 (vs)	
>C=C<	1648–1638 (vs)	Monosubstituted
	~1650 (vs)	*Gauche* isomer
	~1660 (vs)	*Cis* isomer
	1676–1665 (s)	*Trans* isomer
	1678–1664 (vs)	Trisubstituted
	1680–1665 (s)	Tetrasubstituted
Aldimines	1673–1639	Dialkyl substituents at higher frequency; diaryl substituents at lower end of range
	1405–1400 (s)	
Aldoximes and ketoximes	1680–1617 (vs)	
—N=N—	1580–1570 (vs)	Nonconjugated
	1442–1380 (vs)	Conjugated to aryl ring
	1060–1030 (vs)	Aryl compounds

*Abbreviations: vs, very strong; s, strong; m, moderate.

mercial blend of 80% 2,4- and 20% 2,6-toluene diisocyanate. The two spectra illustrate the complementary nature of the two spectroscopic methods. The infrared spectrum provides evidence for the identification and location of substituent groups on the aromatic ring. For example, the highest intensity infrared band near 2280 cm^{-1} is quite characteristic of the —N=C=O group. The Raman spectrum possesses a high-intensity band near 1510 cm^{-1} which is due to a symmetric stretching of the aromatic ring and the —N=C=O group. Although there are numerous coincidences between the two spectra, the differences in relative band intensities can be dramatic. For example, the 2280 cm^{-1} band is very intense in the infrared spectrum but very weak in the Raman spectrum.

Another pair of spectra is shown in Fig. 9.5 for a compound whose molecular weight is 140. There is a carbonyl band at 1787 cm^{-1} that is difficult to characterize initially. Only unsaturated C—H stretch is present. Phenyl ring frequencies

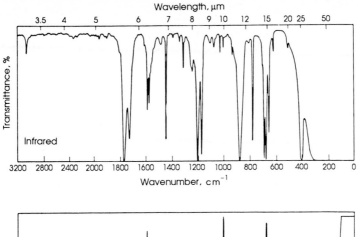

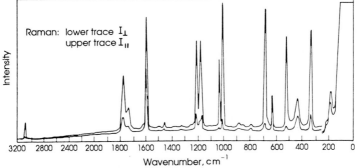

FIGURE 9.5 Infrared and Raman spectra for benzoyl chloride.

are seen at 1600, 1594, and 1426 cm^{-1}, but the substitution pattern is confusing in the infrared. However, characteristic Raman bands at 1025, 1000, and 615 cm^{-1} clearly identify a monosubstituted ring. The symmetrical C—CH$_3$ deformation frequency near 1380 cm^{-1} is absent in both spectra. Something is needed for the terminal group. One has $140 - (77 + 28) = 35$ atomic mass units left, so chlorine is the terminal atom; thus the compound is benzoyl chloride. Now the confusing bands in the ring puckering region near 700 cm^{-1} can be associated with the C(=O)—Cl stretch. Also, aryl chlorides absorb strongly at 1785 to 1765 cm^{-1}.

9.4.1 Polarization Measurements

The depolarization ratio is defined as

$$p = \frac{P_\parallel}{P_\perp} \tag{9.1}$$

where $P_\perp$ is the radiant power observed when the electrical vector of the Raman radiation is in the yz plane and $P_\parallel$ is the power of the Raman radiation polarized in the xy plane. The ratio of the radiant powers is measured by setting the analyzer prism in the path of the Raman scattered radiation at 0° and then at 90°. Since P_i is always greater, the depolarization ratio may vary from near 0 for highly symmetrical types of vibrations to a maximum of 0.75 for totally nonsymmetrical vibrations.

If the incident radiation is polarized in the xy plane (parallel illumination) and then in the xz plane (perpendicular illumination), the depolarization ratio may vary from 0 to a maximum of 0.86. Note the parallel and perpendicular traces of the Raman spectrum in Fig. 9.5.

The instrument operation should be checked with the known Raman bands. The 218-cm^{-1} band of carbon tetrachloride should have the maximum value, and the 459-cm^{-1} band should have a value that is essentially zero (Fig. 9.6).

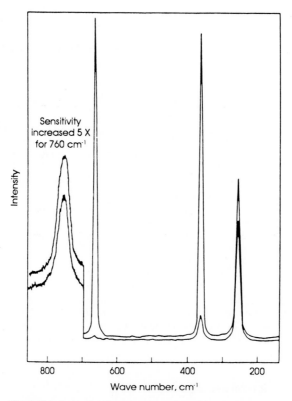

FIGURE 9.6 Partial Raman spectrum of chloroform illustrating depolarization. In the lower trace the direction of polarization of the incident beam is perpendicular to the direction of observation; in the upper trace the incident beam is polarized parallel to the direction of observation.

9.5 QUANTITATIVE ANALYSIS

The radiant power of a Raman line is measured in terms of an arbitrarily chosen reference line, usually the line of carbon tetrachloride at 459 cm^{-1}, which is scanned before and after the spectral trace of the sample. Peak areas on the spectrum are converted to *scattering coefficients* by dividing the area of the sample peak by the average of the areas of the dual traces of carbon tetrachloride. Both standards and samples must be recorded in cells of the same dimensions.

The scattering coefficient based on the area under a recorded peak is directly proportional to the volume fraction of the compound present. Although peak heights may be used for mixtures in which the components all have the same molecular type, peak areas compensate for band broadening.

REFERENCES

1. J. A. Dean, ed., *Handbook of Organic Chemistry*, McGraw-Hill, New York, 1986.

BIBLIOGRAPHY

Baranksa, H., A. Labudzinska, and J. Terpinski, *Laser Raman Spectrometry, Analytical Applications*, Wiley, New York, 1988.

Grasselli, J. G., M. K. Snavely, and B. J. Bulkin, *Chemical Applications of Raman Spectroscopy*, Wiley, New York, 1981.

Parker, F. S., *Applications of Infrared, Raman, and Resonance Raman Spectroscopy in Biochemistry*, Plenum, New York, 1983.

Spiro, T. G., *Biological Applications of Raman Spectroscopy*, Vols. 1 and 2, Wiley, New York, 1987.

Strommen, D. P., and K. Nakamoto, *Laboratory Raman Spectroscopy*, Wiley, New York, 1984.

CHAPTER 10
FLAME SPECTROMETRIC METHODS

10.1 INTRODUCTION

Atomic spectroscopy includes all analytical techniques which employ the emission and/or absorption of electromagnetic radiation by individual atoms. There are three kinds of emission spectra:

1. Continuous spectra which are emitted by incandescent solids.
2. Line spectra, which are characteristic of atoms that have been excited and are emitting their excess energy.
3. Band spectra, which are emitted by excited molecules.

This chapter discusses methods that use combustion flames and a single nonflame method, electrothermal atomic absorption spectrometry. Three techniques can be used to observe the atomic vapor that is produced when a sample solution is nebulized and passed into a flame. These are atomic absorption spectrometry (AAS), flame emission spectrometry (FES), and atomic fluorescence spectrometry (AFS) (Fig. 10.1). Of these, AAS and FES are the most widely used.

Flame spectrometric methods, qualitative and quantitative, can be applied to clinical materials (serum, plasma, and biologic fluids), soils, plant materials, plant nutrients, and samples of inorganic and organic substances. The specific frequency of the radiation (emitted or absorbed) identifies the element. The intensity of emitted (or absorbed) radiation at the specific frequency is proportional to the amount of the element present.

10.2 INSTRUMENTATION FOR FLAME SPECTROMETRIC METHODS

The basic instrumentation for flame atomic emission spectrometric methods requires these items:

1. An atom source, either a flame or an electric furnace
2. A monochromator to isolate the specific wavelength of light to be used

FLAME EMISSION

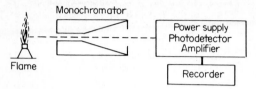

ATOMIC ABSORPTION

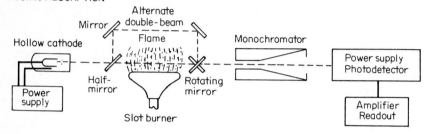

ATOMIC FLUORESCENCE

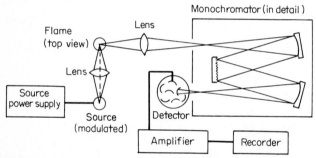

FIGURE 10.1   Schematic diagrams illustrating the several techniques: flame emission, flame atomic absorption, and flame atomic fluorescence.

3. A detector to measure the light transmitted

4. Electronics to treat the signal and a data logging device to display the results

A primary light source must be added, either a hollow cathode lamp or an electrodeless discharge lamp, for work in atomic absorption spectrometry.

10.2.1 Atom Sources

The atom source must produce free atoms of analytical material from the sample. A high-energy acetylene–nitrous oxide flame (2800°C) is usually desired to rapidly desolvate and efficiently dissociate the analyte in the sample in order to minimize interferences. This is particularly true if an oxide of the analyte has a high dissociation energy. A lower temperature acetylene-air flame (2400°C) is sometimes desired when an element is too easily ionized in a hotter flame. A special source, the electrothermal (graphite) furnace, is described in a later section. The sample is introduced as an aerosol into the flame and as a liquid or solid into the furnace.

For FES and AAS a premixed, laminar-flow flame is employed. A slot burner and expansion chamber are shown in Fig. 10.2. The flame burner head or the interior of the tube of the furnace is aligned so that it intersects the light path of the spectrophotometer.

10.2.2 Nebulization

In both FES and AAS, the sample solution is introduced as an aerosol into the flame. Pneumatic nebulization is the technique used in most flame spectrometric determinations. The sample solution is drawn through the inner annulus of two concentric annuli by the pressure differential generated by the high-velocity gas stream (usually the oxidant of the flame gases) passing across or concentric with

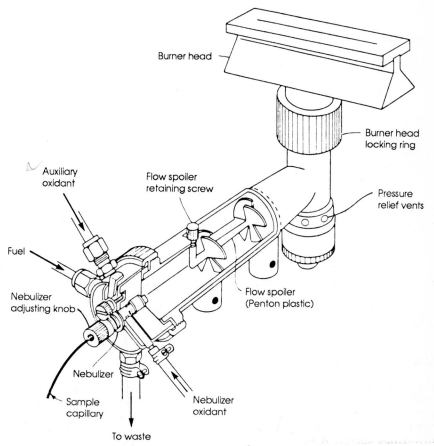

FIGURE 10.2 Slot burner and expansion chamber. *(Courtesy of Perkin-Elmer Corp.)*

the sample orifice. A cloud of droplets of varying diameters is produced within the expansion chamber. The larger droplets are removed from the sample stream upon collision with the wall surfaces. Many nebulizer units break up the original droplets into smaller droplets by impact beads or a counter gas jet. The final aerosol, now a fine mist, is combined with the oxidizer-fuel mixture and carried into the burner (Fig. 10.2). Only a small percentage (usually 2 to 3 percent) of the nebulized analyte solution reaches the flame. Unanticipated changes in viscosity and surface tension are avoided during aerosol generation by keeping the sample and standard matrices identical and by avoiding total acid or salt concentrations greater than about 0.5%.

10.2.3 Atomization

The flame atomization processes for the salt MX are outlined in Fig. 10.3. It is

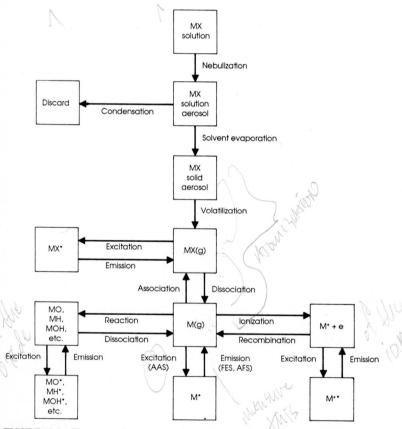

FIGURE 10.3 Flame atomization processes for the salt MX. Asterisk (*) indicates excited state.

complex, and everything leading to the production of free atoms must take place in the few milliseconds that correspond to the upward movement of the molecules and atoms through the flame gases. The atomization step must convert the analyte within the aerosol into free analyte atoms for AAS and flame atomic emission spectrometry.

For AAS the following sequence of events occurs in rapid succession.

1. *Desolvation of the aerosol*:   The water, or other solvent, is vaporized, leaving minute particles of dry salt.

2. *Vaporization of the resulting particles*:   At the high temperature of the flame, the dry salt is vaporized (converted to gaseous molecules).

3. *Dissociation of gaseous molecules*:   Part or all of the gaseous molecules are dissociated to give neutral atoms.

For flame atomic emission, two more events must occur.

4. *Excitation of atoms (and molecules)*:   The neutral metal atoms are excited (and sometimes ionized) by the thermal energy of the flame. (Excitation occurs in AAS by thermal collisions as well as by absorption of radiation from the light source.)

5. *Emission*:   From the excited electronic level(s) of the atom, a reversion takes place to the ground electronic state with the emission of light whose wavelength is characteristic of the element and whose intensity is proportional to the amount of analyte element present.

Unfortunately, some of the metal atoms unite with other radicals or atoms that are present in the flame gases or which are introduced into the flame along with the test element. For example, neutral metal atoms may unite with OH radicals or atomic oxygen present in the flame gases to form MOH, MH, and MO molecular species. These reactions weaken both the atomic absorption signal and the atomic emission signal at the wavelength of the metal emission or absorption line(s). The MOH, MH, and MO molecules may be excited and, if so, will emit characteristic molecular band spectra which clutter the general background. Some molecular band spectra are quite strong and have been used in flame emission as, for example, the band centered at 554 nm due to CaOH and band heads at 606 and 622 nm due to CaO molecules. To eliminate or minimize the interference from molecular bands, the use of hotter flames (acetylene–nitrous oxide) is recommended, along with the addition of an ionization buffer to minimize any ionization of metal atoms.

10.2.4   Spectrophotometers

The spectrophotometers are similar to those used in visible-ultraviolet instruments discussed in Chap. 6. The usual detectors are photomultiplier tubes. A chopper will be added either to modulate the radiation from the sample in FES or to modulate the light source in AAS.

Care must be taken never to exceed the saturation limit of the photomultiplier tube by flooding it with light. This situation will arise when too much radiation emanates from the flame and is often caused by too high a concentration of sample components, even though these signals are eliminated by modulation in the final signal readout.

10.2.4.1 Scale Expansion. When the electronic noise in the detector-amplifier system can be reduced to a negligible level, scale expansion can be used. The zero point is displaced off-scale by applying a potential opposite to the signal arriving from the detector. Thus, the full scale of the readout device can be used as the upper end of a greatly expanded scale. Alternatively, the lower end of the instrument scale can be expanded by adjusting the amplification of the signal.

10.3 FLAME COMPOSITION AND TEMPERATURE

A premixed, laminar-flow flame with well defined zones (Fig. 10.4) has lower background emission and less noise as a result of optical turbulence than a corresponding diffusion flame (total consumption atomizer-burner) where the zones are intermixed. Temperatures of common premixed flames are given in Table 10.1. The exact temperature of the flame depends on fuel-oxidant ratio and is generally highest for a stoichiometric mixture. Generally, no attempt is made to subtly adjust flame temperatures for a particular fuel-oxidant mixture other than the use of lean or fuel-rich flames, as discussed later.

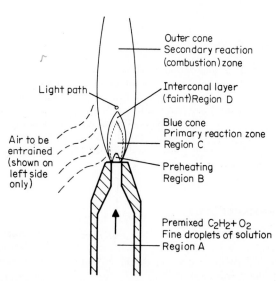

FIGURE 10.4   Schematic structure of a laminar-flow flame. (*From J. A. Dean, Flame Photometry, McGraw-Hill, New York, 1960.*)

TABLE 10.1   Characteristics of Common Premixed Flames

Fuel	Oxidant	Temperature, °C*	Burning velocity,† cm · s⁻¹
Acetylene	Air	2400	160–266 (160)
Acetylene	Nitrous oxide	2800	260
Acetylene	Oxygen	3140	800–2480 (1100)
Hydrogen	Air	2045	324–440
Hydrogen	Nitrous oxide	2690	390
Hydrogen	Oxygen	2660	900–3680 (2000)
Propane	Air	1925	43

*Values are for a stoichiometric mixture.
†Values of burning velocity in parentheses are probably the ones most applicable to laboratory burners.

10.3.1 Burning Velocity

Flame propagation rate or burning velocity is important. If it exceeds approximately 400 cm · s⁻¹, the flame likely will flash back into the mixing chamber and an explosion will result. That is why acetylene-oxygen flames must *never* be used with a premixed burner. When using acetylene–nitrous oxide flames, first ignite the flame with an acetylene-air mixture (turn on the air first). Then override the air with the nitrous oxide gas flow until the flame exhibits the characteristic reddish color in the interconal region. Reverse this sequence when extinguishing the flame; turn off the nitrous oxide while the air is flowing, then turn off the acetylene flow before turning off the air. Special burner heads (5-cm slot length, 0.5 mm width) and control units are required for igniting and extinguishing acetylene–nitrous oxide flame.

10.3.2 Flame Profile

The concentration of excited (Fig. 10.5) and unexcited (Fig. 10.6) atoms in a flame varies in different parts of the flame envelope. Although usually cooler, a fuel-rich flame (ratio of fuel to oxidant exceeds that needed for stoichiometric combustion) provides the reducing atmosphere necessary for the production of a large free-atom population of those elements (alkaline earth elements, Al, B, Sb, and Ti) that have a tendency to form refractory oxides. The free-atom concentration of these metals is negligible unless analyzed in the reducing environment of a fuel-rich acetylene–nitrous oxide flame.

10.3.3 Observation Site

The region that is viewed within the flame is important. For example, the emission lines of boron (249.7 nm) and antimony (259.8 nm) are either absent or very weak in the outer mantle of a stoichiometric flame, but they appear in high concentrations in the reaction zone (blue cone) of a fuel-rich flame (Fig. 10.5). In the illustration mixture, "strength" refers to the actual ratio of oxygen-acetylene as compared to the ratio for a stoichiometric flame. For mixture strengths 0.40 and less, the inner conal gases will be bright green as a result of the high concentra-

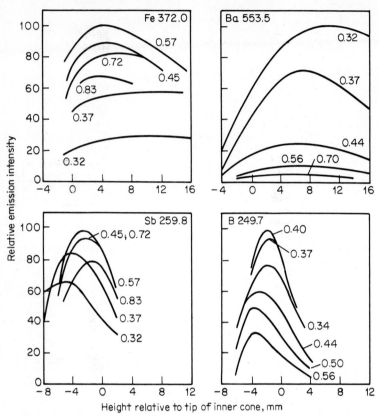

FIGURE 10.5   Flame emission intensity as a function of height of observation with mixture strength as parameter.   [*After J. A. Dean and J. E. Adkins, Analyst, 91:709 (1966). Courtesy of Society for Analytical Chemistry.*]

tion of CH radicals, which provide an excellent reducing atmosphere for refractory boron oxide. Some means of adjusting the burner height relative to the observation light path of the spectrometer will be needed to observe the interconal reaction zone.

10.3.4   Ionization Buffer

When flame temperatures are high enough to cause ionization of the analyte atoms, as shown in Table 10.2 (for example, a pair of calcium ion emission lines appear at 393.3 and 396.8 nm), an ionization buffer must be incorporated into the sample solution. An ionization buffer is an easily ionizable element added to the matrix, but an element which will not add any spectral line interference. Often easily ionized elements such as K, Cs, or Sr are added in approximately 100-fold excess (as compared to the analyte concentration) to suppress the ionization of easily ionized elements—those whose ionization potentials are less than about 8

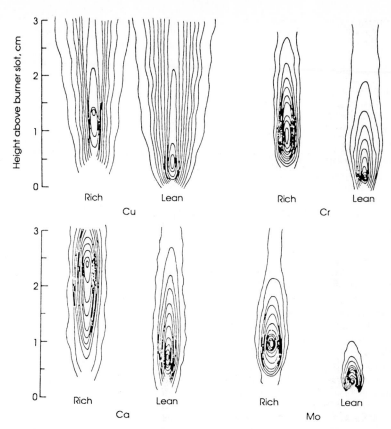

FIGURE 10.6 Distribution of atoms in a 10-cm acetylene-air flame. Fuel-rich and fuel-lean results are shown. Contours are drawn at intervals of 0.1 absorbance unit with maximum absorbance in the center. [*After C. S. Rann and A. N. Hambly, Anal. Chem., 37:879 (1965). Copyright (1965) American Chemical Society.*]

eV. The use of ionization buffers is important when using the acetylene–nitrous oxide flame and desirable when handling samples that contain variable amounts of the alkali and alkaline earth metals which are easily ionized.

10.3.5 Releasing Agents

Releasing agents provide a chemical means for overcoming some vaporization interferences. A releasing agent may either combine with the interfering substance or deny the analyte to the interfering substance by mass action. For example, in calcium determinations a few hundred parts per million of lanthanum or strontium are often added routinely to solutions to minimize interference due to anions (such as phosphate, vanadate, and tungstate). The lanthanum or strontium preferentially binds the anion, thus releasing the calcium atoms. Calcium can also be

TABLE 10.2   Percent Ionization of Selected Elements in Flames

Element	Ionization potential, eV	Acetylene-air, 2200°C	Acetylene–nitrous oxide, 2800°C
Lithium	5.391	0	16
Sodium	5.139	0.3	26
Potassium	4.340	2.5	82
Rubidium	4.177	13.5	90
Cesium	3.984	28	96
Magnesium	7.646		4
Calcium	6.113	0.01	7
Strontium	5.694	0.1	17
Barium	5.211	1.9	

released by complexing the calcium with EDTA. Once in the flame the EDTA is destroyed.

10.4   ATOMIC ABSORPTION SPECTROMETRY

Atomic absorption is the process that occurs when a ground state atom absorbs energy in the form of electromagnetic radiation at a specific wavelength and is elevated to an excited state (Fig. 10.7). The atomic absorption spectrum of an element consists of a series of resonance lines, all originating with the ground electronic state and terminating in various excited states. Usually the transition between the ground state and the first excited state is the line with the strongest absorptivity, and it is the line usually used.

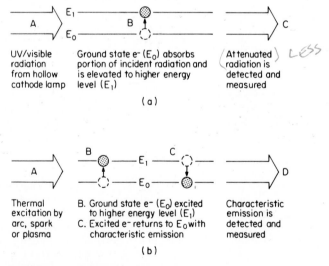

FIGURE 10.7   Mechanism of (a) atomic absorption (flame and carbon furnace) and (b) flame emission spectrometry.

Transitions between the ground state and excited states occur only when the incident radiation from a source is exactly equal to the frequency of a specific transition. Part of the energy of the incident radiation P_0 is absorbed. The transmitted radiation P is given by

$$P = P_0 e^{-(k_v b)} \tag{10.1}$$

where k_v is the absorption coefficient of the analyte element and b is the horizontal path length of the radiation through the flame. Atomic absorption is determined by the difference in radiant power of the resonance line in the presence and absence of analyte atoms in the flame. The width of the line emitted by the light source must be smaller than the width of the absorption line of the analyte in the flame.

The amount of energy absorbed from a beam of radiation at the wavelength of a resonance line will increase as the number of atoms of the selected element in the light path increases. The relationship between the amount of light absorbed and the concentration of the analyte present in standards can be determined. Unknown concentrations in samples are determined by comparing the amount of radiation they absorb to the radiation absorbed by standards. Instrument readouts can be calibrated to display sample concentrations directly.

10.4.1 Light Sources

The schematic diagram of a shielded-type hollow-cathode lamp is shown in Fig. 10.8. The lamp has a Pyrex body with an exit window of quartz. It is filled with an inert gas, usually argon or neon, at a pressure of 4 to 10 torr. Neon gas provides greater intensity of emitted element lines; argon is used only when a neon emission line lies in close proximity to a resonance line of the cathode element. An anode wire is positioned along the outside of the cylindrical cathode which is constructed for (or lined on the interior with) the element to be determined. Multielement lamps are constructed with an alloy or sinter of pressed metal pow-

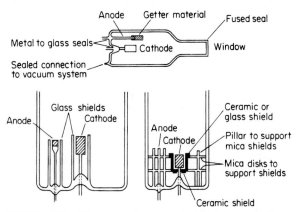

FIGURE 10.8 Schematic diagram of shielded-type hollow-cathode lamp.

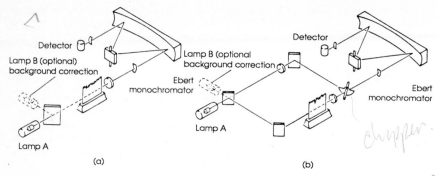

FIGURE 10.9   Optical diagrams of (*a*) a single-beam and (*b*) a double-beam atomic absorption spectrometer.

der. A protective shield (nonconductive) of mica around the outside of the cathode just behind the lip prevents spurious discharges around the outside of the cathode. Lamps are operated at currents below 30 mA.

Electrodeless discharge lamps are made by sealing a metal iodide salt (or small amounts of the metal and iodine) and argon at a low pressure in a small quartz tube. The tube is placed inside a ceramic cylinder on which the antenna from a 2450-MHz microwave generator is coiled.

10.4.2   Atomic Absorption Spectrophotometer

Optical diagrams of (a) a single-beam and (b) a double-beam absorption spectrophotometer are shown in Fig. 10.9. Typical focal length is 0.33 to 0.50 m with a Czerny-Turner or Ebert configuration. Two gratings, mounted back-to-back on a turntable, cover the wavelengths of 190 to 440 nm and 400 to 900 nm, respectively. Wavelengths should be keyboard- or switch-selectable.

A liability of AAS is the need for a different hollow-cathode lamp for each element to be analyzed. Only a few multielement lamps are available. However, hollow-cathode lamps with their narrow emission line widths provide virtual specificity for each element.

10.5   FLAME EMISSION SPECTROMETRY

Flame emission spectrometry (FES) is a simple and convenient method for analyzing solutions of organic and inorganic materials for individual metals. FES is particularly suited to alkali and alkaline earth metals, with 1 to 5 g of a dissolved sample being adequate for several analyses. Extremely low concentrations of metals can be determined (Table 10.3). The quantitative estimation for most metals, without concentration, is on the order of 0.1 to 2.0 $\mu g \cdot mL^{-1}$.

10.5.1   Principle

A solution of the sample is sprayed into a flame possessing the thermal energy required to excite the element to a level at which it will radiate its characteristic

TABLE 10.3 Useful Flame Emission and Atomic Absorption Lines

Element	Wave-length, nm	Flame emission, detection limit, $\mu g \cdot mL^{-1} \cdot (0.1\ mV)^{-1}$	Atomic absorption sensitivity, $\mu g \cdot mL \cdot (1\%\ Abs)^{-1}$
Ag	328.1	0.2	0.13
	338.3	0.2	0.20
Al	394.4	0.3	2
	396.2	0.5	1.2
AlO	484.2	0.5	
As	197.2		1.8
	235.0	2.2	
B	249.8	7	10
BO_2	518.0	3	
Ba	553.5	0.03	2.6
Be	234.9	0.15	0.03
Bi	223.0	6	0.7
Ca	422.7	0.07	0.08
Cd	228.8	4	0.03
	326.1	6	20
Co	240.7		0.19
	345.4	0.03	
Cr	357.9	0.2	0.22
	425.4	0.10	0.6
Cs	852.1	0.008	0.05
Cu	324.8	0.6	0.1
	327.4	2.0	0.2
Fe	372.0	0.03	0.5
Ga	417.2	0.02	3.7
Ge	265.2	1	2
Hg	253.6	2.5	0.001 (cold vapor)
In	410.2	0.14	2.6
K	766.5	0.0005	0.005
LaO	791	0.005	
Li	670.8	0.00003	0.0006
Mg	285.2	0.2	0.008
Mn	279.5	0.02	0.06
	403.1	0.005	0.6
Na	589.0	0.0005	0.002
Nb	405.6	1	28
Ni	232.0		0.16
	341.5	0.03	0.57
HPO	525	13	
Pb	283.3	6	0.6
Pd	244.8		0.3
	340.5	0.1	
Rb	780.0	0.001	0.005
Re	346.1	2.5	0.7
S_2	394	3	
Sb	217.6	13	0.6
	259.8	0.6	
Se	196.0		0.5
Si	251.6	4	2
Sn	225.5	5	1.6

TABLE 10.3   Useful Flame Emission and Atomic Absorption Lines (*Continued*)

Element	Wave-length, nm	Flame emission, detection limit, $\mu g \cdot mL^{-1} \cdot (0.1\ mV)^{-1}$	Atomic absorption sensitivity, $\mu g \cdot mL \cdot (1\%\ Abs)^{-1}$
Sr	460.7	0.09	0.06
Te	214.3	7	0.4
	238.6	2	43
Tl	377.6	0.02	0.03
V	318.4	0.4	0.4
Zn	213.9	50	0.025

Source: J. A. Dean and T. C. Rains, eds., *Flame Emission and Atomic Absorption Spectrometry,* Vol. 3, Dekker, New York, 1975.

line-emission spectrum. For an atom or molecule in the ground electronic state to be excited to a higher electronic energy level, it must absorb energy from the flame via thermal collisions with the constituents of the partially burned flame gases. Upon their return to a lower or ground electronic state, the excited atoms and molecules emit radiation characteristic of the sample components.

Band spectra arise from electronic transitions involving molecules. For each electronic transition there will be a whole suite of vibrational levels involved. This causes the emitted radiation to be spread over a portion of the spectrum. Band emissions attributed to triatomic hydroxides (CaOH) at 554 nm and monoxides (AlO, strongest band at 484 nm) are frequently observed and occasionally employed in FES. The boron oxide system gives very sensitive bands at 518 and 546 nm.

10.5.2 Instrumentation

The number of atoms in an excited state, or of molecules excited, is extremely small. The emitted radiation passes through a monochromator that isolates the specific wavelength for the desired analysis. A photodetector measures the radiant power of the selected radiation which is correlated with the concentration of analyte in the sample and in standards. FES requires a monochromator capable of providing a bandpass of 0.05 nm or less in the first order. The instrument should have sufficient resolution to minimize the flame background emission and to separate atomic emission lines from nearby lines and molecular fine structure. A monochromator equipped with a laminar-flow burner for flame AAS serves equally well for FES. However, in FES it is often desirable to scan a portion of the spectrum.

10.6 ATOMIC FLUORESCENCE SPECTROMETRY

In atomic fluorescence spectrometry (AFS) the exciting source (hollow-cathode or electrodeless discharge lamp) is placed at right angles to the flame (acetylene–nitrous oxide) and the optical axis of the spectrometer. Some of the incident radiation from the light source is absorbed by the free atoms of the analyte which are formed by the thermal energy of the flame. The atoms are raised to the excited state that corresponds to the origin of the emission line of the source. Immediately after this absorption, energy is released as the excited atoms return to the ground state.

Instrumentation resembles that required for AAS except that the light source is placed at right angles to the flame and optical axis of the spectrometer. In principle the fluorescence is measured against a zero background of the flame; in practice there may be a slight amount of scattered primary radiation from particulates in the flame gases. Use of a modulated source and ac detection lessens this problem.

The intensity of the fluorescence is linearly proportional to the exciting radiation flux, thus explaining the need for an intense source. The AFS technique is linear to the analyte concentration over several orders of magnitude and allows simultaneous multielement analyses with little or no spectral interferences. The marriage of the argon inductively coupled plasma as an exceedingly hot atomization system and pulsed hollow cathode lamps as narrow line excitation sources is the heart of a commercial instrument for AFS.

10.7 ELECTROTHERMAL TECHNIQUE

Electrothermal (nonflame) atomizers offer several attractive features that complement flame AAS.

1. Only small amounts (10^{-8} to 10^{-11} g absolute) of analyte are required.

2. Solids can be analyzed directly, often without any pretreatment.

3. Small amounts of liquid samples, 5 to 100 μL, are needed.

4. Background noise is very low.

5. Sensitivity is increased because the production of free analyte atoms is more efficient than with a flame.

On the negative side, matrix effects are usually more severe, and precision, typically 5 to 10 percent, compares unfavorably with that of AAS and FES.

10.7.1 Principle

After insertion of the sample into the electrothermal atomizer, a heating sequence is initiated to take the sample through three steps: dry, ash or char, and atomize.

1. In the drying cycle, the sample is heated for 20 to 30 s at 110°C to evaporate any solvent or extremely volatile matrix components.

2. The ash or char step is performed at an intermediate temperature (often 500°C) to volatilize higher-boiling matrix components and to pyrolyze matrix materials that will crack and carbonize. The latter would be fats and oils. Loss of analyte may occur if the ashing temperature is too high or is maintained for too long. Recommended ash temperatures for selected elements are given in Table 10.4.

3. In the atomization step, maximum power is applied to raise the furnace temperature as quickly as possible to the selected atomization temperature or the

TABLE 10.4   Ash Temperatures with Graphite Furnace

Element	Recommended ash temperature, °C	Ash temperature with Pd modifier, °C
Antimony	800	1400
Arsenic	800*	1500
Bismuth	500	1100
Cadmium	300	550
Chromium	1300	1300
Cobalt	900	1200
Copper	900	1100
Gold	700	1100
Iron	800	1300
Lead	400	1000
Manganese	800	1200
Mercury	120	450
Nickel	900	1200
Selenium	700*	1100
Silver	500	950
Tellurium	500	1300
Tin	800	1300
Thallium	400	1500
Zinc	400	900

*Nickel modifier added.

maximum furnace temperature. The analyte residue is volatilized and dissociated into free atoms which will absorb light from the AAS source. The transient absorption signal must be measured rapidly.

10.7.2   Instrumentation

Three components make up an electrothermal atomizer: the workhead, the power unit, and the controls for the inert gas supply. A cross section of a heated graphite atomizer is shown in Fig. 10.10. It consists of a hollow graphite cylinder 28 mm long and 8 mm in diameter whose interior is coated with pyrolytic graphite. Electrodes at each end of the cylinder are connected to a power supply. A metal housing surrounding the furnace is water-cooled to allow the entire unit to be rapidly restored to ambient temperature after each run. Inert gas, usually argon, enters the cylinder at both ends and exits through the sample introduction port at the center of the cylinder. The gas flow removes the matrix components vaporized during the ashing step (to prevent subsequent vaporization during the atomization step and thereby a large background absorption signal) and prevents oxidation of the graphite cylinder during the heating cycles.

Liquid samples are introduced with a microsyringe through the small opening in the top of the cylinder and placed on a thin graphite plate or platform which is located within the cylinder (Fig. 10.11). Care must be taken not to scratch the pyrolytic graphite surface. Solid samples can be introduced through the end of the cylinder on a microdish made of tungsten. The sample is heated by radiation from the walls of the cylinder which enables the walls and vapor to reach a steady-state temperature before the sample is atomized.

A miniature version of the electrothermal analyzer is the carbon rod atomizer

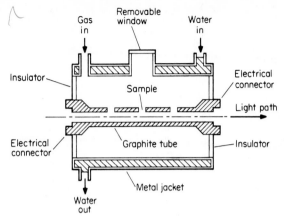

FIGURE 10.10 Cross section of a heated graphite atomizer. (*Courtesy of Perkin-Elmer Corp.*)

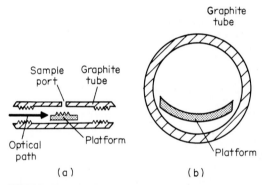

(a)   (b)

FIGURE 10.11 (*a*) The modified L'vov platform. Side view of the platform position within the graphite tube. Tube dimensions, 28 mm by 6 mm (i.d.). Platform dimensions, 7×5 mm. (*b*) The modified L'vov platform, end view. [*After S. R. Koirtyohann and M. A. Kaiser, Anal. Chem., 54:1518A (1982). Copyright (1982) American Chemical Society.*]

(Fig. 10.12). A tube or cup unit is supported between two graphite electrodes whose outside ends are inserted in water-cooled terminal blocks. The tube is 9 mm long and 3 mm in diameter; sample capacity is 10 μL for smooth tubes. The tube can be replaced with a vertical cup for solid samples or samples that require preliminary chemical treatment performed directly in the cup.

10.7.3 Matrix Modifier

Palladium(II) nitrate is a useful chemical modifier which will stabilize many elements to temperatures several hundred degrees higher than those possible other-

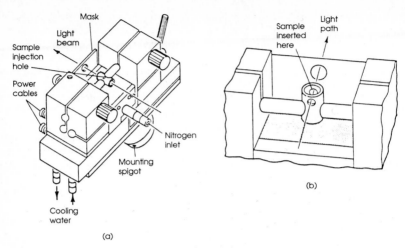

FIGURE 10.12 Carbon rod atomizer: (*a*) horizontal rod version and (*b*) vertical cup version. (*Courtesy of Varian Associates, Inc.*)

wise (Table 10.4). Palladium modifies the analyte element by converting it to a more thermally stable species. This lessens analyte volatility and permits a higher temperature for the ash step which more efficiently removes bulk matrix constituents without loss of analyte element. Thus palladium reduces signal dependency on the sample composition and produces greater consistency in absorption analyte signals. The greatest temperature shifts are achieved for these elements: As, Bi, Ga, Ge, P, Pb, Sb, Se, Te, and Tl.

The palladium concentration is an important parameter, requiring careful investigation in methods development. It should be added as the nitrate (not as chloride). For relatively clean samples, lower levels of palladium (50 to 200 mg $\cdot$ L^{-1}) may be sufficient to shift analyte atomization away from the sample constituents and to cause background and chemical interferences. For more complex samples, higher concentrations of palladium (200 to 1000 mg $\cdot$ L^{-1}) may be necessary. If analyte signals produce late, broad, and irregular peaks, a higher atomization temperature may also be necessary. Hydrogen gas (5% hydrogen in 95% argon) is introduced during the ramp to the ash step from the dry step to assist in producing metallic palladium. Argon is then used for the ash and atomization steps. Magnesium nitrate (1000 mg $\cdot$ L^{-1}) has been combined with palladium(II) nitrate and hydrogen gas for the determination of a variety of elements.

10.7.4 Precautions

The proper temperature and timing parameters must be carefully selected for each step of the electrothermal process. The progress of each step should be observed by monitoring the absorption signal without background correction.

In the drying step the evaporation of solvent must be smooth and gentle. Foaming and splattering will result in loss of analyte. If observed, decrease the heating rate.

In the ashing step, most organic materials pyrolyze at around 350°C. If a residue of amorphous carbon (resembles a spider web in appearance) is formed, a stream of air or oxygen may be introduced into the furnace to convert the carbon into carbon dioxide. Low-boiling elements, such as lead, may be partially lost if this step is prolonged unnecessarily or the final temperature is too high.

The atomization step must be rapid. The analyte should be converted into free atoms almost instantly so that the absorption peak is sharp.

Not every AAS spectrometer possesses a response time adequate to measure the transient absorption signal from an electrothermal analyzer. Lifetime of the free atoms in the optical path is 0.01 s or less.

Operational life of the graphite cylinder surface or tube is finite. A pyrolytic graphite coating prolongs the life and also prevents the sample from soaking into ordinary graphite. Formation of metal carbides by some elements is also prevented.

10.8 CHEMICAL VAPORIZATION

For elements (As, Ge, Sb, Se, Sn, and Te) that form volatile hydrides, the samples can be pretreated with sodium borohydride, dispensed in pellet form, as the reducing agent is added to a 1 to 4 M HCl solution. The gaseous hydride is released and injected into the atomizer in a stream of inert gas. Either a hydrogen-air flame and conventional AAS instrument or a low-temperature tube-type quartz furnace is used to generate the gaseous free metal atoms.

Mercury compounds need only be heated to generate mercury vapor, although chemical reduction is an alternative.

10.9 BACKGROUND CORRECTION TECHNIQUES

Correction for background associated with flames and matrix components is a serious problem in FES and AAS methods. Background radiation from the flame itself arises from hydrogen molecules, OH radicals, and partially burned fuel and solvent molecules. These species plus continua from metals and metal oxides and hydroxides constitute the flame background.

The type of background correction technique used will depend on the element to be determined, the matrix, the desired accuracy, and the availability of a given system.

10.9.1 Background Correction in FES

10.9.1.1 Scanning Methods. A somewhat tedious correction method involves scanning the spectrum on both sides of the analyte emission line (or band). A baseline is drawn beneath the emission line from the background (extrapolated).

After a preliminary wavelength scan has been made, the line plus background signal is measured for a group of samples and standards at the wavelength of the emission line. The wavelength setting is moved to a suitable position on the background only, first on one side of the emission and then on the other. The average reading is subtracted from the line plus background reading. Computer-controlled equipment will make these measurements automatically.

10.9.1.2   Wavelength Modulation.   In one technique a refractor plate which is made to oscillate about the vertical axis of the light beam is placed inside a monochromator after the entrance slit or before the exit slit. The optical beam is refracted and will cause a small oscillating displacement (and therefore spectral distribution) of the light beam leaving the monochromator. Thus, the analyte line plus background and adjacent background are measured alternately and their signals subtracted. Wavelength modulation measures the intensity of a weak line superimposed on an intense and continuum background.

10.9.2   Background Correction in AAS

Background absorption occurs in AAS as molecular absorption and as light scattering by particulate matter (unevaporated droplets, but more likely unevaporated salt particles remaining after desolvation of the aerosol). (Light scattering in AFS would appear as a signal that could not be distinguished from the analyte signal.) The absorption is usually broad in nature, compared to the monochromator band pass and line-source emission widths. The analyte signal could be on a sloping or level background. Background correction is often required for elements with resonance lines in the far-ultraviolet region, and is essential to achieve high accuracy in determining low levels of elements in complex matrices. Electrothermal atomization methods invariably require correction, and the background may change with time.

10.9.2.1   Use of a Continuum Source.   This commonly used form of background correction uses a deuterium lamp for elements in the range of 180 to 350 nm and a tungsten-halide lamp for elements in the range of 350 to 800 nm. Both lamps are readily available. The continuum source is aligned in the optical path of the spectrometer so that light from the continuum source and light from the primary lamp source are transmitted alternately through the flame (or axis of the electrothermal analyzer tube). The continuum source has a broadband emission profile and is only affected by the background absorption whereas the primary lamp source is affected by background absorption and absorption resulting from the analyte in the flame. Subtraction of the two signals (electronically or manually) eliminates or at least lessens the effect of background absorption. This method does not correct for problems of scatter or spectral interferences. Highly structured molecular absorbance causes incorrect results. It is difficult to match the reference and sample beams exactly because different geometries and optical paths exist between the two beams.

10.9.2.2   Smith-Hieftje Method.   The Smith-Hieftje system utilizes the phenomenon of self-reversal which will occur when the light source is operated at high currents. In practice a hollow cathode lamp line source is cycled at periods of low current for several milliseconds and then cycled at high current for several hundred microseconds. The time must be precisely controlled so that the same number of photons reaches the detector during each phase of the cycle. At high currents the absorption signal caused by the analyte atoms is eliminated through self-reversal, but the background is not changed. The background signal can then be subtracted electronically to produce the corrected signal. Optical alignment is simple and matching of two light sources is not required. Special hollow cathode lamps must be employed to avoid burnout at high currents and not every hollow cathode lamp (for some elements) will produce a usable self-reversal effect.

10.9.2.3 Zeeman Effect Method. When under the influence of a magnetic field, the absorption line splits into three components for the normal Zeeman effect, with the wavelength of one component coinciding with the resonance emission line from the light source and other components shifting to shorter and longer wavelengths, respectively. When the light beam from the lamp is passed through a polarizer rotated so that the light is parallel to the central component, the emission line from the lamp is absorbed by any analyte atoms present in the flame, whereas the two wing components are unaffected. As the polarizer is rotated to the perpendicular position, there is no absorption by the atomic vapor of the sample. However, light scattering and broad band molecular absorption are measured in both polarizer configurations. Therefore, by using parallel emission lines as the sample beam and perpendicular emission lines as the reference beam, electronic subtraction of the two absorbances will produce the true absorbance of the analyte.

The Zeeman effect can be produced in the light source or in the atomic vapor (flame or furnace). The magnet placement around the light source imposes no restrictions on the atomizer but requires hollow cathode lamps that are designed to operate in a magnetic field. Baseline drift may be greater than when the magnet is placed around the atomizer. In the latter configuration, the magnetic system must be compatible with the atomizer and will be bulkier. However, the same wavelength is used for the signal and the reference beam, which minimizes spectral interference.

Spectral interferences may be overcome if lines are 0.02 nm or more apart. Analytical calibration curves may be double-valued, that is, two widely different concentrations may give the same absorbance. The polarizer contributes to light loss. Of the three correction systems, it is the most costly.

BIBLIOGRAPHY

Dean, J. A., *Flame Photometry,* McGraw-Hill, New York, 1960.

Dean, J. A., and T. C. Rains, eds., *Flame Emission and Atomic Spectrometry: Theory,* Vol. 1, 1969; *Components and Techniques,* Vol. 2, 1971; *Elements and Matrices,* Vol. 3, 1975, Dekker, New York.

Ebdon, L., *An Introduction to Atomic Absorption Spectroscopy—A Self Teaching Approach,* Heydon, Philadelphia, 1982.

Mayer, B., *Guidelines to Planning Atomic Spectrometric Analysis,* Elsevier, New York, 1982.

Ottaway, J., and A. Ure, *Practical Atomic Absorption Spectroscopy,* Pergamon, New York, 1983.

CHAPTER 11
ATOMIC EMISSION SPECTROSCOPY WITH PLASMA AND ELECTRICAL DISCHARGES

11.1 INTRODUCTION

⟨Atomic emission spectroscopy that utilizes high-temperature atomization sources is a technique for simultaneously determining the concentration of about 70 elements in inorganic and organic matrices⟩This capability is achieved with plasma and electric discharges but with more spectral interferences and without the relative simplicity of atomic absorption spectrometry.⟨A minute amount of sample, often as little as a milligram, is vaporized and thermally excited to the point of atomic emission. For a quantitative determination, 20 mg of inorganic solid sample is usually adequate.⟩

Emission spectroscopic techniques can be applied to almost every type of sample. Areas of investigation include chemicals, minerals, soils, metals and alloys, plastics, agricultural products, foodstuffs, and water analysis. With preconcentration of the sample, most elements can be determined at the low parts-per-billion level (Table 11.1).

11.1.1 Principle

When a substance is excited by a plasma or electric discharge (arc or spark), elements present emit light at wavelengths that are specific for each element. The light emitted is dispersed by a grating or prism monochromator.⟨The spectral lines produced are recorded either on a photographic plate or, in modern systems, by diode arrays or photomultiplier tubes linked directly to computer-driven data processing systems.⟩

Table 11.1   Comparative Detection Limits in $\mu g \cdot L^{-1}$*

Element	ICP-AES	Flame AAS	Plasma AFS
A. Nonrefractory metals			
Ag	3	3	<0.1
As	30	250	10
Au	10	10	0.4
Bi	40	60	7
Ca	0.05	3	<0.1
Cd	2	0.6	<0.1
Co	7	7	0.4
Cr	4	5	0.4
Cu	3	5	0.2
Fe	3	5	0.4
Hg	20	200	3
In	40	50	3
Mg	0.1	0.3	<0.1
Mn	1	3	0.2
Ni	8	8	0.3
Pb	30	20	5
Pd	30	20	2
Sb	30	70	6
Se	60	1000	5
Sr	0.2	6	0.3
Te	40	150	2
Tl	40	30	10
Zn	2	1	<0.1
B. Alkali metals			
K	60	3	0.6
Li	3	2	<0.1
Na	20	0.4	<0.1
C. Refractory metals			
Al	20	40	4
B	2	2000	80
Ba	0.4	20	25
Be	0.2	2	0.5
Ge	40	150	20
Mo	5	40	7
Si	10	300	50
Sn	25	80	20
Ti	2	100	25
V	4	100	20
W	20	1000	200
Y	2	500	50

*Signal-to-noise = 2, 20-s integrations.
Source: Courtesy of Baird Corporation.

11.2 INSTRUMENTATION

Instrumentation for atomic emission spectroscopy (AES) involves these components:

1. The sampling device and source, often as an integral unit
2. The spectrometer
3. The detector and readout device

11.3 SAMPLING DEVICES AND SOURCES

A wide variety of sources are needed to handle the diverse sample matrices. For solid samples, arc excitation is more sensitive, while spark sources are more stable. Plasma sources are the choice for solutions and gaseous samples.

11.3.1 Direct-Current Arcs

The dc arc consists of a high-current low-voltage discharge between two solid electrodes, one of which may be the sample or an electrode supporting the sample while the other is the counter electrode. In the United States the anode is generally the sample-containing electrode, whereas in Europe it is the cathode. High-purity graphite is a popular electrode material because it resists attack by strong acids or redox reagents and, being highly refractory, permits the volatilization of high-boiling sample components. Its emission spectrum contains few lines.

Several electrode configurations are shown in Fig. 11.1. Conducting samples are usually ground flat and used as one electrode with a pointed graphite counter electrode. Powdered samples are mixed with graphite and pressed into a pellet or placed in a pedestal holder. Metal samples can be ground to a point and used with a graphite counter electrode.

Because of its high sensitivity the dc arc is well suited for qualitative survey analyses of both trace and major elements or for semiquantitative analyses. An internal standard will minimize the effects of arc instability and the sample matrix. The element selected as the internal standard must have vaporization and excitation characteristics that closely match the element being determined. The tendency of the dc arc to wander around the sample surface seriously affects quantitative precision, unless a total burn is employed; on the other hand, it offers excellent qualitative surveys.

11.3.2 Alternating-Current Arcs

The ac arc momentarily stops at the end of each half-cycle and then reignites during the next half-cycle when the applied voltage exceeds the voltage required for the dielectric breakdown of the gas between the electrodes. The sampling spot moves rapidly about on the electrode surface (120 times per second for a 60-Hz

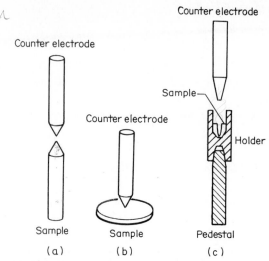

FIGURE 11.1 Electrode configurations: (*a*) point-to-point, (*b*) point-to-plane, and (*c*) carrier distillation.

source). Consequently, reproducibility (precision) is superior to that of the dc arc; however, the sensitivity is less.

11.3.3 High-Voltage Alternating-Current Spark

Although not as sensitive as the dc arc source, the high-voltage, ac spark provides the greatest precision and stability of all the electric discharge sources. It also has a sampling spot that moves rapidly about on the electrode surface. The ac spark is the method of choice for analyses of ferrous metals in industrial operations.

11.3.4 Inductively Coupled Plasma Sources

The inductively coupled argon plasma (ICAP or ICP) torch derives its sustaining power from the interaction of a high-frequency magnetic field and ionized argon gas. The schematic configuration of the ICAP torch is shown in Fig. 11.2. The plasma is formed by a tangential stream of argon gas flowing between the outer two quartz tubes of the torch assembly. Radio-frequency power (27 to 30 MHz, 2 kW) is applied through the induction coil. This sets up an oscillating magnetic field. Power transfer between the induction coil (two-turn primary winding) and the plasma (one-turn secondary winding) is similar to a power transfer in a transformer. An eddy current of cations and electrons is formed as the charged particles are forced to flow in a closed annular path. The fast-moving ions and electrons collide with more argon atoms to produce further ionization and intense thermal energy as they meet resistance to their flow. A long, well-developed tail emerges from the flame-shaped plasma which forms near the top of the torch and

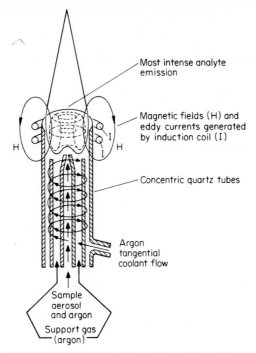

Most intense analyte emission

Magnetic fields (H) and eddy currents generated by induction coil (I)

Concentric quartz tubes

Argon tangential coolant flow

Sample aerosol and argon

Support gas (argon)

FIGURE 11.2 Schematic configuration of an induction-coupled argon plasma torch.

above the induction coils. This tail is the spectroscopic source. Temperatures in the plasma range from 6000 to 10 000 K.

The argon gas stream is initially seeded with free electrons from a Tesla discharge coil. These seed electrons quickly interact with the magnetic field of the coil and gain sufficient energy to ionize argon atoms by collisional excitation.

A second stream of argon provides a vortex flow of gas to cool the inside quartz walls of the torch. This flow also serves to center and stabilize the plasma. The sample is injected as an aerosol through the innermost concentric tube of the torch through the center of the plasma. An ICAP source possesses several advantages.

1. The analytes are confined to a narrow region.
2. The plasma provides simultaneous excitation of many elements.
3. The analyst is not limited to analytical lines involving ground-state transitions but can select from first or even second ionization state lines. For the elements Ba, Be, Fe, Mg, Mn, Sr, Ti, and V, the ion lines provide the best detection limits.
4. The high temperature of the plasma ensures the complete breakdown of chemical compounds (even refractory compounds) and impedes the formation of other interfering compounds, thus virtually eliminating matrix effects.

5. The ICAP torch provides a chemically inert atmosphere and an optically thin emission source.

6. Excitation and emission zones are spatially separated; this results in a low background. The optical window used for analysis lies just above the apex of the primary plasma and just under the base of the flamelike afterglow.

7. This low background, combined with a high signal-to-noise ratio of analyte emission, results in low detection limits, typically in the parts-per-billion range.

11.3.5 Direct-Current Plasma Sources

In the three-electrode direct-current plasma (DCP) source, the argon plasma jet is formed between two carbon anodes and a tungsten cathode in an inverted Y configuration (Fig. 11.3). The sample excitation region and observation zone are centered in the crook of the Y where the temperature is approximately 6000 K. The plasma is initiated by moving the three electrodes into contact with argon-driven pistons and then withdrawing them. Samples are nebulized and the aerosol is introduced into the excitation region of the plasma. The plasma is stable in the presence of aerosols that contain large amounts of dissolved solids, organic solvents, and high concentrations of acids or bases.

As with the ICAP, the emission lines of atoms and ions are observed in a region separated from the main plasma. This leads to a high signal-to-background ratio and low detection limits. However, several limitations exist with the DCP source:

1. The electrodes are consumed and require reshaping after 2 h of continuous operation.

2. The excitation characteristics are greatly affected by high concentrations of easily ionized elements such as Na and Ca.

3. The DCP source cannot be incorporated into totally automated systems.

4. The relatively high consumption rates of argon pose a cost consideration.

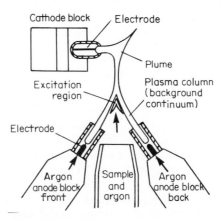

FIGURE 11.3 Schematic of the dc argon plasma source in the Y configuration.

11.4 ATOMIC EMISSION SPECTROMETERS

High-resolution high-luminosity monochromators are necessary to isolate a spectral line from its background without loss of radiant power. Both concave and plane diffraction gratings are used as dispersive elements.

11.4.1 Direct-Reading Instruments

The schematic diagram of a direct-reading (nonscanning) spectrometer is shown in Fig. 11.4. It has a holographic concave grating mounted in a Rowland circuit configuration with a focal length of 1 m. In this configuration the entrance slit, grating, and focal plane lie on the circumference of the Rowland circle (which has a radius of curvature half that of the grating). An array of slits with mirrors projects and focuses the selected spectral lines onto the cathodes of the photomultiplier tubes.

Direct readers are usually built for a specific analytical requirement by care-

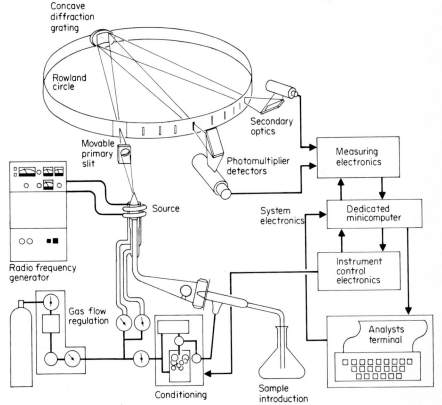

FIGURE 11.4 Schematic diagram of a direct-reader spectrometer using a holographic concave grating in the Rowland circle configuration. (*Courtesy of Applied Research Laboratories.*)

fully positioning the exit slits (and accurate control of the slit width) and photomultiplier tubes to measure specific lines of elements of interest. Instruments generally measure from 8 to 24 elements in a single matrix. The array of receiver systems are often grouped into several bridges, one for each type of matrix, so that the instrument can be used to analyze several types of samples successively.

Integration times are from 25 to 40 s. The radiant power of the spectral line is converted by the photomultiplier tube into current that is used to charge a capacitor-resistor circuit. In integration with constant time, a voltage-to-frequency circuit converts the capacitor voltage into electrical pulses of equal height which are counted to give a number proportional to the radiant power. The output is in digital form and ready for computer processing.

The spectrophotometers are calibrated with a high- and low-concentration standard. The observed line intensities are related directly to concentrations. Typically, sample analysis, data output, and any associated data management functions are completed within 2 min after sample introduction.

11.4.2 Plane-Grating Spectrophotometers

Almost all scanning spectrometers use a plane grating in an Ebert mounting. In this mounting light of all wavelengths is brought to a focus on the detector without changing the detector-to-mirror distance. Wavelengths are easily changed by simply rotating the grating by a computer-controlled stepper motor. This permits the programmed selection of desired wavelengths and the sequential examination of spectral lines. The computer also checks the analytical parameters for each line analyzed (wavelength, slit width), makes the necessary background correction, performs required calculations, and prints the results.

If the monochromator is enclosed in a vacuum, lines of sulfur, phosphorus, and boron can be used.

The spectral lines may be recorded on a photographic plate (Fig. 11.5). The location and intensity of the lines produced by the sample are compared either visually or by means of a photoelectric densitometer with the lines produced by suitable standards.

11.5 ICP ATOMIC FLUORESCENCE SPECTROSCOPY

The use of an inductively coupled plasma as an atomization cell for the observation of atomic fluorescence results in a technique that eliminates most spectral interferences and background shifts often associated with other spectroscopic techniques. The block diagram for an ICP atomic fluorescence spectrometer is shown in Fig. 11.6. An array of up to 12 interchangeable element modules surrounds an ICP. Each optical module, specific for one element, is composed of a specially designed hollow cathode lamp, an appropriate optical filter, and a photomultiplier detector. During a sample measurement, each module's hollow cathode lamp is pulsed at approximately 500 Hz in sequence around the array. Only one element is being excited, with its resultant fluorescence emission being detected at that time.

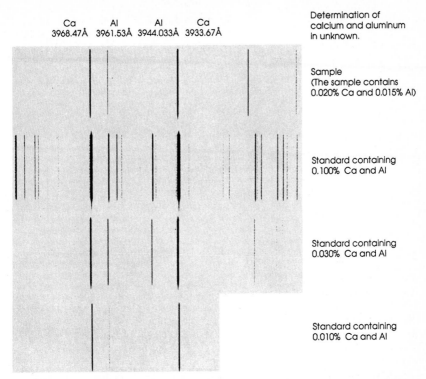

FIGURE 11.5   Location and intensity of spectral lines compared with two standard samples. Lines are the singly ionized doublet of calcium and the atomic doublet of aluminum.

⟨ Light from a given hollow cathode lamp is directed into the central channel of the ICP plasma. Part of the atomic fluorescence produced is collected by the optical module. The instrument recognizes the element module installed and its position in the array, and automatically sets a recommended lamp current.⟩

Introduction of standard and sample solutions into the plasma is accomplished by means of a cross-flow nebulizer mounted on a a glass-horn spray chamber. Solutions can be aspirated directly or fed to the nebulizer by a peristaltic pump.

⌐The atomic fluorescence spectra are simple with no overlapping spectral lines; thus they do not require extensive computer processing, expensive monochromators, or complex scanning mechanisms for correction or isolation of spectral lines. The plasma serves solely as an atomization cell to produce ground-state atoms. Excitation energy comes from an external light source; the fluorescence (reemission) is viewed out of the path of the optical beam. Atomic fluorescence occurs only at the most sensitive atomic resonance lines. The resolution achieved surpasses that of even an echelle emission spectrometer because the effective spectral bandwidth is equivalent to the width of the absorption transition in the plasma. A number of detection limits are listed in Table 11.1.

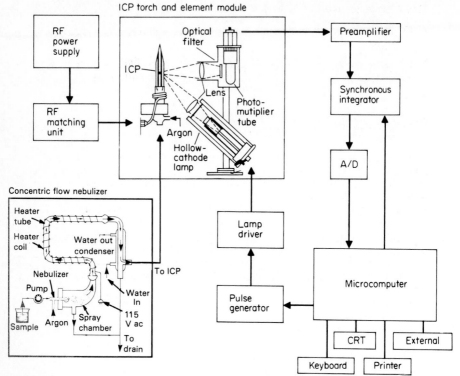

FIGURE 11.6　Block diagram for ICP atomic fluorescence spectrometer.　(*Courtesy of Baird Corp.*)

11.6　COMPARISON OF AAS AND ICP METHODS

From among the several methods discussed in Chaps. 10 and 11, AAS and ICP dominate atomic spectroscopy. What are the relative merits of these two methods?

11.6.1　Advantages of ICP over AAS

1. Large dynamic concentration range (up to 10^5) prevails for many elements.
2. Several elements in a sample can be simultaneously determined without the need for repeated aspirations, adjustment of instrument parameters, and proper tracking of the samples.
3. Sensitivities are better at high concentrations.

11.6.2   Advantages of AAS over ICP

1. Instrumentation is much simpler and less expensive by a factor of one-half to one-fifth.

2. For most elements, ICP cannot match the detection limits attainable with furnace or fluorescence AAS, particularly at low concentrations.

BIBLIOGRAPHY

Barnes, R., ed., *Emission Spectroscopy*, Halsted, New York, 1976.

Boumans, P. W. J. M., ed., *Inductively Coupled Plasma Emission Spectroscopy*, Wiley, New York; Part 1: Methodology, Instrumentation, and Performance, 1987; Part 2, Applications and Fundamentals, 1987.

Keliher, P., W. J. Boyko, R. H. Clifford, J. L. Snyder, and S. F. Zhu, *Anal. Chem.*, 58(5), 334R (1986).

Montaser, A., and D. Golightly, *Inductively Coupled Plasmas in Analytical Atomic Spectroscopy*, VCH Publishers, Deerfield Beach, Fla., 1987.

Sacks, R., "Emission Spectroscopy," *Treatise on Analytical Chemistry*, 2d ed., P. J. Elving, E. J. Meehan, and I. M. Kolthoff (eds.), Chap. 6, Part I, Vol. 7, Wiley, New York, 1981.

Walsh, M., and M. Thompson, *A Handbook of Inductively Coupled Plasma Spectroscopy*, Methuen, New York, 1983.

CHAPTER 12
X-RAY METHODS

X-ray methods involve the excitation of an atom by the removal of an electron from an inner energy level, usually from the innermost K level or from one of the three L levels. Atoms can be excited either by direct bombardment of the sample with electrons (direct emission analysis, electron probe microanalysis, and Auger emission spectroscopy) or by irradiation of the sample with X rays of shorter wavelength (X-ray fluorescence analysis). Electron spectroscopy for chemical analysis (ESCA) measures the energy of the electrons ejected from inner electron levels when the sample is bombarded by a monochromatic X-ray beam. In X-ray absorption the intensity of an X-ray beam is diminished as it passes through material. X rays are also diffracted by the planes of a crystal which provides a useful method for qualitative identification of crystalline phases.

12.1 PRODUCTION OF X RAYS AND X-RAY SPECTRA

When an atom is bombarded by sufficiently energetic electrons or X radiation, an electron may be ejected from one of the inner levels of the target atoms. The place of the ejected electron is promptly filled by an electron from an outer level whose place, in turn, is taken by an electron coming from still farther out. Each transition is accompanied by the release of an X-ray photon, whose energy is characteristic of the element from which it originated. The measurement of the various photon energies produced by sample excitation provides a means of identifying its constituent elements. A count of the photons provides the quantification of each element.

Electron bombardment of the anode in an X-ray tube causes the emission of both a continuum of radiation and the characteristic emission lines of the anode material. When incident electrons (or X-ray photons) of sufficient energy impinge upon an atom, an electron from an inner shell may be photoejected from that atom. The energy required to initiate photoejection is called the absorption-edge energy. The absorption edge will be a sharp discontinuity in the plot of mass absorption coefficient versus wavelength.

There is successive ionization—first of electrons in the outermost levels of the sample or target, then of electrons in the L levels as the three L absorption edges are progressively exceeded, and finally in the K level as the K absorption edge is exceeded (Fig. 12.1).

The energy required to lift a K electron out of the environment of the atom

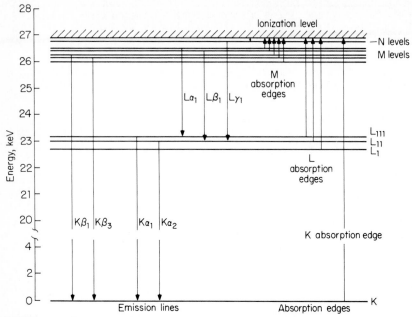

FIGURE 12.1   Energy-level diagram of cadmium for X-ray transitions.

(exceed the ionization limit) must exceed the energy of the K absorption edge. The relationship between the voltage applied across an X-ray tube (or the energy of incident X radiation, in volts) and the wavelength λ, in angstroms (Å), is given by the Duane-Hunt equation:

$$\lambda = \frac{hc}{eV} = \frac{12\ 393}{V} \tag{12.1}$$

where V = X-ray tube voltage, V
   e = charge on electron
   h = Planck's constant
   c = velocity of light

Following spectroscopic selection rules, electrons from outer shells (L and M) will undergo transitions to fill the K-shell vacancy. In so doing, they may emit an X-ray photon. (In a competing process, the energy released may be internally converted in the atom to cause the ejection of a secondary, or Auger, electron. Auger spectroscopy is discussed later.) The energy of the emitted radiation will be characteristic of the element and of the particular transition.

The wavelength of an absorption edge is always shorter than that of the corresponding emission lines. For example, in energy units, the $K\alpha_1$ line represents the difference: K edge minus L_{III} edge. By contrast, the K edge is the difference between the K energy level and the ionization limit. Several characteristic wavelengths of absorption edges and emission lines for selected elements are given in Table 12.1.

The characteristic K or L emission lines (or absorption edges) of each element vary in a regular fashion from one element to another. The characteristic wave-

TABLE 12.1 Characteristic Wavelengths of Absorption Edges and Emission Lines for Selected Elements

Element	Minimum potential for excitation of K lines, kV	K absorption edge, Å	$K\beta$, Å	$K\alpha_1$, Å	L_{III} absorption edge, Å
Magnesium	1.30	9.512	9.558	9.889	247.9
Aluminum	1.559	7.951	7.981	8.337	170
Chromium	5.988	2.070	2.085	2.290	20.7
Manganese	6.542	1.896	1.910	2.102	19.40
Cobalt	7.713	1.608	1.621	1.789	15.93
Nickel	8.337	1.488	1.500	1.658	14.58
Copper	8.982	1.380	1.392	1.541	13.29
Zinc	9.662	1.283	1.295	1.435	12.13
Molybdenum	20.003	0.620	0.632	0.709	4.912

Source:: J. A. Dean, ed., *Lange's Handbook of Chemistry*, 13th ed., McGraw-Hill, New York, 1985.

lengths decrease as the atomic number of the elements increases. Stated more exactly, the frequency of a given type of X-ray line increases approximately as the square of the atomic number of the element involved.

X-ray emission lines or absorption edges are quite simple because they consist of very few lines as compared to optical emission or absorption spectra observed in the visible-ultraviolet region. X-ray spectra are not dependent on the physical state of the sample or on its chemical composition, except for the lightest elements, because the innermost electrons are not involved in chemical binding and are not significantly affected by the behavior of the valence electrons.

12.2   INSTRUMENTATION

Analytically useful X radiation is generated by bombarding the sample with electrons in the range of 5 to 100 keV or with X-ray photons in a similar energy range. Instrumentation associated with X-ray emission methods, and specifically for a plane-crystal spectrometer as used in X-ray fluorescence, is shown in Fig. 12.2. The source of the X-ray photons may be an X-ray tube, a secondary target irradiated with photons from an X-ray tube, or emission from a radionuclide source. The X-ray photons are directed through a collimator onto a single analyzing crystal. The analyzing crystal acts as a diffraction grating. Scanning through the entire angular range of the goniometer permits the radiation at a particular angular position to be correlated with wavelength through the Bragg condition, which is discussed later. The detector is rotated at twice the angular rate of the analyzing crystal.

12.2.1   X-Ray Tube

The X-ray tube is a high-vacuum sealed-off unit and is shown schematically in Fig. 12.3. The target (anode), usually copper or molybdenum, is viewed from a very small glancing angle above the surface for diffraction work and at an angle of

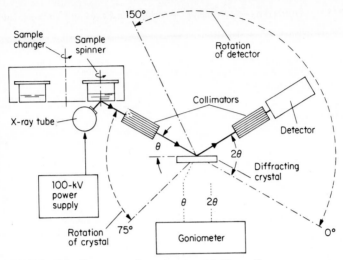

FIGURE 12.2 Geometry of a plane-crystal X-ray fluorescence spectrometer. (*Courtesy of Philips Electronic Instruments.*)

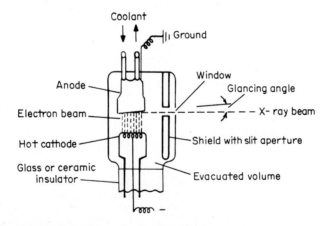

FIGURE 12.3 Schematic of an X-ray tube.

about 20° for fluorescence work. Dissipation of the generated heat is accomplished by water-cooling the target and occasionally by rotating it. The X-ray beam leaves the tube through a thin window of beryllium. For wavelengths of 6 to 70 Å, ultrathin films of aluminum or cast Parlodion are used as window material, and the equipment must be evacuated or flushed with helium.

12.2.2 Collimators and Filters

Radiation from an X-ray tube is collimated by a bundle of tubes, 0.5 mm or

TABLE 12.2 Filters for Common Targets of X-Ray Tubes

Target element	$K\alpha_1$, Å	$K\beta$, Å	Filter	K absorption edge filter, Å	Thickness,* mm	Percent loss of $K\alpha_1$
Mo	0.709	0.632	Zr	0.689	0.081	57
Cu	1.541	1.392	Ni	1.487	0.013	45

*To reduce the intensity of the $K\beta$ line to 0.01 that of the $K\alpha_1$ line.

smaller in diameter and usually a few centimeters long. Collimators are placed between the sample and the analyzer crystal and between the analyzer crystal and the detector.

The K spectra of metals used as target material contain three strong lines, $K\alpha_1$, $K\alpha_2$, and $K\beta_1$. For X-ray diffraction, the $K\beta_1$ radiation can be reduced by using a thin foil filter, usually of the element of next lower atomic number to that of the target element. In general, an element with an absorption edge at a wavelength between two X-ray emission lines may be used as a filter to reduce the intensity of the line with the shorter wavelength; the longer wavelength ($K\alpha$ in our example) lines are transmitted with a relatively small loss of intensity. Filters for the common targets of X-ray tubes are listed in Table 12.2. Filters are placed at the entrance slit of the monochromator (called goniometer in X-ray work).

12.2.3 Analyzing Crystals

The analyzing crystal takes the place of a grating in a monochromator. Virtually monochromatic X radiation is obtained by reflecting X rays from crystal planes. The relationship between the wavelength of the X-ray beam λ, the angle of diffraction θ, and the distance between each set of atomic planes of the crystal lattice d, is given by the Bragg condition:

$$m\lambda = 2d \sin \theta \qquad (12.2)$$

The range of wavelengths usable with various analyzing crystals is governed by the d spacings of the crystal planes and by the geometric limits to which the goniometer can be rotated (see Fig. 12.2). The d value should be small enough to make the angle 2θ greater than 8°, even at the shortest wavelength used. A small d value is favorable for producing a large dispersion of the spectrum to give good separation of adjacent lines. On the other hand, a small d value imposes an upper limit to the range of wavelengths that can be analyzed. The goniometer is limited mechanically to about 150° for a 2θ value. A final requirement is the reflection efficiency.

Lithium fluoride is the optimum crystal for all wavelengths less than 3 Å. Pentaerythritol and potassium hydrogen phthalate are the crystals of choice for wavelengths from 3 to 20 Å. The long-wavelength analyzers are prepared by dipping an optical flat into the film of the metal fatty acid about 50 times to produce a layer 180 molecules in thickness. Table 12.3 gives a list of crystals commonly used.

TABLE 12.3   Typical Analyzer Crystals

Crystal	Reflecting plane	Lattice spacing d, Å	Useful range, Å Maximum	Minimum	Reflectivity
Silicon	111	1.568	3.03	0.219	High
Lithium fluoride	200	2.014	3.89	0.281	High
Calcium fluoride	111	3.16	6.11	0.440	High
Pentaerythritol	002	4.371	8.44	0.610	High
Potassium hydrogen phthalate	1011	13.2	25.50	1.841	Medium

12.3 DETECTORS

12.3.1 Proportional Counters

A proportional counter, shown in Fig. 12.4, consists of a central wire anode surrounded by a cylindrical cathode. The two electrodes are enclosed in a gas-tight envelope typically filled to a pressure of 80 torr of argon plus 20 torr of methane or ethanol (or 0.08 torr of chlorine). Radiation enters a thin window of mica, about 2 to 3 mg · cm^{-2} thick. For work at very long wavelengths, typical window materials are 1-μm aluminum screen dipped in Formvar (usable for sodium and magnesium X rays) and 1-μm cast Formvar or collodion films for oxygen, nitrogen, and boron X rays.

Each ionizing particle that enters the active volume of the counter collides with the filling gas to produce an ion pair (argon cation plus an electron). Under the influence of a potential gradient of 300 to 600 V applied across the electrodes, the initial electron soon acquires sufficient velocity to produce a new pair of ions

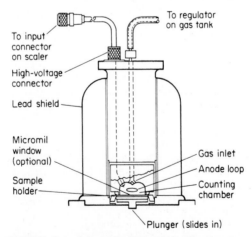

FIGURE 12.4   Schematic diagram of a flow proportional counter mounted in a shield; the very thin window is optional. The sample is inserted into the active volume by means of the lateral slide holder.

by collision with another atom of argon. This gives rise to an avalanche of electrons traveling toward the central anode. The output pulse is proportional to the number of primary pairs produced by the original ionizing particle.

The positive ions, if allowed to reach the cathode, would produce photons that would initiate a fresh discharge. Because the methane, ethanol, or halogen gas molecules have a lower ionization potential than argon, after a few collisions the ions moving toward the cathode consist of only these lower energy particles that are unable to produce photons. When the organic filling gas ions are discharged at the cathode, they decompose into various molecular fragments and eventually the quenching gas is exhausted. Counter life is limited to about 10^{10} counts. Because chlorine atoms merely recombine, a halogen-quenched counter has a life in excess of 10^{13} counts.

The dead time, the time during which the counter will not respond to an entrant ionizing particle, is about 250 ns. Count rates up to 200,000 counts per second are possible; the upper limit is imposed by the associated electronic circuitry. About 30 eV is required for the production of an ion pair in this detector.

12.3.2 Scintillation Counters

Certain substances (called scintillators, phosphors, or fluors) will emit a pulse of visible light or near-ultraviolet radiation when they are subjected to X radiation. The light is observed by a photomultiplier tube (with light-amplification stages), as shown in Fig. 12.5; the combination is called a scintillation counter. For X radiation a sodium iodide crystal doped with 1% thallium(I) is the best scintillator. When such radiation interacts with a NaI(Tl) crystal, iodine atoms are excited. Upon their return to the ground state, the reemitted ultraviolet radiation pulse is promptly absorbed by the thallium atom and, in turn, reemitted as fluorescent light at 410 nm (near the optimum wavelength response of a blue-sensitive photomultiplier tube). The crystal is sealed within an enclosure of aluminum foil which protects the crystal from atmospheric moisture and also serves as an internal reflector.

Dead time is 250 ns. Response is proportional to the energy of the X radiation; 500 eV is required to produce a photoelectron. The scintillation counter is usable throughout the important X-ray region, 0.3 to 2.4 Å, and possibly to 4 Å.

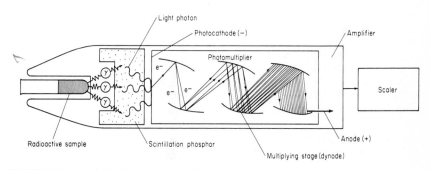

FIGURE 12.5 A scintillation counter.

12.3.3 Lithium-Drifted Semiconductor Detectors

The solid-state Ge(Li) and Si(Li) detectors can be considered as layered structures in which a lithium-diffused active region separates a p-type entry side from an n-type side. Detectors are fabricated by drifting lithium ions (a donor) into and through p-type germanium or silicon until only a layer of p-type material remains. All acceptors within the bulk material are compensated and this high-resistivity (or intrinsic) region becomes the radiation-sensitive region. Under a reverse bias of approximately 800 to 1000 V, the active region serves as an insulator with an electric field gradient throughout its volume. When X radiation enters the intrinsic region, photoionization occurs with an electron-hole pair created for each 3.8 eV of photon energy for Si(Li) and 2.65 eV of photon energy for Ge(Li). The charge produced is rapidly collected under the influence of the bias voltage. The detector must be maintained at 77 K at all times (unless extremely pure germanium crystals, a recent development, are used). The charge collected each time an X-ray photon enters the detector is converted into a digital value representing the photon energy, which is interpreted as a memory address by a computer.

Silicon semiconductor detectors are preferred for X rays longer than 0.3 Å. For shorter wavelengths, Ge detectors are necessary since Si cannot absorb the X rays effectively because the available crystals are not deep enough. The response time is about 10 ns.

12.3.4 Comparison of X-Ray Detectors

For a given amount of energy absorbed, about 10 times as many electron-hole pairs are formed as ion pairs in a gas proportional counter, and about 170 times as many electron-hole pairs as photoelectrons in NaI(Tl) scintillation counters. Since the relative resolution of a detector is proportional to the square root of the signal, the resolution of the Ge(Li) detector is about a factor of 13 better than the NaI(Tl) detector and 3 times better than a gas proportional counter.

12.4 PULSE-HEIGHT DISCRIMINATION

The measurement of pulse height provides the analyst with a tool for energy discrimination. The method is applicable whenever the amplitude of the pulse from an X-ray detector is proportional to the energy dissipation in the detector. All three types of detectors described meet this requirement.

To operate a pulse-height amplifier, adjust the amplifier to produce voltage output pulses of suitable magnitude. Then sort the pulses into groups according to their pulse heights. Set the baseline discriminator to pass only those pulses above a certain amplitude. Finally, adjust the second discriminator (variously called the window width, the channel width, or the acceptance slit) so that all pulses above the sum of the baseline discriminator and the acceptance slit are also rejected. Only the pulses within the confines of these two discriminator settings pass on to the counting stages (Fig. 12.6). A Si(Li) or Ge(Li) detector, because of its narrow pulse amplitude discrimination, can discriminate (resolve) between elements one or two atomic numbers apart. The pulse-height analyzer serves, in effect, as a secondary monochromator. This process permits the accumulation of an emission spectrum and its subsequent display on a video screen.

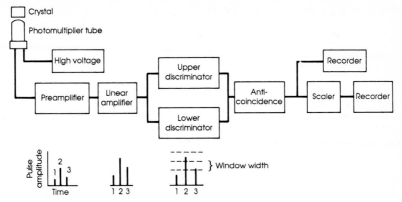

FIGURE 12.6 Block diagram of a single-channel pulse-height analyzer.

An auxiliary electronic circuit, called the pulse pile-up rejection circuit, prevents charge collection from multiple photons that enter in rapid succession.

12.5 X-RAY ABSORPTION METHODS

12.5.1 Principle

Although the experimental equipment required is complex, the measurement is straightforward. The specimen is irradiated with two or more monochromatic X rays while the incident (P_0) and transmitted (P) radiant energies are monitored. Because only one element has changed its mass absorption coefficient at an absorption edge, the following relationship pertains:

$$2.3 \log \frac{P}{P_0} = (\mu'' - \mu')W\rho x \qquad (12.3)$$

where the term in parentheses represents the difference in mass absorption coefficient at the edge discontinuity, W is the weight fraction of the element, and the product ρx is the mass thickness of the sample in grams per square centimeter. There is no matrix effect, which gives the absorption method an advantage over X-ray fluorescence analysis in some cases.

In contrast to absorption measurements in other portions of the electromagnetic spectrum, only a single attenuation measurement is made on each side of the edge because spectrometers that provide a continuously variable wavelength of X radiation are not commercially available. Instead, two X-ray emission lines are used for each edge. Primary excitation from an X-ray tube strikes a target. For example, a crystal of $SrCO_3$ would provide the Sr $K\alpha_1$ line and a crystal of RbCl would provide the Rb $K\alpha_1$ line. The Sr line lies on the short wavelength side of the Br K edge and the Rb line on the long wavelength side of the edge. After suitable standards have been run to calibrate the equipment, bromine can be determined in various materials, such as dibromoethane in gasolines.

For the light elements whose valence electrons may be involved in the X-ray absorption, the exact position of the absorption edge (fine structure) can give the oxidation state of the atom in question.

12.5.2 Radiography

The gross structure of various types of specimens may be examined by absorption techniques using relatively simple equipment. The microradiographic camera fits as an inset in the collimating system of commercial X-ray equipment. Photographic film with an extremely fine grain makes magnification up to 200 times possible. Sample thicknesses vary from a few hundredths to a few tenths of a millimeter; only a few seconds of exposure are necessary.

12.6 X-RAY FLUORESCENCE METHOD

12.6.1 Principle

When a sample is bombarded by a beam of X radiation that contains wavelengths shorter than the absorption edge of the spectral lines desired, characteristic secondary fluorescent X-ray spectra are emitted. Besides X rays, electron bombardment is used in the scanning electron microscope and the electron microprobe. Certain radionuclides are X-ray emitters and can be used as excitation sources, particularly in portable equipment.

The X-ray fluorescence method rivals the accuracy of wet chemical techniques in the analysis of major constituents. However, for trace analyses, it is difficult to detect an element present in less than one part in 10 000. In absolute terms the limit is about 10 ng. The method is attractive for elements that lack reliable wet chemical methods and for the analysis of nonmetallic specimens. X-ray fluorescence is one of the few techniques that offers the possibility of quantitative determinations with about 5 to 10 percent relative error without the use of a suite of standards.

12.6.2 Instrumentation

The general arrangement for exciting, dispersing, and detecting fluorescent radiation with a plane-crystal spectrometer is shown in Fig. 12.2. The sample is irradiated with an unfiltered beam of primary X rays. A portion of the fluorescence X radiation is collimated by the entrance slit of the goniometer and directed onto the surface of the analyzing crystal. The radiations reflected according to the Bragg condition (Eq. 12.2) pass through the exit collimator to the detector.

For elements of atomic number less than 21, a vacuum of 0.1 torr is needed or the system must be flushed with helium. Below magnesium (atomic number 12) the transmission becomes seriously attenuated.

12.6.3 Sample Handling

Samples are best handled as liquids if they can be conveniently dissolved. Sample depth should be at least 5 mm so that the sample will appear infinitely thick to the primary X-ray beam. If possible, solvents that do not contain heavy atoms should be used. Water and nitric acid are superior to hydrochloric or sulfuric acid. Powders are best converted into a solid solution by fusion with lithium borate (both light elements). Powders can also be pressed into a wafer but should be heavily diluted with a material that has a low absorption, such as powdered starch, lithium carbonate, lampblack, or gum arabic to avoid matrix effects (as done by a borax fusion).

12.6.4 Matrix Effects

Matrix effects arise from the interaction of elements in the sample to affect the X-ray emission intensity in a nonlinear manner. They are often negligible when thin samples, which may be collected on a filter, mesh, or membrane, are used for analyses. The most practical way to correct for matrix effects is to use the internal standard technique. Even so, this technique is only valid if the matrix elements affect the reference line and analytical line in exactly the same way. The following list presents the potential problems:

1. If a disturbing element has an absorption edge between the comparison lines, preferential absorption of the line on the short-wavelength side of the edge will occur.
2. If fluorescence from a matrix line lies between the absorption edges of the analytical and reference elements, selective enhancement results for the element whose absorption edge lies at the longer wavelength.
3. Line intensity can be enhanced if a matrix element absorbs primary radiation and then, by fluorescence emits radiation that, in turn, is absorbed by a sample element and causes that latter to fluoresce more strongly.

12.7 ENERGY-DISPERSIVE X-RAY SPECTROMETRY

Energy-dispersive X-ray spectrometry differs from X-ray fluorescence in that X rays emitted from elements in the sample are observed by a solid-state detector and pulse-height analyzer without using a crystal analyzer and collimators with their attendant energy losses. Equipment is very compact. Computer-based multichannel analyzers are then used to acquire a spectrum of counts versus energy and to perform data analysis. As this information is being collated, it can be simultaneously displayed as a spectrum on a video screen. Computer-generated emission lines can be superimposed on the spectrum displayed. Often elements can be identified by the line positions alone. When overlap of the emission lines from one or more elements may be possible, one should use the relative intensities of the lines as well as their positions to confirm the elemental identification. Detection limits in bulk material are typically a few parts per million.

Energy-dispersive X-ray spectrometry offers the advantage of being a simultaneous multielement method for samples with elements separated by one or two atomic numbers. Quantitative determinations can range from the use of simple intensity-concentration standard working curves to computer programs that convert intensity to concentration. Emission lines that have no spectral overlap can be measured from a spectrum using software to perform integration of the net peak intensity above the background for the peak of interest. Cases in which peak overlap is of concern require spectrum-fitting techniques from a library of reference spectra. Care must be taken to maintain the accuracy of the energy calibration of the instrumentation.

Primary X-ray tubes are operated at low power and can be air-cooled. The tube anode material is usually Rh, Ag, or Mo. A filter wheel, typically fitted with six filters, is located between the tube and the sample. The X-ray tube can also be focused on secondary targets to produce specific fluorescent X-ray lines.

Portable instruments are designed around X rays emitted from a radionuclide source. A variety of isotopes are needed to provide radiation over the energy range needed.

Unfortunately, the resolution of an energy dispersion instrument is as much as 50 times less than the wavelength dispersion spectrometer using a crystal analyzer. For precise quantitative measurements, wavelength dispersion followed by energy-dispersive detectors and pulse-height analyzers must be used to get sufficient resolution.

12.8   ELECTRON-PROBE MICROANALYSIS

When materials are bombarded by a high-energy electron beam, as in an electron microscope, X-ray fluorescence radiation is produced. By incorporating an X-ray fluorescence spectrometer directly into the instrument, it is possible to obtain the same sort of elemental data as obtained by normal X-ray fluorescence directly on the area viewed or scanned by the electron beam. It is possible to obtain qualitative and quantitative elemental data from a volume of 0.1 μm^3 for elements of atomic number five (boron) and higher. When the electron beam is scanned across the sample, a point-by-point spatial distribution of elements across the surface (or area) of the sample is produced. Excitation is restricted to thin surface layers because the electron beam penetrates to a depth of only 1 or 2 μm into the specimen.

The electron optical system consists of an electron gun followed by two electromagnetic focusing lenses to form the electron beam probe. The specimen is mounted as the target inside the high-vacuum column of the instrument. If not conductive, the surface of the specimen must treated with a coating to make its surface conductive in order to avoid the problems of specimen charging with electron bombardment.

12.9   ELECTRON SPECTROSCOPY FOR CHEMICAL APPLICATIONS (ESCA)

Electron spectroscopy for chemical applications, or X-ray photoelectron spectroscopy (XPS), is a nondestructive spectroscopic tool for studying the surfaces of solids. Any solid material can be studied and all elements (except hydrogen) can be detected by this technique, usually at 0.1 atomic percent abundance.

12.9.1   Principle

When a specimen is exposed to a flux of X-ray photons of known energy, all electrons whose binding energies E_b are less than the energy of the exciting X rays are ejected. The kinetic energies E_{kin} of these photoelectrons are then measured by an energy analyzer in a high-resolution electron spectrometer. For a free molecule or atom, conservation of energy requires that

$$E_b = h\nu - E_{kin} - \phi \qquad (12.4)$$

where $h\nu$ is the energy of the exciting radiation and ϕ is the spectrometer work function, a constant for a given analyzer. The binding energy is indicative of a specific element and a particular structural feature of electron distribution.

12.9.2 Chemical Shifts

The binding energies of core electrons are affected by the valence electrons and therefore by the chemical environment of the atom. Chemical shifts, as they are called, are observed for every element except hydrogen and helium. Applicability to carbon, nitrogen, and oxygen makes ESCA an important structural tool for organic materials. In general, the ESCA chemical shifts lie in the range 0 to 1500 eV; peak widths vary from 1 to 3 eV. To make assignments, one must refer to a catalog of element reference spectra or to correlation charts.

12.9.3 ESCA Instrumentation

Instrumentation for ESCA involves (1) a source of soft X rays, usually Mg $K\alpha_{1,2}$ and Al $K\alpha_{1,2}$, (2) a device that collects the emitted electrons, counts them, and carefully measures their kinetic energy (an energy analyzer), and (3) a vacuum system capable of providing an operating pressure of about 5×10^{-6} torr. Samples must be low-pressure solids or liquids condensed onto a cryogenic probe. The two alternative sources are needed to distinguish ESCA peaks from Auger peaks. When a different X-ray source is used, the ESCA peaks shift in kinetic energy but the kinetic energies of Auger peaks remain constant and appear at the same energy position in the spectrum.

One type of ESCA spectrometer places the X-ray anode, a spherically bent crystal disperser, and the sample on a Rowland circle. In the energy analyzer an electrostatic field sorts the electrons by their kinetic energies and focuses them at the detector. A continuous channel electron multiplier counts the electrons at each step as the electrostatic field is increased in a series of small steps. A plot of the counting rate as a function of the focusing field yields the spectrum.

12.9.4 Uses

The intensity of a photoelectron line is proportional not only to the photoelectric cross section of a particular element but also to the number of atoms of that particular element present in the sample. Analyses of mixtures are often accurate to ±2 percent.

Because of the unique ability of ESCA to study surfaces, the technique is used to study heterogeneous catalysts, polymers and polymer adhesion problems, and materials such as metals, alloys, and semiconductors. The probe depth is 1 to 2 nm.

12.10 AUGER EMISSION SPECTROSCOPY (AES)

Once an atom is ionized, it must relax by emitting either an X-ray photon or an electron. The latter is the nonradiative Auger process which nature chooses in most instances, and increasingly so for elements of atomic number less than 30. A combination electron gun and energy analyzer unit is shown schematically in Fig. 12.7.

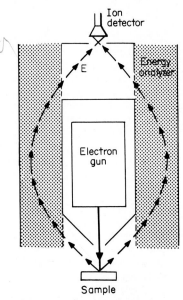

FIGURE 12.7 Schematic diagram of an Auger spectrometer and cylindrical energy analyzer.

A *KLL* Auger transition involves these processes: A K electron undergoes the initial ionization. An L-level electron moves in to fill the K-level vacancy and, at the same time, gives up the energy of that transition to another L-level electron. The latter becomes the ejected Auger electron. The energy of the ejected electron is a function only of the atomic energy levels involved in the Auger transition and is thus characteristic of the atom from which it came. Most elements have more than one intense Auger peak; *LMM* and *MNN* are other transitions. A recording of the spectrum of energies of Auger electrons released from any surface (Fig. 12.8) is compared with the known spectra of pure elements. Typically the sampling depth is about 2 nm.

Shifts from one element to the next are about 25 eV; peak positions can be measured to an accuracy of ±1 eV. Spectra of all the elements lie between 50 and 1000 eV. Both qualitative and quantitative analyses of the elements in the immediate surface atomic layers is possible with Auger emission spectroscopy. When

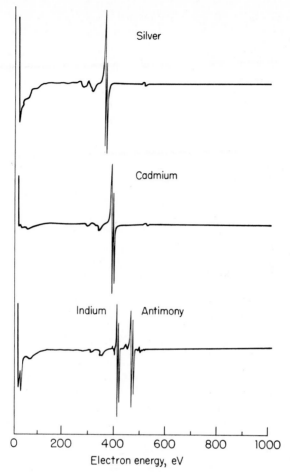

FIGURE 12.8 Auger spectra (differential form) from silver, cadmium, indium, and antimony; all are *MNN* transitions.

combined with a controlled removal of surface layers by ion sputtering, AES provides the means to solve some very important problems involving surfaces.

12.11 X-RAY DIFFRACTION

12.11.1 Principle

When a beam of monochromatic X radiation is directed at a crystalline material, one observes reflection or diffraction of the X rays at various angles with respect to the primary beam. The relationship between the wavelength of the X radiation, the angle of diffraction, and the distance between each set of atomic planes of the crystal lattice is given by the Bragg condition (Eq. 12.2). From the Bragg condition one can calculate the interplanar distances of the crystalline material. The interplanar spacings depend solely on the geometry of the crystal's unit cell while the intensities of the diffracted X rays depend on the types of atoms in the crystal and the location of the atoms in the fundamental repetitive unit, the unit cell.

From the rearranged Bragg equation:

$$\theta = \sin^{-1}\left(\frac{\lambda}{2d}\right) \tag{12.5}$$

the two factors within the parentheses can be seen to control the choice of X radiation. Because the parenthetical term cannot exceed unity, the use of long-wavelength radiation limits the number of reflections that are observed, whereas short-wavelength radiation tends to crowd individual reflections very close together. Furthermore, radiation just shorter in wavelength than the absorption edge of an element in the specimen should be avoided because the resulting fluorescent radiation increases the background. For these reasons, a multiwindow X-ray tube with anodes of Ag, Mo, W, and Cu is used.

The applications of X-ray diffraction can be conveniently considered under three main headings: powder diffraction, polymer characterization, and single-crystal structure studies. In addition there are many specialized uses.

12.11.2   Powder Diffraction

In the powder method, the sample is a large collection of very small crystals, randomly oriented. A polycrystalline aggregate is formed into a cylinder whose diameter is smaller than the diameter of the incident X-ray beam. The diffraction pattern is a series of nonuniformly spaced cones (as intercepted on the photographic film) whose spacings are determined by the prominent planes of the crystallites.

Metal samples may be machined to a cylindrical configuration, plastic materials can often be extruded through suitable dies, and all other samples are best ground to a fine powder (200 to 300 mesh) and shaped into thin rods after mixing with collodion binder or simply tamped into a uniform glass capillary (Fig. 12.9). Liquids must be converted into crystalline derivatives.

The X-ray pattern of a pure crystalline substance can be considered as a "fin-

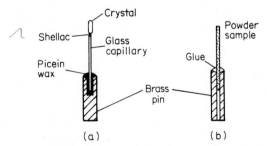

FIGURE   12.9 Specimen   mounts   for   X-ray diffraction: (*a*) single crystal and (*b*) powdered sample.

gerprint" with each crystalline material having, within limits, a unique diffraction pattern. The ASTM has published the powder diffraction patterns of some 50,000 compounds. An identification of an unknown compound is made by comparing the interplanar spacings of the powder pattern of a sample to the compounds in the ASTM file. The d value for the most intense line of the unknown is looked up first in the file. In a single compound, the d values of the next two most intense lines are matched against the file values. After a suitable match for one component is obtained, all the lines of the identified component are omitted from further consideration. The intensities of the remaining lines are rescaled by setting the strongest intensity equal to 100 and repeating the entire procedure. If X-ray fluorescence data are also added, the comparison is even more definitive. A systematic search, either manual or by computer, usually leads to an identification within an hour. Mixtures of up to nine compounds can often be completely identified. The minimum limit of detection is about 1 to 2 percent of a single phase.

In addition to identifying the compounds in a powder, the diffraction pattern can also be used to determine the degree of crystalline disorder, crystalline size, texture, and other parameters associated with the state of the crystalline materials. Differentiation among various oxides such as MnO, Mn_2O_3, MnO_2, and Mn_3O_4, or between materials present in such mixtures as $NaCl + KBr$ or $KCl + NaBr$ is easily accomplished by X-ray diffraction. Identification of various hydrates, such as $Na_2CO_3 \cdot H_2O$ and $Na_2CO_3 \cdot 10H_2O$, is another application.

12.11.3 Polymer Characterization

The following information can be obtained from wide-angle and small-angle X-ray studies of polymers:

1. Degree of crystallinity
2. Crystallite size
3. Degree and type of preferred orientation
4. Polymorphism
5. Microdiffraction patterns
6. Information concerning the macrolattice of the crystallites

Fibers and partially oriented samples show spotty diffraction patterns rather than uniform cones; the more oriented the specimen, the spottier the pattern. The degree of crystallinity and crystallite size can be measured in powders, films, and fibers. This determination usually takes a few hours. The degree of orientation in uniaxially oriented materials, such as fibers, can also be determined in a few hours. If the orientation is other than uniaxial, the type of orientation and an approximate determination of the degree of orientation require similar times.

Polymorphism, that phenomenon where a chemical compound can exist in more than one crystalline form, is often exhibited by polymers. From a study of the diffraction pattern, the presence of polymorphism can be ascertained and the approximate percentages of the polymorphs present determined. Such a study usually requires 1 to 2 days.

Micro X-ray diffraction can be used to identify inclusions in polymers and other materials. It can also be used to study the relationship of the skin of fibers to their interiors with regard to crystallinity, crystallite size, and orientation. Areas as small as 25 μm^2 can be studied. One limitation is that the specimen must be transparent to the X-ray beam. For most materials, a thickness up to 1 mm can be tolerated. Studies require 1 to 2 days.

Small-angle X-ray scattering is used to obtain information about the macrolattice, which can be defined as a periodic array of matter in space which is greater than 5.0 nm in size. In many polymers the chains fold back on themselves forming crystallites. This leads to crystalline and amorphous domains ranging in size from less than 10 nm to almost 100 nm. Small-angle scattering studies are used to determine the size, distribution, and orientation of these domains.

12.11.4   Single-Crystal Structure Determination

Three-dimensional molecular structures can be determined using single-crystal X-ray diffraction procedures. The structures of molecules of considerable complexity (up to about 100 atoms) can be completely elucidated. In addition to the chemical structure, a crystal determination reveals the configuration and conformation of the molecule in the solid state.

The crystal, less than 1 mm in size, is affixed to a thin glass capillary (with shellac) that in turn is fastened to a brass pin which is mounted in the X-ray diffraction unit. In the rotating-crystal method, monochromatic X radiation is incident on the crystal that is rotated about one of its axes. The reflected beams lie as spots on the surface of cones that are coaxial with the rotation axis.

12.11.5   Other Specific Uses

A few of the specialized uses of X-ray diffraction are listed.

1. High-temperature diffractometry.
   a. Determination of expansion coefficients of the crystallographic axes.
   b. Phase diagrams and phase-transition studies.
   c. Studies of order-disorder transitions.
2. Determination of precise crystallographic lattice constants (within $\pm 0.000\ 01$ nm).
3. Studies of the distribution of one molecular species within the host lattice of a second species in solid solutions.
4. Studies of the state of noncrystalline and amorphous solids using computer analysis of time-averaged X-ray data. The results can be used to study the spatial relationship or arrangement between molecules in noncrystalline solids.

More sophisticated applications use software packages that include programs for indexing powder patterns, isostructural searches aiding in solid solutions analysis, preferred orientation calculations, automated texture analysis, incorporation of user-generated patterns into a database, and calculation of lattice parameters to provide extremely accurate distances between atomic planes of the crystal lattice.

BIBLIOGRAPHY

Bertin, E. P., *Principles and Practice of X-Ray Spectrometric Analysis,* 2d ed., Plenum, New York, 1978.

Jenkins, R., R. W. Gould, and D. Gedcke, *Quantitative X-Ray Spectrometry,* Dekker, New York, 1981.

Tertian, R., and F. Claisse, *Principles of Quantitative X-Ray Fluorescence Analysis,* Heyden, London, 1982.

Thompson, M., M. D. Baker, A. Christie, and J. F. Tyson, *Auger Emission Spectroscopy,* Chemical Analysis, Vol. 74, Wiley, New York, 1985.

CHAPTER 13
RADIOCHEMICAL METHODS

13.1 INTRODUCTION

Radioactivity is the spontaneous disintegration of an atom that is accompanied by emission radiation. There are many radioactive elements which are isotopes (have the same atomic number but different atomic mass) of nonradioactive elements. Among these are hydrogen, carbon, iodine, and cobalt. An atom of a radioactive isotope has the same number of orbital electrons as an atom of its nonradioactive counterpart and will, in general, behave chemically and biologically like the nonradioactive species. Therefore experimental and diagnostic as well as analytical procedures can utilize atoms of radioactive isotopes as tracers. The difference between the radioactive and the nonradioactive atoms of identical elements is the number of neutrons in the nucleus, the number of protons and electrons being the same for all. (Some elements have more than two isotopes.)

13.1.1 Particles Emitted in Radioactive Decay

Atomic radiation is generally of three types.

13.1.1.1 Alpha Radiation. The emitted radiation consists of heavy, positively charged particles (helium nuclei or alpha particles) which travel only 5 to 7 cm in air and have very little penetrating effect. A piece of paper, a rubber glove, or the dead outer layer of the human skin stops them. Their energies are very high. As a result, the ionizing power of an alpha particle is high. This type of radiation can generally be distinguished from the other types on the basis of pulse amplitude (see Chap. 12).

13.1.1.2 Beta Radiation. In beta radiation, the emitted particle may be a very energetic negatively charged electron (a beta particle) or a positron (a positively charged electron). A 0.5-MeV beta particle has a range in air of 1 m and produces 60 ion pairs per centimeter on its path. The most energetic can penetrate about 8.5 mm of skin, but almost all are stopped by 3.2 mm of aluminum or 13 mm of Lucite.

13.1.1.3 Gamma and X Radiation. Gamma and X radiation are made up of electromagnetic waves of higher frequency and much greater penetrating ability than those in the ultraviolet spectrum. They cannot be entirely stopped, but they can be attenuated (i.e., their intensity can be reduced) as they pass through lead or concrete barriers. Their ionizing power is less than that of alpha or beta radiation. Substances emitting gamma radiation are dangerous and must be handled with due regard to all safety precautions.

A positron-emitting atom may decay by capturing one of its own orbital K-electrons (*K*-capture or internal conversion). The excess energy is emitted as gamma radiation. The daughter elements (one atomic number less than its parent) has vacant a K-orbital; X radiation that is characteristic of the daughter is emitted when L- and M-level electrons fall into the K-level.

13.1.2 Radioactivity and Isotopes

Each element has a characteristic number of protons and electrons. Isotopes of a particular element are atoms which have the same number of protons, but a different number of neutrons. For example, carbon atoms always contain six protons and six electrons. Its most abundant form (98.892 percent natural abundance), ^{12}C, also contains six neutrons. Carbon's stable isotope (1.108 percent natural abundance), ^{13}C, contains six protons, six electrons, but seven neutrons. The radioactive isotope, ^{14}C contains six protons, six electrons, and eight neutrons.

All isotopes of elements with an atomic number greater than 83 are radioactive. A nucleus with a ratio of neutrons to protons too great for stability undergoes a nuclear transition which reduces the neutron-to-proton ratio. One type of transition (called beta or negatron decay) involves the transformation of a neutron into a proton with the ejection of a negative electron and a neutrino; the atomic number of the product is one higher than its parent.

The atomic number of the parent atom is reduced when the decay process involves a positron emission, electron capture, or alpha decay. The emission of two protons and two neutrons bound together as a helium nucleus is possible for some nuclei with high mass number. The emitted particle is named an α particle and the process is alpha decay.

Radioactive decay of a nucleus often results in the formation of a daughter nucleus in an excited unstable state. As the daughter nucleus changes from the excited state to a state of lower energy, energy is released, usually as electromagnetic radiation in the form of gamma radiation. The change is termed an isomeric transition because the nucleus decays with no change in atomic mass or number.

13.2 UNITS AND CHARACTERISTICS OF RADIOACTIVITY

Activity is expressed in terms of the *Curie* (Ci) where 1 Ci is 3.700×10^{10} disintegrations per second (dps). *Specific activity* is the term used to describe the rate of radioactive decay of a substance whose energies are measured in millions of electronvolts, MeV. It is expressed by disintegrations per second per unit mass or volume, or in units such as microcurie or millicurie per milliliter, per gram, or per millimole. The latter is preferable for labeled compounds.

13.2.1 Radioactive Decay

The decay of a radionuclide follows the first-order rate law:

$$\frac{dN}{dt} = -\lambda N \tag{13.1}$$

where N is the number of radionuclide atoms that remain at time t and λ is the characteristic decay constant. The activity A, the quantity usually observed or computed, is related to N by the equation

$$A = \lambda N \tag{13.2}$$

After integration, the rate equation from which the decay of radioactivity may be calculated is

$$A_t = A_0 e^{-\lambda t} \tag{13.3}$$

where A_0 is the activity at some initial time and A_t is the activity after elapsed time t.

13.2.2 Half-Life

Radioactive substances decay (disintegrate) at a statistical rate which cannot be altered by any chemical or physical treatment of the radionuclide. In order to measure this rate, one can determine how long it takes for one-half of the radionuclide to decay. This is called the *half-life* of the radionuclide, and can be expressed as

$$t_{1/2} = \frac{1}{\lambda} \ln \frac{A}{A/2} = \frac{0.693}{\lambda} \tag{13.4}$$

One-half of the radionuclide will decay in that time; then one-half of what is left will decay in that same time interval; and so on. After 10 half-lives only 0.1 percent of the radionuclide remains. An accurate knowledge of the characteristic decay constant is essential when working with short-lived radionuclides to correct for the decay that occurs while the experiment is in progress.

A selection of radionuclides and their characteristics is given in Table 13.1. Radionuclides of elements occurring naturally in living organisms, such as ^{3}H, ^{14}C, ^{32}P, ^{35}S, and ^{131}I, are used for biologic and medical research as tracers or in labeled compounds wherein one or more radioactive atoms have replaced stable atoms of the same element in the compound.

EXAMPLE 13.1   The activity of a manganese 56 ($t_{1/2} = 2.576$ h) sample at the beginning of an experiment was 3.68×10^6 counts per second. What would be the activity 2 h later at the conclusion of the experiment?

$$A_t = 3.68 \times 10^{10} \exp \left[\frac{-0.693(2.0 \text{ h})}{2.576 \text{ h}} \right]$$

$$= 2.153 \times 10^6 \text{ counts per second}$$

TABLE 13.1   Nuclear Properties of Selected Radioisotopes

Radioisotope	Half-life	Target isotope			Major radiations, energies in MeV (γ intensities, %)
			Natural abundance, %	Thermal neutron cross section, barns	
^{3}H	12.26 yr				$\beta.^-$ 0.0186; no γ
^{14}C	5730 yr	^{13}C	1.108	0.0009	β^- 0.156; no γ
^{15}O	123 s				β^+ 1.74; γ 0.511
^{22}Na	2.62 yr				β^+ 1.820, 0.545; γ 0.511, 1.275(100)
^{24}Na	14.96 h	^{23}Na	100	0.53	β^- 1.389; γ 1.369(100), 2.754(100)
^{28}Al	2.31 min	^{27}Al	100	0.235	β^- 2.85; γ 1.780(100)
^{32}P	14.28 day	^{31}P	100	0.19	β^- 1.710; no γ
^{35}S	87.9 day	^{34}S	4.22	0.27	β^- 0.167; no γ
^{36}Cl	3.08×10^5 yr	^{35}Cl	75.53	44	β^- 0.714; γ 0.511
^{38}Cl	37.29 min	^{37}Cl	24.47	0.4	β^- 4.91; γ 1.60(38), 2.17(47)
^{40}K	1.26×10^9 yr		0.118	70	β^- 1.314; β^+ 0.483; γ 1.460(11)
^{42}K	12.36 h	^{41}K	6.77	1.2	β^- 3.52; γ 0.31, 1.524(18)
^{45}Ca	165 day	^{44}Ca	2.06	0.7	β^- 0.252
^{51}Cr	27.8 day	^{50}Cr	4.31	17	γ 0.320(9); ϵ^- 0.315
^{56}Mn	2.576 h	^{55}Mn	100	13.3	β^- 2.85; γ 0.847(99), 1.811(29), 2.110(15)
^{55}Fe	2.60 yr	^{54}Fe	5.84	2.9	Mn x-rays
^{59}Fe	45.6 day	^{58}Fe	0.31	1.1	β^- 1.57, 0.475; γ 0.143(1), 0.192(3), 1.095(56), 1.292(44)
^{60}Co	5.263 yr	^{59}Co	100	19	β^- 1.48, 0.314; γ 1.173(100), 1.332(100)
^{63}Ni	92 yr	^{62}Ni	3.66	15	β^- 0.067; no γ
^{65}Ni	2.564 h	^{64}Ni	1.16	1.5	β^- 2.13; γ 0.368(5), 1.115(16), 1.481(25)
^{64}Cu	12.80 h	^{63}Cu	69.1	4.5	β^- 0.573; β^+ 0.656; e^- 1.33, γ 0.511, 13.4(1)
^{65}Zn	245 day	^{64}Zn	48.89	0.46	β^+0.327; e^- 1.106; γ 0.511, 1.115(49)
^{69m}Zn	13.8 h	^{68}Zn	18.6	0.10	γ 0.439(95); e^- 0.429

TABLE 13.1 Nuclear Properties of Selected Radioisotopes (*Continued*)

Radio-isotope	Half-life	Target isotope			Major radiations, energies in MeV (γ intensities, %)
			Natural abundance, %	Thermal neutron cross section, barns	
^{76}As	26.4 h	^{75}As	100	4.5	β⁻ 2.97; γ 0.559(43), 0.657(6), 1.22(5), 1.44(1), 1.789, 2.10(1)
^{80}Br	17.6 min	^{79}Br	50.52	8.5	β⁻ 2.00; β⁺ 0.87; γ 0.511, 0.61(7), 0.666(1)
^{80m}Br	4.38 h				γ 0.037(36); e⁻ 0.024, 0.036, 0.047
^{82}Br	35.34 h	^{81}Br	49.48	3	β⁻ 0.444; γ 0.554(66), 0.619(41), 0.698(27), 0.777(83), 0.828(25), 1.044(29), 1.317(26), 1.475(17)
^{90}Sr	27.7 y				β⁻ 0.546; no γ
^{90}Y	64.0 h				β⁻ 2.27; no
^{110m}Ag	255 day	^{109}Ag	48.65	89	β⁻ 1.5; γ 0.658(96), 0.68(16), 0.706(19), 0.764(23), 0.818(8), 0.885(71), 0.937(32), 1.384(21), 1.505(11)
^{122}Sb	2.80 day	^{121}Sb	57.25	6	β⁻ 1.97; β⁺ 0.56; γ 0.564(66), 1.14(1), 1.26(1)
^{124}Sb	60.4 day	^{123}Sb	42.75	3.3	β⁻ 2.31; γ 0.603(97), 0.644(7), 0.72(14), 0.967(2), 1.048(2), 1.31(3), 1.37(5), 1.45(2), 1.692(50), 2.088(7)
^{128}I	24.99 min	^{127}I	100	6.4	β⁻ 2.12; γ 0.441(14), 0.528(1), 0.743, 0.969

TABLE 13.1 Nuclear Properties of Selected Radioisotopes (*Continued*)

Radio-isotope	Half-life	Target isotope			
			Natural abundance, %	Thermal neutron cross section, barns	Major radiations, energies in MeV (γ Intensities, %)
^{137}Cs	30.0 yr				β^- 0.511, 1.176; γ 0.662(85)
^{137m}Ba	2.554 min				γ 0.662(89); e$^-$ 0.624, 0.656
^{198}Au	2.697 day	^{197}Au	100	98.8	β^- 0.962; γ 0.412(95), 0.676(1), 1.088
^{204}Tl	3.81 yr	^{203}Tl	29.5	11	β^- 0.766

Source: J. A. Dean, ed., *Lange's Handbook of Chemistry*, 13th ed., McGraw-Hill, New York, 1985.

13.3 MEASUREMENT OF RADIOACTIVITY

The random nature of nuclear disintegrations requires that a large number of individual disintegrations be observed to obtain a counting rate or total sample count with a desired statistical significance. Several problems must be considered in measurement of radioactivity.

1. The radionuclide emission may be absorbed in the air path or in the walls of the detector.
2. The dead time (unresponsive period between radioactive events) may lead to uncounted events.
3. The energy of the ionizing particle may be insufficient to produce an ion in the sensitive volume of the detector.

13.4 DETECTORS

Detectors described in Chap. 12 are also suitable for measurement of radioactivity. When resolution is the highest priority, measurement should be made with a Ge(Li) or Si(Li) detector coupled to a multichannel, pulse-height analyzer system. If sensitivity is paramount, a proportional counter in conjunction with a scintillation crystal or liquid scintillation system is the choice.

13.4.1 Scintillation Counters

A good match should exist between the emission spectrum of the scintillator and the response curve of the photocathode of the multiplier phototube. The decay time is 250 ns for a sodium iodide crystal doped with 1% thallium(I) iodide, a crystal useful for counting beta particles and gamma radiation. When radiation interacts with a NaI(Tl) crystal, the transmitted energy excites the iodine atoms. Upon their return to the ground electronic state, this energy is reemitted in the form of a light pulse in the ultraviolet which is promptly absorbed by the thallium atom and reemitted as fluorescent light at 410 nm (see Fig. 12.5). The well-type arrangement of the scintillation crystal increases the counting efficiency to approximately 100 percent by surrounding the sample with the detector crystal. This arrangement is best for counting gamma radiation.

13.4.2 Proportional Counters

A sample of radioactive material can be placed inside the active volume of a flow proportional counter, thus avoiding losses due to window absorption (see Fig. 12.4). The chamber is purged with a rapid flow of counter gas (P-10 gas, a mixture of 10% methane in argon) and the flow maintained during counting of samples. Counter life is virtually unlimited. Such a counter is particularly suited for distinguishing and counting low-energy alpha and beta particles.

13.5 STATISTICS OF RADIOACTIVITY MEASUREMENTS

Because of the randomness of radioactive decay, the true average count of any sample is not known. However, the relative error of a particular number of counts can be estimated. We need to know the counting rate N_s/t_s or the total counts N_s, where t_s is the counting time. There is always some background activity (background counts N_b and counting time for background only t_b) that the detector registers along with the sample activity.

13.5.1 Fractional Error

The fractional error F_y of the sample, after correction for background ($N_s - N_b$), is given by

$$F_y = \frac{K}{N_s - N_b} \sqrt{\frac{N_s}{t_s} + \frac{N_b}{t_b}} \qquad (13.5)$$

where K is the number of standard deviations (Table 13.2) for a particular error or confidence limit.

13.5.2 Distribution of Counting Time

The optimum distribution of counting time between background and sample is given by

$$\frac{t_b}{t_s} = \sqrt{m\left(\frac{N_b}{N_s}\right)} \qquad (13.6)$$

TABLE 13.2   Table of Constants of Relative Error

Error (confidence limit)	Probability of occurrence, %	K
Probable	50.0	0.675
Standard deviation (one sigma)	31.7	1.000
Ninety-five hundredths	5.00	1.960
Two sigmas	4.55	2.000
Ninety-nine hundredths	1.00	2.576
Three sigmas	0.27	3.000

where m is the number of experimental samples measured in that batch for a single radionuclide, not including the background measurements.

13.5.3 Preset Time or Preset Count

The counting time required for a sample to achieve a predetermined precision is given by the expression

$$t_s = \frac{K^2}{F_y^2}\left[\frac{N_s + N_b/c}{(N_b - N_s)^2}\right] \tag{13.7}$$

where $c = mN_b/N_s$. The time required to count the background is obtained from Eq. 13.6. For a preset count, the counting time (Eq. 13.7) is multiplied by the counting rate N_s/t_s.

13.6 LIQUID SCINTILLATION COUNTING

Liquid scintillation counting has as its primary application the counting of weak beta emitters, such as tritium, carbon 14, and phosphorus 32.

13.6.1 Liquid Scintillators

Samples are dissolved in a "liquid cocktail" that contains a suitable solvent, a primary scintillator and perhaps a secondary scintillator, and additives to improve water miscibility and to permit counting at low temperatures. The emitted beta particles collide with the solvent molecules which become excited. In turn, the excited solvent molecules transfer their energy to the scintillator molecules, a portion of which is then emitted in the form of photons of visible light whose wavelength lies in the responsive range of a photomultiplier detector. A block diagram of the energy transfer in liquid scintillation counting is shown in Fig. 13.1.

Solvents, such as toluene, xylene, or 1,4-dioxane, must efficiently transfer energy to a scintillator molecule. 1,4-Dioxane is used when large amounts of water are involved, or a nonionic surfactant such as Triton X-100 can be incorporated into a toluene-based system.

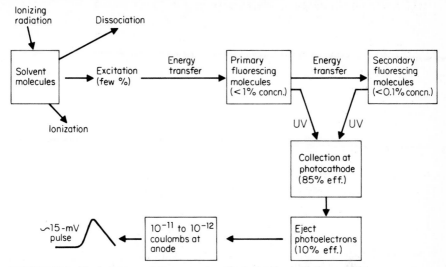

FIGURE 13.1 Block diagram of the energy transfer in liquid scintillation counting.

The scintillator must be capable of absorbing light at the wavelength emitted by the bulk solvent molecules and reemitting it at a longer wavelength that matches the spectral sensitivity of the blue-sensitive photomultiplier tubes. The signal is amplified and shaped in the amplifier circuitry; it is then sent to the pulse-height analyzer where pulses of a certain energy range are then accumulated by the scaler and processed by the timer.

13.6.2 Sample Vials

Sample vials should have a low background count from naturally occurring potassium 40, for example, less than 10 counts per minute when determining tritium. High-density polyethylene vials are economical and are used for counting samples. Borosilicate glass vials are best for sample storage because they are impervious to toluene, xylene, and 1,4-dioxane.

13.6.3 Color and Chemical Quenching

Color quenching, in which the sample absorbs a portion of the fluorescent photons emitted by the scintillator, is common in colored specimens (such as blood or plant tissues). It is best eliminated by treating the sample prior to counting; the sample can be digested using a mixture of hydrogen peroxide and perchloric acid or completely combusted to water (for tritium) or to a soluble carbonate (for carbon 14) or phosphate (for phosphorus 32).

In chemical quenching, some of the energy released by the emission of the beta particle during decay of the radionuclide is absorbed by the components of the liquid cocktail. The method of standard additions is helpful.

13.7 APPLICATIONS OF RADIONUCLIDES AS TRACERS

Radioactive tracers are used to follow the behavior of atoms or groups of atoms in any analytical scheme or other chemical reaction, in an industrial system, or in a biologic process. A radioactive tracer unequivocally labels the particular atoms to be traced. The specificity and sensitivity of detection of radionuclides is extremely good; less than 10^{-18} g can often be detected. However, a tagging radionuclide must not be exchangeable with similar atoms in other compounds.

A chemical separation of inactive, and occasionally active, contaminants generally precedes the activity measurement. Ordinary analytical techniques and reactions form the basis for most radiochemical separations, which might involve carrying a microconstituent by coprecipitation with a macroconstituent (called a carrier), solvent extraction, volatilization, adsorption, ion exchange, electrodeposition, and chromatography.

The active material must be spread in a uniform layer over a definite area unless 4-π geometry (such as a well-type scintillation counter) is employed. Electrodeposition onto a flat surface gives excellent deposits for many metals; a thin film of plastic sprayed with gold can serve as the active electrode. The sample may be spread as a slurry in a solvent which is later evaporated; deposits that are not coherent can be stabilized with a binder such as collodion.

A few typical examples of applications using radioactive tracers are:

1. Study of polymer synthesis mechanisms.
2. Study of catalyst poisoning, including decomposition and interactions.
3. Measurement of homogeneous and heterogeneous reaction rates.
4. Studies of biochemical metabolism.
5. Studies of solubility, coprecipitation, and separation.
6. Measurement of nutrient uptake in vegetation.
7. Measurement of permeation rates through thin films and metal-metal self-diffusion studies at elevated temperatures.
8. Studies of surface corrosion.
9. Studies of the stability of solvents and additives during long-time storage.
10. Identification and location of molecules, such as DNA, within cells to study rates of cell division and tissue growth.
11. Study of the concentration of different radionuclides in body organs; the size, outline, and areas of increased or decreased radioactive uptake can be determined.

13.8 ISOTOPE DILUTION ANALYSIS

Isotope dilution analysis is useful where no quantitative procedure is known, such as in radiocarbon dating of archaeological specimens and in the analysis of complex biochemical mixtures. The procedure is as follows:

1. Add to the sample a known weight W_1 of the compound that is in the sample, but the added compound is tagged with the radioactive element.
2. Determine the specific activity A_1 of the tagged compound separately.
3. Isolate from the mixture a small amount of pure compound that is sufficient for an accurate determination.
4. Measure the specific activity A of the isolated material.

The amount W of inactive material present is given by

$$W = W_1\left(\frac{A_1}{A} - 1\right)$$ (13.8)

13.9 ACTIVATION ANALYSIS

13.9.1 Principle

Activation analysis is an extremely sensitive elemental analysis in which various elements present in samples are made radioactive by bombardment with suitable nuclear particles. The induced radionuclides are then identified or quantitatively measured. Neutron activation analysis is the most common and widely used form of the activation analysis methods. It involves bombarding the sample with neutrons, usually thermal (slow) neutrons generated in a nuclear reaction, but in some cases fast neutrons generated by an accelerator. In neutron activation analysis, the sample is kept in a neutron flux for a length of time that is sufficient to produce radionuclide product in amounts that can be measured with the desired statistical precision. At the termination of the irradiation, the activity A_0 will be

$$A_0 = \Phi\sigma N\left[1 - \exp\left(-\frac{0.693t}{t_{1/2}}\right)\right]$$ (13.9)

where Φ = number of bombarding particles, or flux, in $cm^{-2} \cdot s^{-1}$)
    σ = reaction cross section, cm^2 per target atom (10^{-24} cm^2 per nucleus, also denoted as barns)
    N = number of target nuclei available
    $t_{1/2}$ = half-life of radionuclide
    t = duration of irradiation period

The induced activity will reach 98 percent of the maximum value for irradiation periods equal to six half-lives. In fact, at $t/t_{1/2}$ ratios of 1, 2, 3, 4, 5, and 6, the corresponding values of the term within the brackets of Eq. 13.8 are 0.5, 0.75, 0.87, 0.94, 0.97, and 0.98, respectively.

EXAMPLE 13.2   What is the activity for a 10.0-mg sample (from a painting) of pigment containing 1.50% chromium after a 24.0-h irradiation in a flux of 1×10^{14} neutrons $cm^{-2} \cdot s^{-1}$. Chromium 51 has a half-life of 27.8 days; chromium 50, the target isotope, has a natural abundance of 4.31 percent and a thermal neutron cross section of 17 barns.

$$A_0 = \frac{(0.0150)(0.0100 \text{ g})(0.0431)(6.02 \times 10^{23} \text{ nuclei} \cdot \text{mol}^{-1})}{52.00 \text{ g} \cdot \text{mol}^{-1}}$$

$$\times (1 \times 10^{14} \text{ neutrons} \cdot \text{cm}^{-2} \cdot \text{s}^{-1})(17 \times 10^{-24} \text{ cm}^2 \cdot \text{nuclei}^{-1})$$

$$\times \left\{ 1 - \exp \left[\frac{-0.693(1.00 \text{ day})}{27.8 \text{ days}} \right] \right\}$$

$$= 3.13 \times 10^6 \text{ disintegrations per second}$$

Longer-lived activities, such as chromium 52, are enhanced by the use of a longer irradiation period, followed, if necessary, by an appreciable delay for the decay of interfering short-lived activities, before counting the chromium 52. The reverse is true for short-lived activities.

13.9.2 Instrumental (Nondestructive) Methods

The instrumental or nondestructive approach is preferred whenever applicable. A set of samples and standards are irradiated under exactly the same flux conditions. After the irradiation step the samples are transferred to a multichannel gamma-ray spectrometer and counted for a suitable period of time. Gamma rays are emitted at characteristic energies. An illustrative gamma-ray spectrum (or pulse-height spectrum) is shown in Fig. 13.2. The shaded area indicated under the photopeak at 0.847 MeV would be used for quantitative work. The two less intense photopeaks at 1.811 and 2.11 MeV aid in qualitative identification of manganese. The X rays of iodine arise from the NaI(TlI) crystal detector used to ob-

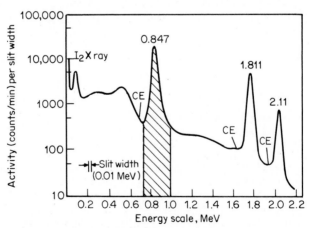

FIGURE 13.2 Gamma spectrum of manganese 56. The area indicated under the photopeak at 0.847 MeV would be used in quantitative work. CE indicates the Compton edges for each photopeak.

tain the ^{56}Mn spectrum. Observation of the decay rates using sequential measurements of the gamma spectrum over time will supply the half-life of the nuclide and confirm the identification.

For the analysis of mixtures, spectrum stripping is performed. The lower-energy portion of the gamma-ray spectrum of the most energetic full-energy peak is subtracted from the total spectrum. This step is repeated for each successively less energetic peak. These operations are performed automatically on the multichannel analyzer from standard curves stored in the memory with the aid of appropriate software programs.

13.9.3 Radiochemical Separations

If interferences from other induced activities cannot be removed by either a suitable decay period or use of filters (for beta activities), a postirradiation chemical separation will be needed. The sample (and standard) is dissolved and chemically equilibrated with an accurately known amount of carrier (usually about 10.0 mg) for each element of interest. Then the analyte is separated and purified by any suitable separation procedure and finally counted. The amount of carrier element, if added as a hold-back carrier of the analyte radionuclide, is recovered and measured quantitatively. Results are normalized to 100 percent recovery. Complete recovery is unnecessary; however, the higher the recovery, the better the counting statistics.

13.10 RADIATION SAFETY

13.10.1 Labeling of Radioactive Substances

All radioactive substances are supplied in approved containers which are designed to specific standards and which must pass extreme testing for resistance to damage from fire and shock. There are type A and B packages used for small and large quantities. Only a specified quantity of the radionuclide can be shipped in a particular package. Labeling is standardized (Fig. 13.3); in the case of an emergency the hazard can be quickly assessed.

The following labeling procedures have been adopted:

Category I	White label	The dose rate at the surface of the package is 0.5 mSv/h^{-1} or less. (The abbreviation mSv stands for millisievert.)
Category II	Yellow label	The dose rate at the surface of the package does not exceed 10 mSv/h^{-1} and the dose rate 1 m from the center of the package does not exceed 0.5 mSv/h^{-1}.
Category III	Red label	The dose rate does not exceed 200 mSv/h^{-1} at the surface of the package, and and the dose rate 1 m from the center of the package does not exceed 10 mSv/h^{-1}.

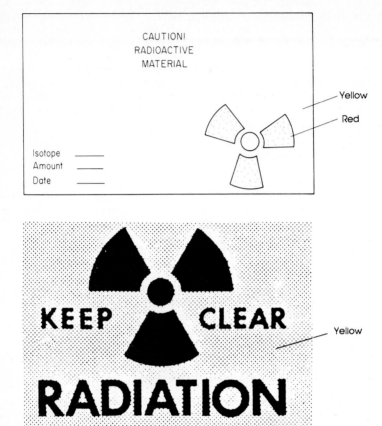

FIGURE 13.3   Standard radiation warning labels and signs.

13.10.2   Handling of Radioactive Substances

Small quantities of various radioactive isotopes are available without license because such quantities are considered to be harmless under general laboratory conditions. All radioactive substances must be handled so that any unnecessary or excessive exposure is prevented. Usually the greatest exposure occurs during the transfer of the radionuclide from its container to the experimental equipment. The type of shielding for personnel depends on the type of radioactive emission. Alpha emitters do not require any shielding. Beta emitters can be shielded by sheets of plastic such as transparent methacrylate. Gamma emitters require lead or steel shielding.

13.10.3 Storage of Radioactive Substances

The storage requirements of radioactive materials depend upon the type of emitter and the quantity involved. Normally, laboratory quantities of radionuclides may not require specialized storage areas, but the containers should be positioned away from any counting areas so as to prevent stray radiation from affecting and increasing the counter background.

Disposal regulations for radioactive wastes can be obtained from the Nuclear Regulatory Commission.

13.10.4 Radiation Hazards and Safeguards

Gamma and beta radiation are hazardous both inside and outside the body; proper shielding and precautions must be followed. Unless adequate precautions are taken, radioactive material can get into the body by inhalation, by ingestion, or through cuts and abrasions on exposed skin surfaces. Some materials can even be absorbed through the intact skin. Safeguards against dust are of prime importance. The laboratory area should be kept clean, and special air monitors and respiratory protection should be provided wherever they are needed.

When shielding or other control measures are being planned, the complete range of radiations of the original material and all of the decay products must be considered. For example, ruthenium 106 has a very low-energy beta emission but its "daughter," rhodium 106, emits both high-energy beta and gamma radiation.

The equipment in which radioactive materials is used should be maintained as a closed system. Special markings or tags should be used to warn all persons who come in contact with radioactive materials. All apparatus and containers used in each stage of the laboratory procedure from storage to disposal should be plainly marked with the date, isotope, decay products, and the type of radiation emitted. Protective gloves and shoe covers, even complete clothing changes, may be necessary. Areas in which radioactive materials are being used should be clearly marked and restricted. No smoking, eating, or applying cosmetics should be permitted in these areas. Decontamination facilities may be needed for personnel leaving restricted areas. Hands, feet and clothing should be checked upon leaving the area. Personal exposure can be checked with pocket monitoring devices.

BIBLIOGRAPHY

Barkouskie, M. A., "Liquid Scintillation Counting," *Am. Lab.*, **8**:101 (May 1976).

Bernstein, K., "Neutron Activation Analysis," *Am. Lab.*, **12**:151 (September 1980).

Finston, H. L., "Radioactive and Isotopic Methods of Analysis," in *Treatise on Analytical Chemistry*, Chap. 96, I. M. Kolthoff and P. J. Elving, eds., Part I, Vol. 9, Wiley-Interscience, New York, 1971.

Friedlander, G., J. W. Kennedy, E. S. Macias, and J. M. Miller, *Nuclear and Radiochemistry*, 3d ed., Wiley, New York, 1981.

Katz, S. A., "Neutron Activation Analysis," *Am. Lab.*, **17**:16 (June 1985).

CHAPTER 14
NUCLEAR MAGNETIC RESONANCE SPECTROSCOPY

14.1 INTRODUCTION

Nuclear magnetic resonance (NMR) spectroscopy is a powerful method for elucidating chemical structures. The spectra obtained answer many questions such as (referring to specific nuclei)[1]:

1. Who are you?
2. Where are you located in the molecule?
3. How many of you are there?
4. Who and where are your neighbors?
5. How are you related to your neighbors?

The result is often the delineation of complete sequences of groups or arrangements of atoms in the molecule. The sample is not destroyed in the process.

NMR can also be used for a particular facet of a structure, such as chain length or moles of ethoxylation, and in the study of polymer motion by relaxation measurements. Kinetic studies of reactions at temperatures in the range from −150 to 200°C are another application. Useful information can also be obtained from complex mixtures, such as, for example, the total aldehydic content of a perfume, the phenyl-methyl ratio in polysiloxanes, the monoester-diester ratios in emulsifiers, and the alcohol-water ratios in colognes and after shaves. In solvent systems an understanding of the basic types of solvents present can be gained from scanning the product in the NMR region.

Integration of areas under the absorption peaks and the peak of the internal standard enables quantitative analysis to be performed.

14.2 BASIC PRINCIPLES

The nuclei of certain isotopes have an intrinsic spinning motion around their axes which generates a magnetic moment along the axis of spin. The simultaneous application of a strong external magnetic field H_0 and the radiation from a second and weaker radio-frequency (rf) source H_1 (applied perpendicular to H_0) to the

14.1

nuclei results in transitions between energy states of the nuclear spin (Fig. 14.1). Absorption occurs when these nuclei undergo transition from one alignment in the applied field to an opposite one. The energy needed to excite these transitions can be measured. The resonance frequency v that causes the transitions between energy levels is given by

$$\Delta E = hv = \frac{\mu H_0}{I} \tag{14.1}$$

where μ = magnetic moment of the nucleus
I = spin quantum number, h/π
h = Planck's constant

Nuclei with $I = 1/2$ give the best resolved spectra because their electric quadrupole moment is zero. These nuclei include ^{1}H, ^{13}C, ^{19}F, ^{29}Si, and ^{31}P.

As indicated by Eq. 14.1, the frequency of the resonance condition varies with the value of the applied magnetic field. For example, in a magnetic field of 14 092 G, protons require a frequency of 60 MHz to realign their spins. In a field of 23 490 G, the frequency rises to 100 MHz. Higher values of field strength lead to a stronger absorption signal (roughly in proportion to the square of the magnetic field strength). Nuclear properties of selected elements are given in Table 14.1.

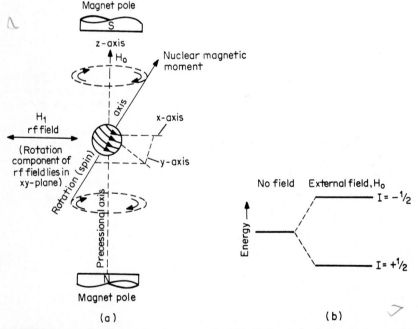

FIGURE 14.1 (a) Spinning nucleus in a magnetic field. (b) Energy-level diagram for a nucleus with a spin of 1/2.

TABLE 14.1 Nuclear Properties of Selected Elements

Nuclide	Natural abundance, %	Spin I	Sensitivity at constant field relative to ^1H	NMR frequency for a 1000-G field,* MHz
^1H	99.985	1/2	1.000	4.257 60
^{13}C	1.108	1/2	0.015 9	1.070 54
^{15}N	0.365	−1/2	0.001 04	0.431 5
^{19}F	100	1/2	0.834	4.005 43
^{29}Si	4.71	−1/2	0.078 5	0.845 8
^{31}P	100	1/2	0.066 4	1.723 8

*NMR frequency (H_1) at any magnetic field (H_0) is the entry in column 5 multiplied by the value of the magnetic field in kilogauss. For example, in a magnetic field of 23.4924 kG, protons will precess at 4.25760 × 23.4924 = 100.00 MHz and carbon 13 will precess at 1.07054 × 23.4924 = 25.1496 MHz.
Source: J. A. Dean, ed., *Handbook of Organic Chemistry*, McGraw-Hill, New York, 1986, p. 6-68.

14.2.1 Relaxation Processes

Energy absorbed and stored in the higher energy state must be dissipated and the nuclei returned to the lower energy state. Otherwise, the rf system equalizes the populations of the energy states, the spin system becomes saturated, and the absorption signal disappears.

The *spin-lattice relaxation* involves interaction of the spin with the fluctuating magnetic fields produced by the random motions of neighboring nuclei (called the "lattice"). Nuclei lose their excess energy as thermal energy to the lattice. The relaxation process is first order and decreases exponentially with time. T_1 is a rate constant called the spin-lattice relaxation time. It depends on the average distance of the nucleus from magnetic neighboring nuclei and on the types of molecular motion that the functional group is undergoing. Internal rotations and segmental motions in polymers are detected in this way. In typical organic liquids and dilute solutions, the time is in the range 0.01 to 100 s.

The *spin-spin relaxation* process involves a second time constant T_2. Nuclei exchange spins with neighboring nuclei through the interaction of their magnetic moments, and they thereby lose phase coherence. T_2 can be calculated from the dispersion mode signal as a function of time.

Automated T_1 and T_2 measurements are done with computer software. The time constants aid in understanding physical properties of polymers.

14.3 INSTRUMENTATION

NMR instrumentation (Fig. 14.2) involves these basic units:

1. A magnet to separate the nuclear spin energy states.
2. (a) One rf channel for field or frequency stabilization, which produces stability for long-term operation, (b) one rf channel to furnish irradiating energy to the sample, and (c) a third channel which may be added for decoupling nuclei.
3. A sample probe houses the sample and also coils for coupling the sample with

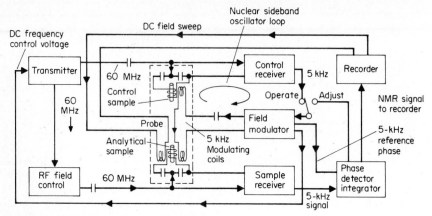

FIGURE 14.2   Schematic diagram of an NMR spectrometer.   (*Courtesy of Varian Associates.*)

the rf transmitter and the phase-sensitive detector. It is inserted between the pole faces of the magnet.

4. A detector to collect and process the NMR signals.

5. A sweep generator for sweeping the rf field through the resonance frequencies of the sample. Alternatively, the magnetic field may be swept and the rf field held constant.

6. A recorder to display the spectrum.

14.3.1   Sample Handling

The liquid sample is contained in a cylindrical, thin-walled precision-bore glass tube (purchased from supply houses) that has an outer diameter of 5 mm. The tube is filled until the length-diameter ratio is about 5. An air-bearing turbine rotates the sample tube at a rate of 20 to 40 revolutions per second. For proton spectra, a few micrograms to a few milligrams of sample is dissolved in a deuterated solvent such as chloroform-d, acetone-d_6, or benzene-d_6.

The deuterium replaces the hydrogen nuclei that would generate a background solvent signal that would overwhelm the proton signals from the sample. Also, the NMR response of the deuterium nuclei is used to lock the ratio of H_0/ν of the NMR spectrometer over long periods of time. The same solvents are used to dissolve samples when the ^{13}C spectrum is desired. Because the isotopic abundance of ^{13}C is only 1.3 percent, the sample size is larger than that required for proton spectra—between a few milligrams and a few hundred milligrams.

14.3.2   The Magnet

Less expensive instruments may use permanent magnets. These are simple and inexpensive to operate, but they do require extensive shielding and must be thermostatted to $\pm0.001°C$.

Electromagnets require elaborate power supplies and cooling systems. They offer the opportunity to use different field strengths to disentangle chemical shifts from multiplet structures and to study different nuclei. The upper limit of magnet strength is 23.49 kG.

For higher field strengths, one must turn to cryogenic superconducting solenoids. Instruments with fields of 93.9 and 117.4 kG permit rf fields of 400 and 500 MHz, respectively.

14.3.3 Pulsed (Fourier Transform) NMR

In Fourier transform NMR all the spectral lines are excited simultaneously. A strong pulse of energy (H_1) is applied to the sample for a very short time (1 to 100 μs). In the interval between pulses, the free precession of the magnetic moments of the nuclei gradually decay. This free induction decay requires several spin-lattice time constants ($3T_1$ to $5T_1$). The decay signal following each repetitive pulse (about 0.82 s for 8192 data points) is digitized by a fast analog-to-digital converter, and the successive digitized transient signals are coherently added in the computer until an adequate signal-to-noise ratio is obtained. Using the Cooley-Tukey algorithm, the computer then performs a fast Fourier transformation to the frequency domain and plots a normal spectral presentation of the NMR spectrum in 10 to 20 s.

Pulsed Fourier transform NMR makes possible the study of less sensitive nuclei, such as ^{13}C, ^{14}N, ^{15}N, ^{17}O, ^{31}P, and even unstable, radioactive nuclei with spins. Proper selection of pulse sequences using computer control allows measurement of the various T_1 values.

14.3.4 Magic Angle Spinning and Cross Polarization

Line broadening from a polycrystalline material can largely be removed by rotating the sample very rapidly (faster than 2 kHz) about an axis oriented at an angle of approximately 54.7° (the magic angle) with respect to the external magnetic field H_0. The averaging that occurs is similar to tumbling in liquids.

Solids are machined into rotors or packed as powders into hollow rotors. The rotor is dropped into the probe.

For ^{13}C studies, cross polarization techniques overcome the problems of long spin-lattice relaxation times. Details are given by Miknis and colleagues.[2]

14.4 SPECTRA AND MOLECULAR STRUCTURE

For most purposes, NMR spectra are described in terms of *chemical shifts* and *coupling constants*. The spin-lattice (T_1) and spin-spin (T_2) relaxation times have already been discussed.

14.4.1 Chemical Shifts

On the basis of nuclear properties alone, all protons would precess at exactly the same frequency in a fixed magnetic field and would be indistinguishable from each other. Fortunately, in different chemical environments, shifts in resonant frequency arise from partial shielding of the nuclei from the applied magnetic field by the electron cloud around it. The density of this cloud varies with the number and nature of the neighboring atoms. The specific location of the shifted resonant frequencies indicates the chemical nature of the nuclei and provides some information concerning their neighbors. The resonance position of proton spectra is always stated with respect to the resonance of the protons in a reference compound. This overcomes the problem of NMR spectrometers with different field strengths. The magnitude of the chemical shift is expressed in parts per million.

$$\delta = \frac{H_r - H_s}{H_r} \times 10^6 \qquad (14.2)$$

where H_r and H_s are the positions of the absorption lines for the reference and sample, respectively, expressed in magnetic units.

For proton spectra in nonaqueous media, the reference material is tetramethylsilane, $(CH_3)_4Si$, abbreviated TMS, whose position is assigned to exactly 0.0 on the delta (δ) scale. TMS is also a suitable reference material for ^{13}C and ^{29}Si.

The location of proton resonances for C—H bonds may be summarized as follows:

1. When only aliphatic groups are substituents, the range is $\delta = 0.9$ to 1.5.

2. Protons in CH_2 and CH groups appear slightly further downfield (in that order) from protons in CH_3 groups.

3. An adjacent, unsaturated bond (whether alkene or aryl) shifts the resonance position of CH_3 to $\delta = 1.6$ to 2.7.

4. An adjacent nitrogen atom causes the proton resonance to lie in the range 2.1 to 2.3 for alkyl—N—CH groups, 2.8 to 3.0 for aryl—N—CH groups, and 2.8 to 3.1 for amide groups.

5. An adjacent oxygen atom markedly shifts proton signals downfield to $\delta = 3.2$ to 3.4 to aliphatic entities and to $\delta = 3.6$ to 3.9 for aryl—O—CH situations.

6. Aryl protons are found at $\delta = 7.2$.

7. Aldehyde protons in aliphatics lie in the range $\delta = 9.4$ to 10.0 and those for aryls at $\delta = 9.7$ to 10.5.

Table 14.2 gives some proton chemical shifts. Other more complete tables for proton, carbon 13, nitrogen 15 (or 14), fluorine 19, silicon 29, and phosphorus 31 chemical shifts and coupling constants can be found in handbooks edited by Dean.[3,4]

Figure 14.3 shows an NMR spectrum consisting only of singlets in distinctively different environments. Resonance absorption confined to a single field position (no multiplets) implies that each proton-containing group is isolated from the others by at least one intervening entity that contains no protons. From left to right we have aryl protons (the ring connected to a carbon), a methylene group that is flanked by an aryl group on one side and an —O—(C=O)— group on the other, and a methyl group connected to a carbonyl group. An —O— or —CO— adjacent to the phenyl ring would create a multiplet out of the phenyl ring protons. The alkyl group adjacent to an oxygen would lie around $\delta = 3.6$; the group adjacent to

TABLE 14.2 Proton Chemical Shifts

Values are given on the δ scale. Abbreviations used in the table: R, alkyl group and Ar, aryl group

Substituent group	Methyl protons	Methylene protons	Methine protons
HC—C—CH$_2$	0.95	1.20	1.55
HC—C—NR$_2$	1.05	1.45	1.70
HC—C—C≡C	1.00	1.35	1.70
HC—C—Ar	1.15	1.55	1.80
HC—C—OH (or —OR)	1.20	1.50	1.75
HC—C—OAr	1.30	1.55	2.00
HC—C—O(C=O)R	1.30	1.60	1.80
HC—C—NO$_2$	1.60	2.05	2.50
HC—C—O(C=O)Ar	1.65	1.75	1.85
HC—C—Br	1.80	1.85	1.90
HC—C=C	1.60	2.05	
HC—C˙C	1.70	2.20	2.80
HC—(C=O)OR	2.00	2.25	2.50
HC—SR	2.05	2.55	3.00
HC—(C=O)R	2.10	2.35	2.65
HC—C˙N	2.15	2.45	2.90
HC—CHO	2.20	2.40	
HC—Ar	2.25	2.45	2.85
HC—NR$_2$	2.25	2.40	2.80
HC—(C=O)Ar	2.40	2.70	3.40
HC—NRAr	2.60	3.10	3.60
HC—SO$_2$R [HC—(SO)R]	2.60	3.05	
HC—Br	2.70	3.40	4.10
HC—NH(C=O)R	2.95	3.35	3.85
HC—Cl	3.05	3.45	4.05
HC—OH and HC—OR	3.20	3.40	3.60
HC—NH$_2$	3.50	3.75	4.05
HC—O(C=O)R	3.65	4.10	4.95
HC—OAr	3.80	4.00	4.60
HC—O(C=O)Ar	3.80	4.20	5.05
HC—F	4.25	4.50	4.80
HC—NO$_2$	4.30	4.35	4.60
Cyclopropane		0.20	0.40
Cyclopentane		2.45	
Cyclohexane		1.50	1.80

Substituent group	Proton shift	Substituent group	Proton shift
HC˙C	2.35	HO—C=O	0–12
HC˙CAr	2.90	HO—SO$_2$	11–12
HC˙C—C=C	2.75	HO—AR	0.5–5.6
HAr	7.20	HO—R	0.5–4.5
HCO—O	8.1	HS—Ar	2.8–3.6
HCO—R	9.4–10.0	HS—R	1–2
HCO—Ar	9.7–10.5	HN—Ar	3–6
HO—N=C (oxime)	9–12	HN—R	0.5–5

Source: J. A. Dean, ed., *Handbook of Organic Chemistry*, McGraw-Hill, New York, 1986, pp. 6-70 to 6-71.

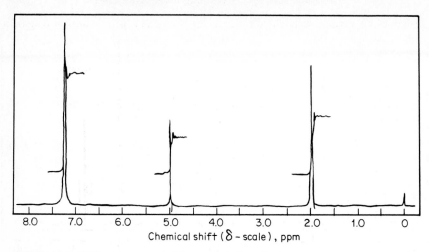

8.0   7.0   6.0   5.0   4.0   3.0   2.0   1.0   0

Chemical shift (δ - scale) , ppm

FIGURE 14.3 NMR spectrum containing only singlets. The integrator trace is superimposed on each resonance peak. Each resonance position (and flanking groups) is aryl(—C—), (aryl)—CH$_2$—(OC=O), and CH$_2$—(C=O).

a carbonyl group would be around $\delta = 2.0$. Chemical shifts settle the question (combined with the integrator traces which are discussed in a later section).

14.4.2 Spin-Spin Coupling

The real power of NMR derives from its ability to define complete sequences of groups or arrangements of atoms in the molecule. The absorption band multiplicities (splitting patterns) give the spatial positions of the nuclei. These splitting patterns arise through reciprocal magnetic interaction between spinning nuclei in a molecular system facilitated by the strongly magnetic binding electrons of the molecule in the intervening bonds. This coupling, called *spin-spin coupling,* causes mutual splitting of the otherwise sharp resonance lines into multiplets (Fig. 14.4). The strength of the spin-spin coupling, denoted by J, is given by the spacings between the individual lines of the multiplets, which will be the same in both coupled multiplets—an aid in structure elucidation. In Fig. 14.4 note the identical spacing between the quartet centered around $\delta = 4.0$ and the triplet centered around $\delta = 1.4$. Also note the identical spacings for the two doublets at $\delta = 6.85$ and $\delta = 7.40$.

Proton-proton couplings in aliphatic organic compounds are usually transmitted through only two or three bonds. Longer-range coupling occurs between protons separated by unsaturated systems. Couplings also depend on geometry. Adjacent axial-axial protons of cyclohexane are strongly coupled, whereas axial-equatorial and equatorial-equatorial protons are coupled only moderately. Coupling constants of *trans* and *cis* protons on olefinic double bonds have a ratio of approximately 2, which is useful in assigning structures of geometric isomers.

The number of lines in multiplets from protons is $n + 1$ lines, where n is the number of nuclei producing the splitting. Thus, one neighboring proton splits the observed resonance of a proton on an adjacent group into a doublet whose intensities are in the ratio 1:1, two produce a triplet (1:2:1), three a quartet (1:3:3:1),

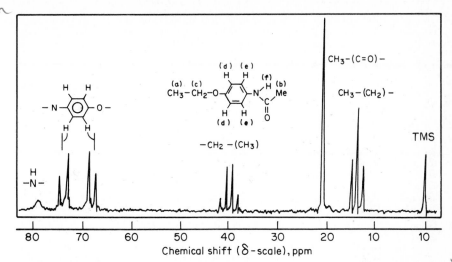

FIGURE 14.4 NMR spectrum of phenacetin, illustrating the origin of resonance positions and multiplet patterns.

four a quintet (1:4:6:4:1), and so on. The peaks within a multiplet are proportional to the coefficients of the binomial expansion. These features are illustrated in Fig. 14.4. The relative intensity of each of the multiplets, integrated over the whole multiplet, is proportional to the number of nuclei in the group—useful for quantitative work and for structural assignments.

As the external magnetic field strength increases, the multiplets move farther apart but the spacing of the peaks within each multiplet remains the same. This is one distinct advantage in the use of high-field strengths obtainable with 300-, 400-, and 500-MHz NMR spectrometers. The ratio J/δ is crucial and determines the appearance of the spectrum. When J/δ is 0.05 or less, the spectrum consists of well-separated multiplets with the theoretical intensities of the peaks. When $J/\delta \hbar 0.15$, the intensities are no longer binomial and the spectrum deviates noticeably from ideal appearance. When the chemical shift difference vanishes, the multiplet collapses to a singlet. For example, a C—O—CH$_2$—CH$_2$—O—C structure would show only a single peak and not a pair of triplets; the chemical shift for each methylene group is identical.

Other nuclei with spins of 1/2 interact with protons (and each other) and cause observable spin-spin couplings. Only significant numbers of fluorine and phosphorus atoms occur naturally; the presence of one of these elements may be deduced from an otherwise unexplained coupling effect in a proton spectrum.

14.4.3 Integration

Under proper operating conditions, the area under an absorption signal is directly proportional to the number of protons producing that signal. Thus, we have available a proton counter to supplement the mass spectrometer as a carbon counter. The integrator allows the protons to be assigned within each resonance group if the total is known. If the total is not known, quite often a particular singlet or multiplet can be tentatively assigned protons, from which the proportional numbers in the remaining groups can be found. If the numbers are rational in each multiplet, the assignments can be accepted. Referring again to Fig. 14.3, the heights of the integrals are in the proportion (in millimeters from the trace) 28: 11:17. Either the downfield singlet—assumed to be a benzene ring—can be tentatively assigned five protons or the upfield singlet can be assigned three protons on the assumption it is a methyl group. Both assumptions lead to a proton ratio of 5:2:3, respectively, which provides a clue that the molecule contains a monosubstituted phenyl ring plus isolated methylene and methyl groups.

14.5 QUANTITATIVE ANALYSIS

The area under an absorption band (or multiplet) is proportional to the number of nuclei responsible for the absorption. A device for integrating the absorption signal is a standard item on commercial NMR spectrometers. Accuracy is typically within ±2 percent.

For quantitative analysis, a known amount of a reference compound can be included with the sample. Ideally, the signal of the reference compound should be a strong singlet lying in a region of the NMR spectrum unoccupied by sample peaks. Quantitative analysis requires that at least one resonance band from each component in a mixture be free from overlap by other absorption signals. The amount of the unknown is calculated by

$$W_{unk} = W_{std}\frac{N_{std}M_{unk}A_{unk}}{N_{unk}M_{std}A_{std}} \tag{14.3}$$

where A's = two peak areas
N's = numbers of protons in groups giving rise to absorption peaks
M's = molecular weights of unknown and standard
W's = weights of unknown and internal standard

Whenever the empirical formula is known, the total height divided by the number of protons yields the increment of height per proton. If the assignment of a particular absorption band has been deduced, the increment per height per proton is calculated from the height for the assigned group divided by the number of protons in the particular group. The same procedure would be used for other nuclei such as carbon 13 and nitrogen 15.

14.6 ELUCIDATION OF NMR SPECTRA

Application of NMR to structure analysis is based primarily on the empirical correlation of structure with observed chemical shifts and coupling constants. Extensive NMR spectra and surveys have been published.[5-8] An unique advantage of NMR is that spectra can often be interpreted by using reference data from structurally related compounds. Internal simulation programs the NMR spectrum of an unknown to be compared against a postulated structure.

EXAMPLE 14.1   Let us complete the structural elucidation of the compound whose NMR spectrum is shown in Fig. 14.4.

1. The singlet at $\delta = 2.1$ has to be from an isolated methyl group; its downfield position indicates that it is adjacent to a carbonyl group.
2. The triplet-quartet patterns are coupled with each other; the triplet is a methyl group split by two adjacent protons, which in turn are split into a quartet by the three methyl protons.
3. The pair of doublets in the aryl region imply strong coupling between protons *ortho* to each other, that is, a *para*-substituted aryl ring. Furthermore, different atoms must be connected directly to the ring. After consulting Table 14.2, we see that nitrogen causes the downfield shift to $\delta = 7.4$ and that oxygen causes the upfield shift to $\delta = 6.85$. Unperturbed benzene protons would appear at $\delta = 7.2$.
4. The "hump" at $\delta = 7.9$ can now be assigned to a proton connected to a nitrogen atom; the smearing of the proton signal is caused by the electric quadruple moment of nitrogen.

Now our pieces are:

$$CH_3—CH_2— \qquad CH_3—C(=O)— \qquad {}^\wedge NH— \qquad {}^\wedge N—C_6H_4—O \ (para\text{-})$$

The methylene group is far downfield; an adjacent oxygen would place it around $\delta = 3.6$. Something in the beta position is causing it to be further downfield; a likely candidate is the phenyl ring's double bonds. Putting the pieces together, we have

$$CH_3—CH_2—O—C_6H_4—NH—C(=O)—CH_3$$

which is phenacetin, often used in headache preparations.

14.6.1 Other Techniques Useful with Complex Spectra

14.6.1.1 Shaking with Deuterium Oxide. Brief vigorous shaking with a few drops of D_2O results in a complete exchange of deuterium for the labile protons attached to O, N, and S and the collapse of their absorption signal. In Fig. 14.5, note the collapse of the —OH signal and the conversion of the alpha-methylene quartet to a triplet. As a consequence of spin coupling to the hydroxyl proton and the adjacent methylene group, the signal had been a quartet. The other signals in this group are the six lines from the methine proton, shown as light lines over the trace and overlapping the quartet (and the triplet after deuteration).

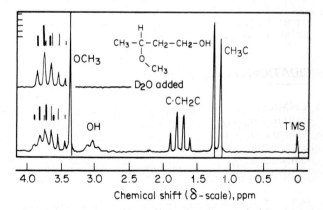

FIGURE 14.5 Example of hydroxyl deuteration.

14.6.1.2 Spin Decoupling. Figure 14.6 shows how spectra can be simplified and how spin-spin coupling partners can be identified. In homonuclear decoupling, the sample is simultaneously irradiated with a second, relatively strong rf field that is perpendicular to H_0 and that is at resonance with the nuclei to be decoupled; these nuclei collapse into single lines (or to the multiplicity from the normal coupling with a third nucleus). Irradiation of the methyl group simplifies the complex pattern of the vinyl protons from a quartet of doublets into simple doublets. If the aldehyde proton were to be irradiated, the small doublets (due to the weak coupling with the aldehyde proton) on the other multiplet patterns would disappear.

Heteronuclear decoupling involves irradiating the sample at the NMR frequency characteristic of the heteroatom. One could disentangle the coupling between proton and fluorine spins or the perturbations caused by nitrogen in the spectrum of dimethylformamide.

14.6.1.3 Shift Reagents. Often fluorinated β-diketones of europium(III) or praseodymium(III) will spread proton resonance peaks across a broader range of magnetic field strength, thereby improving the J/δ ratio. Increments of shift reagent are added until sufficient resolution is attained (Fig. 14.7). Eu(III) chelates induce downfield shifts (away from the TMS reference position) whereas those of Pr(III) induce upfield shifts.

14.6.1.4 Solvent Influence. Through screening of particular protons, solvents will cause shifts. For example, the addition of 50 percent benzene to a sample of 4-butyrolactone dissolved in CCl_4 causes the ring protons to appear at a higher field as well-separated multiplets.

14.6.2 Troubleshooting

In comparison to other spectroscopic methods, NMR is highly dependent on the performance of the spectrometer, the sample, the sample tube, and solute conditions. To obtain the appropriate information from spectra and to understand the basic instrumental problems, the discussion by Tchapla et al.[9] should be consulted.

REFERENCES

1. J. A. Dean, "Use of Nuclear Magnetic Resonance in Determination of Molecular Structure of Urinary Constituents of Low Molecular Weight," *Clin. Chem.,* **14**:326 (1968).

2. F. P. Miknis, V. J. Bartuska, and G. E. Maciel, "Crosspolarization ^{13}C NMR with Magic-Angle Spinning," *Am. Lab.,* **11**:19 (November 1979).

3. J. A. Dean, ed., *Lange's Handbook of Chemistry,* 13th ed., McGraw-Hill, New York, 1985.

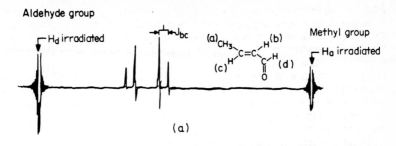

(a)

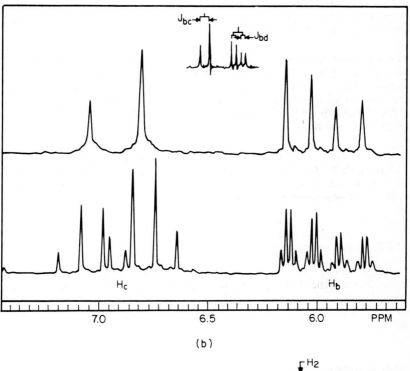

(b)

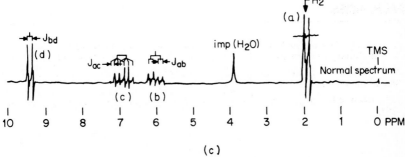

(c)

FIGURE 14.6 Double resonance spectrum of crotonaldehyde. (*a*) After irradiation of both the methyl and aldehyde protons; (*b*) after irradiation of the methyl protons; (*c*) normal NMR spectrum.

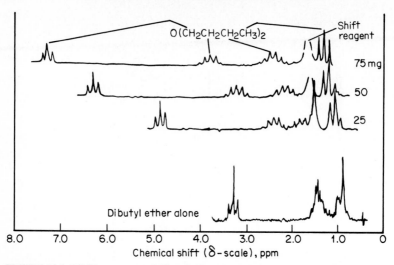

FIGURE 14.7 NMR spectrum of dibutyl ether alone (*lower trace*) and with increasing amounts of europium shift reagent added to the sample.

4. J. A. Dean, ed., *Handbook of Organic Chemistry*, McGraw-Hill, New York, 1986.
5. *The Aldrich Library of NMR Spectra*, Edition 7, Vols. 1 and 2, Aldrich Chemical Co., Milwaukee, 1983.
6. Sadtler Research Laboratories, *Nuclear Magnetic Resonance Spectra*, Philadelphia, a continuously updated subscription service.
7. R. M. Silverstein, G. C. Bassler, and T. C. Morrill, *Spectrometric Identification of Organic Compounds*, 4th ed., Wiley, New York, 1981.
8. Asahi Research Center Co., Ltd., Japan, *Handbook of Proton-NMR Spectra and Data*, Academic, Orlando, Fla., 1985.
9. A. Tchapla, G. Emptoz, A. Aspect, and A. Nahon, "Troubleshooting in Proton NMR," *Am. Lab.*, **17**:38 (November 1985).

BIBLIOGRAPHY

Allerhand, A., and S. R. Maple, "Ultra-High Resolution NMR," *Anal. Chem.*, **59**:441A (1987).

Becker, C. D., *High Resolution NMR*, Academic, New York, 1980.

Harris, R. K., *Nuclear Magnetic Resonance Spectroscopy*, Wiley, New York, 1986.

Levy, G. C., R. Lichter, and G. Nelson, *Carbon-13 Nuclear Magnetic Resonance Spectroscopy*, 2d ed., Wiley, New York, 1980.

Macomber, R. S., "A Primer on Fourier Transform NMR," *J. Chem. Educ.*, **62**:213 (1985).

Sanders, J. K. M., and B. K. Hunter, *Modern NMR Spectroscopy: A Guide for Chemists*, Oxford Univ. Press, New York, 1986.

Williams, D. A. R., and D. J. Mowthorpe, *Nuclear Magnetic Resonance Spectroscopy*, Wiley, New York, 1986.

CHAPTER 15
MASS SPECTROMETRY

Mass spectrometry provides qualitative and quantitative information about the atomic and molecular composition of inorganic and organic materials. As an analytical technique it possesses distinct advantages:

1. Increased sensitivity over most other analytical techniques because the analyzer, as a mass-charge filter, reduces background interference.
2. Excellent specificity from characteristic fragmentation patterns to identify unknowns or confirm the presence of suspected compounds.
3. Information about molecular weight is provided.

15.1 INSTRUMENT DESIGN

Functionally, all mass spectrometers have these components (Fig. 15.1):

1. An inlet sample system
2. An ion source
3. An ion acceleration system
4. A mass (ion) analyzer
5. An ion-collection system, usually an electron multiplier detector
6. A data handling system
7. A vacuum system connected to components 1 through 5

To provide a collision-free path for ions once they are formed, the pressure in the spectrometer must be less than 10^{-6} torr.

Calibration of a mass spectrometer must be done in the electron-impact mode, Perfluoroalkanes are often used as markers because they provide a peak at intervals of masses corresponding to CF_2 groups.

15.1.1 Inlet Sample Systems

Gas samples are transferred from a vessel of known volume (3 mL), where the pressure is measured, into a reservoir (3 to 5 L). Volatile liquids are drawn through a sintered disk into the low-pressure reservoir where they are vaporized

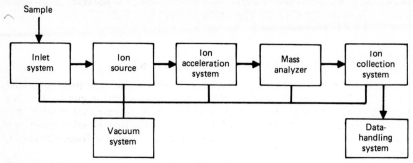

FIGURE 15.1 Components of a mass spectrometer.

instantly. Oftentimes a nonvolatile compound can be converted into a derivative that has sufficient vapor pressure.

The gaseous sample enters the source through a pinhole in a piece of gold foil. For analytical work, molecular flow (where the mean free path of gas molecules is greater than the tube diameter) is usually preferred. However, in isotope-ratio studies viscous flow (where the mean free path is smaller than the tube diameter) is employed to avoid any tendency for various components to flow differently from the others.

15.2 IONIZATION METHODS IN MASS SPECTROMETRY

Ionization methods in mass spectrometry are divided into gas-phase ionization techniques and methods that form ions from the condensed phase inside the ion source. All ion sources are required to produce ions without mass discrimination from the sample and to accelerate them into the mass analyzer. The usual source design, shown in Fig. 15.2 specifically for an electron-impact ion source, has an ion withdrawal and focusing system. The ions formed are removed electrostatically from the chamber. Located behind the ions is the repeller which has the same charge as the ions to be withdrawn. A strong electrostatic field between the

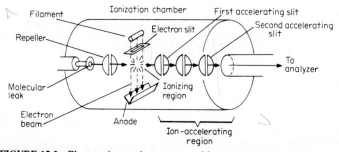

FIGURE 15.2 Electron-impact ion source and ion accelerating system.

first and second accelerating slits of 400 to 4000 V, which is opposite in charge to the ions, accelerates the ions to their final velocities.

15.2.1 Electron-Impact Ionization

The electron-impact source (Fig. 15.2) is the most commonly used ionization method. The ionizing electrons from the cathode of an electron gun located perpendicular to the incoming gas stream collide with the sample molecules to produce a molecular ion. A source operating at 70 V, the conventional operating potential, has sufficient energy to also cause the characteristic fragmentation of sample molecules.

Some compounds do not give a molecular ion in an electron-impact source. This is a disadvantage of this source.

15.2.2 Chemical Ionization[1]

Chemical ionization results from ion-molecule chemical interactions that involve a small amount of sample with an exceedingly large amount of a reagent gas. The source must be tightly enclosed with an inside pressure of 0.5 to 4.0 torr. The pressure outside the source is kept at about 4 orders of magnitude less than the inside by a differential pumping system (Fig. 15.3).

Often the primary reason for using this technique is to determine the molecular weight of a compound. For this purpose a low-energy reactant, such as $tert$-$C_4H_9^+$ (from isobutane) is frequently used. In the first step the reagent gas is ionized by electron-impact ionization in the source. Subsequent reactions between the primary ion and additional reagent gas produce a stabilized reagent gas plasma. When a reagent ion encounters a sample molecule (MH), several products may be formed:

MH_2^+ by proton transfer

M^+ by hydride abstraction

MH^+ by charge transfer

Practically all the spectral information will be clustered around the molecular ion, or one mass unit larger or smaller, with little or no fragmentation. This type of ionization is desirable when an analysis of a mixture of compounds is needed and the list of possible components is limited. The general absence of carbon-carbon cleavage reactions for the chemical ionization spectra means that they provide little skeletal information.

Negative chemical ionization[2] can be conducted with hydroxide and halide ions. For these studies the charges on the repeller and accelerating slits in the ion source are reversed, with the repeller having a negative charge.

15.2.3 Other Ionization Methods

The less frequently used ionization methods receive only brief mention here. For more details consult the references cited.

$Field$ $ionization$[3] and $field$ $desorption$[4] are techniques used for studying sur-

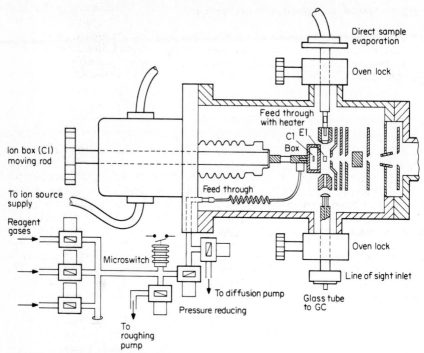

FIGURE   15.3 Combination   chemical   ionization   and   electron-impact   ionization source. (*Courtesy of Varian Associates.*)

face phenomena, such as adsorbed species and trapped samples, and the results of chemical reactions on surfaces; they are also suitable for handling large lipophilic polar molecules.

Fast atom bombardment[5] and *californium 252 desorption*[6] techniques deal rather effectively with polar substances (usually of higher molecular weight) and salts. Samples may be bulk solids, liquid solutions, thin films, or monolayers.

15.3   MASS ANALYZERS

The function of the mass analyzer is to separate the ions produced in the ion source according to their different mass-charge ratios. The analyzer section is continuously pumped to a very low vacuum so that ions may pass through without colliding with the gas molecules. The energies and velocities v of the ions moving into the mass analyzer are determined by the accelerating voltage V from the ion source slits and the charge z on the ions of mass m:

$$\frac{1}{2}m_1v_1^2 = \frac{1}{2}m_2v_2^2 = \frac{1}{2}m_3v_3^2 = \cdots = zV \tag{15.1}$$

15.3.1 Magnetic-Deflection Mass Analyzer

In a single-focusing magnetic-sector mass analyzer, the ion source, the collector slit, and the apex of the sector shape (usually 60°) are collinear, as shown in Fig. 15.4. Upon entering the magnetic field, the ions are classified and segregated into beams, each with a different m/z ratio.

$$\frac{m}{z} = \frac{H^2 r^2}{2V} \tag{15.2}$$

where H is the strength of the magnetic field and r is the radius of the circular path followed by the ions. Since the radius and the magnetic field strength are fixed for the particular sector instrument, only ions with the proper m/z ratio will pass through the analyzer tube without striking the walls, where they are neutralized and pumped out of the system as neutral gas molecules. Focusing is accomplished by changing either the electrostatic accelerating voltage or the magnetic field strength; often the former is allowed to diminish while the spectrum is scanned. Each m/z ion from light to heavy is successively swept past the detector slit at a known rate. The detector current is amplified and displayed on a strip-chart recorder. Since the ion paths are separated from one another, the recorder signal will fall to the baseline and then rise as each mass strikes the detector. The height of the peaks on the chart will be proportional to the number of ions of the corresponding mass-charge ratio.

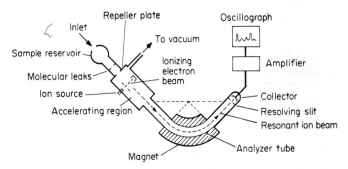

FIGURE 15.4   Schematic diagram of a Nier 60° sector mass spectrometer.

A magnetic-sector mass analyzer has a mass range of 2500 at 4-kV ion accelerating voltage. Mass resolution is continuously variable up to 25 000 (10 percent valley definition). Metastable peaks that aid in structural elucidation are recorded.

15.3.2 Double-Focusing Sector Spectrometers

Because single-focusing mass analyzers are not velocity focusing for ions of a given mass, their resolving power is limited. In double-focusing mass spectrometers an electrostatic deflection field is incorporated between the ion source and the magnetic analyzer. Resolving power lies in the range of 100 000. In the spectrometer shown in Fig. 15.5, additional focusing is achieved with quadrupole

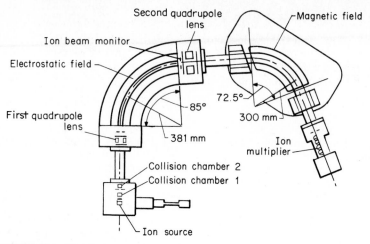

FIGURE 15.5   Ultrahigh resolution double-focusing mass spectrometer.   (*Courtesy of Jeol, Ltd.*)

lenses placed before the electrostatic field and between the electrostatic and magnetic fields.⟩

15.3.3   Quadrupole Mass Analyzer

In the quadrupole mass analyzer, ions from the ion source are injected into the quadrupole array, shown in Fig. 15.6.
⌐ Opposite pairs of electrodes are electrically connected, one pair at $+U_{dc}$ volts and the other pair at $-U_{dc}$ volts. An rf oscillator supplies a signal to each pair of

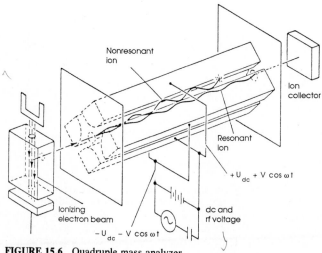

FIGURE 15.6   Quadruple mass analyzer.

rods, but the signal to the second pair is retarded by 180°. When the ratio U_{dc}/V_{rf} is controlled, the quadrupole field can be set to pass ions of only one m/z ratio down the entire length of the quadrupole array. When the dc and rf amplitudes are changed simultaneously, ions of various mass-charge ratios will pass successively through the array to the detector and an entire mass spectrum can be produced.

Registration of negative ions, as from a chemical ionization source, is possible with two electron multipliers, one for positive and one for negative ions.

Scan rates can reach 780 daltons · s^{-1} before resolution is significantly affected. The quadrupole mass analyzer is ideal for coupling with a gas chromatograph.

15.3.4 Time-of-Flight Spectrometer

In the time-of-flight (TOF) mass spectrometer (Fig. 15.7), the ions leave the source as discrete ion packets by pulsing the voltage on the accelerating slits at the exit of the ion source. Upon leaving the accelerating slits, the ions enter into the field-free region (drift path) of the flight tube, 30 to 100 cm long, with whatever velocity they have acquired (Eq. 15.1). Because their velocities are inversely proportional to the square roots of their masses, the lighter ions travel down the flight tube faster than the heavier ions. The original ion packet becomes separated into "wafers" of ions according to their mass/charge ratio. The wafers are collected sequentially at the detector.

A spectrum can be recorded every 10 s. This makes the TOF mass spectrometer suitable for kinetic studies and for coupling with a gas chromatograph to examine effluent peaks. Resolution is limited to masses below 1000 daltons.

15.3.5 Resolving Power

The most important parameter of a mass analyzer is its resolving power. Using the 10 percent *valley definition,* two adjacent peaks (whose mass differences are

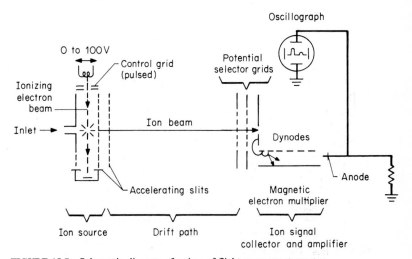

FIGURE 15.7   Schematic diagram of a time-of-flight mass spectrometer.

Δm) are said to be separated when the valley between them is 10 percent or less of the peak height (and the peak heights are approximately equal). For this condition, $\Delta m/m$ equals the peak width at a height that is 5 percent of the individual peak height.

A resolution of 1 part in 800 adequately distinguishes between m/z 800 and m/z 801 so long as the peak intensity ratio is not greater than 10 to 1. However, if one wanted to distinguish between the parent peaks of 2,2-naphthylbenzothiophene (260.0922) and 1,2-dimethyl-4-benzoylnaphthalene (260.1201), the required resolving power is

$$\frac{m}{\Delta m} = \frac{260}{260.1201 - 260.0922} = 9319$$

15.4   DETECTORS

After leaving a mass analyzer, the resolved ion beams sequentially strike some type of detector. The electron multiplier, either single or multichannel, is most commonly used.

15.4.1   Electron Multiplier

The electron multiplier resembles the photomultiplier tubes described in Chap. 6. The ion beam strikes a conversion dynode which converts the ion beam to an electron beam. Further multiplication is done in two ways. A discrete dynode multiplier has 15 to 18 individual dynodes arranged in a venetian blind configuration and coated with a material that has high secondary electron emission properties. A magnetic field forces the secondary electrons to follow circular paths, causing them to strike successive dynodes.

Continuous dynode multipliers are in the configuration of a curved hollow tube to form the multiplier channel. The walls are coated with the secondary emitter. A linear configuration is illustrated in Fig. 15.8 for a channel electron multiplier array. Pore diameters range from 10 to 25 μm.

15.5   CORRELATION OF MASS SPECTRA WITH MOLECULAR STRUCTURE

15.5.1   Molecular Identification

In the identification of a compound, the most important information is the molecular weight. The mass spectrometer is able to provide this information, often to four decimal places. One assumes that no ions heavier than the molecular ion form when using electron-impact ionization. The chemical ionization spectrum will often show a cluster around the nominal molecular weight.

Several relationships aid in deducing the empirical formula of the parent ion (and also molecular fragments). From the empirical formula, hypothetical molec-

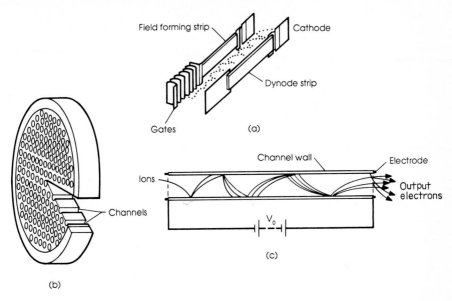

FIGURE 15.8 Electron multiplier detector. (*a*) Schematic of a continuous dynode multiplier (not as a bent cone). (*b*) Multichannel array. (*c*) Detail of electron amplification.

ular structures can be proposed using the entries in the formula indices of Beilstein and *Chemical Abstracts*.

15.5.2 Natural Isotopic Abundances

The relative abundances of natural isotopes produce peaks one or more mass units larger than the parent ion (Table 15.1*a*). For a compound $C_wH_xO_zN_y$, there is a formula that allows one to calculate the percent of the heavy isotope contributions from a monoisotopic peak P_M to the P_{M+1} peak:

$$100 \frac{P_{M+1}}{P_M} = 0.015x + 1.11w + 0.37y + 0.037z$$

Tables of abundance factors have been calculated for all combinations of C, H, N, and O up to mass 500.[7]

Compounds that contain chlorine, bromine, sulfur, or silicon are usually apparent from prominent peaks as masses, 2, 4, 6, and so on, units larger than the nominal mass of the parent or fragment ion. For example, when one chlorine atom is present, the $P + 2$ mass peak will be about one-third the intensity of the parent peak. When one bromine atom is present, the $P + 2$ mass peak will be about the same intensity as the parent peak. The abundance of heavy isotopes is treated in terms of the binomial expansion $(a + b)^m$, where a is the relative abundance of the light isotope, b is the relative abundance of the heavy isotope, and m is the number of atoms of the particular element present in the molecule. If two bromine atoms are present, the binomial expansion is

TABLE 15.1

(a) Abundances of some polyisotopic elements, %					
Element	% Abundance	Element	% Abundance	Element	% Abundance
^{1}H	99.985	^{16}O	99.76	^{33}S	0.76
^{2}H	0.015	^{17}O	0.037	^{34}S	4.22
^{12}C	98.892	^{18}O	0.204	^{35}Cl	75.53
^{13}C	1.108	^{28}Si	92.18	^{37}Cl	24.47
^{14}N	99.63	^{29}Si	4.71	^{79}Br	50.52
^{15}N	0.37	^{30}Si	3.12	^{81}Br	49.48

(b) Selected isotope masses			
Element	Mass	Element	Mass
^{1}H	1.0078	^{31}P	30.9738
^{12}C	12.0000	^{32}S	31.9721
^{14}N	14.0031	^{35}Cl	34.9689
^{16}O	15.9949	^{56}Fe	55.9349
^{19}F	18.9984	^{79}Br	78.9184
^{28}Si	27.9769	^{127}I	126.9047

Source: J. A. Dean, ed., *Lange's Handbook of Chemistry*, 13th ed., McGraw-Hill, New York, 1985.

$$(a + b)^2 = a^2 + 2ab + b^2$$

Now substituting the percent abundance of each isotope (^{79}Br and ^{81}Br) into the expansion:

$$(0.505)^2 + 2(0.505)(0.495) + (0.495)^2$$

gives

$$0.255 + 0.500 + 0.250$$

which are the proportions of $P:(P + 2):(P + 4)$, a triplet that is slightly distorted from a 1:2:1 pattern. When two elements with heavy isotopes are present, the binomial expansion $(a + b)^m(c + d)^n$ is used.

Sulfur 34 enhances the $P + 2$ peak by 4.2 percent; silicon 29 enhances the $P + 1$ peak by 4.7 percent and the $P + 2$ peak by 3.1 percent.

15.5.3 Exact Mass Differences

If the exact mass of the parent or fragment ions is ascertained with a high-resolution mass spectrometer, this relationship is often useful for combinations of C, H, N, and O (Table 15.1b):

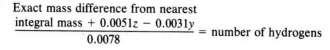

Exact mass difference from nearest
$$\frac{\text{integral mass} + 0.0051z - 0.0031y}{0.0078} = \text{number of hydrogens}$$

One substitutes integral numbers (guesses) for z (oxygen) and y (nitrogen) until the divisor becomes an integral multiple of the numerator within 0.0002 mass unit.

For example, if the exact mass is 177.0426 for a compound containing only C, H, O, and N (note the odd mass which indicates an odd number of nitrogen atoms), thus

$$\frac{0.0426 + 0.0051z - 0.0031y}{0.0078} = 7 \text{ hydrogen atoms}$$

when $z = 3$ and $y = 1$. The empirical formula is $C_9H_7NO_3$ since

$$\frac{177 - 7(1) - 1(14) - 3(16)}{12} = 9 \text{ carbon atoms}$$

15.5.4 Number of Rings and Double Bonds

The total number of rings and double bonds can be determined from the empirical formula ($C_wH_xO_zN_y$) by the relationship

$$\frac{1}{2}(2w - x + y + 2)$$

when covalent bonds make up the molecular structure. Remember that the total number for a benzene ring is 4 (one ring and three double bonds); for a triple bond it is 2.

15.5.5 General Rules

1. If the nominal molecular weight of a compound containing only C, H, O, and N is even, so is the number of hydrogen atoms it contains.
2. If the nominal molecular weight is divisible by 4, the number of hydrogen atoms is also divisible by 4.
3. When the nominal molecular weight of a compound containing only C, H, O, and N is odd, the number of nitrogen atoms must be odd.

15.5.6 Metastable Peaks

If the mass spectrometer has a field-free region between the exit of the ion source and the entrance to the mass analyzer, as does the sector instrument shown in Fig. 15.4, metastable peaks m^* may appear as a weak, diffuse (often humped-shaped) peak, usually at a nonintegral mass. The one-step decomposition process takes the general form:

$$\text{Original ion} \rightarrow \text{daughter ion} + \text{neutral fragment}$$

The relationship between the original ion and daughter ion is given by

$$m^* = \frac{(\text{mass of daughter ion})^2}{\text{mass of original ion}}$$

For example, a metastable peak appeared at 147.9 mass units in a mass spectrum with prominent peaks at 65, 91, 92, 107, 108, 155, 172, and 200 mass units. Try all possible combinations in the above expression. The fit is given by

$$147.9 = \frac{(172)^2}{200}$$

which provides this information:

$$200^+ \rightarrow 172^+ + 28$$

The probable neutral fragment lost is either $CH_2{=}CH_2$ or CO.

15.6   MASS SPECTRA AND STRUCTURE

The mass spectrum is a fingerprint for each compound because no two molecules are fragmented and ionized in exactly the same manner on electron-impact ionization. When mass spectra are reported, the data are normalized by assigning a value of 100 to the most intense peak (denoted as base peak). Other peaks are reported as percentages of the base peak.

A very good general survey for interpreting mass spectral data is given by Silverstein et al.[8]

15.6.1   Initial Steps in Elucidation of a Mass Spectrum

1. Tabulate the prominent ion peaks, starting with the highest mass.
2. Usually only one bond is cleaved. In succeeding fragmentations a new bond is formed for each additional bond that is broken.
3. When fragmentation is accompanied by the formation of a new bond as well as by the breaking of an existing bond, a rearrangement process is involved. The resulting mass peak will be even when only C, H, and O are involved. The migrating atom is almost exclusively hydrogen; six-membered cyclic transition states are most important.
4. Tabulate the probable groups that (a) give rise to the prominent charged ion peaks and (b) list the neutral fragments.

15.6.2   General Rules for Fragmentation Patterns

1. Bond cleavage is more probable at branched carbon atoms: tertiary > secondary > primary. The positive charge tends to remain with the branched carbon.
2. Double bonds favor cleavage beta to the carbon (but see rule 6).
3. A strong parent peak often indicates a ring.
4. Saturated ring systems lose side chains at the alpha carbon. Upon fragmentation, two ring atoms are usually lost.
5. A heteroatom induces cleavage at the bond beta to it.

6. Compounds that contain a carbonyl group tend to break at this group; the positive charge remains with the carbonyl portion.
7. For linear alkanes, the initial fragment lost is an ethyl group (never a methyl group), followed by propyl, butyl, and so on. An intense peak at mass 43 suggests a chain longer than butane.
8. The presence of Cl, Br, S, and Si can be deduced from the unusual isotopic abundance patterns of these elements. These elements can be traced through the positively charged fragments until the pattern disappears or changes as a result of the loss of one of these atoms to a neutral fragment.
9. When unusual mass differences occur between some ion fragments, the presence of F (mass difference 19), I (mass difference 127), or P (mass difference 31) should be suspected.

15.6.3 Characteristic Low-Mass Fragment Ions

Mass 30	Primary amines
Masses 31, 45, 59	Alcohol or ether
Masses 19 and 31	Alcohol
Mass 66	Monobasic carboxylic acid
Masses 77 and 91	Benzene ring

15.6.4 Characteristic Low-Mass Neutral Fragments from the Molecular Ion

Mass 18 (H_2O)	From alcohols, aldehydes, ketones
Mass 19 (F) and 20 (HF)	Fluorides
Mass 27 (HCN)	Aromatic nitriles or nitrogen heterocycles
Mass 29	Indicates either CHO or C_2H_5
Mass 30	Indicates either CH_2O or NO
Mass 33 (HS) and 34 (H_2S)	Thiols
Mass 42	CH_2CO via rearrangement from methyl ketone or an aromatic acetate or an aryl-$NHCOCH_3$ group
Mass 43	C_3H_7 or CH_3CO
Mass 45	—COOH or OC_2H_5

15.7 SECONDARY ION MASS SPECTROMETRY

Secondary ion mass spectrometry (SIMS) is used for the analysis of surface layers and their composition to a depth of 1 to 3 nm. A focused ion beam strikes the sample surface and releases secondary ions, which are detected by a mass spectrometer. Typical instrumentation might involve a plasma discharge source coupled with a quadrupole mass analyzer. The plasma discharge also serves as the sputtering device to remove successive layers of sample for profiling the material. The SIMS technique affords qualitative identification of all surface elements and permits identification of isotopes and the structural elucidation of molecular compounds present on a surface. Detection sensitivity is in parts per million. SIMS is also useful for analyzing nonvolatile and thermally labile molecules, including polymers and large biomolecules.

REFERENCES

1. B. Munson, "Chemical Ionization Mass Spectrometry," *Anal. Chem.*, **49:**772A (1977).
2. R. C. Dougherty, "Negative Chemical Ionization Mass Spectrometry," *Anal. Chem.*, **53:**625A (1981).
3. M. Anbar and W. H. Aberth, "Field Ionization Mass Spectrometry," *Anal. Chem.*, **46:** 59A (1974).
4. W. D. Reynold, "Field Desorption Mass Spectrometry," *Anal. Chem.*, **51:**283A (1979).
5. M. Barber et al., "Fast Atom Bombardment Mass Spectrometry," *Anal. Chem.*, **54:** 645A (1982).
6. R. D. Macfarlane, "Californium-252 Plasma Desorption Mass Spectrometry," *Anal. Chem.*, **55:**1247A (1983).
7. J. H. Beynon and A. E. Williams, *Mass and Abundance Tables for Use in Mass Spectrometry,* Elsevier, Amsterdam, 1963.
8. R. M. Silverstein, G. C. Bassler, and T. C. Morrill, *Spectrometric Identification of Organic Compounds,* 4th ed., Wiley, New York, 1981.

BIBLIOGRAPHY

McLafferty, F., and D. Stauffer, *Registry of Mass Spectral Data,* 2d ed., Wiley, New York, 1988.
Watson, J. T., *Introduction to Mass Spectrometry,* Raven, New York, 1985.

CHAPTER 16
INTRODUCTION TO ELECTROCHEMICAL METHODS

16.1 INTRODUCTION

Many chemical reactions can be classified as oxidation-reduction reactions (redox reactions) and can be considered to be the result of two reactions, one oxidation and the other reduction. An element is said to have undergone oxidation if it loses electrons or if its oxidation state has increased; that is, it has attained a more positive charge. An element is said to have undergone reduction if it gains electrons or if its oxidation state has been reduced; that is, it has attained a more negative charge. Atoms of elements in their elemental state have a 0 charge.

A typical redox reaction occurs when metallic zinc reacts with a solution of lead(II) nitrate. If a piece of metallic zinc is dropped into a solution of lead(II) nitrate, the zinc atoms lose electrons to become zinc ions with a 2+ charge; they are oxidized. Lead ions (with a 2+ charge) become metallic lead with a 0 charge; they are reduced.

$$Zn(s) + Pb^{2+}(aq) \rightarrow Zn^{2+}(aq) + Pb(s) \tag{16.1}$$

The overall reaction can be considered the result of two half-reactions:

$$Zn(s) \rightarrow Zn^{2+}(aq) + 2e^- \quad \text{oxidation reaction} \tag{16.2}$$

$$Pb^{2+}(aq) + 2e^- \rightarrow Pb(s) \quad \text{reduction reaction} \tag{16.3}$$

Lead is a *stronger oxidizing agent* than zinc because, in the reaction above, the lead has a stronger attraction for the electrons and takes them away from the zinc. By the same reasoning, zinc is a stronger reducing agent than the lead because it releases its electrons more easily. Each substance has its own characteristic affinity for electrons, and that affinity is the basis for constructing the electrochemical-reaction or electromotive series table (Table 16.1) which reflects that affinity in terms of volts. The table reflects the difference in affinity for electrons between an element and its ion or between two intermediate oxidation states of an element. The values given in Table 16.1 are reduction potentials, where $E°$ is the single electrode potential when each substance involved in the oxidation-reduction reaction is at unit "activity." If a particular substance is more easily oxidized than hydrogen, its $E°$ is assigned a negative value. If a substance is not oxidized as easily as hydrogen its $E°$ is assigned a positive value.

As one proceeds downward in Table 16.1, the oxidizing agents decrease in

TABLE 16.1   Potentials of Selected Half-Reactions at 25°C

A summary of oxidation-reduction half-reactions arranged in order of decreasing oxidation strength and useful for selecting reagent systems.

Half-reaction		$E°$, V
$F_2(g) + 2H^+ + 2e^-$	$= 2HF$	3.06
$O_3 + 2H^+ + 2e^-$	$= O_2 + H_2O$	2.07
$S_2O_8^{2-} + 2e^-$	$= 2SO_4^{2-}$	2.01
$Ag^{2+} + e^-$	$= Ag^+$	2.00
$H_2O_2 + 2H^+ + 2e^-$	$= 2H_2O$	1.77
$MnO_4^- + 4H^+ + 3e^-$	$= MnO_2(s) + 2H_2O$	1.70
$Ce(IV) + e^-$	$= Ce(III)(in\ 1\ M\ HClO_4)$	1.61
$H_5IO_6 + H^+ + 2e^-$	$= IO_3^- + 3H_2O$	1.6
$Bi_2O_4(bismuthate) + 4H^+ + 2e^-$	$= 2BiO^+ + 2H_2O$	1.59
$BrO_3^- + 6H^+ + 5e^-$	$= ½Br_2 + 3H_2O$	1.52
$MnO_4^- + 8H^+ + 5e^-$	$= Mn^{2+} + 4H_2O$	1.51
$PbO_2 + 4H^+ + 2e^-$	$= Pb^{2+} + 2H_2O$	1.455
$Cl_2 + 2e^-$	$= 2Cl^-$	1.36
$Cr_2O_7^{2-} + 14H^+ + 6e^-$	$= 2Cr^{3+} + 7H_2O$	1.33
$MnO_2(s) + 4H^+ + 2e^-$	$= Mn^{2+} + 2H_2O$	1.23
$O_2(g) + 4H^+ + 4e^-$	$= 2H_2O$	1.229
$IO_3^- + 6H^+ + 5e^-$	$= ½I_2 + 3H_2O$	1.20
$Br_2(liq) + 2e^-$	$= 2Br^-$	1.065
$ICl_2^- + e^-$	$= ½I_2 + 2Cl^-$	1.06
$VO_2^+ + 2H^+ + e^-$	$= VO^{2+} + H_2O$	1.00
$HNO_2 + H^+ + e^-$	$= NO(g) + H_2O$	1.00
$NO_3^- + 3H^+ + 2e^-$	$= HNO_2 + H_2O$	0.94
$2Hg^{2+} + 2e^-$	$= Hg_2^{2+}$	0.92
$Cu^{2+} + I^- + e^-$	$= CuI$	0.86
$Ag^+ + e^-$	$= Ag$	0.799
$Hg_2^{2+} + 2e^-$	$= 2Hg$	0.79
$Fe(III) + e^-$	$= Fe^{2+}$	0.771
$O_2(g) + 2H^+ + 2e^-$	$= H_2O_2$	0.682
$2HgCl_2 + 2e^-$	$= Hg_2Cl_2(s) + 2Cl^-$	0.63
$Hg_2SO_4(s) + 2e^-$	$= 2Hg + SO_4^{2-}$	0.615
$H_3AsO_4 + 2H^+ + 2e^-$	$= HAsO_2 + 2H_2O$	0.581
$Sb_2O_5 + 6H^+ + 4e^-$	$= 2SbO^+ + 3H_2O$	0.559
$I_3^- + 2e^-$	$= 3I^-$	0.545
$Cu^+ + e^-$	$= Cu$	0.52
$VO^{2+} + 2H^+ + e^-$	$= V^{3+} + H_2O$	0.337
$Fe(CN)_6^{3-} + e^-$	$= Fe(CN)_6^{4-}$	0.36
$Cu^{2+} + 2e^-$	$= Cu$	0.337
$UO_2^{2+} + 4H^+ + 2e^-$	$= U^{4+} + 2H_2O$	0.334
$BiO^+ + 2H^+ + 3e^-$	$= Bi + H_2O$	0.32
$Hg_2Cl_2(s) + 2e^-$	$= 2Hg + 2Cl^-$	0.2676
$AgCl(s) + e^-$	$= Ag + Cl^-$	0.2223
$SbO^+ + 2H^+ + 3e^-$	$= Sb + H_2O$	0.212
$CuCl_3^{2-} + e^-$	$= Cu + 3Cl^-$	0.178
$SO_4^{2-} + 4H^+ + 2e^-$	$= SO_2(aq) + 2H_2O$	0.17
$Sn^{4+} + 2e^-$	$= Sn^{2+}$	0.154
$S + 2H^+ + 2e^-$	$= H_2S(g)$	0.141
$TiO^{2+} + 2H^+ + e^-$	$= Ti^{3+} + H_2O$	0.10
$S_4O_6^{2-} + 2e^-$	$= 2S_2O_3^{2-}$	0.08
$AgBr(s) + e^-$	$= Ag + Br^-$	0.071

TABLE 16.1 Potentials of Selected Half-Reactions at 25°C (*Continued*)

A summary of oxidation-reduction half-reactions arranged in order of decreasing oxidation strength and useful for selecting reagent systems.

Half-reaction		$E°$, V
$2H^+ + 2e^-$	$= H_2$	0.0000
$Pb^{2+} + 2e^-$	$= Pb$	$- 0.126$
$Sn^{2+} + 2e^-$	$= Sn$	$- 0.136$
$AgI(s) + e^-$	$= Ag + I^-$	$- 0.152$
$Mo^{3+} + 3e^-$	$= Mo$	$- 0.2$
$N_2 + 5H^+ + 4e^-$	$= H_2NNH_3^+$	$- 0.23$
$Ni^{2+} + 2e^-$	$= Ni$	$- 0.246$
$V^3 + e^-$	$= V^{2+}$	$- 0.255$
$Co^{2+} + 2e^-$	$= Co$	$- 0.277$
$Ag(CN)_2^- + e^-$	$= Ag + 2CN^-$	$- 0.31$
$Cd^{2+} + 2e^-$	$= Cd$	$- 0.403$
$Cr^{3+} + e^-$	$= Cr^{2+}$	$- 0.41$
$Fe^{2+} + 2e^-$	$= Fe$	$- 0.440$
$2CO_2 + 2H^+ + 2e^-$	$= H_2C_2O_4$	$- 0.49$
$H_3PO_3 + 2H^+ + 2e^-$	$= H_3PO_2 + H_2O$	$- 0.50$
$U^{4+} + e^-$	$= U^{3+}$	$- 0.61$
$Zn^{2+} + 2e^-$	$= Zn$	$- 0.763$
$Cr^{2+} + 2e^-$	$= Cr$	$- 0.91$
$Mn^{2+} + 2e^-$	$= Mn$	$- 1.18$
$Zr^{4+} + 4e^-$	$= Zr$	$- 1.53$
$Ti^{3+} + 3e^-$	$= Ti$	$- 1.63$
$Al^{3+} + 3e^-$	$= Al$	$- 1.66$
$Th^{4+} + 4e^-$	$= Th$	$- 1.90$
$Mg^{2+} + 2e^-$	$= Mg$	$- 2.37$
$La^{3+} + 3e^-$	$= La$	$- 2.52$
$Na^+ + e^-$	$= Na$	$- 2.714$
$Ca^{2+} + 2e^-$	$= Ca$	$- 2.87$
$Sr^{2+} + 2e^-$	$= Sr$	$- 2.89$
$K^+ + e^-$	$= K$	$- 2.925$
$Li^+ + e^-$	$= Li$	$- 3.045$

Source: J. A. Dean, ed., *Lange's Handbook of Chemistry*, 13th ed., McGraw-Hill, 1985.

strength and the reducing agents increase. In general, if the two half-reactions are represented by the following equations:

$$(\text{Oxidizing agent})_1 + ne^- = (\text{reducing agent})_1 \qquad (16.4)$$

$$(\text{Oxidizing agent})_2 + ne^- = (\text{reducing agent})_2 \qquad (16.5)$$

and the second equation occurs lower in the table than the first equation, then a reaction may occur between oxidant$_1$ and reductant$_2$, whereas no reaction is possible between oxidant$_2$ and reductant$_1$. No predictions concerning the rate of the reaction are possible.

If two or more reactions between two substances are possible, usually the reaction that involves half-reactions which are farthest apart in the table will occur first.

16.2　ELECTROCHEMICAL CELL

We cannot determine absolute electrode potentials; therefore all electrode potentials are based on an arbitrary standard, the potential of the standard hydrogen electrode (SHE). Thus, when the electrode potential of an oxidation-reduction reaction is mentioned, what is actually implied is the electromotive force (emf) of the electrochemical cell, which is made up of two half-cells:

$$\text{Pt, } H_2(1 \text{ atm}) \,|\, H^+(m = 1.228) \,||\, M^{n+}(a = 1), M^\circ$$

standard hydrogen electrode　　　individual half-cell

liquid junction

where the partial pressure of hydrogen gas at the surface of the platinum electrode is 1 atm (equal to 1.01325×10^5 Pa or 1.013 25 bar), m is the molal concentration of hydrogen ions, and a is the activity of the M^{n+} species. Other standard notations used in the half-cell are a vertical line to represent a phase boundary, a comma to separate two components in the same phase, and a double vertical line to indicate a phase boundary with an associated liquid-junction potential.

To achieve the largest possible surface area, the platinum electrode is coated with a layer of finely divided platinum, known as platinum black. At all temperatures the standard hydrogen electrode is assigned a value of 0.

By an international convention adopted by the International Union of Pure and Applied Chemistry (IUPAC), all standard potentials, such as listed in Table 16.1, are written as reductions.

16.3　ACTIVITY SERIES

16.3.1　Relative Reactivities of Different Metals

Metals vary in their activity, or in the ease with which they give up electrons to form ions. Some metals are extremely difficult to change to their corresponding ions, while others are extremely easy to convert to ions. Lithium is the most active metal. The activity decreases with each succeeding metal. Metals below hydrogen in Table 16.1 yield hydrogen when treated with an acid. Those above hydrogen are unreactive with acids (unless the acids are oxidizing agents). Potassium, sodium, and calcium are so active that they react with cold water. Aluminum, zinc, and iron react with water if it is in the form of steam. Any metal which is below another metal will displace that metal in solution, as was shown in Eq. 16.1.

16.4　OXIDATION NUMBERS

The oxidation number is sometimes called the *oxidation state*. The basic facts to remember are:

1. Elements in the free state have an oxidation number of *zero*.

2. In simple ions, which contain one atom, the oxidation number is equal to the charge on the ion.

3. In the majority of compounds which contain oxygen (except peroxides), the oxidation number of each oxygen atom is -2. In peroxides the oxidation number of oxygen is -1.

4. In the majority of compounds which contain hydrogen, the oxidation number of hydrogen is $+1$, except in hydrides, where it is -1.

5. In molecules which are neutral, the sum of the positive charges equals the sum of the negative charges.

6. In radicals and complex ions, the sum of all the oxidation numbers making up the radical or complex ion must equal the sum of the charge on the ion.

7. Among the nonmetals, the oxidation number of the element in anions usually changes in steps of two electrons; for example, sulfur is $+6$ in SO_4^{2-}, $+4$ in SO_3^{2-}, $+2$ in $S_2O_3^{2-}$ (the average), 0 in free sulfur, and -2 in S^{2-}. Likewise, chlorine is $+7$ in $HClO_4$, $+5$ in $HClO_3$, $+3$ in $HClO_2$, $+1$ in $HClO$, and -1 in Cl^-. Also, nitrogen is $+5$ in HNO_3, $+3$ in HNO_2, $+1$ in N_2O, -1 in NH_2OH, and -3 in NH_4^+.

16.5 EQUIVALENT WEIGHT IN REDOX REACTIONS

In redox reactions the equivalent weight (or mass) involves the transfer of 1 mol of electrons. Thus, the redox equivalent weight of a substance can be calculated by dividing its molecular (formula) weight by the number of electrons which it loses or gains in the half-reaction which takes place.

A 1 N oxidizing solution or a 1 N reducing solution has one equivalent weight of the reactant in 1 L of solution. Thus, a 1 N solution of $KMnO_4$ (when reduced to Mn^{2+}) would be 0.2 M.

16.6 EFFECT OF CONCENTRATION ON ELECTRODE POTENTIALS

The potential E of any electrode (half-cell) for an oxidation-reduction system (here abbreviated ox/red), such as Eqs. 16.4 and 16.5, is given by the generalized form of the Nernst equation:

$$E = E° - \frac{RT}{nF} \ln \frac{a_{red}}{a_{ox}} = E° - \frac{2.3026RT}{nF} \log \frac{a_{red}}{a_{ox}} \qquad (16.6)$$

where
$E°$ = standard electrode potential
R = molar gas constant
T = absolute temperature
n = number of electrons transferred in electrode reaction
a_{ox}, a_{red} = activities of oxidized and reduced forms involved in electrode reaction, respectively

Stating the Nernst expression in terms of concentrations and common logarithms and assuming the temperature is 25°C (298 K), Eq. 16.6 becomes

$$E = E^{\circ\prime} - \frac{0.059\ 16}{n} \log \frac{[\text{red}]}{[\text{ox}]} \tag{16.7}$$

where E° is the formal electrode potential defined in terms of concentration (moles per liter).

16.7 EFFECT OF COMPLEX FORMATION ON ELECTRODE POTENTIALS

Reagents that react with one or both participants of an electrode process can significantly affect the electrode potential. Two cases will be examined. In the presence of significant concentrations of ammonia, the zinc-zinc(II) system will be converted essentially to a zinc-zinc(II) tetrammine system. The formation of the zinc tetrammine complex is represented by the equilibrium:

$$Zn^{2+} + 4NH_3 \rightleftharpoons Zn(NH_3)_4^{2+} \tag{16.8}$$

for which the formation constant is written as

$$K_f = \frac{[Zn(NH_3)_4^{2+}]}{[Zn^{2+}][NH_3]^4} = 2.9 \times 10^9 \tag{16.9}$$

For the half-reaction involving the zinc ion–zinc system, the electrode potential is expressed by

$$E = E^{\circ} + 0.059\ 16 \log [Zn^{2+}] \tag{16.10}$$

Combining Eqs. 16.9 and 16.10 yields the potential of a zinc electrode in aqueous ammonia systems (ignoring any intermediate complexes):

$$E = E^{\circ} + 0.059\ 16 \log \frac{1}{K_f[NH_3]^4} + 0.059\ 16 \log [Zn(NH_3)_4^{2+}] \tag{16.11}$$

The second term on the right-hand side of Equation 16.11 contains the effect of the complexing agent.

For a redox couple involving two oxidation states in solution, such as the aquo-cobalt species,

$$Co^{3+} + e^- = Co^{2+} \qquad E^{\circ} = 1.84V \tag{16.12}$$

In the presence of aqueous ammonia, both the cobalt(II) hexammine and the cobalt(III) hexammine species predominate. The respective formation constants are

$$\frac{[Co(NH_3)_6^{2+}]}{[Co^{2+}][NH_3]^6} = K_f' = 10^5 \tag{16.13}$$

$$\frac{[Co(NH_3)_6^{3+}]}{[Co^{3+}][NH_3]^6} = K_f'' = 10^{34} \tag{16.14}$$

Substitution of these values into the Nernst equation for the cobalt system gives

$$E = E° + 0.0592 \log \frac{K_f'}{K_f''} + 0.0592 \log \frac{[\text{Co(NH}_3)_6^{3+}]}{[\text{Co(NH}_3)_6^{2+}]} \tag{16.15}$$

Here the shift in potential is a function of the ratio of the formation constants for each of the electroactive species. Generally, the compound with the higher oxidation state will form the more stable complex, and if it does, the shift in electrode potential will be in the negative direction. In the example above, the shift is significant, being 1.72 V.

16.8 ELECTROANALYTICAL METHODS OF ANALYSIS

Each basic electrical measurement of current (i), resistance (R), and potential (V) has been used alone or in combination for analytical purposes.

16.8.1 Steady-State Methods

Steady-state or static methods entail measurements under equilibrium conditions existing throughout the bulk of the solution. Time is eliminated as a variable, and equilibrium is assured by vigorously stirring the solution. Each of the following steady-state methods is discussed in succeeding chapters.

Ion-selective potentiometry: Electrode potential is measured at zero current.

Null-point potentiometry: Analyte additions or subtractions are done to null any potential difference at zero current.

Potentiometric titrations: Electrode potential is measured as a function of volume of titrant; current is zero.

Amperometric titrations: Current is measured as a function of the volume of titrant; potential is kept constant.

16.8.2 Transient Methods

We can impose our analytical will on many electrochemical systems by subjecting them to impressed current or voltage signals and observing qualitatively and quantitatively how they respond. This is done by applying a variety of voltage-time patterns to a microelectrode in a sample solution and analyzing the resultant currents. These transient methods constitute the field of *voltammetry* and its subset *polarography*. The concentration of the electroanalyte is the variable.

16.8.3 Controlled Potential Methods

The weight of a separated phase or the integrated current, that is current multiplied by time or coulombs, is a measure of the total amount of material converted to another electrochemical form. The potential at the working electrode is strictly controlled. In *controlled-potential electroanalysis* the weight of the separated phase is measured. In *controlled-potential coulometry* the total number of coulombs flowing through the system is measured.

16.8.4 Charge Transport by Migration

There is a group of methods that depend upon charge transport by migration. Electron-transfer reactions are unimportant. These methods are:

Conductance measurements: The conductance $(1/R)$ is measured as a function of concentration. *Conductometric titrations:* Conductance versus the volume of titrant is measured.

BIBLIOGRAPHY

Bard, A. J., and L. R. Faulkner, *Electrochemical Methods, Fundamentals and Applications,* Wiley, New York, 1980.

Lingane, J. J., *Electroanalytical Chemistry,* 2d ed., Interscience, New York, 1958.

CHAPTER 17
MEASUREMENT OF pH

17.1 ACIDS AND BASES

An acid is a proton donor. It can donate a proton to a substance capable of accepting the proton. In aqueous solutions, water is always available as a proton acceptor, so the ionization of an acid HA can be written as

$$HA + H_2O \; 6 \; H_3O^+ + A^- \tag{17.1}$$

where a water molecule accepts a proton to form a hydronium ion and the conjugate base A^- of the acid HA. Since most of this chapter involves ionization of acids in aqueous solution, Eq. 17.1 will be written

$$HA \; 6 \; H^+ + A^- \tag{17.2}$$

The equilibrium constant for the ionization of an acid is

$$\frac{[H^+]\,[A^-]}{[HA]} = K_a \tag{17.3}$$

An acid may be (1) a neutral molecule, as illustrated, (2) a positively charged species, such as

$$NH_4^+ \pm NH_3 + H^+ \tag{17.4}$$

or (3) a negatively charged species, such as

$$HC_2O_4^- \pm C_2O_4^{2-} + H^+ \tag{17.5}$$

Remember that H^+ really stands for a hydrated proton, or a proton associated with one or more water molecules.

The strength of an acid is related to the extent that the dissociation reactions proceed to the right, or to the magnitude of the acidic equilibrium constant K_a.

17.1.1 Water

Water is an amphiprotic substance. It can either lose or gain a proton. Pure water ionizes according to the reaction:

$$H_2O + H_2O \; 6 \; H_3O^+ + OH^- \tag{17.6}$$

or in an abbreviated form

$$H_2O \rightleftharpoons H^+ + OH^- \tag{17.7}$$

The equilibrium constant for Eq. 17.7 at 25°C is given by the expression:

$$\frac{[H^+][OH^-]}{[H_2O]} = K = 1.8 \times 10^{-16} \tag{17.8}$$

Because the concentration of water in dilute aqueous solutions is approximately constant and equal to 55.5 mol · L, Eq. 17.8 may be written

$$[H^+][OH^-] = K_w = 1 \times 10^{-14} \tag{17.9}$$

This new constant K_w is called the ion-product constant for water. In any aqueous solution, the product of the concentrations of the hydrogen ion and the hydroxyl ion will always be equal to the constant K_w.

In a neutral solution

$$[H^+] = [OH^-] = \sqrt{K_w} = 1.0 \times 10^{-7} \tag{17.10}$$

K_w is a function of temperature (Table 17.1).

In an acid solution the hydronium ion concentration is greater than the hydroxyl ion concentration, and the hydronium ion concentration is greater than 1×10^{-7}.

17.1.2 Solvents

Solvents can be classified into four groups depending on their behavior with respect to proton transfer: (1) basic, (2) acidic, (3) amphiprotic, and (4) aprotic. Of course, these classifications cannot be rigid, and they hold only to a degree. That is, one cannot always say that a solvent is exclusively acidic with absolutely no basic properties or that a solvent is definitely aprotic. The position of the autoprotolysis ranges of solvents, relative to the intrinsic strength of index acids, is indicated schematically in Figure 17.1. The autoprotolysis constant is 14 in water, varies from 15 to 19 in alcohols, and is about 14 in glacial acidic acid.

TABLE 17.1 Ionic Product Constant of Water

Temperature, °C	pK_w	Temperature, °C	pK_w
0	14.944	50	13.262
5	14.734	55	13.137
10	14.535	60	13.017
15	14.346	65	12.908
20	14.167	70	12.800
25	13.997	75	12.699
30	13.833	80	12.598
35	13.680	85	12.510
40	13.535	90	12.422
45	13.396	95	12.341

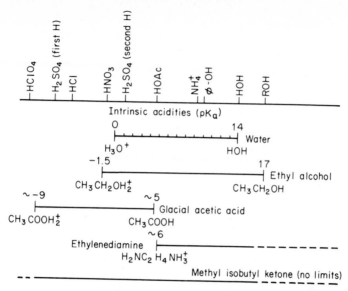

FIGURE 17.1 Schematic representation of autoprotolysis ranges of selected solvents in relation to the intrinsic strength of certain index acids. Influence of dielectric constant is not included.

17.1.2.1 Basic Solvents.

Basic solvents are those which have a pronounced tendency to accept protons from solutes, forming the solvated proton as one product of the reaction. Examples of common protophilic solvents used in titrations are the amines, for example, pyridine, butylamine, ethylenediamine, and piperidine. Basic solvents enhance the properties of weak acids. Phenols and carboxylic acids produce distinctive end points in butylamine or ethylenediamine.

The titrant is usually sodium aminoethoxide or tetrabutylammonium hydroxide in benzene-methanol (10 or 25% alcohol). The titrant is standardized against benzoic acid.

17.1.2.2 Acidic Solvents.

Acidic solvents are those which have a tendency to donate protons to solutes. The most common ones in acid-base titrations are glacial acetic acid and acetic anhydride. By repeated freezing and distillation, glacial acetic acid can be freed of water. Alternatively, acetic anhydride can be added to remove residual water; the amount added depends on the estimated water content.

Amino acids yield sharp end points because once the carboxylic acid group is swamped, the zwitterion equilibrium is removed. The titrant is a solution of perchloric acid in 1,4-dioxane or glacial acetic acid which has been standardized with potassium hydrogen phthalate or sodium carbonate.

17.1.2.3 Amphiprotic Solvents.

Amphiprotic solvents are those which can act as both proton donors and proton acceptors; that is, they have both acidic and basic properties. Water and the low-molecular-weight alcohols are common examples.

Suitable titrants for basic solutes are perchloric acid or p-toluene sulfonic acid dissolved in the same solvent as the solute. Tetrabutylammonium hydroxide in 2-propanol is a suitable basic titrant.

17.1.2.4 Aprotic Solvents. Aprotic solvents have no acidic or basic properties. If their dielectric constant is low, they have low ionizing power. These solvents have certain advantages over the other classes of solvents. Because they do not interact with dissolved solutes, no leveling action is exerted. Except for the influence of the dielectric constant, each solute will exhibit its intrinsic acidic or basic strength. Aprotic solvents include aromatic and aliphatic hydrocarbons, carbon tetrachloride, and 4-methyl-2-pentanone. The last named is particularly suitable; acidic impurities are removed by passing the solvent through a column of activated alumina.

Having no autoprotolysis limits, the range of applicability of aprotic solvents is limited only by the strength of the acid or base titrant. The former is usually perchloric acid or *p*-toluene sulfonic acid; the latter is a quaternary ammonium base. If suitably spaced, a series of successive (electrometric) end points can be achieved in a solvent such as methyl isobutyl ketone. Distinct end points can be achieved in the order enumerated for each of the following solutes when they are all together in a mixture and the titrant is tetrabutylammonium hydroxide: perchloric acid, hydrochloric acid, salicylic acid, acetic acid, and phenol. The first and second protons of sulfuric acid give very sharp end points when titrated in the same system.

17.2   CONCEPT OF pH

The term pH is defined as the negative logarithm (base 10) of the hydrogen-ion concentration expressed in molarity, that is,

$$pH = -\log [H^+] \tag{17.11}$$

Unfortunately, pH involves a single ion activity which is not measurable.

17.2.1 Operational Definition of pH

It is universally agreed that the definition of pH difference is an operational one. The pH difference is ascertained from the emf of an electrochemical cell of the following design:

Electrode reversible to hydrogen ions	Unknown (x) or standard (s) buffer solution	Salt bridge	Reference electrode

The two bridge solutions may be any molality of potassium chloride greater than 3.5 m. To a good approximation, the pH-responsive electrodes in both cells may be replaced by other hydrogen-ion-responsive electrodes. In most measurements, a single glass electrode-reference electrode probe assembly is transferred between the cells. E_x is the emf of the probe assembly when immersed in the solution of unknown pH_x, and E_s is the emf of the probe when immersed in a standard reference material whose value is pH_s. pH_x is given by the relationship

$$pH_x = pH_s + \frac{E_x - E_s}{2.303RT/F} \qquad (17.12)$$

where T = temperature, K
R = the gas constant
F = the Faraday constant

17.2.2 Standard Reference Solutions

To utilize Eq. 17.12, a value of pH at each temperature must be assigned to solutions selected as primary pH standards, as has been done in Table 17.2 over the temperature range 0 to 95°C from measurements on cells without liquid junctions. pH values for operational standard reference solutions have also been designated; the latter reference solutions include a liquid-junction factor, where the liquid junctions are formed within vertical 1-mm capillary tubes. Two solutions from the latter set are included in Table 17.2 to cover the extremes of the pH scale; they are the potassium tetraoxalate and the calcium hydroxide solutions.

As fabricated, glass-reference electrode assemblies exhibit variations in the reproducibility of the internal reference electrode within the glass electrode, in the liquid-junction potential between the filling solution of the reference electrode and the test solution, and the asymmetry potential associated with glass electrodes. These differences should all be eliminated in the standardizing procedures that use standard reference pH buffers (*see* R. G. Bates[1]). However, at pH values less than 2 or greater than 12, and for ionic strengths greater than 0.1, the reproducibility of the liquid-junction potential is seriously impaired and errors as large as several tenths of a pH unit can result.

The operator of pH equipment should be aware of the following limitations in the use of pH reference buffer standards:

1. The uncertainty in pH_s values is ±0.005 pH unit at 0 to 60°C and ±0.008 pH unit at 60 to 95°C.
2. The operational definition of pH is valid (a) for only dilute solutions and (b) for the pH range 2 to 12.
3. The use of the pH probe assembly in the very acid or very basic region requires two secondary standards (potassium tetraoxalate and calcium hydroxide solutions), which are included in Table 17.2.
4. The temperature must be known to ±2°C for an accuracy of ±0.01 pH unit. Don't overlook the fact that dissociation equilibria and junction potentials have significant temperature coefficients. Solutions should be measured at their operating temperatures and not cooled to room temperature. In an industrial process it is often sufficient to know that at a certain stage a particular pH value is maintained.

17.3 POTENTIOMETRIC pH MEASUREMENTS

The direct measurement of an electrode potential using suitable indicator and reference electrodes allows the hydrogen-ion concentration to be determined. A glass membrane acts as a half-cell, generating a potential proportional to the log-

TABLE 17.2 National Bureau of Standards Reference pH_s Buffer Solutions*

Temp., °C	Secondary standard, 0.05 M K tetroxalate	KH tartrate (sat'd. at 25°C)	0.05 M KH_2 citrate	0.05 M KH phthalate	0.025 M each KH_2PO_4 Na_2HPO_4	0.008695 M KH_2PO_4 0.03043 M Na_2HPO_4	0.01 M $Na_2B_4O_7$	0.025 M each $NaHCO_3$ Na_2CO_3	Secondary standard, $Ca(OH)_2$ (sat'd. at 25°C)
0	1.666	—	3.863	4.003	6.984	7.534	9.464	10.317	13.423
5	1.668	—	3.840	3.999	6.951	7.500	9.395	10.245	13.207
10	1.670	—	3.820	3.998	6.923	7.472	9.332	10.179	13.003
15	1.672	—	3.802	3.999	6.900	7.448	9.276	10.118	12.810
20	1.675	—	3.788	4.002	6.881	7.429	9.225	10.062	12.627
25	1.679	3.557	3.776	4.008	6.865	7.413	9.180	10.012	12.454
30	1.683	3.552	3.766	4.015	6.853	7.400	9.139	9.966	12.289
35	1.688	3.549	3.759	4.024	6.844	7.389	9.102	9.925	12.133
40	1.694	3.547	3.753	4.035	6.838	7.380	9.068	9.889	11.984
50	1.707	3.549	3.749	4.060	6.833	7.367	9.011	9.828	11.705
60	1.723	3.560	—	4.091	6.836	—	8.962	—	11.449
70	1.743	3.580	—	4.126	6.845	—	8.921	—	—
80	1.766	3.609	—	4.164	6.859	—	8.885	—	—
90	1.792	3.650	—	4.205	6.877	—	8.850	—	—
95	1.806	3.674	—	4.227	6.886	—	8.833	—	—
Buffer value, β	0.070	0.027	0.034	0.016	0.029	0.016	0.020	0.029	0.09
Dilution value, $\Delta pH_{1/2}$	+0.186	+0.049	+0.052	+0.052	+0.080	+0.07	+0.01	+0.079	−0.28

*Numbers given are "conventional" pH values. Properties of these buffer solutions are included at the foot of each column.
Source: R. G. Bates, *J. Res. Natl. Bur. Std.*, 66A, 179–183 (1962); B. R. Staples and R. G. Bates, *J. Res. Natl. Bur. Std.*, 73A, 37 (1969).

17.6

arithm of the analyte activity (concentration). This potential is measured relative to a reference electrode that is also in contact with the sample. Essentially no current flows during the measurements to avoid disturbing the equilibrium at the sample-membrane interface or generating an *iR* drop (potential gradient across the system).

17.3.1   Reference Electrodes

A reference electrode is an oxidation-reduction half-cell of known and constant potential at a particular temperature. It consists of three parts (Fig. 17.2):

1. An internal element
2. A filling solution which also constitutes the salt-bridge electrolyte.
3. An area in the tip of the electrode that permits a slow, controlled flow of filling solution to escape the electrode and maintain electrical conductance with the remainder of the electrochemical cell.

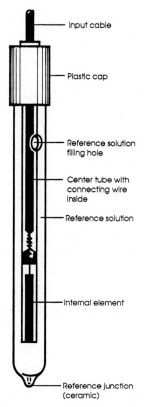

FIGURE 17.2   Schematic of a reference electrode.

17.3.1.1   Internal Elements.   There are two choices for a reference electrode: (1) the calomel or mercury–mercury(I) chloride half-cell and (2) the silver–silver chloride half-cell as the internal element. Both electrodes are anion reversible.

The *silver–silver chloride electrode* is often a silver wire coated with silver chloride that is immersed in a filling solution of potassium chloride of known concentration, usually 1.0 M, and saturated with silver chloride. In the cartridge form the metal is in contact with a paste of the salt moistened with electrolyte, which is enclosed in an inner glass tube.

One equilibrium involved is

$$Ag^+ + e^- \leftrightharpoons Ag \tag{17.13}$$

The expression for the electrode potential is

$$E = 0.799 - 0.059\ 16 \log [Ag^+] \tag{17.14}$$

In addition, there is a chemical equilibrium,

$$AgCl(s) \leftrightharpoons Ag^+ + Cl^-, \qquad K_{sp} = 1.8 \times 10^{-10} \tag{17.15}$$

Combining Eqs. 17.14 and 17.15 gives the Nernst expression for the chloride ion:

$$E = 0.799 + 0.059\ 16 \log K_{sp} - 0.059\ 16 \log [Cl^-] \tag{17.16}$$

This simplifies to

$$E = 0.2222 - 0.059\ 16 \log [Cl^-] \tag{17.17}$$

Thus when the electrode system is immersed in a filling solution that contains a constant (and known) amount of the chloride ion, the electrode potential is constant (and known). Table 17.3 lists potentials of common reference electrodes as a function of temperature.

Calomel electrodes have a construction that is similar to the cartridge type of silver–silver chloride electrodes. The electrode comprises a platinum wire in contact with mercury and a paste of mercury(I) chloride, mercury, and potassium chloride, moistened with the filling solution. The wire is enclosed in an inner glass tube and makes contact with a filling solution of potassium chloride (usually 0.1 M, 1.0 M, or saturated, 4.2 M) through a porous plug.

17.3.1.2   Comparison of Reference Electrodes.   Calomel electrodes are easy to prepare and maintain. They must never be used at temperatures higher than 80°C. The saturated calomel electrode (SCE), although easiest to prepare, reaches equilibrium more slowly following temperature changes.

Silver–silver chloride electrodes can be used up to the boiling point of water (and higher under special conditions). The temperature coefficient of the electrode potential is less than for calomel electrodes. These electrodes are also more stable over long periods of time.

17.3.2   Liquid Junctions

Electrical contact between the sample and the reference electrode is established by a slow but continuous leak of filling solution through a constricted orifice. The liquid junction must be clean and free flowing, and must make good contact with the sample. Liquid junctions come in several physical forms: (1) a sleeve or ta-

TABLE 17.3 Potentials of Reference Electrodes in Volts as a Function of Temperature
(Liquid-junction potential included)

Temperature, °C	0.1 M KCl calomel*	Sat'd. KCl calomel*	1.0 M KCl Ag/AgCl†
0	0.3367	0.25918	0.23655
5			0.23413
10	0.3362	0.25387	0.23142
15	0.3361	0.2511	0.22857
20	0.3358	0.24775	0.22557
25	0.3356	0.24453	0.22234
30	0.3354	0.24118	0.21904
35	0.3351	0.2376	0.21565
38	0.3350	0.2355	
40	0.3345	0.23449	0.21208
45			0.20835
50	0.3315	0.22737	0.20449
55			0.20056
60	0.3248	0.2235	0.19649
70			0.18782
80		0.2083	0.1787
90			0.1695

*R. G. Bates et al., *J. Res. Natl. Bur. Std.*, **45**, 418 (1950).
†R. G. Bates and V. E. Bower, *J. Res. Natl. Bur. Std.*, **53**, 283 (1954).

pered ground-glass junction (Fig. 17.3*a*), (2) a porous ceramic or quartz-fiber junction (Fig. 17.2), and (3) a double chamber salt bridge (Fig. 17.3*b*).

1. *Sleeve junctions* have an electrolyte flow of about 0.1 mL $\cdot$ h^{-1}, a self-cleaning facility. This junction is useful when working with slurries, emulsions, suspensions, pastes, gels, and nonaqueous solvent systems.

2. *Porous ceramic* or *fiber junctions* have a very small flow rate of filling solution (about 8 μL $\cdot$ h^{-1}). This type of junction should be used when contamination of the sample by the filling solution must be avoided.

3. *Double junction* salt bridges overcome all problems with leakage of electrolyte into the sample or compatibility of filling and sample solution. The leakage from the internal filling solution is retained in the outer (chamber) salt bridge, which contains an innocuous electrolyte and can be flushed frequently. This type of junction must be used when the test solution contains ions that would precipitate or complex with silver or mercury(II) ions found in the filling solution or with strong reducing agents that might reduce silver ions to silver metal at the junction.

17.3.3 Maintenance of Reference Electrodes

Reference electrodes should be stored separately in a dilute KCl solution of approximately 0.1 M. To prevent back flow and possible contamination of the electrolyte, always maintain the level of the filling solution in the reference electrode

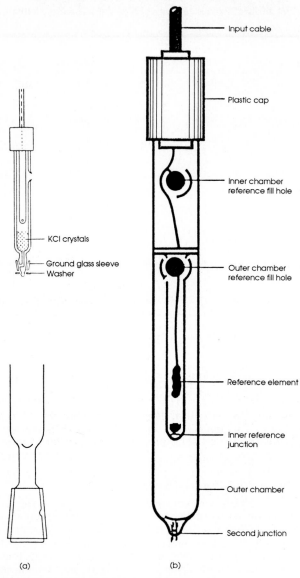

FIGURE 17.3 Liquid junctions: (*a*) sleeve or tapered ground glass junction and (*b*) double chamber construction.

above the level of both soaking and sample solutions and above the internal elements.

17.3.3.1 Blocked or Clogged Liquid Junctions. A blocked or clogged liquid junction can be detected in several ways:

1. Apply air pressure to the filling hole and observe the dried junction—a bead of electrolyte should form readily.
2. Measure the junction resistance with an ohmmeter—it should read less than 20 000 Ω.
3. Place the glass-reference electrode assembly into 250 mL of deionized water (with stirring) and note the reading. Add 50 mg of ultrapure solid KCl and note the reading. The reading should change less than 0.08 pH.

The junction of the reference electrode can be unblocked as follows:

1. Gel-filled electrodes need only to be immersed in warm water for several minutes or overnight in 0.1 M KCl.
2. Sleeve-type junctions can be unblocked by simply loosening the sleeve to drain the electrolyte, rinsing the cavity with distilled water, then refilling with fresh KCl solution.
3. Porous ceramic or cracked bead junctions should be immersed in warm distilled water to clear the junction. Apply air pressure to the fill hole to reestablish the electrolyte flow.
4. For porous plugs, try boiling the junction in dilute KCl for 5 to 10 min or soak overnight in 0.1 M KCl. If this fails, carefully sand or file the porous plug.

17.3.3.2 Contamination on Tip and Surface.   Contamination on the surface and tip can be removed as follows:

1. For protein layers, wash with pepsin or 0.1 M HCl.
2. For inorganic deposits, wash with EDTA or dilute acids.
3. For grease films, wash with acetone, methanol, or diethyl ether.

17.3.3.3 Faulty Internal Elements or Contaminated Filling Solutions.   Faulty internal elements, contaminated filling solution, or too-great junction potentials can be checked by plugging the suspect electrode into the reference jack of your pH meter and a reference electrode known to be working properly into the pH indicating electrode jack (pin-jack adapter needed). Immerse both electrodes in a saturated KCl solution. The observed pH meter reading should be 0 ± 5 mV for similar electrodes, or about +44 ± 5 mV if a calomel reference is tested against a silver–silver chloride reference electrode.

17.3.4 Glass-Indicating Electrodes

Typical pH-sensitive glass membranes are either sodium–calcium silicate or lithium silicates with lanthanum and barium ions added to retard silicate hydrolysis and lessen sodium ion mobility. All glass electrodes must be conditioned for a time by soaking in water or in a dilute buffer solution. Upon immersion of the membrane in water, the surface layer becomes involved in an ion-exchange process between the hydrogen ions in the solution and the sodium or lithium ions of the membrane.

The glass pH electrode (Fig. 17.4) has an internal reference electrode immersed in a chloride salt buffer solution that is sealed within the tip or bulb of the electrode. Often a phosphate buffer at pH 7 is used. An external reference electrode completes

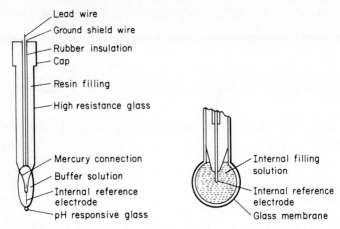

FIGURE 17.4   Construction of a glass mambrane, pH-responsive electrode.

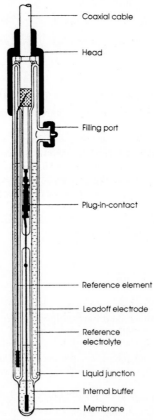

FIGURE 17.5 Combination of a pH-responsive glass and a reference electrode.

the probe assembly. The combination electrode contains both the pH-sensing and reference electrodes in a single probe body, as shown in Fig. 17.5.

Glass electrodes are available in a variety of configurations. For medical studies, special electrodes are available that can be inserted into blood vessels, the stomach, and other parts of the body. Syringe and capillary electrodes require only one or two drops of solution, whereas others penetrate soft solids (such as leather) or pastes. The normal-size electrode operates with a solution volume of 1 to 5 mL. Only the pH-sensitive membrane needs to be completely immersed; the remainder of the glass electrode is made of rugged, inert glass.

The chemical composition of the glass membrane may dictate its usage. Universal glass electrodes have a resistance of about 100 $M\Omega$ at 25°C; they can be used down to 0°C. The rugged membrane withstands abuse and rough handling typically found in many industrial applications. Glass membranes may suffer chemical attack by the test solution, so chemical durability is important. For longer membrane life at extreme pH values, a full-range high-pH glass electrode should be used. This is the membrane that should be used in solutions with high alkali metal content and high pH values; however, its lower temperature limit is 10°C.

17.3.5 Precautions When Using Glass Electrodes

1. When not in use, keep the pH electrode assembly immersed in a buffer solution. For long-term storage, carefully dry the electrode and place it in a protective container.

2. Thoroughly wash the electrode with distilled water after each measurement.

3. Rinse the electrode with several portions of the test solution before making the final measurement.

4. Vigorously stir poorly buffered solutions during measurement; otherwise, the stagnant layer of solution at the glass-solution interface tends toward the composition of the particular membrane.

5. Wipe suspensions and colloidal material from the membrane surface with a soft tissue. Avoid scratching the delicate membrane.

6. Never use glass electrodes in acid fluoride solutions because the membrane will be attacked chemically.

17.3.6 Troubleshooting Tips

When poor or suspect readings are encountered and the reference electrode has been eliminated as a problem, (1) inspect the electrode for visible cracks or scratches on the membrane surface. If either exists, the electrode must be discarded. (2) Perform a standard calibration for one buffer, then transfer the probe assembly to a second buffer. A reading accurate to within ±0.05 pH unit from the standardization value should be obtained within 30 s. If this condition is not met, the electrode should be rejuvenated.

Rejuvenation of the sensing membrane may succeed after trying the following procedures:

1. Immerse the membrane into 0.1 M HCl for about 15 s. Rinse with distilled

water, immerse into 0.1 *M* KOH for 15 s, and rinse with distilled water. Cycle the membrane through these solutions several times.

2. Immerse the membrane into 20% ammonium hydrogen fluoride for exactly 3 min. Thoroughly rinse the electrode with distilled water, dip into concentrated hydrochloric acid, and rinse with distilled water. Soak the electrode in pH 4 buffer solution for 24 h. If proper electrode function has not been restored, discard the electrode.

17.4   pH METER

As shown in Fig. 17.6, an electrometric pH measurement system (pH meter) consists of

1. pH-responsive electrode
2. Reference electrode
3. Potential-measuring device—some form of high-impedance electronic voltmeter

pH-responsive glass electrodes and reference electrodes have been discussed earlier.

Electronic pH meters are simply voltmeters with scale divisions in pH units which are equivalent to the values of $2.3026RT/F$ (in millivolts) per pH unit. Values of this function at several temperatures are given in Table 17.4. To achieve a reproducibility of ±0.005 pH unit, the pH meter must be reproducible to at least 0.2 mV. The current drawn from the probe assembly should be 10^{-12} A or less.

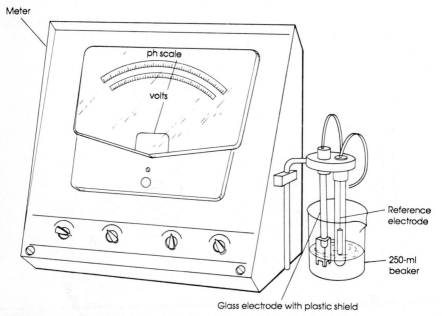

FIGURE 17.6   pH meter and probe assembly.

TABLE 17.4   Values of $2.3026RT/F$ at Several Temperatures in Millivolts

t, °C	Value	t, °C	Value	t, °C	Value
0	54.197	35	61.141	70	68.086
5	55.189	38	61.737	75	69.078
10	56.181	40	62.133	80	70.070
15	57.173	45	63.126	85	71.062
18	57.767	50	64.118	90	72.054
20	58.165	55	65.110	95	73.046
25	59.157	60	66.102	100	74.038
30	60.149	65	67.094		

The schematic circuit diagram of a pH meter, based on an operational amplifier, is shown in Fig. 17.7. The operational controls on a pH meter are best understood by reference to Fig. 17.8. The relationship for the emf of a pH probe assembly is

$$E = k - KT(\text{pH}) \tag{17.18}$$

This is the equation of a straight line with slope $-KT$ and a zero intercept of k on the E axis. The glass electrode with its internal reference electrode and filling solution (usually pH 7.00) must have an isopotential point at 0 V.

17.4.1   Operating Procedure

Proceed as follows when making pH measurements:

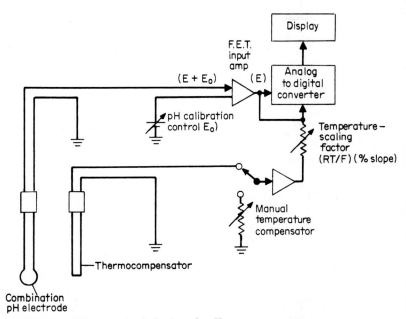

FIGURE 17.7   Schematic circuit diagram of a pH meter.

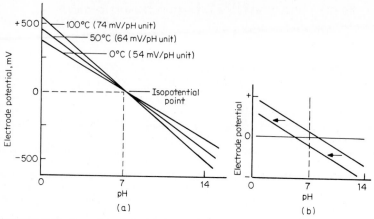

FIGURE 17.8 Operation of the (*a*) slope and (*b*) intercept controls shown schematically.

1. Turn on the pH meter. Different manufacturers designate the neutral position of the selector switch with various markings such as Standby, Bal, and so on.

2. Raise the pH probe assembly from the storage solution.

3. Rinse the electrodes thoroughly with distilled water and then with the standard reference solution.

4. Immerse the pH probe assembly in the standard reference buffer that is nearest the expected pH value of the sample.

5. Adjust the temperature control to the temperature of the solution. This amounts to adjusting the *KT* factor in Eq. 17.20 (and enumerated in Table 17.4) by means of the emf/pH slope control about the isopotential point.

6. Rotate the selector switch to pH.

7. Bring the meter reading into juxtaposition with the pH_s value by adjusting the intercept (variously named Zero, Standardization, Calibration, or Asymmetry) control.

8. Remove the pH probe assembly, rinse with distilled water and then with portions of the test solution. Blot the electrodes between rinses with absorbent, lint-free tissue.

9. Lower the pH probe assembly into the test solution. With the selector switch still at pH, read the pH of the test solution.

After standardization with the standard reference buffer, it is always wise to check the functionality of the probe system by measuring the pH of a second reference buffer; the two reference buffers should bracket the expected pH of the sample. Faulty glass electrodes or plugged external reference electrodes may seem to operate correctly upon initial standardization, but they will not give the proper pH reading for the second pH reference buffer.

Microcomputer-based pH meters measure the output of the electrode assembly and subject it to appropriate algorithms based on Eq. 17.18. A prescribed operating protocol must be carefully followed for calibration and standardization.

17.5 COLORIMETRIC pH DETERMINATION

17.5.1 Indicators

Indicators are substances which, in solution, change color according to the pH of the solution. In many instances they will be incompletely dissociated acids or bases in which the dissociated form differs in structure and color from the undissociated form.

Ideally an indicator should exhibit a sharp color change over a narrow pH range, preferably about 1.5 pH units. Most indicators have two colors, one in the acidic form and the other in the basic form. Examples are members of the sulfonephthalein series. Some indicators have only one visible color and lose it completely when in the conjugate form; phthaleins and nitrophenols are examples.

Indicators should be brilliant in color so that only a few drops are needed; such a small amount will not affect the pH of the test solution. Indicators should be soluble in water. The perception of one form of an indicator in the presence of another colored form depends upon the observer's power to detect changes of tone or shade in mixtures of two colors. The actual transformation interval to an average observer assumes no less than 9 percent of one form together with 91 percent of the other form. In pH units, this encompasses a range of 2 pH units which is centered about the pK_{HIn}. Indicators and their pH ranges are listed in Table 17.5. Test strips of individual indicators are available.

17.5.2 Universal Indicators

Mixtures of indicators, called universal indicators, can be prepared. One such mixture is composed of phenolphthalein, methyl red, dimethylaminoazobenzene, bromothymol blue, and thymol blue. The mixture goes through a series of color changes, starting with red at pH 2 and passing through orange at pH 4, yellow at

TABLE 17.5 Indicators and Their pH Ranges

Indicator	pH range	pK_{HIn}	Color change (acid-alkaline)
o-Cresol red (acid range)	0.2–1.8		Red-yellow
Thymol blue (acid range)	1.2–2.8	1.75	Red-yellow
2,4-Dinitrophenol	1.3–3.2		Red-yellow
Methyl orange	3.1–4.4	3.40	Red-orange
Bromophenol blue	3.0–4.6	4.05	Yellow-blue
Bromocresol green	4.0–5.6	4.68	Yellow-blue
Methyl red	4.4–6.2	4.95	Red-yellow
Bromocresol purple	5.2–6.8	6.3	Yellow-purple
Chlorophenol red	5.4–6.8	6.0	Yellow-red
Bromothymol blue	6.2–7.6	7.1	Yellow-blue
Phenol red	6.4–8.0	7.9	Yellow-red
Cresol red	7.2–8.8	8.2	Yellow-red
Thymol blue	8.0–9.6	8.9	Yellow-blue
Phenolphthalein	8.0–10.0	9.4	Colorless-red
Thymolphthalein	9.4–10.6	10.0	Colorless-blue

pH 6, green at pH 8, and blue at pH 10. Although accuracy never exceeds 1 pH unit, a universal indicator is useful for rough preliminary testing. Test papers impregnated with mixed indicators are available.

17.5.3 Methods of Color Comparison

Test strips are immersed in the sample, and the resulting color is compared with the colors on the dispenser. Reliability is usually within 0.2 to 0.3 pH unit.

In the direct reading method, the sample solution containing a known concentration of indicator is compared with solutions of known pH value, each containing the same quantity of indicator. In the slide comparator, the solutions for comparison are put up in small glass vials. Each set of solutions covers a range of pH values in steps of 0.2 pH unit. The solutions to be tested and the standard are examined together by transmitted light. When the color is matched, the test solution has the same pH as the standard. When not matched exactly, the color can be adjudged to lie between two standard solutions and, perhaps, the position in the interval estimated within 0.05 pH unit. A number of sets of vials are needed to cover the entire pH range.

More permanent artificial color standards are available in the form of glass color filters. A comparative color wheel contains a series of color standards which duplicate in intervals of 0.2 pH unit the colors of one indicator. The entire unit is held up to a light source, and the operator views the sample and color standard together through either a split field eyepiece or two round holes with an opal screen as background. The wheel is rotated until the color of one of the glass standards matches the test solution. The latter is contained in a rectangular glass cell, to which has been added a specified amount of indicator solution.

REFERENCES

1. R. G. Bates, *Determination of pH: Theory and Practice,* 2d ed., Wiley, 1973.

BIBLIOGRAPHY

Fisher, J. E., "Measurement of pH," *Am. Lab.,* **16:**54 (June 1984).

Rothstein, F., and J. E. Fisher, "pH Measurement: The Meter," *Am. Lab.,* **17:**124 (September 1985).

CHAPTER 18
POTENTIOMETRY

This chapter discusses the direct measurement of an electrode potential from which the activity of an ion may be derived. Because the measurement of pH is such a common method and also involves colorimetric procedures, the use of pH-responsive glass electrodes was discussed in Chap. 17. The novel ion-selective electrodes, other than pH-responsive electrodes, and potentiometric titrations form the subject matter of this chapter.

18.1 ION-SELECTIVE ELECTRODES

An ion-selective electrode (ISE) is a potentiometric probe whose output potential, when measured against a suitable reference electrode, is logarithmically proportional to the activity of the selected ion in the test solution. Ion activities are the thermodynamically effective free ion concentrations. In dilute solutions, ion activity approaches the ion concentration. Activity measurements are valuable because the activities of ions determine rates of reactions and chemical equilibria.

ISEs have several valuable features which make them superior to other methods of analysis.

1. Easy to use.
2. Rapid response.
3. Benchtop analysis typically requires 1 to 2 min.
4. Many analyses can be performed directly without the need for time-consuming sample preparation, such as centrifugation and filtration.
5. Solution color or turbidity does not affect results.
6. Constant precision throughout their linear measuring range, which can be as large as six decades.
7. Rugged, compact, and lightweight; their portability allows them to be used in field work.
8. Unresponsive to oxidation-reduction couples in the solution.

There are three types of membrane sensors: glass, solid state, and solid matrix (liquid ion exchange). Gas-sensing and biocatalytic electrodes are merely special designs that incorporate one of the three types into the system.

18.1.1 Glass Membrane Electrodes

The glass electrode construction is the same as shown in Fig. 17.4 for pH-responsive electrodes. Substitution of aluminum for part of the silicon in the alkali metal–silicate glasses produces cation-selective glass membranes. Glasses of the composition 11% Na_2O:18% Al_2O_3:71% SiO_2 are highly sodium-selective with respect to other alkali metal ions. The sodium-responsive glass electrode has found extensive use in clinical work.

18.1.2 Solid-State Sensors

The electrode body is composed of an inert epoxy formulation. Bonded to the electrode body is the sensing membrane, a solid-state ionic conductor. Completing the electrode is an internal reference electrode, usually silver–silver chloride, and an internal filling solution. A cross-sectional view is shown in Fig. 18.1. The lower limits of useful response are imposed by the solubility of the sensor's membrane material in the sample solution.

The active membrane of the fluoride electrode has a single crystal of LaF_3 doped with europium(II) and an internal solution that is 0.1 M in NaF and in NaCl. The fluoride ion activity controls the potential of the inner surface of the LaF_3 membrane, and the chloride ion activity fixes the potential of the internal Ag–AgCl wire reference electrode. When in contact with the sample at 25°C, the external surface of the membrane responds to the fluoride ion activity in the sample:

$$E = \text{constant} - 0.0592 \log [F^-] \tag{18.1}$$

The fluoride-responsive electrode follows a logarithmic response when fluoride concentrations are as low as 10^{-5} M and a useful response to at least 10^{-6} M fluoride ion.

A group of ISEs based on the Ag_2S membrane are available. By itself, such a membrane can be used to detect silver ions or to measure sulfide ion levels. The

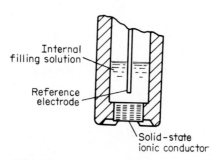

FIGURE 18.1 Cross-sectional view of solid-state sensor. (*Courtesy of Orion Research, Inc.*)

dynamic range of the electrode extends from saturated solutions down to silver and sulfide levels on the order of 10^{-8} M. For sulfide measurements, a special buffer must be mixed with the samples to raise the pH of the sample solution, to free sulfide bound to hydrogen, to fix the total ionic strength, and to retard oxidation of the sulfide ion.

If the silver sulfide is altered by dispersing within it another metal sulfide, the corresponding metal-selective electrode is obtained. Two solid-phase equilibria must now be established—the solubility product equilibria of silver sulfide and the metal sulfide, which must be larger than that of the silver sulfide. Useful copper-, cadmium-, and lead-selective electrodes have been prepared from CuS, CdS, and PbS, respectively, dispersed in a silver sulfide matrix.

Other more soluble silver salts, dispersed within the silver sulfide matrix, make up the anion-selective electrodes of chloride, bromide, iodide, and thiocyanate, respectively. The halide electrodes exhibit useful ranges (in pI or $-\log$ [ion] units) up to 5, 6, and 7, respectively, for chloride, bromide, and iodide ions.

18.1.3 Solid or Liquid Matrix Electrodes

In the solid matrix electrode, the sensing membrane is an ion exchanger permanently embedded in a plastic material that is sealed to the inert electrode body. The construction is shown in Fig. 18.2. In the older liquid matrix electrodes, the ion exchanger was held within the internal structure of a porous membrane at the tip of the electrode where contact was made with the test solution.

Some examples of solid and liquid matrix electrodes are:

The *calcium*-selective electrode uses the calcium salt of bis(2-ethylhexyl)phosphoric acid as the ion exchanger; the aqueous internal filling solution consists of a fixed concentration of calcium and chloride ions.

The *perchlorate*-sensing membrane uses the iron(III) ion-pair association complex of 1,10-phenanthroline as the ion exchange site group.

The *nitrate* (and *fluoroborate*) electrode uses a nickel(II)1,10-phenanthroline site group.

A *potassium*-selective electrode incorporates a neutral carrier, the valino-

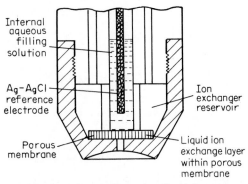

Internal aqueous filling solution

Ag–AgCl reference electrode

Porous membrane

Ion exchanger reservoir

Liquid ion exchange layer within porous membrane

FIGURE 18.2 Construction of a liquid ion-exchange electrode. (*Courtesy of Orion Research, Inc.*)

mycin molecule. This molecule is a doughnut-shaped complex with an electron-rich pocket in the center into which potassium ions are selectively bound through ion-dipole interactions to replace the hydration shell around potassium.

18.1.4 Gas-Sensing Electrodes

Gas-sensing electrodes are available for the measurement of ammonia, carbon dioxide, nitrogen oxide, and sulfur dioxide. A gas-permeable membrane isolates the analyte from the sample, and a thin buffer layer traps the analyte and converts it to some ionic species that can be detected with a pH or ion-selective electrode. The response and recovery time is relatively long—often 30 s to 5 min.

In the ammonia-selective electrode, dissolved ammonia from a sample diffuses through a fluorocarbon membrane into the internal filling solution where the ammonia reacts with water to form hydroxide and ammonium ions. Either the ammonium ion concentration can be measured with an ammonium ion–selective glass electrode or the pH of the system can be monitored with a pH-responsive glass electrode. Standards and samples are adjusted to a fixed pH or to a pH greater than 11. Sensitivity extends from 10^{-6} to 1 M.

The carbon dioxide sensor possesses a microporous Teflon membrane that separates the sample from an internal electrolyte of sodium hydrogen carbonate. An internal pH glass electrode monitors the hydrogen carbonate–carbonate equilibrium as it is affected by the carbon dioxide dissolved in the sample.

The sulfur dioxide–hydrogen sulfite ratio and the nitrous acid–nitrite equilibria can be followed similarly by using an internal pH glass electrode.

18.1.5 Biocatalytic Membrane Electrodes

The membranes in biocatalytic electrodes are multilayered composites containing one or more biocatalysts (often enzymes) immobilized in a gel layer that coats a conventional ion-selective electrode. Numerous enzyme-substrate combinations are possible. One example will suffice. An urea-responsive electrode has the enzyme urease fixed in a layer of acrylamide gel held in place around the pH glass electrode bulb by a nylon mesh or a thin cellophane film. The urease acts specifically upon urea in the sample solution to yield ammonium ions (and hydroxide ions) which diffuse through the gel layer and are sensed by the internal pH electrode (or ammonium ion-selective electrode).

18.2 ION ACTIVITY EVALUATION METHODS

For an electrode responsive to the activity of ion X, a_x, the electrode response is

$$E_x = \text{constant} + S \log a_x \tag{18.2}$$

where E_x is the potential reading on a pION meter and S is the slope of the calibration plot (approximately 60 mV for a monovalent ion and 30 mV for a divalent ion). Calibration procedures must use solutions of known activity or concentration, depending on which parameter is required. At less than 10^{-3} to 10^{-4} M the two quantities are practically indistinguishable.

18.2.1 Direct Calibration Plot

In this method several measurements of E are made in solutions of known activity (or concentration). These solution are prepared by serial dilution of a concentrated standard. The recommended ionic strength buffer is added to each standard and to each unknown. One buffer for fluoride ion determination consists of 0.25 M acetic acid, 0.75 M sodium acetate, 1 M sodium chloride, and 1 mM sodium citrate [for masking aluminum(III) and iron(III), which interfere by complexing with the fluoride ion]. Appropriate ionic strength (perhaps combined with a pH buffer) is added for other selective ion determinations. The calibration plot of electrode potential versus activity will be linear over several orders of magnitude; the concentration plot will deviate from linearity above 10^{-4} to 10^{-3} M. The potentials of unknown sample solutions can then be measured, and their activities (or concentrations) can be read directly from the calibration plot. The lower limit of detection reflects the experimental difficulty of preparing extremely dilute solutions without extensive adsorption on and desorption from the surfaces of containment vessels and the electrodes. Solubility equilibria and ion-pair association equilibria also affect the lower limit of electrode utility.

18.2.2 Method of Additions

In the method of additions (standard addition or subtraction), the approximate concentration (or activity) of the sample is first estimated from the observed electrode potential (rough calibration plot):

$$E_1 = \text{constant} + S \log C_u \tag{18.3}$$

A known volume of the standard solution (which should be about 10 times the estimated sample activity or concentration) is added. After correction for dilution of the original sample by the addition, and the corresponding correction of the concentration of the added standard:

$$E_2 = \text{constant} + S \log [C'_u + C'_s] \tag{18.4}$$

Combining equations, one gets

$$\Delta E = E_2 - E_1 = S \log \frac{C'_u + C'_s}{C_u} \tag{18.5}$$

Volume corrections are

$$C'_u = C_u \frac{V_u}{V_u + V_s} \quad \text{and} \quad C'_s = C_s \frac{V_s}{V_u + V_s}$$

The concentration of the unknown can be explicitly determined by solving the equation:

$$C_u = C_s \left(\frac{V_s}{V_u + V_s}\right) \left[10^{\Delta E/S}\left(\frac{V_u}{V_u + V_s}\right)\right]^{-1} \tag{18.6}$$

Sample addition is the inverse of the standard addition method. It is useful for samples that are small, highly concentrated, or dirty. It cannot be used when the unknown species is complexed. With this technique the sample is added to a known volume or a standard solution. The unknown concentration is determined by solving this equation:

$$C_u = C_s \left[10^{\Delta E/S} \left(\frac{V_u + V_s}{V_u} \right) \left(\frac{V_s}{V_u} \right) \right]$$ (18.7)

Known addition and subtraction methods are particularly suitable for samples with a high unknown total ionic strength. Where the species being measured is especially unstable, known subtraction is preferred over known addition.

18.3 SELECTIVITY AND INTERFERENCE

As their name implies, ion selective electrodes are selective rather than specific for the particular ion. Thus a potassium selective electrode responds not only to the activity of potassium ions in solution but also to some fraction of the sodium ions present. In a first approximation, the effect of foreign cations on the electrode potential may be fitted by an extended Nikolsky equation:

$$E = \text{constant} + \frac{RT}{z_i F} \ln \left[a_i + \sum K_{ij} a_j^{(z_i/z_j)} \right]$$ (18.8)

where a_i = activity of primary ion with charge of z_i
a_j = activity of interfering ion with charge of z_j
K_{ij} = selectivity ratio characteristic a given membrane

The value of the constant in the equation depends on the choice of the external reference electrode.

The simplest way to determine the selectivity ratio is shown in Fig. 18.3. Potential measurements are made in a series of solutions, with the activity of the ion

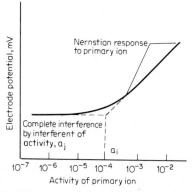

FIGURE 18.3 A method for calculating selectivity ratios.

to be measured varying in the presence of a constant background level of the interference. The region over which the electrode response is linear with respect to the test ion is clearly demarcated; this is the slope factor (S), here equal to RT/z_iF. The intercept of the extension of the response line (slope) with the horizontal line in the region of high interference defines a particular intercept activity for the primary ion a_i. The particular activity is related to the selectivity ratio by the second term in the brackets of Eq. 18.8:

$$a_i = K_{ij}a_j^{(z_i/z_j)} \qquad\qquad (18.9)$$

where a_j is the constant background level of the interference. Knowing the activities of a_i and a_j, one can solve for the selectivity ratio. The selectivity (the inverse of the selectivity ratio) is quoted for various ion-selective electrodes in Table 18.1. From the foregoing discussion, it should be obvious that selectivity ratios are not constant but vary with the concentration of both the primary and interference ion and, therefore, are often stated at a particular concentration.

EXAMPLE 18.1 Valinomycin membranes show excellent selectivity for potassium—about 3800 times that of sodium and 18 000 times that of hydrogen ions. The selectivity ratio ($1/3800 = 2.632 \times 10^{-4}$) indicates that a 0.1 M sodium solution will give the same response as a 2.6×10^{-5} M potassium solution; that is, a 100 percent error in the apparent potassium concentration. To lower the error due to the presence of sodium ions to 2 percent, the sodium ion concentration should not exceed 0.02×0.1 M, or 0.002 M, for this level of potassium ion concentration. If the potassium concentration were tenfold greater, the sodium ion concentration could also be tenfold greater and the error in the potassium readings would be the same.

  Continuing the example, but for hydrogen ion interference, the hydrogen ion selectivity ratio is $1/18\ 000 = 5.56 \times 10^{-5}$. Thus, the pH of the sample should not exceed 6.26 if the potassium readings in the 1 M range are to be accurate within 1 percent.

In solid-state sensors, the interference arises from surface reactions that can convert one of the components of the solid membrane to a second insoluble compound. As a result, the membrane loses sensitivity to the test ion. For example, chloride ion can interfere with bromide ion measurements if the following reaction takes place:

$$Cl^- + AgBr(s) \rightarrow AgCl(s) + Br^- \qquad\qquad (18.10)$$

It will occur if the ratio of chloride ion activity to bromide ion activity exceeds the ratio of the solubility product of silver chloride to the solubility product of silver bromide, or

$$\frac{1}{K_{ij}} = \frac{a_{Cl}}{a_{Br}} = \frac{1.8 \times 10^{-10}}{5.0 \times 10^{-13}} = 360 \qquad\qquad (18.11)$$

An expected response also fails to occur when the concentration of a primary ion approximates the solubility of the membrane material. This is the basis for the statement made earlier that the bromide ion–selective electrode exhibited a useful response down to pBr = 6 (or, more precisely, $K_{sp} = 6.15$).

ited by the sharpness of the potential change at the end point and by the accuracy with which the volume of titrant can be delivered. Generally, solutions more dilute than 10^{-3} *M* do not give satisfactory end points. This is a limitation of potentiometric titrations.

18.5.1 Location of the Equivalence Point

The equivalence point may be calculated or it can be located by inspection from the inflection point of the titration curve. It is the point that corresponds to the maximum rate of change of cell emf per unit volume of titrant added (usually 0.05 or 0.1 mL). A typical experimental titration curve and its derivatives are shown in Fig. 18.5. Appropriate software algorithms relieve the analyst from the tedium of calculating the derivatives.

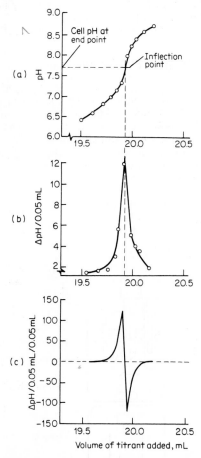

FIGURE   18.5 Potentiometric   titration curves: (*a*) experimental titration curve, (*b*) first derivative curve, and (*c*) second derivative curve.

18.6 AUTOMATIC TITRATORS

An entire titration can be performed automatically by a titrator equipped with a microcomputer and analog-to-digital converters, and using dedicated software. Digitally controlled stepper motors deliver the titrant in precise increments, often slowing the delivery rate as the end point is approached. A fully automatic unit comes equipped with a sample turntable. After each titration the turntable rotates, indexes the next sample beneath the electrode assembly, and actuates the titration switch. Automatic titrators are ideal for performing multiple analyses in which the analytical procedure remains fixed, as in a quality control situation.

18.6.1 End Point Titration Systems

Routine titrations in the laboratory are often performed as end point titrations; that is, as titrations where the addition of titrant is stopped at a predetermined potential or pH value. Instruments for this purpose must not only be able to stop the titration at the end point, but they must be able to gradually reduce the titrant delivery before the end point is reached, so equilibrium in the sample is assured and overshooting of the end point is prevented.

An end point titration system comprises the usual titration assembly (Fig. 18.4) except that an autoburet replaces the manually operated buret. The titrator receives a signal proportional to the potential or pH value of the solution from the mV or pH meter and controls the titrant delivery of the autoburet.

The selected end point is set on the titrator together with a "proportional band" and a "delay sec." The setting of the proportional band corresponds to the pH or mV span prior to the end point wherein the titrant delivery is gradually reduced until the end point is reached. On some instruments this is known as the "anticipator circuit." The titrator reduces the titrant delivery within the proportional band by starting and stopping the autoburet, so the titrant is delivered in small increments. When the time interval between two successive increments exceeds the preselected "delay sec.," the titration is concluded and the volume of titrant is read from the autoburet.

The choice of setting for the proportional band depends on the strength of the inflection on the normal titration curve. The setting of the "delay sec" depends on the reaction rate of the titration, the stirring efficiency, and the possible presence of unwanted side reactions (such as the absorption of carbon dioxide from the air). For the titration of a strong acid with a strong base, the solution at the start of the titration must be pH 3 or less; the end point is set for pH 7 and the proportional band is set at 2 pH units. For the first part of the titration and until pH 5 is reached, the autoburet is running continuously. After pH 5 is reached, the titrant is added in steadily decreasing increments until pH 7 is attained. (Because of the absorption of carbon dioxide, the end point should be selected slightly on the acid side of neutrality.)

18.6.2 Recording Titrators

It is possible to record titration curves automatically. With the same titration assembly (Fig. 18.4), the recorder receives the signal from the electronic voltmeter for the X axis. The consumption of titrant from the buret is electronically linked to the Y axis of the recorder. A complete titration curve is then obtained without preselecting the stopping point. Reassessment of the correct end point can be done at a later time because all data have been recorded.

18.6.3 Derivative Titrators

When a derivative titration is conducted, the pH or mV signal must be differentiated electronically with respect to the volume of titrant added. A differentiation module usually has a selector switch for selecting up or down scale titrations and an inflection selector switch which serves as a combined filter and amplifier. For poorly defined end points, the inflection switch is set in the "weak" position, which provides for maximum amplification of the meter signal and maximum rejection of noise. The "medium" position is for intermediate cases, and the "abrupt" position is used when the end points on the titration curve are well defined. (Nomenclature on switches varies with the instrument manufacturer.)

A microcomputer computes the first and second derivative of the titration curve. At the point where the sign of the second derivative changes from a positive to a negative value (or vice versa) the computer calculates the end point by interpolating to the zero point on the second derivative curve.

This type of titrator requires no prior knowledge of the number or location of end point(s); that is, a predetermined end point is not set and a proportional band or anticipator-type circuit is not used. The computer does all the work. However, the reaction rate must be reasonable, or the titration will require a long period of time to reach an end point.

18.6.4 Automatic Burets

Burets with digital display of readings from 0.001 to 50.00 mL are ideal; they are available as semiautomatic or fully automatic versions. The tedious manipulation of stopcocks, filling of burets, and meniscus reading problems are eliminated. Buret volume is divided into 100 000 pulses which assures accurate and precise addition of titrant.

BIBLIOGRAPHY

Bailey, P. L., *Analysis with Ion Selective Electrodes,* 2d ed., Heyden, London, 1980.

Freiser, H., ed., *Ion-Selective Electrodes in Analytical Chemistry,* Vol. 2, Plenum, New York, 1980.

Koryta, J., *Ions, Electrodes and Membranes,* Wiley, New York, 1982.

Rechnitz, G. A., "Bioanalysis with Potentiometric Membrane Electrodes," *Anal. Chem.,* **54:**1194A (1982).

Serjeant, E. P., *Potentiometry and Potentiometric Titrations,* Wiley, New York, 1984.

CHAPTER 19
VOLTAMMETRIC METHODS

Voltammetric techniques can be used to study solution composition through current-potential relationships in an electrochemical cell and with the current-time response of a microelectrode at a controlled potential. These methods rank among the most sensitive analytical techniques available. They are routinely used for the determination of electroactive inorganic elements and organic substances (Table 19.1) in nanogram and even picogram amounts; often the analysis takes place in seconds. Voltammetric techniques can distinguish between oxidation states, which is important information because the oxidation state may affect a substance's reactivity and toxicology.

The large repertoire of voltammetric methods (Table 19.2) demands flexibility in the format of instrumentation for electrochemical excitation and observation. After consideration of the instrumentation that is basic to all voltammetric methods, the individual techniques are discussed.

TABLE 19.1   Reducible Organic Functional Groups

Aldehydes	Disulfide
Alkene	Heterocylcic double bond
Alpha-halogenated ketone or	Hydroxylamine
aryl methane	Nitrate
Aryl alkyne	Nitrite
Aryl halides	Nitro
Azo	Nitroso
Azoxy	Peroxy
Dibromides	Polynuclear aromatic ring systems

19.1   INSTRUMENTATION

19.1.1   The Dropping Mercury Microelectrode

The dropping mercury microelectrode involves forcing a stream of mercury through a glass capillary (0.05 to 0.08 mm i.d.) under the pressure of an elevated reservoir of mercury connected to the capillary by flexible tubing (Fig. 19.1). Mercury issues from the capillary at the rate of one drop every 2 to 5 s. A dropping mercury electrode (dme) has these advantages:

TABLE 19.2   Repertoire of Voltammetric Methods

Linear potential sweep (dc) voltammetry
   Classical dc polarography at the dropping mercury electrode
   Current-sampled (Tast) voltammetry
Potential-step methods
   Normal pulse voltammetry
   Differential pulse voltammetry
   Square-wave voltammetry
Cyclic potential sweep voltammetry
   Cyclic voltammetry
Stripping voltammetry
   Anodic stripping voltammetry
   Cathodic stripping voltammetry
Controlled potential in flowing systems
   Amperometric titrations
   Constant potential amperometric detection (in HPLC)

---

1. Its surface area is reproducible with a given capillary.
2. The constant renewal of the drop surface (as it expands) eliminates poisoning effects.
3. Hydrogen exhibits a very high overpotential on mercury ($\cong 1.2$ V) which extends the usefulness of the dme to approximately -2.7 V versus the SCE in aqueous solutions with a tetrabutylammonium salt as the supporting electrolyte.
4. The diffusion current assumes a steady value immediately and is reproducible.
5. Mercury forms amalgams with many metals and thereby lowers their reduction potential.

To achieve drop synchronization, essential for reproducible measurements in many voltammetric methods, a drop of mercury is detached from the capillary at regular time intervals. One type delivers a sharp knock to the capillary. These intervals are less than the normal drop life. Each drop is permitted to grow until its area changes the least, usually after the first 1.5 to 2.0 s of drop life.

19.1.2   Solid Electrodes

Solid electrodes may be fabricated from platinum, gold, or glassy carbon. At periodic intervals, as selected by the operator, the stationary electrode tip is rapidly raised and lowered.

19.1.3   Rotating Disk Electrodes

A simple form of a rotating disk electrode consists of a platinum wire sealed in an insulating material with the sealed end ground smooth and perpendicular to the rod axis. In the laboratory, rotating disk electrodes are used primarily for ultra trace analysis and electrode kinetics research.

19.1.4   Thin-Film Mercury Electrode

This type of electrode is prepared by depositing a thin film of mercury on a glassy carbon electrode. The thin film limits the use of this type of electrode to analyte

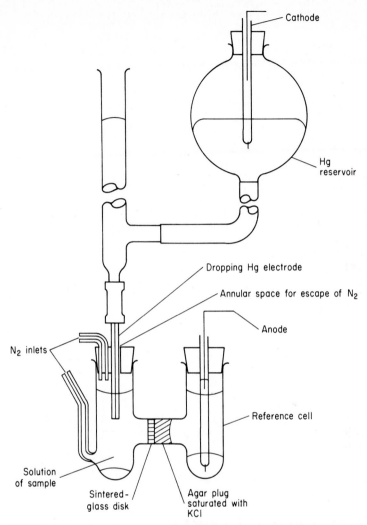

FIGURE 19.1 A dropping mercury electrode and voltammetric cell.

concentrations of less than 10^{-7} M. These electrodes are used only when maximum sensitivity is required as for stripping voltammetry.

19.1.5 Deaeration of Solutions

Oxygen must be removed from analyte solutions by bubbling pure nitrogen through the solution via a fritted glass dispersion frit for 1 to 2 min. The nitrogen gas should be presaturated with solvent vapor to prevent any change in the sample concentration. During experiments the bubbling is stopped; however, a slight positive pressure of nitrogen should be maintained over the solution. Oxygen in solution is reduced in two steps: the first starting at -0.1 V versus SCE and the second at about -0.9 V versus SCE.

19.1.6   Three-Electrode Potentiostat

Potentiostatic control of the working electrode potential accompanied by the measurement of the current at that electrode is the basis of voltammetric methods. The schematic diagram of a three-electrode potentiostat is shown in Fig. 19.2. It consists of a working (indicator) electrode, a reference electrode positioned as close as possible to the working electrode, and an auxiliary (counter) electrode to complete the electrochemical cell. If the working electrode is the cathode, then the auxiliary electrode will be the anode. The function of the potentiostat is to observe the potential of the reference-working electrode pair through a circuit that draws essentially no current. Since the potential of the reference electrode is constant, the circuit really senses the potential of the working electrode without the need to correct for any iR drop across the cell between the anode and cathode. This makes it possible to use nonaqueous solvents of high resistance and quite dilute aqueous electrolytes. As a consequence, distortion of wave shapes is less pronounced, if not entirely eliminated.

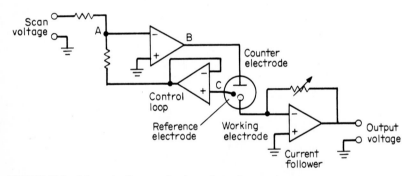

FIGURE 19.2   Schematic diagram of a three-electrode potentiostat.

19.2   LINEAR POTENTIAL SWEEP (DC) VOLTAMMETRY

In the classical case (polarography), a linear dc potential ramp is applied between two electrodes, one small and easily polarized working electrode (the dme) and the other relatively large and relatively constant in potential (often a pool of mercury). Before proceeding further, some terms require definition.

19.2.1   Diffusion Current

An electroactive species undergoing an electron exchange at a microelectrode may be brought to the electrode surface by three modes:

1. *Migration* of charged ions or dipolar molecules capable of undergoing directed movement in an electric field.
2. *Convection* due to thermal currents and by density gradients; convection occurs whenever the solution is stirred into the path of the electrode (hydrodynamic transport).
3. *Diffusion* under the influence of a concentration gradient.

The effect of electrical migration can be eliminated by adding some salt ("supporting electrolyte") in a concentration at least 100-fold greater than the analyte. Usually a potassium or quaternary ammonium salt is added; their discharge potential is more negative than most cations. In the presence of a supporting electrolyte, the resistance of the solution, and thus the potential gradient (iR drop) through it, is made desirably small.

Convection effects can be minimized by using quiescent (unstirred) solutions during current measurements.

Under these conditions, the maximum diffusion current (or limiting current) is the current flowing as a result of an oxidation-reduction process at the electrode surface. The limiting current is controlled entirely by the rate at which the reactive substance can diffuse through the solution to the working electrode. The diffusion current is given by the Ilkovic equation:

$$i_d = 708nCD^{1/2}m^{2/3}t^{1/6} \qquad (19.1)$$

where i_d = diffusion current, μA
D = diffusion coefficient of analyte, $cm^2 \cdot s^{-1}$
C = concentration of analyte, mM
m = mass of mercury flowing through capillary tip, $mg \cdot s^{-1}$
t = drop life, s

In classical polarography, measurements are made just at the end of the drop's life when the diffusion current is greatest. Reproducibility is improved by using an electromechanical drop dislodger.

19.2.2 Current-Potential Curves

The potential of an oxidation-reduction system under conditions of diffusion control is given by

$$E = E^\circ - \frac{0.059\ 16}{n} \log \frac{i}{i_d - i} + \frac{0.059\ 16}{n} \log \left(\frac{D_{red}}{D_{ox}}\right)^{1/2} \qquad (19.2)$$

A current-potential curve (Fig. 19.3) shows the damped sawtooth pattern that arises from the growth and detachment of each mercury drop. As the potential is slowly increased (the dc potential ramp) in the negative direction, the voltammogram (polarogram) exhibits a sigmoidal curve for each electroactive species in the solution. By definition, the *half-wave potential* is the point where the current is one-half the limiting diffusion current (Fig. 19.4). At this point the first logarithmic term in Eq. 19.2 becomes zero, and

$$E_{1/2} = E^\circ + \frac{0.059\ 16}{n} \log \left(\frac{D_{red}}{D_{ox}}\right)^{1/2} \qquad (19.3)$$

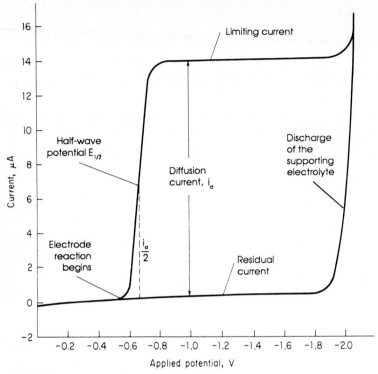

FIGURE 19.3 Current-voltage curve.

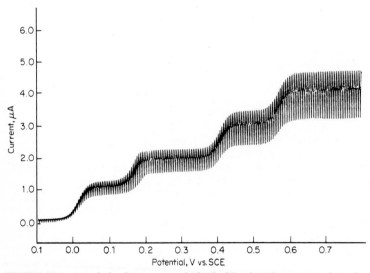

FIGURE 19.4 A typical voltammogram obtained with a dropping mercury electrode.

Many pertinent organic and inorganic half-wave potentials are listed in the *Handbook of Organic Chemistry*.[2]

To obtain the true diffusion (faradaic) current, a correction must be made for the charging or capacitive current, often called the *residual current*. Detection limits are about 10^{-5} M. The charging current arises from the current that flows to charge each fresh mercury drop as it grows. The correction may be made in two ways:

1. The residual current of the supporting electrolyte alone (see Fig. 19.4) is evaluated in a separate voltammogram.

2. Extrapolate the residual current portion of the voltammogram immediately preceding the rising part of the current-potential curve. Take the difference between this extrapolated line and the current-potential plateau as the faradaic (limiting) current. This method is questionable at low concentrations (less than 10^{-4} M).

Maxima may arise from convection around the growing mercury drop. Sometimes they are recognizable peaks, but other types of maxima simply enhance the plateau current without visibly distorting the wave shape. As a precaution, surfactants, such as gelatin or Triton X-100, are routinely added in small quantities (0.005 to 0.01%) to all test solutions.

19.2.3 Current-Sampled (Tast) Voltammetry

In current-sampled voltammetry the current is measured (sampled) at a fixed time (and for a short period of time) near the end of the drop life. This sampled current measurement is held and recorded until the same point in time is reached on the succeeding drop when the measurement is updated to the new measured value of current. A voltammogram is free of drop growth oscillations. The voltammogram resembles a series of steps whose width is the drop time (Fig. 19.5). To control drop life and current sampling time, a synchronized time circuit

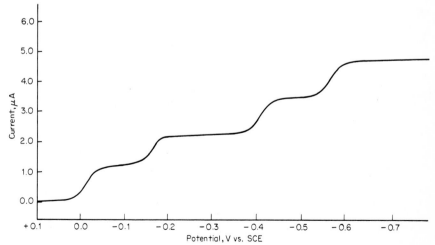

FIGURE 19.5 Current-sampled (Tast) voltammogram. Each step, not recognizable on the graph, represents the current during the lifetime of a mercury drop.

and drop knocker are required. Detection limits are near 10^{-6} M. The improvement in detection limits over those of classical polarography is largely a result of freedom from the current fluctuations caused by drop growth and fall.

19.3   POTENTIAL STEP METHODS[3]

Potential step methods are based on the measurement of current as a function of time after applying a potential. These methods seek to optimize the ratio of faradaic to charging current by applying a sudden change (pulse) in applied potential and sampling the faradaic current just before the drop is detached but after the capacitative current has largely decayed. The potential step methods discriminate against the charging current by delaying the current measurement until close to the end of the pulse.

19.3.1   Normal Pulse Voltammetry

In normal pulse voltammetry a series of square-wave voltage pulses of successively increasing magnitude is superimposed upon a constant dc voltage signal (Fig. 19.6). Near the end of each pulse (perhaps 50 ms in duration) and before the drop is dislodged, the current is sampled for perhaps 17 ms. The sampled current is presented to a recorder as a constant signal until the current sample taken in the next drop lifetime replaces it. A staircase plot traces the potential-current curve. The time delay between pulses must be long enough to restore all concentration gradients at a dropping mercury electrode or a solid microelectrode to their original state before the next potential pulse is applied. Each potential pulse is made a few millivolts higher for each drop (or for each pulse when a solid electrode is used).

19.3.2   Differential Pulse Voltammetry

In differential pulse voltammetry a series of potential pulses of fixed but small amplitude (10 to 100 mV) is superimposed on a constant dc voltage ramp (Fig. 19.7) near the end of the drop life and after the drop has attained the bulk of its growth. The current is sampled immediately before applying the potential pulse (perhaps 17 ms) and again (for 17 ms) just before the drop is dislodged. Subtraction (instrumentally) of the first current sampled from the second provides a stepped, peak-shaped derivative voltammogram.

Differential pulse voltammetry discriminates effectively against the capacitive component of the current signal. Detection limits are improved by the return of the signal to the baseline after each peak.

19.3.3   Square-Wave Voltammetry[4]

In square-wave voltammetry a symmetrical square-wave potential pulse E_{sw} is superimposed on a staircase waveform where the forward pulse (point 1 in Fig. 19.8) of the square wave is coincident with the staircase step. The current is sam-

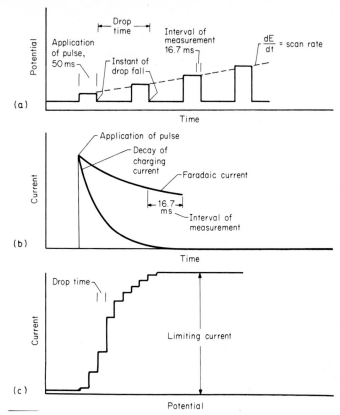

FIGURE 19.6 Normal pulse voltammetry. (*a*) Schematic showing the pulse application, drop time, and interval of current measurement; (*b*) the variation of current during the pulse; (*c*) stepped voltammogram.

pled twice during each square-wave cycle, once at the end of the forward pulse and again at the end of the reverse pulse (point 2). The difference between the current observed at point 1 (forward current) and that at point 2 (reverse current) is the measured signal. Because the amplitude of the square-wave modulation is so large, the reverse pulse causes reoxidation of the product produced on the forward pulse back to the original state with the resulting anodic current. Concentration levels of parts per billion are easily detectable.

19.4 CYCLIC POTENTIAL SWEEP VOLTAMMETRY[5]

Cyclic voltammetry consists of cycling the potential of a stationary electrode immersed in a quiescent solution and measuring the resulting current (Fig. 19.9). An isosceles triangular or staircase potential waveform is used. Scans can be initi-

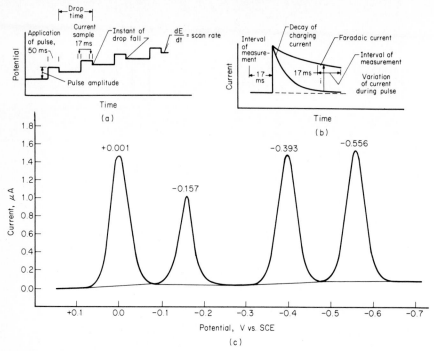

FIGURE 19.7   Differential pulse voltammetry. (*a*) A linearly increasing scan voltage on which a 35-mV pulse is superimposed during the last 50 ms of the drop life. (*b*) Variation of current during the pulse application. (*c*) The net signal observed between the two intervals of current measurement, for four electroactive substances. Individual steps not shown.

ated either in the cathodic or anodic direction. In the example, the initial potential (+0.150 V versus Ag/AgCl) is chosen to avoid any electrolysis of electroactive species in the sample when the cycling is initiated. Then the potential is scanned in the negative direction at 20 mV · s^{-1} until the desired negative potential is reached. For the first electroactive species present, a cathodic current commences to flow at about +0.07 V. The cathodic current increases rapidly until the surface concentration of oxidant at the electrode surface approaches zero, which is signaled by the current peaking at −0.029 V. The current then decays as the solution surrounding the electrode is depleted of oxidant. Additional rises occur as additional electroactive species undergo reaction or the system reaches the discharge potential of the supporting electrolyte. At the selected point (−0.800 V) and about 45 s after initiation of the forward scan, the potential is switched to scan in the positive direction. Briefly, reduction of oxidant continues to occur until the potential becomes sufficiently positive to bring about oxidation of the reductant that had been accumulating adjacent to the electrode surface. An anodic current begins to flow and increases rapidly until the surface concentration of the accumulated reductant approaches zero, at which point the anodic current peaks. The anodic current then decays as the solution surrounding the electrode is depleted of reductant formed during the forward scan. The anodic peaks grow and decay for all other reversible electrode reactions.

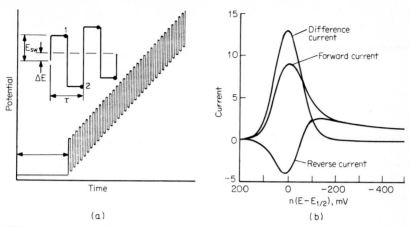

FIGURE 19.8 (*a*) Excitation signal for square-wave voltammetry. (*b*) Forward, reverse, and difference current.

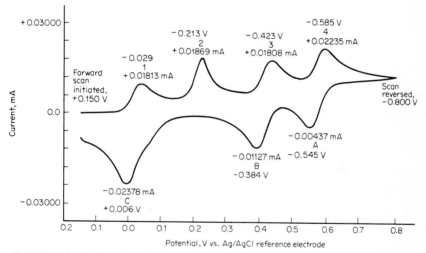

FIGURE 19.9 Cyclic voltammogram of four electroactive species.

Although peak currents can be measured and are linearly related to the analyte concentration, it is sometimes difficult to establish the correct baseline. Cyclic voltammetry finds its greatest use in the study of electrochemical reversibility, kinetics, and transient intermediates. The number of electrons transferred in the electrode reaction can be calculated if the diffusion coefficient is known (or if n is known, the estimation of the diffusion coefficient). Cathodic and anodic peak potentials are separated by $57/n$ mV for a reversible electrode reaction (provided that the switching potential is more negative than $100/n$ mV of the reduction peak). Also, for a reversible wave, the peak potential is independent of the scan rate (usually 20 to 100 mV $\cdot$ s^{-1}) and the peak current is proportional to the square root of the scan rate. Quasi-reversibility is characterized by a separation of cathodic and anodic peak potentials larger than $57/n$ mV and irreversibility by the disappearance of a reverse peak (as shown by the disappearance of the anodic wave for the species whose cathodic peak occurred at -0.213 V).

19.5   STRIPPING VOLTAMMETRY[6]

Stripping voltammetry is the most sensitive electrochemical technique presently available. The technique is applicable to analytes that oxidize or reduce reversibly at a solid (thin-film mercury) electrode or that form an insoluble species with the electrode material, which can subsequently be removed electrochemically.

Basically, stripping voltammetry is a two-step operation. During the first step the ion or ions of interest are electrolytically deposited on the working electrode by controlled potential electrolysis (see Chap. 20). After a quiescent period, a reverse potential scan, or stripping step, is applied in which the deposited analyte(s) is removed from the electrode. The preconcentration or electrodeposition step provides the means for substantially improving the detection limit for the stripping step, often by a factor of up to 1 million.

19.5.1   Anodic Stripping Voltammetry

Anodic stripping voltammetry (ASV) is used to determine the concentration of trace amounts of metal ions that can be preconcentrated at an electrode by reduction to the metallic state. Very electropositive metal ions, such as mercury(II), gold(III), silver, and platinum(IV), are deposited on glassy carbon electrodes. Other metal ions are deposited on a thin-film mercury electrode or a mercury drop electrode.

In the plating step, a deposition potential is chosen that is more negative than the half-wave potential of the most electronegative metal to be determined; in Fig. 19.10 the plating step was done at -0.900 V versus SCE. Potential-current voltammograms are helpful in selecting the deposition potential which should lie on the diffusion current plateau of the most electronegative metal system. The deposition step is seldom carried to completion; often the plating time is 60 s. Usually only a fraction of the metal ions need to be deposited. Thus during the plating step, the temperature and rate of stirring must be kept as constant and reproducible as possible, and the deposition time must be strictly controlled so that the same fraction of metal ion is removed during each experiment with samples and standards. Interference from dissolved oxygen is eliminated by purging the system with purified nitrogen for 2 to 10 min and maintaining a blanket of nitrogen gas over the solution during the entire operation.

Following the deposition step, stirring is halted for a period of 30 to 60 s to allow convection currents to decrease to a negligible level and to allow time for any amalgam to stabilize. The deposition potential is still applied to the working electrode. Then, in the stripping step, the potential is scanned in the positive direction (from -0.9 to $+0.28$ V in Fig. 19.10) using usually the differential pulse voltammetry. The potentials of the stripping peaks identify the respective metals, and the area under each current peak is proportional to the concentration of the metal species.

Standards and a blank are carried through identical plating and stripping steps.

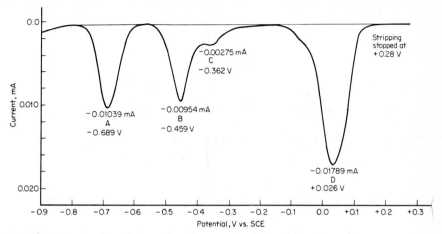

FIGURE 19.10   Anodic stripping voltammogram using differential pulse voltammetry during the stripping cycle.

The method of standard addition is often used for evaluation. If desirable, the supporting electrolyte may be changed after the deposition step to one better suited for the stripping process. Prior to each experiment, the supporting electrolyte must be conditioned (or purified) by applying a potential of 0.0 V versus SCE for 1 to 2 min to clean the electrode by removing metal contaminants.

19.5.2 Cathodic Stripping Voltammetry

Cathodic stripping voltammetry follows the same steps as outlined for anodic stripping voltammetry: (1) preconcentration by applying an controlled oxidation potential to an electrode, (2) a quiescent period, and (3) subsequent stripping via a negative potential scan using differential pulse voltammetry. Cathodic stripping voltammetry is used to determine anions that form insoluble mercury(I) salts at the mercury surface or insoluble silver salts on a silver electrode. A partial list of species that can be determined include arsenate, chloride, bromide, iodide, chromate, sulfate, selenide, sulfide, mercaptans, thiocyanate, and thio compounds.

19.6 AMPEROMETRIC TITRATIONS

In amperometric titrations the potential is maintained at some constant value, usually on a diffusion current plateau, and the current is measured and plotted against the volume of titrant. Either the indicator (working) electrode is rotated at a constant rate or the solution is stirred at a constant velocity past a stationary solid electrode. The reference electrode should be large in area so that when appreciable currents pass through the electrochemical cell its potential will remain constant. Typical equipment is shown in Fig. 19.11.

Only one of the reactants need be electroactive at the working potential. To avoid corrections for dilution, the titrant should be tenfold more concentrated

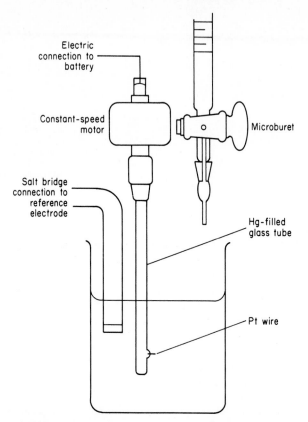

Electric connection to battery

Constant-speed motor

Microburet

Salt bridge connection to reference electrode

Hg-filled glass tube

Pt wire

FIGURE 19.11   Schematic of equipment for an amperometric titration with a rotating platinum electrode.

than the solution being titrated. The shape of the titration curve can be predicted from hydrodynamic voltammograms of the analyte obtained at various stages of the titration (Fig. 19.12). Only three or four experimental points need to be accumulated to establish each branch of the titration curve (Fig. 19.13). Since amperometric titration curves are linear on either side of the equivalence point, the equivalence point is found by extrapolation of data points away from the vicinity of the equivalence point. Data points selected between 0 and 50 percent and between 150 and 200 percent of the end point volume will lie in regions where the common ion effect suppresses dissociation of complexes and solubility of precipitates.

Several shapes of titration curves are possible. An L-shaped curve is obtained when only the analyte is electroactive; after the end point the only signal will be the residual current. If both analyte and titrant are electroactive, the current will decrease up to the end point and then increase again as unreacted titrant is added—a V-shaped curve is obtained. If the analyte is not electroactive but the titrant is, a horizontal line is obtained that rises after the end point is obtained. When possible, the latter shape is preferred because the titrant can be added

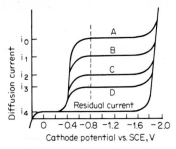

FIGURE 19.12 Successive current-voltage curves of lead ion made after increments of sulfate ion were added.

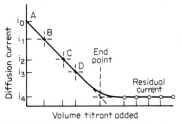

FIGURE 19.13 Amperometric titration curve for the reaction of lead ions with sulfate ions. See Fig. 19.12 for the corresponding current-voltage curves. Performed at −0.8 V versus SCE.

rather rapidly until the diffusion current starts to increase; then three or four data points are taken and the titration is done.

Automation is easy when the titration curve has the "reversed" L-shape. The titrator can be programmed to shut off when a specified current level is reached. Titrant is run into a blank supporting electrolyte until the specified current level is reached, sample is added, and the titrant flow continued until the specified current is attained once again.

19.7 VOLTAMMETRIC METHODS COMPARED

A very effective way to compare some of the voltammetric techniques discussed is to examine the voltammograms obtained using the same solution. The solution selected contained 0.0004 M each of copper(II), lead(II), and cadmium plus 0.00028 M bismuth(III), all in an acetate-tartrate buffer.

A classical dc polarogram of this solution is shown in Fig. 19.4. The polarogram shows individual waves, in order of increasing negative half-wave potentials, for copper, bismuth, lead, and cadmium. This technique has the poorest detection limits (about 10^{-5} M) of any of the voltammetric methods. The culprit is the capacitive current whose compensation is difficult at low concentrations.

A current-sampled (Tast) voltammogram is shown in Fig. 19.5. While the same information is obtained from the dc polarogram, the Tast voltammogram is substantially free from the fluctuating pattern (sawtooth waves) of the classical dc polarogram. Consequently, detection limits are near 10^{-6} M.

A differential pulse voltammogram is shown in Fig. 19.7. While the current scale here is about one-third that for the two preceding voltammograms, the relative peak heights and separation from the residual (background) current are greater. The current is not sampled until the capacitive current has decayed to a negligible value. Also, with a stepped dc ramp the charging current value attributable to the dc potential scan is eliminated. Reversibly reducible ions can be detected at concentrations down to 10^{-8} M. Because the current signal is peak-shaped and returns to the baseline, the selectivity is improved.

Square-wave voltammetry shortens the observation time to about 6 s compared with perhaps 3 min with differential pulse voltammetry. Peak heights are also somewhat larger in square-wave voltammetry. Its rapid response makes it a useful detector in chromatography.

The merits of cyclic voltammetry are largely confined to qualitative and diagnostic experiments. Cyclic voltammetry is capable of rapidly generating a new oxidation state (or reduction state) during the forward scan and then probing its fate on the reverse scan (Fig. 19.9). Systems that exhibit a wide range of rate constants can be studied, and transient species with half-lives of milliseconds are readily detected.

The major advantage of stripping voltammetry is the preconcentration (by factors of 100 or more) of the analyte(s) on or in the small volume of a microelectrode before a voltammetric analysis. Combined with differential pulse voltammetry in the stripping step (Fig. 19.10), solutions as dilute at 10^{-11} M can be analyzed.

REFERENCES

1. J. B. Flato, "The Renaissance in Polarographic and Voltammetric Analysis," *Anal. Chem.*, **44**:75A (1972).

2. J. A. Dean, ed., *Handbook of Organic Chemistry*, McGraw-Hill, New York, 1987.

3. S. A. Borman, "New Electroanalytical Pulse Techniques," *Anal. Chem.*, **54**:698A (1982).

4. J. G Osteryoung and R. A. Osteryoung, "Square Wave Voltammetry," *Anal. Chem.*, **57**:101A (1985).

5. W. R. Heineman and P. T. Kissinger, "Cyclic Voltammetry: Electrochemical Equivalent of Spectroscopy," *Am. Lab.*, **14**:27 (June 1981).

6. W. W. Peterson and R. V. Wong, "Fundamentals of Stripping Voltammetry," *Am. Lab.*, **13**:116 (November 1981).

BIBLIOGRAPHY

Bard, A. J., and L. R. Faulkner, *Electrochemical Methods*, Wiley-Interscience, New York, 1980.

Bond, A. M., *Modern Polarographical Methods in Analytical Chemistry*, Dekker, New York, 1980.

Kissinger, P. T., and W. R. Heineman, eds., *Laboratory Techniques in Electroanalytical Chemistry*, Dekker, New York, 1984.

Koryta, J., and J. Dvorak, *Principles of Electrochemistry*, Wiley, New York, 1987.

CHAPTER 20
ELECTROSEPARATION, COULOMETRY, AND CONDUCTANCE METHODS

20.1 ELECTROSEPARATIONS

When electrodes are immersed in an electrolyte and the applied potential (emf) exceeds the cell emf, an electrochemical reaction is initiated. Oxidation takes place at the anode (positive electrode), and reduction occurs at the cathode (negative electrode). The cathode will react most completely with the strongest oxidizing agent in the solution. In a solution of ionic salts of various metals, such as silver, copper, and lead, the silver ion will be reduced first. It is the strongest oxidizing agent, and all the silver will be converted to the metal form. Copper is next in line (refer to Table 16.1), and the copper ions will be converted into metallic copper. Finally, the lead, which is the least powerful of these three, will be reduced. However, to selectively plate out each metal quantitatively before the next would commence to plate, the potential of the cathode must be controlled during the electrolysis.

20.1.1 Equipment for Controlled-Potential Electroseparations

A schematic arrangement of the equipment needed for controlled potential electrolytic separations is shown in Fig. 20.1. In addition to the platinum gauze anode and cathode, an auxiliary reference electrode is positioned as close as possible to the working electrode (either cathode or anode). The auxiliary reference electrode–working electrode is connected to a vacuum-tube voltmeter (for manual control) or a potentiostat (Fig. 19.2). In manual control the potential of the working electrode is achieved by adjusting the voltage applied to the electrolytic cell (dc output voltage) by manual adjustment of the autotransformer (Variac) at the input of commercial electroanalyzers (Fig. 20.1). A variety of electronic circuits are available for decreasing or increasing automatically the applied emf in order to maintain a constant working electrode potential.

While deposition is in progress, the solution is vigorously stirred with a magnetic stirrer or the anode is rotated. When deposition is complete, the stirring is discontinued but the electric circuit remains connected while the electrodes are

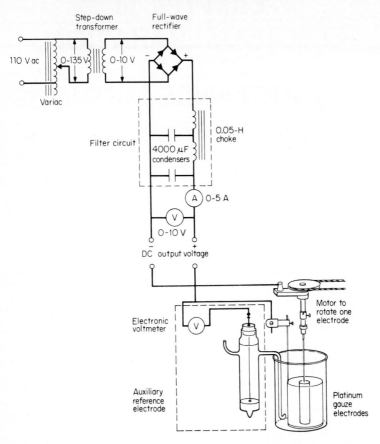

FIGURE 20.1 Equipment for electrodeposition. Enclosed within the dashed lines is the additional equipment required for measuring the working electrode potential.

slowly raised from the solution and washed thoroughly with a stream of distilled water.

20.1.2 Separations with Controlled Potential Electrolysis

An approximate value of the limiting potential of the working electrode can be calculated from the Nernst equation, but lack of knowledge concerning the overpotential term(s) for a system severely limits its usefulness. A more reliable method utilizes the information obtained from current-potential curves (Chap. 19) at solid electrodes.

EXAMPLE 20.1 What range of cathode potentials is needed to deposit silver quantitatively (that is, lower the silver concentration to at least 10^{-6} M) from a solution 0.0100 M in silver nitrate?

Plating of silver will commence when the cathode potential is

$$E = 0.799 - 0.059\ 16 \log \frac{1}{10^{-2}} = 0.681 \text{ V}$$

The removal of silver ions would be considered complete at

$$E = 0.799 - 0.059\ 16 \log \frac{1}{10^{-6}} = 0.444 \text{ V}$$

Thus, the deposition would begin at 0.681 V and would be completed at 0.444 V, an interval of 0.355 V. When a saturated calomel electrode (0.245 V versus SHE) is used as the auxiliary reference electrode, the range would be from 0.436 to 0.199 V.

EXAMPLE 20.2 Would the silver removal be complete before the copper would commence to plate from a solution 0.0100 M in copper(II) ions?
  Copper(II) ions would commence to deposit at

$$E = 0.337 - \frac{0.059\ 16}{2} \log \frac{1}{10^{-2}} = 0.278 \text{ V} \qquad \text{(or 0.033 V versus SCE)}$$

The removal of silver would be complete at 0.199 V (versus SCE), whereas copper would start to deposit at 0.033 V versus SCE. Controlling the cathode potential somewhere in the interval between 0.2 and 0.05 V versus SCE should be satisfactory. The current through the system steadily decreases as the deposition proceeds. However, the maximum permissible current is used at all times so the electrolysis proceeds at the maximum rate.

Many electroseparations are accompanied by the possible evolution of hydrogen gas. Fortunately, the overpotential of hydrogen gas evolution on some electrode material is significant and delays the discharge of hydrogen (Table 20.1). Any overpotential term is added to the Nernst expression and takes the sign of the electrode.

TABLE 20.1 Hydrogen Overpotential on Various Metals

Electrolyte is 1 M H_2SO_4. Overpotential given in volts

Metal	First visible gas bubbles	Current density, 0.01 A $\cdot$ cm^{-2}
Antimony	0.23	0.4
Bismuth	0.39	0.4
Cadmium	0.39	0.4
Copper	0.19	0.4
Lead	0.40	0.4
Mercury	0.80	1.2
Platinum (bright)	0.0	0.09
Silver	0.10	0.3
Tin	0.48	0.5
Zinc	0.48	0.7

EXAMPLE 20.3 Consider the quantitative removal of cadmium from a solution 0.0100 M in cadmium ions. What pH must the solution be adjusted to in order to prevent interference from the evolution of hydrogen gas?

In a 0.1 M nitric acid solution, hydrogen would commence to evolve at a platinum electrode when

$$E = 0.000 - \frac{0.059\ 16}{1} \log \frac{1}{10^{-1}} = -0.059 \text{ V}$$

However, cadmium would not commence plating until

$$E = -0.403 - \frac{0.059\ 16}{2} \log \frac{1}{10^{-2}} = -0.444 \text{ V}$$

The separation is impossible at pH 1 and using platinum electrodes at which the hydrogen overpotential is only about 0.09 V. However, two alterations in procedure will allow cadmium to be deposited. Acetate ions are added to adjust the pH of the solution to pH 5, which will shift the electrode potential at which hydrogen evolves to a more negative value.

$$E = 0.000 - \frac{0.059\ 16}{1} \log \frac{1}{10^{-5}} = -0.296 \text{ V}$$

Second, use an electrode that has been precoated with copper because the hydrogen overpotential is about 0.4 V on copper. Now hydrogen will not interfere until the cathode potential reaches about -0.7 V. At this potential, the residual cadmium concentration remaining in solution is theoretically

$$-0.7 = -0.403 - \frac{0.059\ 16}{2} \log \frac{1}{[\text{Cd}^{2+}]}$$

and
$$[\text{Cd}^{2+}] \leq 1.4 \times 10^{-10}$$

Actually, as soon as a layer of cadmium metal completely covers the electrode surface, the overpotential value of hydrogen gas on a cadmium surface should now be used to calculate the value at which hydrogen would begin to evolve. The overpotential value on cadmium is the same value as on copper. However, if one were plating zinc, the overpotential would increase substantially as soon as the electrode were covered with zinc metal.

From the foregoing examples, several facts emerge.

1. The formal potentials of two oxidation-reduction systems must be $0.178/n$ mV or greater in difference to be able to completely deposit the metal from the more positive oxidation-reduction system before the metal from the next system would commence to plate.

2. A pH change of one unit alters the potential of hydrogen evolution by 0.059 V.

3. The hydrogen overpotential on certain metal surfaces, particularly on mercury and zinc, can be very advantageous.

The addition of a masking agent to form a complex with a metal ion usually shifts the deposition potential in a more negative direction. Tin(II) and lead(II) have almost identical deposition potentials in a chloride medium, but oxidation of tin to the tetravalent state and then complexation of tin(IV) with tartrate ions at pH 5 causes lead to be deposited in the presence of tin. After the solution is acid-ified to destroy the tartrate complex of tin(IV), the tin can be deposited.

20.1.3 Factors Governing Current

During constant-potential electrolysis the current decreases logarithmically with time. The decadic change in concentration of reactant is independent of the initial concentration of reactant. The concentration (or current) at any time t is a func-tion of the initial concentration C_0 (or current):

$$C_t = C_0 e^{-kt} \qquad \text{or} \qquad 2.3 \log \frac{C_0}{C_t} = kt \qquad (20.1)$$

The constant k is given by $DA/V\delta$, where A is the area of the working electrode (in cm^2), D is the diffusion coefficient of the electroactive species (in cm$^2 \cdot$ min^{-1}), V is the volume of the solution (in cm^3), and δ is the thickness of the diffusion layer (in cm). From the terms in the constant k, several factors should be considered in order to optimize an electrodeposition:

1. Use as small a volume of electrolyte as needed to cover the electrodes; a high-form conical beaker is recommended for this purpose.
2. Keep the area of the working electrode as large as possible; here a gauze elec-trode is superior to a perforated metal electrode.
3. Stir the solution because it increases the mass transfer of analyte up to the electrode-solution interface and thereby decreases the thickness of the diffu-sion layer.
4. An elevated temperature, perhaps about 60 °C, usually increases the rate of mass transfer.

Electrolysis is usually discontinued when the current has diminished to 10 to 20 mA. A logarithmic plot of current versus time will enable the value of k to be calculated from the slope of the curve; then the time required to diminish the ini-tial concentration to 10^{-6} M can be estimated.

20.1.4 The Mercury Cathode

A widely used type of electroreduction consists of a pool of mercury (35 to 50 mL) to which a constant potential is applied. The cell shown in Fig. 20.2 is in common use. The anode is a platinum wire formed into a flat spiral. Agitation is accomplished with a mechanical stirrer whose impeller blades are only partially immersed in the mercury. The supporting electrolyte is usually a 0.1 to 0.5 M solution of sulfuric acid or perchloric acid.

This technique finds extensive use for the removal of elements, particularly major constituents of certain samples that interfere in various methods for the minor constituents. Iron in steel and cast iron samples is frequently removed in this way; an alternative method is the use of solvent extraction. No attempt is made to determine any of the metals deposited.

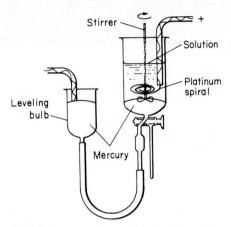

FIGURE 20.2 Mercury cathode cell. [*Reprinted with permission from A. D. Melaven. Ind. Eng. Chem., Anal. Ed., 2:180 (1930). Copyright 1930 American Chemical Society.*]

20.2 CONTROLLED-POTENTIAL COULOMETRY

Coulometric procedures require operation at 100 percent current efficiency and must involve only one overall reaction of known stoichiometry. The method is applicable in the range from milligram quantities down to microgram quantities. Sensitivity is usually limited only by problems of end point detection.

Coulometric methods offer unique advantages:

1. The need for burets and balances is eliminated.
2. The preparation, storage, and standardization of standard solutions is eliminated.
3. Reagents difficult to use or unstable reagents present no problems. They are produced in situ and immediately consumed.
4. The electron becomes the primary standard.[1]

20.2.1 General Principles

Controlled-potential coulometry employs the same equipment as used in controlled-potential electrolysis with the addition of a coulometer. A mercury pool is often used for reduction processes; oxidations can be performed at a cylindrical platinum electrode. The potential of the working electrode is controlled within 1 to 5 mV of the control value by a potentiostat. Current-potential diagrams must be determined for the analyte system and for any possible interfering system. The necessary data can be obtained in two ways:

1. Set the potentiostat to one cathode-reference potential after another in sequence, allowing only enough time at each setting for the current indicator to balance.
2. Perform the coulometric analysis in the usual manner. Periodically adjust the

potential to a value that stops the current flow. Note the net charge transferred up to each adjustment point; a plot of number of coulombs versus the potential provides a coulogram.

A controlled-potential coulometric electrolysis is like a first-order reaction, with the concentration and the current decaying exponentially with time during the electrolysis (Eq. 20.1) and eventually attaining the residual current of the supporting electrolyte. The concentration limits vary from about 2 meq down to about 0.05 meq (set by the magnitude of the residual current).

Advantages of controlled-potential coulometry are:

1. No indicator electrode system is necessary.
2. It proceeds virtually unattended with automatic instruments.
3. Optimum conditions for successive reactions are easily obtained.

As in voltammetry, it is necessary to prereduce the supporting electrolyte in a coulometric reduction, add the sample, and deaerate the system before the reduction is started. Standards should be run under the same conditions that will be used with the samples.

20.2.2 Applications

In a mixture of uranium and chromium, both metals can be prereduced at -0.15 V to uranium(III) and chromium(II). Now, if electrolysis is carried out at -0.55 V, only uranium(III) is oxidized to uranium(IV), not with 100 percent current efficiency but completely. Chromium(II) is then determined by oxidation to chromium(III) at -0.15 V.

Mixtures of two reversible oxidation-reduction states are handled as follows for vanadium(IV)-vanadium(V). At 0.7 V versus SCE, vanadium(IV) is oxidized to vanadium(V); the reduction of vanadium(V) will occur quantitatively at 0.3 V versus SCE. In a mixture of vanadium(IV) and vanadium(V), control the anode potential at 0.75 V and measure the number of coulombs involved in the oxidation of vanadium(IV). Reverse the working electrode potential, control the cathode potential at 0.3 V versus SCE, and measure the coulombs required for the reduction of original and generated vanadium(V). The difference in number of coulombs between the cathodic reduction and the anodic oxidation gives the original vanadium(V) concentration.

20.3 CONSTANT-CURRENT COULOMETRY

Chemical reagents are generated within the supporting electrolyte in constant-current coulometry (often called coulometric titrations). A constant current is maintained through the electrochemical cell throughout the reaction period. The quantity of unknown present is given by the number of coulombs (the product of current and time) of electricity used. The problem is to find electrode reactions that proceed with 100 percent current efficiency and end point detection systems.

20.3.1 Primary Coulometric Titrations

Only electrodes of silver metal, mercury, or mercury amalgams, or electrodes coated with silver–silver halide are suitable sources of the electrogenerated species. For example, the silver ions generated at a silver anode will react with mercaptans dissolved in a mixture of aqueous methanol and benzene to which aqueous ammonia and ammonium nitrate are added to buffer the solution and supply the supporting electrolyte. The end point is determined amperometrically; excess silver ions will generate a signal at a platinum indicator electrode. Before the mercaptan sample is added, free silver ion is generated to a predetermined amperometric (current) signal. The sample is added and the generation continued until the same amperometric signal is attained again. Chloride ion in biologic samples is determined in a similar manner. Combustion in an oxygen flask precedes the titration step for nonionic halides in organic compounds.

20.3.2 Secondary Coulometric Titrations

Secondary coulometric titrations are the most frequently used coulometric technique. These conditions must be met:

1. An active intermediate ion must be generated from an oxidation-reduction buffer (titrant precursor) added in excess to the supporting electrolyte.
2. The intermediate must be generated with 100 percent efficiency.
3. The intermediate must react rapidly and stoichiometrically with the substance being determined.
4. The standard potential of the titrant precursor must lie between the potential "window" of the unknown redox system and the potential at which the supporting electrolyte or another sample constituent undergoes a direct electrode reaction.
5. An end point detection system must be available to indicate when the coulometric generation should be discontinued.

An example will aid the discussion. Consider the coulometric titration of iron(II) to iron(III). The direct coulometric method will not succeed with 100 percent efficiency unless the potential is carefully controlled. When a finite current is forced to flow through the electrochemical cell, the current transported by the iron(II) ions soon falls below that demanded by the imposed current flow. However, if excess cerium(III) ions are added as the titrant precursor, they will begin and continue to transport the current. At the anode the cerium(III) ions will be oxidized to cerium(IV) ions, which will immediately react with unoxidized iron(II) ions. The reaction is stoichiometric and cerium(III) is re-formed. The total coulombs ultimately required will be the sum needed for the direct oxidation of iron(II) and for the indirect oxidation via cerium(IV). Because there is an inexhaustible supply of cerium(III), the anode potential is stabilized at a value less positive than the oxidation of water, which would destroy the coulometric efficiency required. The end point is signaled either potentiometrically with a platinum-reference electrode pair or spectrophotometrically at the wavelength where the first excess of unused cerium(IV) absorbs strongly.

20.3.3 Instrumentation

The instrumentation required consists of an operational amplifier to force a constant current through the generator cell (Fig. 20.3a) and some means to measure

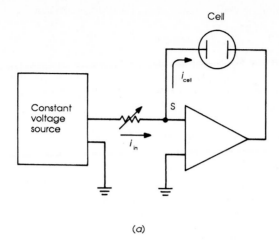

(a)

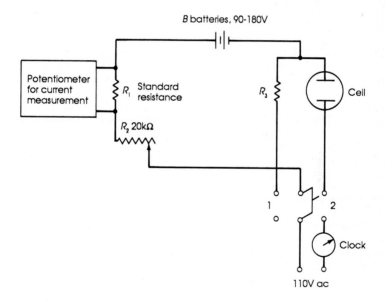

(b)

FIGURE 20.3 Equipment for constant-current coulometry. (a) Operational amplifier (plus a timer needed); (b) manual circuit; R_3 is dummy resistor to be placed in circuit to maintain steady conditions when the cell is disconnected.

the electrogeneration time. A manual circuit (Fig. 20.3b) can be assembled easily. The current source can be a heavy-duty dry cell (6 V) or several B-batteries connected in series with an adjustable rheostat (to control the current level), a precision resistor with a potentiometer connected across its terminals (to measure the current), and the generator electrodes. If electrolytic products generated at the counterelectrode interfere with the reactions at the working electrode, the counterelectrode must be isolated from the remainder of the electrochemical cell by a porous glass frit or other type of salt bridge. The electrochemical cell is stirred throughout the titration.

20.3.4 Applications

Secondary coulometric methods cause uncommon, but useful, titrants such as chromium(II), copper(I), chlorine, bromine, titanium(III) and uranium(V) to be generated in situ. Ordinarily these solutions would be difficult or impossible to prepare and store as standard solutions. Even electrolytic generation of hydroxyl ion offers the advantage that very small amounts are prepared for the determination of very dilute acid solutions, such as would result from adsorption of acidic gases and in a carbonate-free condition.

Internally generated halogens, particularly bromine, have widespread application, especially in organic analysis. Sodium and lithium bromides are quite soluble in various organic solvents in which bromination can be conducted.

The coulometric Karl Fischer titration allows the determination of microgram amounts of water in organic liquids and of moisture in gases. Iodine is generated from an iodide salt in anhydrous methanol plus amine solvents that contain sulfur dioxide.

Azo dyestuffs can be titrated with titanium(III) generated externally and delivered to the hot dye solution via a capillary delivery tube. External generation guarantees that an optimum set of generation and reaction conditions prevail for each step. A double-arm electrolytic cell with separate anode and cathode delivery tubes (one to the sample and other to waste) is used for external generation.

The thickness of corrosion or tarnish films can be measured coulometrically. The specimen is made the cathode, and the film is reduced with a constant, known current to the metal. By following the cathode potential, the end point is taken as the point of inflection of the voltage-time curve. Anodic dissolution is used to determine the successive coatings on a metal surface. For example, the thickness of a tin undercoating and a copper-tin surface layer on iron can be measured because the two coatings exhibit individual step potentials.

20.4 CONDUCTANCE METHODS

Ultra-pure water is a nonconductor of electricity; however, when certain substances are added to water, it does conduct electric current. These substances are called electrolytes. They form positive and negative ions which will carry the electric current—an action called electrolytic conduction. Those substances which remain nonionized in solution are called nonelectrolytes.

20.4.1 Electrolytic Conductivity

Solutions of electrolytes conduct an electric current because the ions migrate under the influence of a potential gradient applied to two electrodes immersed in the solution. The positive ions (cations) are attracted to the negative electrode (cathode) while the negative ions (anions) are attracted to the positive electrode (anode). The flow of current depends upon the magnitude of the applied potential and the resistance of the solution between the electrodes, as expressed by Ohm's law.

The reciprocal of the resistance $1/R$ is called the *conductance S* and expressed in reciprocal ohms or mhos (in SI nomenclature the unit is siemens and the abbreviation is S). The conductance is directly proportional to the cross-sectional area A of the electrodes and inversely proportional to the distance d between them:

$$\frac{1}{R} = S = \kappa\frac{A}{d} \tag{20.2}$$

where κ is the *specific conductance* expressed in $\Omega^{-1} \cdot cm^{-1}$ or $S \cdot cm^{-1}$). For a given electrolytic cell with fixed electrodes, the ratio d/A is a constant, called the *cell constant*, Θ. From Eq. 20.2, it follows that

$$\kappa = S\Theta \tag{20.3}$$

The d/A ratio is determined by measuring the resistance of a standard solution of known specific resistance (Table 20.2), and the cell constant is then computed.

TABLE 20.2 Standard Solutions for Calibrating Conductivity Vessels

Grams KCl per kilogram solution	Conductivity in $S \cdot cm^{-1}$ at		
	0°C	18°C	25°C
71.135	0.065 14	0.097 79	0.111 29
7.419	0.007 134	0.011 161	0.012 850
0.745 3*	0.000 773 3	0.001 219 9	0.001 408 1

*Virtually 0.0100 M.

EXAMPLE 20.4 When a particular conductance cell was filled with a standard solution containing 0.745 263 g KCl per kilogram of solution, which has a specific conductance at 25°C of 0.001 408 $S \cdot cm^{-1}$, it had a resistance of 142.0 Ω. The cell constant is

$$\Theta = (0.001\ 408\ S \cdot cm^{-1})(142.0\ \Omega) = 0.200\ cm^{-1}$$

A historical survey of quantitative electrolytic conductivity has been written by Stork.[2]

20.4.2 Conductance Cells

A conductivity cell with large electrodes very close together has a low cell constant, as in Example 20.4. If the distance is increased and/or the area of the electrodes is decreased, the cell constant increases. For example, a cell with a constant of 10 cm^{-1} would have electrodes perhaps 0.5 cm^2 in area and spaced 5 cm apart. This arrangement would be suitable for measuring the conductance of 0.005 to 2.0% sulfuric acid solutions whose specific conductance ranges from about 0.000 44 to 0.176 S · cm^{-1}. The resistance readings would range from 22 700 Ω for the dilute acid to 57 Ω for the 2% acid solution.

Several types of conductivity cells are shown in Fig. 20.4. The dip-type cell is simplest to use whenever the liquid to be tested is in an open container. Whatever the configuration of the cell, the test solution must completely cover the electrodes. Pipet cells (not shown) permit measurements with as little as 0.01 mL of solution. A pair of individual square platinum electrodes on glass wands is useful in conductometric titrations.

20.4.3 Conductivity Meters

In the classical mode of conductance measurements, resistance measurements are made using some variation of a Wheatstone bridge. A simplified schematic of a conductivity bridge is shown in Fig. 20.5. To balance the capacitive effects in the conductance cell, the bridge circuit must also contain a variable capacitance (8 to 200 pF) in parallel with the balancing resistor. A built-in generator provides bridge current at frequencies of 100, 1000, and 3000 Hz. A lower frequency is

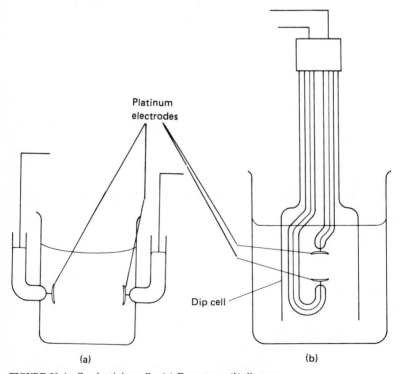

Platinum electrodes

Dip cell

(a)   (b)

FIGURE 20.4 Conductivity cells: (*a*) Freas type, (*b*) dip type.

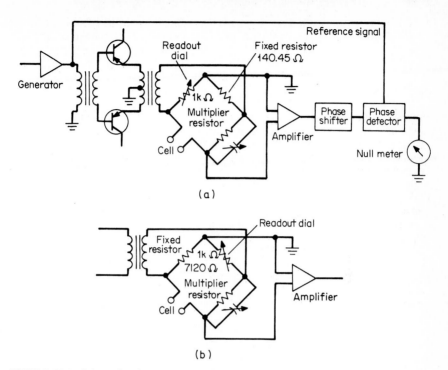

(a)

(b)

FIGURE 20.5 Schematic of a conductivity bridge: (*a*) resistance mode and (*b*) conductance mode. (*Courtesy of Beckman Instruments, Inc.*)

preferred when the measured resistance is high and the higher frequency when the measured resistance is low. Use of an alternating current eliminates the effects of faradaic processes; that is, the deposition potential is not exceeded. The cell constant of the conductivity cell should be selected to maintain the measured resistance between 100 Ω and 1.1 MΩ. Any smooth metal surface can serve as an electrode at an operating frequency of 3000 Hz. Stainless steel electrodes are frequently used for industrial on-line applications.

The Wheatstone bridge can be replaced by operational amplifier circuitry (see Chap. 1), as shown in Fig. 20.6. For conductance measurements, the cell is connected in place of R_1; the result is a current $i = E/R_1$, which is proportional to the

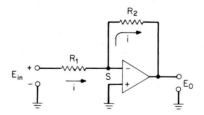

FIGURE 20.6 Operational amplifier used for resistance and conductance measurements.

conductance of the cell. For resistance measurements, a fixed resistance R_1 is used in the input current-generating circuit and the cell is connected in place of R_2.

20.4.4 Conductometric Titrations

In conductometric titrations the variation of the electrical conductivity of a solution during the course of a titration is followed. A conductometric titration is devised so that the ionic species to be determined can be replaced by another ionic species of significantly different conductance. The end point is obtained by the intersection of two straight lines that are drawn through a suitable number of points (usually four for each linear branch) obtained by measurement of the conductivity after each addition of titrant. The titrant should be at least ten times as concentrated as the analyte in order to keep the volume change small. If necessary a correction may be applied. All conductance readings are multiplied by the ratio

$$\frac{V + v}{V}$$

where V is the initial volume and v is the volume of titrant added up to the particular conductance reading.

The major applications are to acid-base titrimetry. Titrant concentrations can be as low as $0.0001\ M$. Under optimum conditions the end point can be located with a relative error of approximately 0.5 percent. Typical acid-base titrations will be considered. Limiting equivalent ionic conductances in aqueous solutions at 25°C are given in Table 20.3. Additional values are available in the literature.[3]

20.4.4.1 Strong Acid Titrated With Strong Base.   Consider the titration of a 0.001 M solution of HCl with 0.1 M NaOH. During the formation of water, the highly conducting hydronium ion ($\lambda_+ = 350$) is replaced by a less highly conducting sodium ion ($\lambda_+ = 50$). The conductivity falls linearly, as shown in Fig. 20.7, reach-

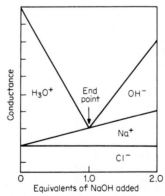

FIGURE 20.7   Titration of hydrochloric acid with sodium hydroxide; also detailed is the conductance of the individual ions.

TABLE 20.3   Limiting Equivalent Ionic Conductances in Aqueous Solutions at 25°C*

Ion	Limiting equivalent ionic conductance, mho-cm$^2 \cdot$ eq^{-1}	Ion	Limiting equivalent ionic conductance, mho-cm$^2 \cdot$ eq^{-1}
Ag$^+$	61.9	Br^{-1}	78.1
Ba^{2+}	63.9	Cl$^-$	76.4
Ca^{2+}	59.5	ClO$_4^-$	67.9
Cu^{2+}	55	F$^-$	54.4
Fe^{2+}	54	Fe(CN)$_6^{4-}$	111
Fe^{3+}	68	Fe(CN)$_6^{3-}$	101
H$^+$	349.8	HCO$_3^-$	44.5
K$^+$	73.5	H$_2$PO$_4^-$	57
Li$^+$	38.7	HPO$_4^{2-}$	33
Mg^{2+}	53	HSO$_3^-$	50
NH$_4^+$	73.5	HSO$_4^-$	50
Na	50.1	I$^-$	76.8
Pb^{2+}	71	NO$_2^-$	71.8
Sr^{2+}	59.5	NO$_3^-$	71.4
UO$_2^{2+}$	32	OH$^-$	198.6
Zn^{2+}	52.8	PO$_4^{3-}$	69.0
Ethylammonium	47.2	Acetate	40.9
Diethylammonium	42.0	Benzoate	32.4
Triethylammonium	34.3	HC$_2$O$_4^-$	40.2
Piperidinium	37.2	C$_2$O$_4^{2-}$	74.2

*Except for H$^+$ (0.0139 deg^{-1}) and OH$^-$ (0.018 deg^{-1}), a temperature coefficient of 0.02 deg^{-1} is applicable to the cations and anions.
Source: J. A. Dean, ed., *Lange's Handbook of Chemistry*, 13th ed., McGraw-Hill, New York, 1985, pp. 6-34 and 6-35.

ing a minimum when the solution consists of only NaCl. Unused NaOH and previously formed NaCl constitute the conductance of the rising branch of the titration curve. The conductance of the solution at any point on the descending branch of the titration curve is given by the expression

$$\frac{1}{R} = \frac{1}{1000\Theta}(C_H\lambda_H + C_{Na}\lambda_{Na} + C_{Cl}\lambda_{Cl}) \tag{20.4}$$

This equation can be expressed in terms of the initial concentration of HCl C_i and the fraction of the acid titrated f:

$$C_H = C_i(1 - f) \qquad C_{Na} = C_i f \qquad C_{Cl} = C_i$$

Substituting these values into Eq. 20.4,

$$\frac{1}{R} = \frac{C_i}{1000\Theta}[\lambda_H + \lambda_{Cl} + f(\lambda_{Na} - \lambda_H)] \tag{20.5}$$

The term within the parentheses in Eq. 20.5 expresses the steepness of the drop in conductivity up to the end point.

20.4.4.2   Incompletely Dissociated Acids (or Bases).

Titration of incompletely dissociated acids (or bases) is somewhat more difficult. Initially the solution has a

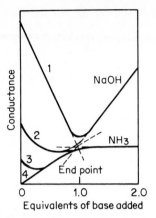

FIGURE 20.8 Acids of different strengths titrated with either sodium hydroxide or aqueous ammonia. The numbered curves are (1) HCl, (2) an acid of $pK_a = 3$, (3) an acid of $pK_a = 5$, and (4) acids whose $pK_a = 7$ or greater.

low conductivity which is due to the few ionized ions present. As neutralization proceeds, the common ion formed represses the dissociation so that an initial fall in conductivity may occur. Consequently, the shape of the initial portion of these titration curves will vary with the strength of the weak acid (or base) and its concentration, as indicated in Figs. 20.8 and 20.9. Pronounced hydrolysis in the vicinity of the end point makes it necessary to select the experimental points for the construction of the two branches considerably removed from the end point.

Sometimes no linear region is obtained preceding the end point. A clever stratagem overcomes the problem. For example, in the titration of a weak acid, sufficient aqueous ammonia is added to neutralize about 80 percent of the weak acid. Then the titration is carried out with standard NaOH. After the remaining weak acid has been neutralized, the titration involves the reaction of the ammonium ion (formed during the neutralization of the weak acid) with NaOH to form ammonia, water, and sodium ions. The conductance falls owing to the replacement of the ammonium ion ($\lambda_+ = 73$) by the sodium ion ($\lambda_+ = 50$). When the replacement is complete, the conductivity increases abruptly because of the unused NaOH, as shown in Fig. 20.10.

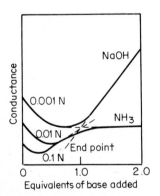

FIGURE 20.9 Effect of concentration upon the shape of the titration curve for an acid whose $pK_a = 4.8$.

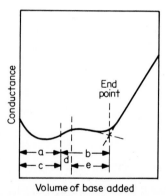

FIGURE 20.10 Titration of a weak acid employing the preliminary addition of aqueous ammonia followed by titration with sodium hydroxide. (*a*) Volume of aqueous ammonia added, (*b*) volume of sodium hydroxide added up to the end point, amount of acid neutralized (*c*) by ammonia and (*d*) by sodium hydroxide, and (*e*) displacement of ammonia from ammonium ions formed in (*c*).

20.4.4.3 *Mixtures of Different Acid Strength.*

A mixture of a strong and a weak acid can be determined in one titration, as illustrated in Fig. 20.11 for the titration of the first and second protons in oxalic acid. The curve would be similar for the titration of a mixture of HCl and acetic acid (or other carboxylic acid). A more distinct second end point is obtained when the titrant is aqueous ammonia (or propyl amine). Weak acids can be titrated only if the product of the ionization constant and the acid concentration exceeds 10^{-11}.

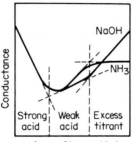

FIGURE 20.11   Titration of a mixture of a strong acid and a weak acid; the specific example involves the two protons of oxalic acid.

The conductometric titration technique is also useful in the titration of the conjugate base of a weakly ionized acid and vice versa. This extends the technique to organic salts such as acetates, benzoates, nicotinates, and so on. One caveat—the ionization constant of the displaced acid or base divided by the original salt concentration must not exceed 5×10^{-3}.

REFERENCES

1. G. W. Ewing "Titrate with Electrons," *Am. Lab.*, **13**:16 (June 1981).
2. J. T. Stork, "Two Centuries of Quantitative Electrolytic Conductivity," *Anal. Chem.*, **56**:561A (1984).
3. J. A. Dean, ed., *Lange's Handbook of Chemistry*, 13th ed., p. 6-36, McGraw-Hill, New York, 1985.

BIBLIOGRAPHY

Bard, A. J., ed., *Electroanalytical Chemistry, a Series of Advances*, Dekker, New York; a series of monographs usually published yearly commencing in 1966.

Kissinger, P. T., and W. H. Heineman, eds., *Laboratory Techniques in Electroanalytical Chemistry*, Dekker, New York, 1984.

Lingane, J. J., *Electroanalytical Chemistry*, 2d ed., Wiley-Interscience, New York, 1958.

Plambeck, J. A., *Electroanalytical Chemistry, Basic Principles and Applications*, Wiley, New York, 1982.

CHAPTER 21
THERMAL ANALYSIS

21.1 INTRODUCTION

Thermal analysis includes a group of techniques in which specific physical properties of a material are measured as a function of temperature. The techniques include the measurement of temperatures at which changes may occur, the measurement of the energy absorbed (*endothermic* transition) or evolved (*exothermic* transition) during a phase transition or a chemical reaction, and the assessment of physical changes resulting from changes in temperature. Most of the thermal analysis techniques in general use can be found in the following categories:

1. *Differential scanning calorimetry* (DSC) measures the amount of energy absorbed or emitted by a sample as a function of either temperature or time. In addition to these direct energy measurements, the precise temperature of the sample material at any point during the experiment is also monitored.

2. *Differential thermal analysis* (DTA) measures the temperature difference between a sample and reference material (as a function of temperature) when both are subjected to a controlled temperature program.

3. *Thermogravimetric analysis* (TGA) monitors the change in mass of a substance as a function of temperature or time while the sample is subjected to a controlled temperature program.

4. *Thermomechanical analysis* (TMA) measures either the dimension or deformation of a substance under a nonoscillatory load and as a function of temperature.

5. *Dynamic mechanical analysis* (DMA) measures the dynamic modulus and/or damping of a substance under an oscillatory load as a function of temperature.

6. *Enthapimetric analysis,* which includes the titrimetric and calorimetric modes, utilizes the temperature change in a system while a titrant is gradually added or measures the thermal energy released during a controlled reaction of the specimen.

Various environments (vacuum, inert, or controlled gas composition) and heating rates from 0.1 to 500°C · min^{-1} are available for temperatures ranging from -190 to 1400°C. The analysis of gas(es) released by the specimen as a function of temperature is possible. The versatility of thermal analysis is thereby increased. A block diagram of a thermal analysis system is shown in Fig. 21.1.

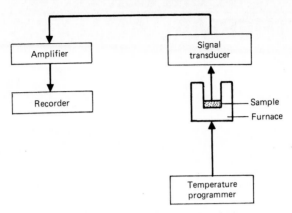

FIGURE 21.1 Block diagram of a thermal analysis system.

21.2 DIFFERENTIAL SCANNING CALORIMETRY AND DIFFERENTIAL THERMAL ANALYSIS

Differential scanning calorimetry (DSC) and differential thermal analysis (DTA) are used to investigate the thermal properties of inorganic and organic materials. DTA detects the physical and chemical changes which are accompanied by a gain or loss of heat in a substance as the temperature is altered. DSC provides quantitative information about these heat changes, including the rate of heat transfer.

When a thermal transition occurs in the sample, thermal energy is added to either the sample or the reference holders in order to maintain both the sample and reference at the sample temperature. Because the energy transferred is exactly equivalent in magnitude to the energy absorbed or evolved in the transition, the balancing energy yields a direct calorimetric measurement of the transition energy at the temperature of the transition.

21.2.1 Instrumentation for DSC

A cell designed for quantitative DSC measurements is illustrated in Fig. 21.2. Two separate sealed pans, one containing the material of interest and the other containing an appropriate reference, are heated (or cooled) uniformly. The enthalpic difference between the two is monitored either at any one temperature (isothermal) or the temperature can be raised (or lowered) linearly. If maximum calorimetric accuracy is desired, the sample and reference thermocouples must be removed from direct contact with the materials.

The gradient temperature experiments can be run slowly ($0.1°C \cdot min^{-1}$) or rapidly (up to $300°C \cdot min^{-1}$). The electronic circuitry detects any change in heat flow in the sample versus the reference cell. This event in turn is sent as an analog signal to an output device such as a strip-chart recorder, digital integrator, or a computer. Information may be obtained with samples as small as 0.1 mg. However, quantitative studies usually require at least 1 mg of sample.

The DSC cell uses a constantan disk as the primary means of transferring heat to the sample and reference holders, and it is also one element of the

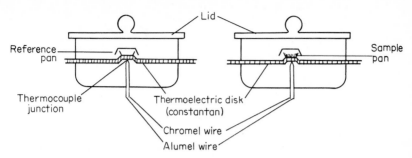

FIGURE 21.2 Arrangement of sample and reference containers with their individual heaters in differential scanning calorimetry.

temperature-sensing thermoelectric junction. Samples in powder, sheet, film, fiber, crystal, or liquid form are placed in disposable aluminum sample pans of high thermal conductivity and weighed on a microbalance. The sample is placed in one sample holder and an empty sample holder serves as the reference. Sample sizes range from 0.1 to 100 mg.

The differential heat flow to the sample and reference through the disk is monitored by the chromel-constantan thermocouples formed by the junction of the constantan disk and the chromel wafer covering the underside of each platform. Chromel and alumel wires connected to the underside of the wafers form a chromel-alumel thermocouple, which is used to monitor the sample temperature. Thin-layer large-area sample distribution minimizes thermal gradients and maximizes temperature accuracy and resolution.

Generally, holders are covered with domed, aluminum sample-holder covers. The covers are essentially radiation shields which enhance baseline linearity and reproducibility from run to run. An air-tight sample-holder enclosure isolates the holders from external thermal disturbances and allows vacuum or controlled atmosphere operation. A see-through window permits observation of physical changes in unencapsulated samples during a scan. Side ports allow addition of catalysts and seeding of supercooled melts.

Three different DSC cells are available: standard (as described), dual sample, and pressure capability. The dual sample cell provides an immediate twofold increase in sample throughput, while the pressure DSC cell provides for pressures from vacuum to 7 MPa (1000 lb · in^{-2}).

21.2.2 Instrumentation for DTA

In DTA a thermocouple is inserted into the center of the material in each sample holder (Fig. 21.3). The sample is in one holder and reference material is placed in the other sample holder. The sample blocks are then heated. The difference in temperature between sample and reference thermocouples is continuously measured. Furnace temperature is measured by an independent thermocouple. Any transition that the sample undergoes results in liberation or absorption of energy by the sample with a corresponding deviation of its temperature from that of the reference. A plot of the differential temperature versus the programmed temper-

Dual thermocouples

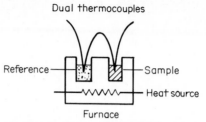

Furnace

FIGURE 21.3 Arrangement of cells and sensors for differential thermal analysis.

ature indicates the transition temperature(s) and whether the transition is exothermic or endothermic.

For high-temperature studies (1200 to 1600°C), the sample holders are fabricated from platinum-iridium.

21.2.3 Applications of DSC and DTA

Applications include melting-point and glass-transition measurements, curing studies, oxidative stability testing, and phase-transition studies.

21.2.3.1 Glass-Transition Temperature. The DSC trace of polyethylene terephthalate (Fig. 21.4) shows an endothermal shift according to the sudden increase in specific heat at the glass-transition temperature (84°C). The glass transition determines the useful lower temperature range of an elastomer, which is normally used well above this temperature to ensure the desired softness. In thermoset materials the glass-transition temperature is related to the degree of cure, with the glass-transition temperature steadily increasing as the degree of cure increases. Additionally, the glass-transition temperature is often used to determine the storage temperature of uncured or partially cured material, since cur-

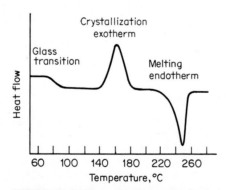

FIGURE 21.4 Three transitions of polyethylene terephthalate obtained on a DSC trace. Sample size: 5.1 mg. Program rate 20°C · min^{-1} Atmosphere: nitrogen.

ing does not begin until the material is heated above its glass-transition temperature.

21.2.3.2 Crystallization and Fusion. Continuing with the DSC trace in Fig. 21.4, the exothermal crystallization enthalpy (at 168°C) as well as the endothermal heat of fusion (at 250°C) can be evaluated by integrating curve areas. In a semicrystalline polymer the crystallinity can be calculated from the measured heat of fusion, which is a characteristic property and widely used for quality control.

21.2.3.3 Curing Reactions. During a curing reaction, energy is released from the sample as crosslinking occurs and a large exothermic peak follows the glass transition.

21.2.3.4 Curie Point. An interesting application of DSC is the determination of the Curie point temperatures of ferromagnetic materials. The Curie point is discussed in more detail in Sec. 21.3.

21.2.3.5 Oxidative Stability. DSC is a useful tool for generating data on oxidative stability of fats and oils. In either the isothermal or the temperature-programmed mode, the onset of a deviation from the baseline can be related to the oxidative induction period. For a sample of polypropylene, the system would be brought to 200°C and held isothermally in a nitrogen atmosphere. Oxygen is then introduced and the time necessary to give the first indication of oxidation (exotherm) is noted. DSC can differentiate quickly between stable and unstable systems and determine which antioxidant system would best preserve the specimen.

21.3 THERMOGRAVIMETRIC ANALYSIS

Thermogravimetric analysis (TGA) is a technique in which the mass of a substance is monitored as a function of temperature or time as the sample specimen is subjected to a controlled temperature program in a controlled atmosphere. Samples are in milligram quantities. Only those events associated with a change in mass will be observed.

21.3.1 TGA Instrumentation

To perform TGA, the equipment must be capable of both simultaneous heating and weighing. Instrumentation requirements are shown in Fig. 21.5; they include the following:

1. A sensitive recording microbalance.
2. A furnace and an appropriate enclosure for heating the sample specimen.
3. A temperature programmer, heat control circuitry, and associated electronics. Linear heating rates from 5 to 10°C · min^{-1} are typical although much higher rates are available. Many important TGA procedures involve isothermal monitoring of the mass loss from a material.

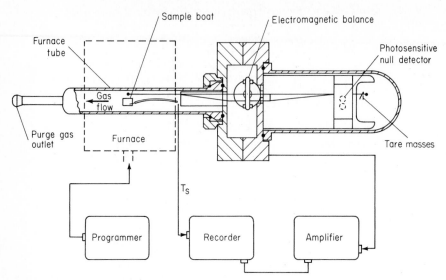

FIGURE 21.5 Schematic of a thermogravimetric analyzer. (*Courtesy of DuPont Clinical and Instruments System Division.*)

4. A pneumatic system for dynamic purging of the furnace and sample chamber.

5. A data acquisition system.

21.3.2 Applications of TGA

Thermogravimetry provides the laboratory chemist with a number of important testing applications. The most important applications include compositional analysis and decomposition profiles at varying temperatures and atmospheric conditions which can be tailored and switched at any point during the experiment. Other important applications include the rapid proximate analysis of coal, quantitative separation of the main sample components in multicomponent mixtures, the determination of the volatile and moisture components in a sample material.

The proximate analysis of coal is performed automatically (Fig. 21.6) as described below.

1. A coal specimen is rapidly heated to 110°C under flowing nitrogen and held at that temperature for 5 min until all of the moisture is lost.

2. The sample is then heated dynamically from 100°C · min⁻¹ to 950°C; the temperature is then held until all of the volatile matter has been liberated.

3. The purge gas is switched from nitrogen to oxygen. The final weight-loss step assigns the percentage of fixed carbon for the coal specimen.

4. The remaining weight in the pan is read directly as the ash value.

The quantitative determination of the main components of an industrial elastomer system follows a similar series of steps, as shown in Fig. 21.7 for a sample of polytetrafluoroethylene.

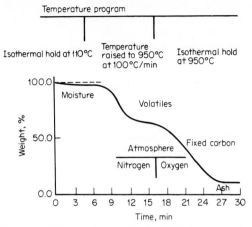

FIGURE 21.6 Proximate analysis of bituminous coal by thermogravimetric analysis.

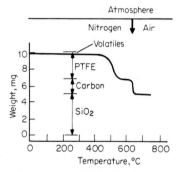

FIGURE 21.7 Thermogravimetric analysis of silica- and carbon-filled polytetrafluoroethylene (PTFE) run at a heating rate of 5 °C · min⁻¹.

1. A sample of 3 to 5 mg is heated from 50°C to 750°C at a rate of 30°C · min⁻¹. Nitrogen is used as an inert purge gas up to 550°C. Softening agents and moisture volatilize up to approximately 300°C. Rubber and elastomers are pyrolyzed up to 550°C.

2. At 550°C, the purge gas is switched to air. Carbon black is burned off in air up to 750°C.

3. The residue represents the amount of inorganic fillers, which in this sample was silicon dioxide.

Curie point measurements by TGA provide an accurate method for the calibration of the temperature axis of the thermogram since the Curie point temperature of many materials is well known and characterized. The ferromagnetic material is suspended in a magnetic field which is oriented such that a vertical component of magnetic force acts on the sample. This magnetic force acts as an equivalent magnetic mass on the TGA microbalance beam to indicate an apparent sample weight. When the sample is heated through its Curie point, the magnetic mass will be lost and the microbalance will indicate an apparent weight loss.

Moisture determination is an important application of TGA. In many industries even small amounts of moisture have serious consequences. When the sample is rapidly heated to 105°C and held at this temperature, any moisture present in the sample is lost. Moisture levels at 0.5 percent, and often below, can be determined.

Determination of the temperature of oxidation of a sample is another TGA application. If magnesium powder is heated from 300 to 900°C in an oxidizing (air) atmosphere, at approximately 682°C a sharp increase in sample weight is noted which corresponds to the rapid oxidation of the material.

21.4   THERMOMECHANICAL ANALYSIS

Thermomechanical analysis (TMA) is used to measure changes in the physical properties of sample materials, such as compression, coefficients of expansion, softening points, heat deflection temperature measurements, and modulus and creep studies. Useful information is provided with regard to behavior of the material at either elevated or reduced temperatures while under an external load, which can be varied. However, very little chemical information can be determined by TMA.

21.4.1   TMA Instrumentation

The schematic diagram (Fig. 21.8) indicates the essential design features of the analyzer. The instrument measures the change in dimension, as a function of temperature and stress. One of several selectable probes makes contact with a sample. The probe touches the upper surface of the sample and applies a well-defined stress. As the sample expands, contracts, or softens, the position of the probe will change. This position is monitored by a linear variable differential transformer (LVDT) which provides a signal proportional to the probe displacement.

Several probe configurations are available for TMA. In the expansion, penetration, and hemispherical configurations, the outer member is the quartz platform and is fixed. The inner members are movable and connected to the LVDT core. The tension and fiber probes put the sample in tension by pulling instead of pushing. The dilatometer probe is designed to measure volume changes.

21.4.2   Applications of TMA

In the penetration and expansion modes, the sample rests on a quartz stage surrounded by the furnace. Under no load, expansion with temperature is observed (Fig. 21.9). At the glass-transition temperature, there is a discontinuous change in expansion coefficient, as evidenced by the elbow in the TMA scan. The thermal coefficient of linear expansion is calculated directly from the slope of the resulting curve. In the penetration mode the penetration of the probe tip into the sam-

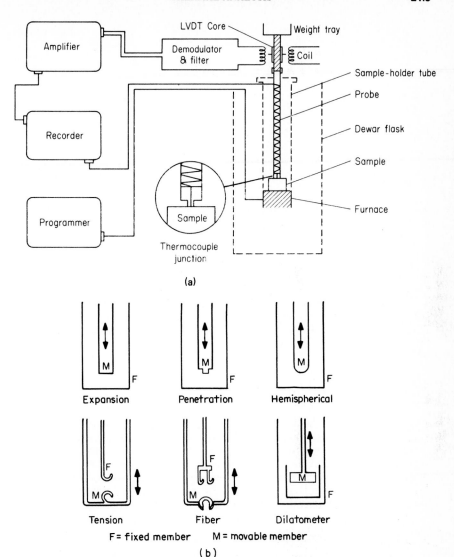

FIGURE 21.8 (*a*) Thermomechanical analyzer. (*b*) TMA probe configurations. (*Courtesy of E. I. DuPont de Nemours, Inc.*)

ple is observed under a fixed weight placed in the weight tray. Sample sizes may range from a 0.1-mil coating to a 0.5-in. thick solid; sensitivities down to a few microinches are observable. In the penetration mode, TMA is a sensitive tool for the characterization and quality control of thin films and coatings. For the measurement of samples in tension, the sample holder consists of stationary and movable hooks constructed of fused silica. This permits extension measurements

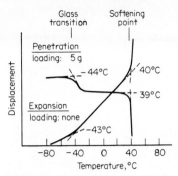

FIGURE 21.9   Glass transitions and softening points of neoprene as detected by penetration and expansion measurements. Heating rate: 5 °C · min⁻¹.

on films and fibers; these measurements are related to the tensile modulus of a sample.

For a swelling measurement, the sample is placed on the bottom of a small cup and is covered by a small disk of aluminum oxide. In an isothermal experiment, the cup is filled with solvent at time zero. The swelling of the sample increases its thickness, which is measured by the probe.

21.5   DYNAMIC MECHANICAL ANALYSIS

Dynamic mechanical analysis (DMA) involves measuring the resonant frequency and mechanical damping of a material forced to flex at a selected amplitude. Two parallel, balanced sample-support arms are free to oscillate around flexure pivots. A specimen material of known dimensions is clamped between the two arms. The force modulator is an electromagnet used in place of the weights required in the static TMA mode. The magnet receives its current impulses through the circuit with an integrated function generator and power amplifier; this oscillates the sample-arm-pivot system. The frequency and amplitude of this oscillation are detected by a linear variable differential transformer positioned at the opposite end of the active arm.

DMA allows the user to observe elastic and inelastic deformations of materials, providing information relative to changes in moduli of shear and elasticity, as well as changes in loss modulus. It also makes possible the detection of glass transitions generally before they are detectable by other means.

21.6   ENTHALPIMETRIC ANALYSIS

The main types of enthalpimetric analysis are thermometric enthalpy titrations and direct injection enthalpimetry. In practical terms they can be differentiated by the way the reactant is introduced to the adiabatic cell.

21.6.1 Thermometric Enthalpimetric Titrations

Thermometric enthalpimetric titrations (TET) are characterized by the continuous addition of the titrant to the sample under effectively adiabatic conditions. The total amount of heat evolved (if the reaction is exothermic) or absorbed (if the reaction is endothermic) is monitored using the unbalance potential of a Wheatstone bridge circuit, incorporating a temperature sensitive semiconductor (thermistor) as one arm of the bridge (Fig. 21.10). Simple styrofoam insulated reaction cells will maintain pseudoadiabatic conditions for the short period of a titration.

The heat capacity of the system will remain essentially constant if the change in volume of the solution is minimized and if the titrant and titrate are initially at the same temperature (usually room temperature). The TET enthalpogram (a thermometric titration curve), shown in Fig. 21.11, illustrates an exothermic titration reaction. The baseline AB represents the temperature-time blank, recorded prior to the start of the actual titration; B corresponds to the beginning of addition of titrant; C is the end point; and CD is the excess reagent line. In order to minimize variations in heat capacity during titrations, it is customary to use titrant 50 to 100 times more concentrated than the specimen being titrated. Thus the volume of the titrate solutions is maintained virtually constant, but the titrant is diluted appreciably. Correction for the latter is conveniently made by linear back extrapolation CB'. Under these conditions, the extrapolated ordinate height BB', represents a measure of the change of temperature due to the titration reaction.

In contrast to most analytical procedures which depend on a property related solely to equilibrium constants (i.e., free energy methods), TET depend on the heat of the reaction as a whole, viz.:

$$\Delta H = \Delta F + T\Delta S \tag{21.1}$$

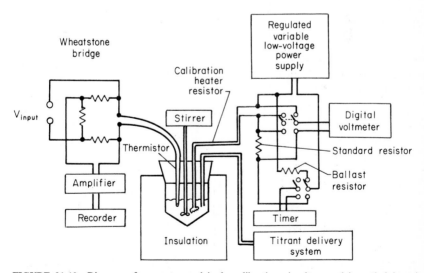

FIGURE 21.10 Diagram of apparatus and joule calibration circuitry used in enthalpimetric analysis.

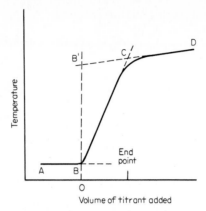

FIGURE 21.11   Typical thermometric titration curve.

Consequently, thermometric titration curves may yield a well-defined end point when all free energy methods fail if the entropy term in Eq. 21.1 is favorable. This is indeed the case in many alkalimetric titrations of weak acids. The corresponding enthalpograms for the titration of hydrochloric acid and boric acid are strikingly similar, because the heats of neutralization are comparable: -56.5 and -42.7 $J \cdot mol^{-1}$, respectively, for HCl and boric acid. For boric acid ($pK_a = 9.24$) the direct free energy (potentiometric titration) method fails to provide a sharp end point.

21.6.1.1 Applications.   Since heat of reaction is the most general property of chemical processes, thermometric titrations have a very wide range of applicability in quantitative analysis. Nonaqueous systems are well suited for this method. Thermometric titrations are very useful in titrating acetic anhydride in acetic acid–sulfuric acid acetylating baths, water in concentrated acids by titration with fuming acids, and free anhydrides in fuming acids. Precipitation and ion-combination reactions such as the halides with silver and cations with ethylenediaminetetraacetate are other possibilities. Even halide titrations in fused salts have been done.

21.6.2   Direct Injection Enthalpimetry

Direct injection enthalpimetry (DIE) involves the (virtually) instantaneous injection of a single "shot" of reactant into the solution under investigation, the reagent being in stoichiometric excess. The corresponding heat evolved (or absorbed) is directly proportional to the number of moles of analyte reacted. Circuitry is identical to that employed with TET. The amount of heat evolved or absorbed is calculated from Joule heating calibration experiments. Precise volume measurements are not a prerequisite of DIE.

The calorimeter is calibrated by passing a constant current i through a calibration heater resistor (which is immersed in the solution) and a standard resistor in the external circuit (Fig. 21.10). The heat dissipated within the calorimeter may then be calculated from the expression:

$$Q \text{ (J)} = \frac{i^2 R_H t}{1} = \frac{V_S V_H t}{R_S} \tag{21.2}$$

where V_S and V_H are potential drops measured across the standard resistor R_S and the calibration heater R_H.

Continuous-flow enthalpimetry is utilized for the on-line analysis of industrial process streams. The technique consists of passing two reactant solutions at a constant rate through a mixing chamber and continuously monitoring the heat output of the product stream. The reagent must be in stoichiometric excess.

BIBLIOGRAPHY

Brennan, W. P., R. B. Cassel, and M. P. DiVito, "Materials and Process Characterization by Thermal Analysis," *Am. Lab.*, **20**:32 (January 1988).

DiVito, M. P., W. P. Brennan, and R. L. Fyans, "Thermal Analysis: Trends in Industrial Applications," *Am. Lab.*, **18**:82 (January 1986).

Gibbons, J. J., "Applications of Thermal Analysis Methods to Polymers and Rubber Additives," *Am. Lab.*, **19**:33 (January 1987).

Hogarth, A. J. C. L., and J. D. Stutts, "Thermometric Titrimetry: Principles and Instrumentation," *Am. Lab.*, **13**:18 (January 1981).

Jordan, J., et al., "Enthalpimetric Analysis," *Anal. Chem.*, **48**:427A (1976).

Kolthoff, I. M., P. J. Elving, and C. Murphy, eds., *Treatise on Analytical Chemistry*, 2d ed., Part I, Vol. 12, Thermal Methods, Wiley, New York, 1983.

Levy, P. F., "Thermal Analysis, an Overview," *Am. Lab.*, **2**:46 (January 1970).

Svehla, G., ed., *Wilson and Wilson's Comprehensive Analytical Chemistry*, Thermal Analysis, Vol. XII, Elsevier, New York, 1984.

Wendlandt, W., *Thermal Methods of Analysis*, 3d ed., Wiley, New York, 1986.

CHAPTER 22
AUTOMATED ANALYSIS

22.1 INTRODUCTION

The classical combustion-tube methods involve tedious, time-consuming microanalytical techniques which require expensive, special laboratories and highly skilled technicians. Automated elemental analyzers offer multisample and unattended operation. The combustion operation is completely automated and is followed by an on-line measurement of the components in the combustion gases. Computerization permits extensive data reduction, calculation, reporting, and storage capabilities.

22.2 MICRODETERMINATION OF CARBON, HYDROGEN, AND NITROGEN

Since organic compounds are characterized by the fact that they contain carbon and usually hydrogen, it can be seen that the ability to measure these elements accurately is of extreme importance. The microcombustion technique is the principal means for determining carbon, hydrogen, and nitrogen. The technique involves several steps.

1. In the purge mode, the weighed sample is dropped into the loading head, which is then sealed and all the interfering gases are purged from the combustion path.

2. In the burn mode, the sample is moved onto a ceramic crucible and into the furnace for combustion in a flowing stream of pure oxygen at 900°C (see Fig. 22.1). The sample boat can subsequently be removed for weighing any residue. Alternatively, the sample can be mixed with cobalt(III) oxide [or a mixture of manganese dioxide and tungsten(VI) oxide] to provide the oxygen and heated to the same combustion temperature. Carbon dioxide, water vapor, nitrogen and some oxides of nitrogen, and oxides of sulfur are possible products of combustion of an organic compound. The burn time is 10 to 12 min.

3. Removal of interfering elements.

4. Measurement of the carbon dioxide, nitrogen, and water vapor formed.

22.2.1 Removal of Interfering Substances

The interfering substances encountered in the determination of carbon and hydrogen are sulfur, the halogens, and nitrogen.

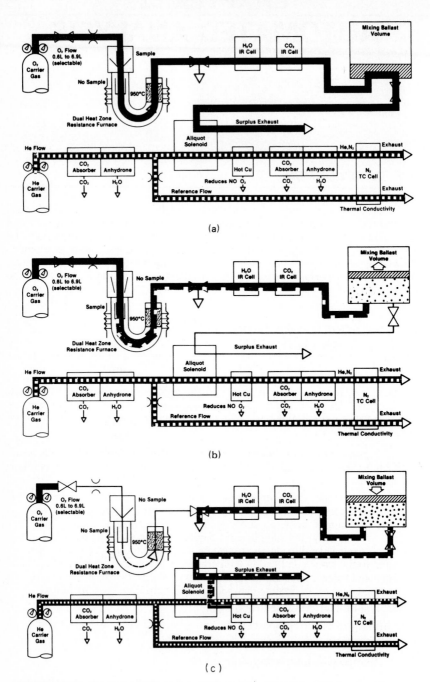

FIGURE 22.1   Flow scheme of a carbon-hydrogen-nitrogen elemental analyzer.   (*Courtesy of Perkin-Elmer Corporation.*)

Hot copper at 550 to 670°C reduces the nitrogen oxides to nitrogen and removes residual oxygen. In carbon-hydrogen analyzers, oxides of nitrogen are removed with manganese dioxide.

Copper oxide converts any carbon monoxide to carbon dioxide.

A magnesium oxide layer in the middle of the furnace removes fluorine.

A silver-wool plug at the exit removes chlorine, iodine, and bromine, and also any sulfur or phosphorus compounds that result from the combustion of the sample.

Calcium oxide removes oxides of sulfur in a secondary combustion zone so that water vapor cannot combine to form sulfurous or sulfuric acid.

22.2.2 Measurement of Combustion Gases

A variety of techniques have been used to separate and measure the components in the combustion gases. These include gas chromatography (Chap. 3), thermal conductivity (Chap. 3), infrared spectrometry (Chap. 8), and coulometry (Chap. 20).

22.2.2.1 Gas Chromatographic Method. In one approach excess oxygen is removed from the combustion gases and the nitrogen oxides are reduced to nitrogen with copper. Helium, used as carrier gas, sweeps the carbon dioxide, water, and nitrogen onto a chromatographic column for separation. Signals from the three chromatographic peaks are integrated to ascertain the quantities present in the sample. In this method relatively small samples must be used so that the combustion products represent a "slug" injection on the chromatographic column (see Chaps. 2 and 3).

In another approach the gases pass through a charge of calcium carbide where water vapor is converted to acetylene. A nitrogen cold trap freezes the sample gases and isolates them in a loop of tubing. A valve seals off the combustion train which is then ready for another sample. The chromatographic separation is begun by removing the cold trap and heating the loop. Another stream of dry helium gas carries the gases from the trap into the chromatographic column where the three gases (nitrogen, carbon dioxide, and acetylene) are completely separated. The chromatographic separation requires about 10 min.

22.2.2.2 Thermal Conductivity Detection. Three pairs of thermal conductivity detectors are used in a differential manner (Fig. 22.1). Specific absorbents are placed between the detectors. The helium carrier gas fills a mixing volume to a specified pressure. The mixed gas is passed through the sample side of detector 1. Water is then removed by a magnesium perchlorate trap from the gases which are then passed through the reference side of the same detector. Similarly carbon dioxide is determined by passing the effluent from the first detector into the sample side of detector 2, removing carbon dioxide from the gas with a soda-asbestos trap, and passing the stripped gas through the reference side of detector 2. Nitrogen is determined by detector 3 which compares the effluent gas from the second detector after removal of carbon dioxide (and water in the earlier measurement) with pure helium.

22.2.2.3 Infrared Detection Methods. Dispersive and nondispersive infrared detectors are available for water and carbon dioxide.

22.2.2.4   Coulometric Detection.   Coulometric detectors for carbon dioxide provide 100 percent efficiency and an absolute digital readout in terms of micrograms of carbon. Hydrogen can be determined by trapping the water from the combustion step on calcium chloride. The water is desorbed by heating and passed through a proprietary material which quantitatively converts water to carbon dioxide for measurement by coulometry.

22.3   TOTAL CARBON, TOTAL ORGANIC CARBON ANALYZER

The total carbon (TC) and total organic carbon (TOC) analyzers are designed for natural and wastewater samples, and seawater. The flow schematic for TC and TOC is shown in Fig. 22.2. Periodically (often every 2.5 min for TC and 5 min for TOC) an aspirated sample is injected into a high-temperature (900°C) reaction chamber through which flows a nitrogen carrier gas with a constant level of oxygen. All oxidizable components are combusted to their stable oxides and all inorganic and organic carbon in the aqueous sample are converted to carbon dioxide. The carbon dioxide generated is transferred by the carrier gas through a scrubber to removed corrosive impurities and interferences. Then the carrier gas flows through a nondispersive infrared analyzer.

To differentiate between total carbon and total organic carbon, a separate sample is drawn and mixed with acid at a constant sample-to-acid ratio. Inorganic carbon in the sample is converted to carbon dioxide and is removed by nitrogen sparging. Then the sample no longer containing any inorganic carbon is treated as described above for total carbon to determine the total organic carbon.

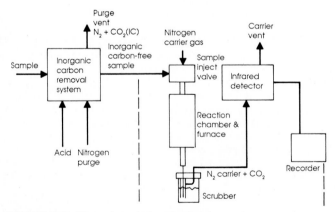

FIGURE 22.2   Flow schematic for a total carbon and a total organic carbon analyzer. (*Courtesy of Ionics, Inc.*)

22.4 KJELDAHL DETERMINATION OF NITROGEN

In the Kjeldahl method the sample is digested with sulfuric acid and a catalyst. Organic material is destroyed, and the nitrogen is converted to ammonium hydrogen sulfate. The heating is continued until the solution becomes colorless or light yellow. Selenium, copper, and mercury, and salts of each, have been used as the catalyst. Potassium sulfate added with the catalyst raises the temperature and thereby speeds the decomposition. A blank should be carried through all the steps of the analysis.

After the digestion is complete, allow the flask to cool. Cautiously dilute with distilled water and cool to room temperature. Arrange a distillation apparatus with a water-jacketed condenser whose adapter tip extends just below the surface of the solution in the receiver. Carefully pour a concentrated solution of NaOH down the side of the digestion flask so that there is little mixing with the solution in the flask. Add several pieces of granulated zinc and a piece of pH test paper. Immediately connect the flask to a spray trap and the condenser. Swirl the solution until mixed; the test paper should indicate an alkaline value. Bring the solution to a boil and distill at a steady rate until only one-third of the original solution remains.

When the reaction mixture is made alkaline, ammonia is liberated and removed by steam distillation. (1) In the classical method, the distillate is collected in a known excess of standard HCl solution. The unused HCl is titrated with a standard solution of NaOH, using as indicator methyl red or bromocresol green. These indicators change color at the pH that corresponds to a solution of ammonium ions. (2) In the boric acid method, the distillate is collected in an excess of boric acid (crystals). The borate ion formed is titrated with a standard solution of HCl, using methyl red or bromocresol green as the indicator.

22.4.1 Special Situations

Compounds containing N—O or N—N linkages must be pretreated or subjected to reducing conditions prior to the Kjeldahl digestion. The N—O linkages are reduced with zinc or iron in acid. There is no general technique for the N—N linkages.

Samples containing very high concentrations of halide can in some instances cause trouble because of the formation of oxyacids known to oxidize ammonia to nitrogen.

For nitrate-containing compounds, salicylic acid is added to form nitrosalicylic acid which is reduced with thiosulfate. Then the digestion can proceed as described.

22.4.2 Automated Kjeldahl Method

The Technicon AutoAnalyzer utilizes a procedure based on the Kjeldahl digestion. The sample is digested using a mixture containing selenium dioxide, sulfuric acid, and perchloric acid. The digest is then made up to a given volume and placed in the autoanalyzer. The ammonium hydrogen sulfate in the digest is automatically sampled and treated with sodium hydroxide. The ammonia liberated is mixed with a phenol-hypochlorite reagent to produce a blue color which is then measured with a filter photometer.

22.5   DETERMINATION OF SULFUR

The determination of sulfur follows the same basic steps outlined earlier for the determination of carbon and hydrogen. A solid sample up to 1 g is weighed in a combustion boat on the integral electronic balance. Coal or coke samples are then covered with a layer of vanadium pentoxide powder. The vanadium pentoxide acts as a flux to moderate sample combustion. The boat with sample is inserted through the open port combustion tube of the resistance furnace where it is correctly positioned under the oxygen inlet by a mechanical stop (Fig. 22.3). Oxidative combustion converts the organic material into carbon dioxide, water, sulfur dioxide, and sulfur trioxide. Raising the temperature to 1350°C ensures the production of sulfur dioxide with no sulfur trioxide. Moisture and dust are removed by appropriate traps.

Liquids, such as petroleum samples, are loaded dropwise on a bed of vanadium pentoxide contained in a crucible with cover. The use of a crucible cover serves to retard the combustion of volatile samples.

22.5.1   Measuring the Sulfur in the Combustion Products

Various methods exist for determining sulfur after oxidation. A simple, straightforward method measures the sulfur dioxide gas by a selective, solid state, infrared detector.

In the amperometric titration method, the sulfur dioxide is pumped from the furnace to a reaction vessel in the analyzer via a heated manifold. The sulfur dioxide is bubbled through a specially formulated diluent and determined directly through an iodometric titration. An electric current is preset for a platinum elec-

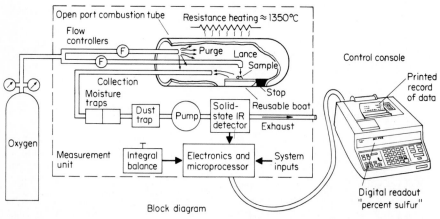

FIGURE 22.3   Schematic diagram of a sulfur elemental analyzer.   (*Courtesy of Leco Corporation.*)

trode in the diluent. As the diluent absorbs sulfur dioxide, the current decreases. The current decrease triggers an integral buret which automatically releases a precisely monitored quantity of titrant to restore the current in the diluent to the preset level. A sensitivity of 0.005% sulfur can be achieved for a nominal 100 mg sample. The specificity of the method eliminates any interference from chlorine, organic nitrogen, phosphorus, and lead anti-knock compounds.

22.5.2 Tube Combustion (Manual)

In the manual method the sample is burned with the aid of a vanadium(V) oxide [or tungsten(VI) oxide] catalyst and pure oxygen in an alundum tube maintained at 1000°C. The combustion products pass successively through magnesium perchlorate, 8-hydroxyquinoline, and free copper (heated at 840°C) which remove water, the halogens, and oxygen. The residual gases are absorbed in neutral hydrogen peroxide which converts all the sulfur oxides to sulfuric acid, which is determined by titration with standard base.

22.5.3 Schöninger Combustion

In the Schöninger combustion technique, the sulfur is converted by oxidation to sulfur dioxide and sulfur trioxide and subsequently oxidized to sulfuric acid with hydrogen peroxide. The methodology is described in Chap. 24. This method is useful for nonvolatile compounds only.

22.6 DETERMINATION OF HALOGENS

Here the term "halogen" refers only to chlorine, bromine, and iodine.

22.6.1 Decomposition of the Organic Material

The sample is decomposed in an oxygen atmosphere at 700°C in the presence of a platinum catalyst. The evolved gases are absorbed in a sodium carbonate solution with hydrazine present.

22.6.2 Measurement of the Halides by Amperometric Titration

Iodide, bromide, and chloride can be successively titrated in mixtures with silver nitrate, using a rotating microelectrode (see Chap. 19, Amperometric Titrations). In a 0.1 to 0.3 M solution of ammonia only silver iodide precipitates. The indicator electrode is held at -0.2 V versus SCE. During the titration of iodide, the current remains constant at zero, or nearly so, until the iodide ions are consumed, and then it rises. After three or four points have been recorded past the end point, the solution is acidified to make it 0.8 M in nitric acid. Immediately the silver ions added in excess and now released from the silver ammine complex combine with the bromide ions and precipitate as silver bromide, and the current drops to zero. The indicator electrode is held at +0.2 V versus SCE. A second rise in the current indicates the end point of the bromide titration. A chloride end point can be obtained by adding sufficient gelatin to make the solution 0.1% in gelatin. Gelatin suppresses the current due to silver chloride, and the titration is continued until the current again rises after the chloride end point.

22.6.3   Separation of the Halides by Ion Exchange

The anion-exchange separation of the halides is carried out on a column of Dowex 1-X10 in the nitrate form by elution with sodium nitrate solutions. For example, a Dowex 1-X10 column, 3.7 cm^2 × 7.4 cm, is eluted at 1.0 mL · min^{-1} with 0.5 M sodium nitrate for chloride ion which elutes within 50 mL. The eluant is then changed to 2.0 M sodium nitrate. Bromide elutes in the next 50 mL. Finally, iodide elutes as an extended peak extending over the volume from 75 to 275 mL of the stronger nitrate solution.

The individually separated halides may be determined by potentiometric titration with silver nitrate.

22.7   OXYGEN DETERMINATION

For oxygen determination a quartz pyrolysis tube that contains platinized carbon is employed. This is followed by a tube that contains copper oxide. The operating temperature is 900°C and a helium atmosphere is used. Any oxygen in the sample forms carbon monoxide which is converted in the copper oxide tube to carbon dioxide. The carbon dioxide is measured as previously described for the carbon determination.

22.8   CONTINUOUS FLOW ANALYSIS

Continuous flow analysis (CFA) refers to any procedure which places the sample(s) to be analyzed in a continuously flowing stream. Often a color-forming reaction and colorimetric measurement of the product serve as the determinative step.

In CFA the samples are successively aspirated from individual vials into flexible tubing by an autosampler. The samples become part of a continuously moving stream in which color development occurs. Eventually the samples pass through the detector which responds to the color developed. Quantitation is based on the absorbance of the samples versus standards run periodically. The movement of the fluid in the tubing and aspiration of the samples is controlled by a peristaltic pump. The stream is segmented with evenly spaced air bubbles which virtually eliminates sample dispersion while creating a scrubbing effect on the tube walls. These two effects reduce cross contamination; however, when a high-reading sample follows or precedes a low-reading sample, both samples should be rerun.

A segmented continuous flow analysis system is shown in Fig. 22.4. Samples, reagents, air, and wash fluid are introduced into the system by the autosampler.

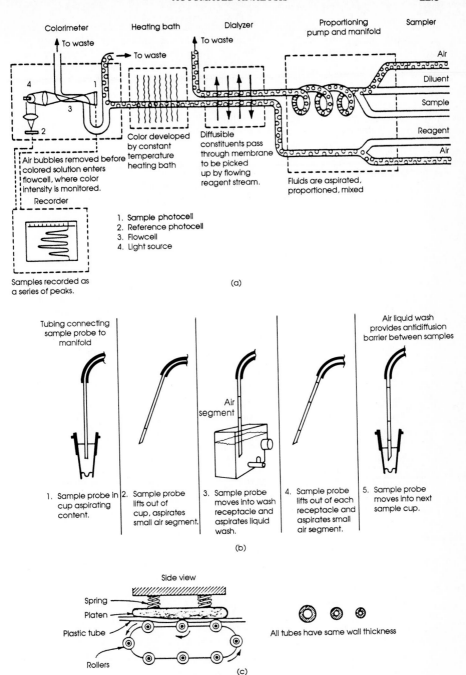

FIGURE 22.4 (*a*) Typical single-channel continuous flow analyzer schematic, (*b*) details of sampler, and (*c*) proportioning pump. (*Courtesy of Technicon Corporation.*)

Variations in modules include mixing with one or more reactants, adding buffers or diluents, dialysis through a membrane into another stream, liquid-liquid separations, heating/cooling coils, and filtering. Modules can be interchanged and rearranged for different analytical methods. Continuous flow analysis is preferred for analysis times longer than 2 min and the sequential addition of three or more reagents.

The heart of the analyzer is the proportionating pump. It can deliver 12 or more separate fluids simultaneously while varying their flow rates in any ratio from 1:1 to 79:1. Variable flow rates are obtained by using flexible tubing with different inside diameters.

An important difference between automatic and manual systems is that in the automatic analyzer reactions do not have to be carried to completion. Since conditions in a given automatic analyzer are unvarying and known, standard samples are subjected to the same treatment as samples of unknown concentration.

22.9 FLOW INJECTION ANALYSIS

Flow injection analysis (FIA) is based on the introduction of a precisely defined volume of sample as a "plug" into a carrier or reagent stream. Sample introduction is via a syringe or valve, the latter actuated by a microprocessor. The result is a sample plug bracketed by carrier; it is a nonsegmented stream in contrast to continuous flow analysis. The absence of air segmentation leads to a higher sample throughput. There is no need to introduce and remove air bubbles, and an expensive high-quality pump is not necessary.

The carrier stream is merged with a reagent stream to bring about a chemical reaction between sample and reagent (Fig. 22.5). The total stream then flows through a detector. Experimental conditions are held constant for both standards and samples in terms of residence time, temperature, and dispersion. The sample concentration is evaluated against appropriate standards treated identically. Sample sizes are in the range of 10 to 300 μL. The flow of reagents is in the range of 0.5 to 9 mL $\cdot$ min^{-1}.

The flow injection analysis uses the laminar flow present in narrow-bore tubing to mix the sample with the reagent, thus eliminating the necessity of air bubble partitioning. A tubing of 0.5-mm i.d. is best. Larger diameter tubing leads to increasing dispersion, whereas the smaller diameter tubing may easily be blocked

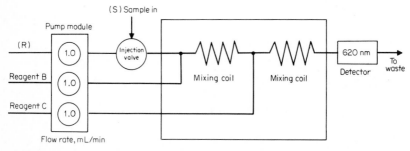

FIGURE 22.5   Schematic diagram of a flow injection analyzer.

and requires high-pressure pumps for operation. FIA is best suited for analyses that require less than 30 s and the sequential addition of only one or two reagents.

BIBLIOGRAPHY

Betteridge, D., "Flow Injection Analysis," *Anal. Chem.,* **50**:832A (1978).

Mottola, H. A., "Continuous Flow Analysis Revisited," *Anal. Chem.,* **53**:1313A (1981).

Ruzicka, J., "Flow Injection Analysis," *Anal. Chem.,* **55**:1041A (1983).

Snyder, L., et al., "Automated Continuous Flow Analysis," *Anal. Chem.,* **48**:942A (1976).

Stewart, K. K., "Flow Injection Analysis. New Tool for Old Assays," *Anal. Chem.,* **55**: 931A (1983).

CHAPTER 23
DETERMINATION OF PHYSICAL PROPERTIES

23.1 DENSITY

The density of any substance can be found by dividing the mass of that substance by the volume that it occupies. The units are grams per cubic centimeter. Water attains its maximum density of $0.999\ 973\ \text{g} \cdot \text{cm}^{-3}$ at 3.98°C. The density of water in terms *milliliters* and *cubic centimeters* is strictly only interchangeable at 3.98°C. At 20°C the density of water is $0.998\ 20\ \text{g} \cdot \text{cm}^{-3}$. For water, the temperature (t_m, °C) of maximum density at different pressures (p) in atmospheres is given by

$$t_m = 3.98 - 0.0225(p - 1) \tag{23.1}$$

Remember that 1 atm = 101 325 Pa (SI system).

23.1.1 Irregularly Shaped Solids

To find the density of irregularly shaped solids, proceed as follows:

1. Determine the mass of the object.
2. Determine the volume by water displacement (Fig. 23.1).
   a. Use a graduated cylinder containing a measured amount of water (original volume).
   b. Submerge the weighed solid completely in the water, and record the final volume.
   c. Subtract the original volume from the final volume and obtain the volume of the object.
3. Divide the mass of the object by the volume.

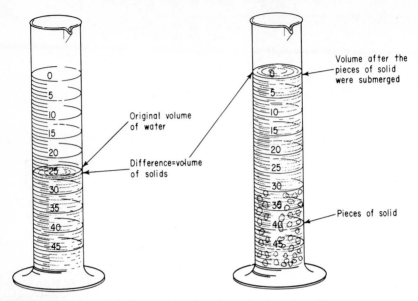

FIGURE 23.1   A method of determining the volume of an irregular solid.

23.1.2 Liquids

23.1.2.1   Density Bottle Method.   Proceed as follows:

1. Use a calibrated-volume liquid container fitted with a thermometer or determine the volume of the container by using distilled water and reference tables listing the density of water at different temperatures.
2. Determine the mass of the empty container (Fig. 23.2).
3. Fill the container to the top of the capillary with the liquid. Place the density bottle in a thermostatically controlled bath. Determine the mass of filled container after removing from the bath and wiping it dry.
4. Obtain the mass of the liquid by subtraction.
5. Divide the mass by the volume of the calibrated container.

Caution   Take care to exclude air bubbles.

23.1.2.2   Fisher-Davidson Gravimeter Method.   The height to which a liquid will rise is inversely proportional to its density. In this method the density of an unknown liquid is determined by comparing the height to which it rises to that of a liquid of known density. After placing the unknown and known liquids in the instrument, a slight vacuum is applied equally to both liquids. Determine the ratio of the heights to which the two liquids will rise. Calculate the density, which will be units of grams per milliliter.

23.1.2.3   Westphal Balance Method.   The Westphal balance (Fig. 23.3) is based upon two concepts:

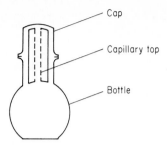

FIGURE 23.2 Density bottle.

1. The mass of a floating object is equal to the mass of the liquid that it displaces.
2. If a body of constant mass is immersed in different liquids, then the corresponding apparent losses in weight of the body are proportional to the masses of the equal volumes of liquid that the body displaced. If one of the liquids is water, the density (specific gravity) of the liquid can be calculated.

Proceed as follows:

1. Standardize the instrument with distilled water by fully immersing the plummet in the water and balancing the beam at the index end.
2. The container of water is removed, and the plummet is carefully dried.
3. Substitute the liquid to be tested. Fully immerse the plummet in the liquid.
4. Adjust the riders so that the indexes of the beam and frame are level.
5. Read the density by the position of the riders.

Note: The 5-g rider is left on the same hook as the plummet for liquids which are heavier than water. It is moved to index numbers on the beam for liquids which are lighter than water.

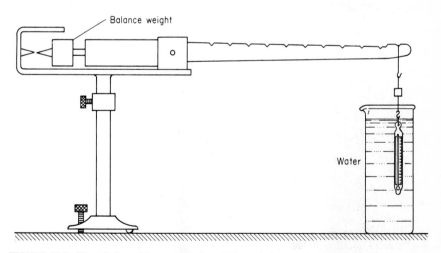

FIGURE 23.3 Westphal specific-gravity balance (shown being standardized with water).

23.1.2.4   Float Method.   A group of weighted, hermetically sealed glass floats can be used to determine the density of a liquid. When such floats are dropped into a liquid and thoroughly wetted, they will float in a more dense liquid, sink in a less dense liquid, or remain suspended in a liquid which has equal density. Density can be determined within ± 0.01 g $\cdot$ mL^{-1}.

23.1.3   Gases

The Dumas method involves the direct determination of the mass of a known volume of gas in a calibrated sphere of known volume. The sphere is fitted with a capillary tube and stopcock, whereby the gas is introduced into the flask. Proceed as follows:

1. Verify the volume of the sphere by filling it with a liquid of known density and determining the mass of the filled sphere.
2. Empty the flask; then clean, dry, and evacuate it.
3. Introduce the gas to be tested, after flushing the sphere with the gas, and determine the mass of the filled sphere. The density calculated is the density of the gas at the temperature at which the determination is made.

23.2   SPECIFIC GRAVITY

Specific gravity is the mass of a substance divided by the mass of an equal volume of water. Specific gravity is expressed by a number; since it is a ratio, it has no units.

23.2.1   Pyknometer Method

Proceed as follows:

1. Measure the mass of a pyknometer (Fig. 23.4).
2. Measure the mass when the pyknometer is filled with distilled water. Subtract the mass obtained in step 1 from this value to obtain the mass of the water.
3. Repeat the procedure when the pyknometer is filled with the unknown liquid. Subtract the mass obtained in step 1 to obtain the mass of the unknown liquid.
4. Divide the mass obtained in step 3 by the mass obtained in step 2 to get the specific gravity. (All mass measurements should be made at the same temperature.)

23.2.2   Hydrometer Method

A hydrometer is a glass container, weighted at the bottom, that has a slender stem calibrated to a standard. Such hydrometers are available in a variety of sizes and ranges for different materials. The depth to which the hydrometer will sink in

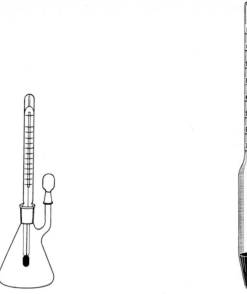

FIGURE 23.4 Pyknometer (can also be used as a density bottle).

FIGURE 23.5 Glass hydrometer for specific-gravity determinations.

a liquid is a measure of the specific gravity of the liquid. Specific gravity is read directly from the calibrated scale on the stem (Fig. 23.5).

To convert specific gravity into other units, use these formulas for degrees Baumé (Bé) on the National Institute of Standards and Technology (U.S) scale, degrees A.P.I. (American Petroleum Institute), or degrees Brix (also called Fisher):

$$°Bé = \frac{140}{\text{sp gr } 15.56°C/15.56°C} - 130 \tag{23.2}$$

$$°A.P.I. = \frac{141.5}{\text{sp gr } 70°F/60°F} - 131.5 \tag{23.3}$$

$$°Brix = \frac{400}{\text{sp gr } 60°F/60°F} - 400 \tag{23.4}$$

23.3 VOLUME

The true volume of powders can be measured by the pressure difference when a gas, preferably helium, under slight pressure is allowed to flow from the sample container into a precisely known reference volume (20 or 135 cm^3). The difference between the initial and final pressure in the sample cell is measured by a precision capacitance pressure transducer to the nearest 0.001 lb · in.$^{-1}$. The above process is repeated for a steel ball of known volume. Accuracy is about 0.2 percent.

23.4   MELTING POINT

The melting point of a crystalline solid is the temperature at which the solid substance begins to change into a liquid. Pure compounds have sharp melting points. Contaminants usually lower the melting point and extend it over a longer range. The melting-point range is the temperature range between which the crystals begin to collapse and melt and the material becomes completely liquid.

As a rule, samples which melt at the same temperature and whose melting point is not depressed by admixture are usually considered to be the same compound.

Narrow-range melting points are indicative of the relative purity of a compound. Acceptably pure compounds have a 1°C range; normal commercially available (organic) compounds have a 2 to 3°C range. Extremely pure compounds have a 0.1 to 0.3°C range.

Some substances will decompose, discolor, soften, and shrink as they are being heated. This behavior can be distinguished from the true melting point of a compound by differential scanning calorimetry and thermomechanical analysis; these methods are discussed in Chap. 21.

23.4.1   Capillary-Tube Methods

In one procedure the melting point is determined by introducing a tiny amount of the compound into a small capillary tube (commercially available) attached to the stem of a thermometer (with a rubber band above the liquid level of the bath) which is centered in a hot-oil bath. Pack tightly to a height of 3 to 4 mm. Use a fresh capillary for each run. Proceed as follows:

1. Heat the oil bath quickly to about 5°C below the melting point (as ascertained on a rough preliminary run), stirring the oil bath continuously.
2. Now heat slowly at about $1°C \cdot min^{-1}$, mixing continuously.
3. Record the temperature when fusion is observed; also record the melting-point range.

An alternative method uses the Thiele melting-point apparatus (Fig. 23.6). When heat is applied to the side arm, that heat is distributed to all parts of the vessel by convection currents in the heating fluid. No stirring is required.

23.4.2   Electric Melting-Point Apparatus

An electric melting-point apparatus is a metal block equipped with a thermometer inserted into a close-fitting hole bored into the block, which is heated by an electric current controlled by a variable transformer. The temperature reading of the thermometer indicates the temperature of the metal block on which the finely

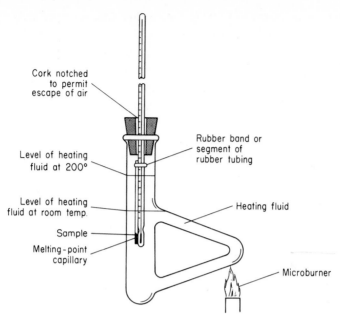

Cork notched
to permit
escape of air

Level of heating
fluid at 200°

Level of heating
fluid at room temp.

Sample

Melting-point
capillary

Rubber band or
segment of
rubber tubing

Heating fluid

Microburner

FIGURE 23.6 Thiele melting-point apparatus.

powdered material melts when placed on the designated area. Heat quickly to within about 5°C below the melting point and then increase the heat slowly.

Always clean the metal surface scrupulously after each use.

The exact melting point can be determined automatically with instruments which use a beam of light to survey the process and a photocell to signal the instant that the sample melts. As the substance melts, the light transmission increases.

23.5 BOILING POINT

The boiling point of a liquid is indicated when bubbles of its vapor arise in all parts of the volume. This is the temperature at which the pressure of the saturated vapor of the liquid is equal to the pressure of the atmosphere under which the liquid boils. Normally, boiling points are determined at the prevailing atmospheric pressure and corrected to standard conditions of 760 mm Hg (1 atm) or 101 325 Pa (pascals). As the pressure decreases, the boiling point will drop.

The correction C_c, which is to be made on the observed temperature t of the boiling point for the actual atmospheric pressure P, is given by the following equation for the centigrade scale:

$$C_c = K(760 - P)(273 + t) \qquad (23.5)$$

where the value of the constant K for several different compounds is given below:

Hydrocarbons, halogen derivatives, ethers, and aldehydes	0.000 125
Esters and ketones	0.000 121
Amines	0.000 118
Alcohols	0.000 100

The variation in the case of acids is too wide to permit of a single value for K.

23.5.1 During Distillation

When a liquid is distilled, the boiling point of the distilling liquid can be read from the thermometer in the distilling heat, which is constantly in contact with the vapors.

23.5.2 Test-Tube Method

The procedure follows:

1. Clamp a test tube containing 2 to 3 mL of the liquid on a stand.
2. Suspend a thermometer with the bulb of the thermometer 2.5 cm above the surface of the liquid.
3. Apply heat gently until the condensation ring of the boiling liquid is 2.5 cm above the bulb of the thermometer.
4. Record the temperature when the reading is constant.

23.5.3 Capillary-Tube Method

Proceed as follows:

1. Seal one end of a piece of 5-mm glass tubing. Attach to a thermometer with a rubber band.
2. Use a pipet to introduce a few milliliters of liquid into the tubing.
3. Drop in a short piece of narrower capillary tubing (sealed at one end) so that the open end is down.
4. Heat until a continuous, rapid, and steady flow of vapor bubbles emerges from the open end of the piece of capillary tubing. Then stop the heating. *Note:* If heating is stopped below the boiling point, the liquid will enter the capillary. If this happens, insert a new capillary tube and restart the procedure.
5. The flow of bubbles will stop, and the liquid will start to enter the capillary tube. Record this temperature; it is the boiling point.

23.5.4 Electronic Methods

The boiling points of very small amounts of liquids can be determined with electronic sensing equipment using photocells. The tube is illuminated from the bottom by dark-field illumination, and as long as no bubbles are present, no light passes through the liquid to reach the photocell sensor. When the boiling point is reached and bubbles begin to rise, the bubbles reflect light to the photocell which triggers the readout indicator. Accuracy is very good: ±0.3°C.

23.6 VISCOSITY

Viscosity is the internal friction or resistance to flow that exists within a fluid, either liquid or gas. Fluid flow in a line has a greater velocity at the center than next to the surfaces, partly because of the friction between the fluid and the boundary surfaces. This causes the adjacent layers to move more slowly.

In the study of oils and organic liquids, viscosity is very significant, because, in industry, "heavier" oils and liquids have higher viscosities, not greater densities.

The unit of viscosity is the pascal-second (Pa · s) in the SI system. This is equal to 10 P (poise) in the cgs system, where formerly viscosities were tabulated in centipoises (cP; 1 cP = 1 mPa · s).

Fluidity is the reciprocal of viscosity. The old unit was called the rhe (1/poise) which in SI units is 0.1 Pa · s.

Kinematic viscosity is equal to the absolute viscosity divided by the mass density. The old unit of kinematic viscosity is the stokes and is equal to 1 mL · s^{-1}.

23.6.1 Saybolt Viscometer Method

The Saybolt viscometer (Fig. 23.7) has a container for liquids with a capacity of 60 mL, fitted with a short capillary tube of special length and diameter. The liquid flows through the tube under a falling head, and the time required for the liquid to pass through is measured in seconds. If temperature is a critical factor, the viscometer is kept at constant temperature in a temperature-controlled bath.

23.6.2 Falling-Piston Viscometer Method

The liquid to be tested is placed in the test cylinder, and the falling piston is raised to a fixed, measured height (Fig. 23.8). The time required for the piston to fall is a measure of the viscosity. The higher the viscosity, the more time it takes for the piston to drop to the bottom.

23.6.3 Rotating-Concentric-Cylinder Viscometer Method

Two concentric cylinders, which are separated by a small annular space, are immersed in the liquid to be tested (Fig. 23.9). One cylinder rotates with respect to the other, and liquid in the space rotates in layers. A viscous force tends to retard

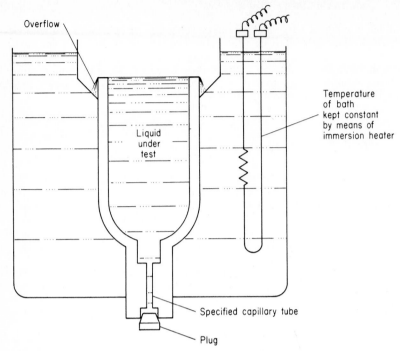

FIGURE 23.7 Saybolt viscometer.

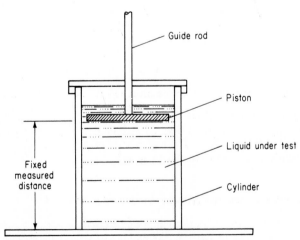

FIGURE 23.8 Falling-piston viscometer.

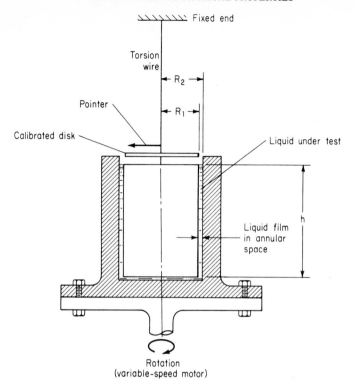

FIGURE 23.9 Rotating-concentric-cylinder viscometer.

the rotation of the cylinder when the viscometer is in motion. The torque on the inner cylinder, which is caused by the viscous force retarding the rotation, is measured by a torsion wire from which the cylinder is suspended.

23.6.4 Fixed-Outer-Cylinder Viscometer Method

The inner cylinder rotates at a constant speed, actuated by a wire wrapped around a drum on the shaft. Force is exerted by the weights of different mass at each end of the wire (Fig. 23.10). The number of revolutions of the cylinder is counted by a revolution indicator. The viscosity of liquids is determined by timing a definite number of revolutions; the time is proportional to the viscosity of the liquid under test and is the basis of comparison.

23.6.5 Falling-Ball Viscometer Method

A ball will fall slowly through a viscous liquid. At first the ball accelerates, but then will fall with a constant velocity. One measures the time required for the ball to fall a known distance, after the condition of uniform viscosity has been achieved. The cylinder (Fig. 23.11) must have a large enough diameter so that

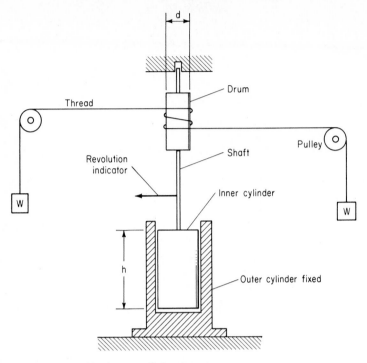

FIGURE 23.10 Fixed-outer-cylinder viscometer.

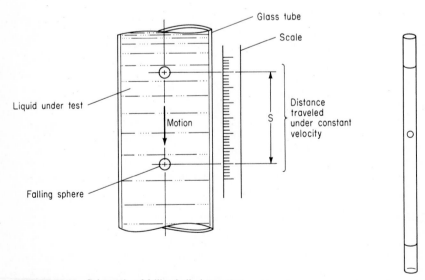

FIGURE 23.11 Schematic of falling-ball viscometer.

(1) no eddy currents are set up and (2) the cylinder surface will not affect the fall of the ball. Commercial falling-ball viscometers are fitted with a fall-release device, and the time required for a steel ball to descend through a precision-bore brass tube is measured with a stopwatch.

23.6.6   Ostwald Viscometer Method

Proceed as follows:

1. Introduce sufficient distilled water into the large round bulb or reservoir (Fig. 23.12).
2. Allow the distilled water to attain thermal equilibrium in a constant temperature bath.
3. Apply suction with a rubber tube to the upper part of the viscometer. This is best done by inverting the viscometer. Draw the liquid up into the tube with the two bulbs to a level above the second bulb.
4. Clock the time needed for the level of the water to pass the signal markings. Make several determinations, using a stopwatch.
5. Drain the viscometer and dry it completely.
6. Add an appropriate volume of the test solution to the reservoir. Then repeat steps 3 through 5.

23.6.7   Shearometers and Rheological Measurements

Often it is desirable to measure separately the two independent variables which dictate the flow behavior of a substance. These are *shear stress*, which is a mea-

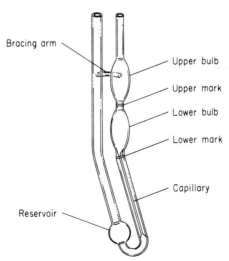

FIGURE 23.12   Ostwald-Cannon-Fenske capillary viscometer.

surement of the forces causing flow, and *shear rate*, which is the rate of deformation of the material. Viscosity is the ratio of shear stress to shear rate.

An electronically controlled drive system produces a known, selected shear rate, and a torque measuring head measures the shear stress required to maintain the shear rate. The drive control system can be programmed to change the rate of shear continuously so that the flow behavior of the material can be plotted automatically as a rheogram (shear stress versus shear rate). Alternatively, the shear rate can be held constant and a plot recorded of viscosity against time. Examples are the structure breakdown of a thixotropic paint or the cure of an adhesive.

23.7 SURFACE TENSION

The cohesion of the molecules of a liquid is manifested in the phenomenon called surface tension. The molecules in the interior of the liquid are subjected to balanced forces between them. The molecules at the surface of the liquid are subjected to unbalanced forces because they are attracted to the molecules below them. As a result, the surface of the liquid appears to resemble an elastic membrane, causing liquid surfaces to contract. This inward pull on those surface molecules results in surface tension, the tendency of a liquid to form drops, and the resistance to expansion of the surface area.

Liquid molecules also have attraction for other substances; this property is called adhesion. When there is an adhesive force between liquids and the surface of a container, the liquid is said to "wet" the surface. This property is called capillarity and is related to surface tension.

23.7.1 Capillary-Rise Method

Liquid will rise in a capillary tube until the gravitational force or pull on the column of liquid is exactly equal to the wetting force. Surface tension of a liquid can be calculated by measuring its capillary rise in a tube and comparing it with the rise of known standards, such as water. The following formula can be used for calculation.

$$\tau = \frac{rh\rho g}{2} \tag{23.6}$$

where τ = surface tension, $dyn \cdot cm^{-1}$
ρ = density of liquid, $g \cdot mL^{-1}$
h = height of column, cm
r = internal radius of tube, cm
g = acceleration due to gravity ($980\ cm \cdot s^{-2}$)

Proceed as follows:

1. Set up the apparatus as shown in Fig. 23.13 in a thermostat.
2. Standardize the apparatus with water. A drop of water-soluble dye facilitates reading the height of the liquid.

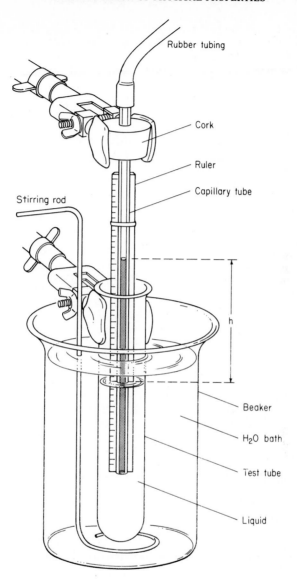

FIGURE 23.13 Apparatus for measuring surface tension by the capillary-rise method (*h* is the height to which the liquid rises).

3. Apply gentle suction to raise the water to the top of the capillary tube, release the suction, and allow the water to fall to equilibrium. Measure the height of the water column. Repeat this step.

4. Clean and dry the equipment. Substitute the test liquid and repeat step 3.

23.7.2 DuNouy Torsion-Wire Tensiometer Method

A stainless-steel torsion wire which is attached to a torsion head extends through the scale and carries a vernier readable to 0.1 dyn. The torsion ring is lowered into the liquid, and as the torsion knob is turned, the torsion is transmitted to the head. The scale is read at the point when the ring breaks the surface of the sample, giving the interfacial tension (surface tension) directly.

23.8 OPTICAL ROTATION

When ordinary white light which is vibrating in all possible planes is passed through a Nicol prism, two polarized beams of light are generated. One of these beams passes through the prism, while the other beam is reflected so that it does not interfere with the plane-polarized transmitted beam.

If the beam of plane-polarized light is also passed through another Nicol prism, it can pass through only if the second Nicol prism has its axis oriented so that it is parallel to the plane-polarized light, If its axis is perpendicular to that of the plane-polarized light, the light will not pass through.

Some substances have the ability to bend or deflect a plane of polarized light; that phenomenon is called optical rotation. Only symmetric molecules can do this. This magnitude and direction of the deflection of the plane of polarized light which passes through an asymmetric substance in solution can be measured in a polarimeter.

23.8.1 Polarimeter

Unpolarized light from the light source passes through a polarizer. Only polarized light is transmitted. This polarized light passes through the sample cell. If the sample does not deflect the plane of light, its angle is unchanged. If an optically active substance is in the sample tube, the light is deflected. The analyzer prism is rotated to permit maximum passage of light and is then said to be aligned. The degree (angle) of rotation is measured (Fig. 23.14). Factors that affect the angle of rotation are:

1. Concentration—the greater the concentration, the greater the angle of rotation.

2. Solvent.

3. Temperature.

4. Wavelength of the polarized light.

5. Nature of the test material.

6. Length of the sample tube.

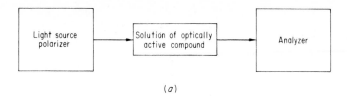

(a)

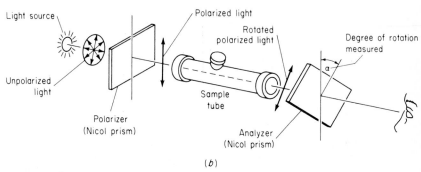

(b)

FIGURE 23.14 Components of a polarimeter.

The direction of rotation is indicated by a (+) for dextrorotation (to the right) and a (−) for levorotation (to the left). Use of the designations d and l is discouraged.

23.8.2 Specific Rotation

Specific rotation is the number of degrees of rotation observed if a 1-dm tube is used and the compound being examined is present to the extent of 1 g per 100 mL. The density for a pure liquid replaces the solution concentration.

$$\text{Specific rotation} = [\alpha] = \frac{\text{observed rotation (degrees)}}{\text{length (dm)} \times \text{(grams per 100 mL)}} \qquad (23.7)$$

The temperature of the measurement is indicated by a superscript and the wavelength of the light employed by a subscript written after the bracket; for example, $[\alpha]_{590}^{20}$ implies that the measurement was made at 20°C using 590 nm radiation.

23.8.3 Using the Polarimeter

Proceed as follows:

1. In a 10-mL volumetric flask, prepare a solution of the compound whose optical activity is to be determined. Filter solution if dust or solid particles are present.
2. Clean, dry, and partially assemble a polarimeter tube.
3. With a dropper fill the tube to overflowing.

4. Slide an end glass on the tube so that no air is entrapped. Close the end of the polarizer tube. Do not screw the cover on too tightly.

5. Position the polarimeter tube in the polarimeter.

6. Turn on the light source, and allow it to warm up.

7. Rotate the analyzer until the two halves of the image, which is viewed through the eyepiece, match exactly.

8. Read the dial and record the magnitude of the rotation.

9. Obtain a blank solvent reading by repeating steps 2 through 8 with pure solvent in a clean polarimeter tube.

10. The difference between readings 8 and 9 is the specific rotation.

23.9   REFRACTIVE INDEX

The refractive index of a liquid is the ratio of the velocity of light in a vacuum to the velocity of light in the liquid. The angle of refraction varies with the wavelength of the light used. Usually the yellow sodium doublet lines are used; they have a weighted mean of 589.26 nm and are symbolized by D. A typical refractive index (η) would be expressed as

$$\eta_D^{20} = 1.4567$$

where the superscript indicates the temperature and the subscript indicates the wavelength of the light source.

When only a single refractive index is available, approximate values over a small temperature range may be calculated using a mean value of 0.000 45 per degree for $d\eta/dt$, and remembering that η decreases with an increase in temperature. If a transition point lies within the temperature range, extrapolation is not reliable.

23.9.1   Specific Refraction

The specific refractive r_D is independent of the temperature and pressure and may be calculated by the Lorentz and Lorenz equation:

$$r_D = \frac{\eta_D^2 - 1}{\rho(\eta_D^2 + 2)} \tag{23.8}$$

where ρ is the density at the same temperature as the refractive index.

The empirical Eykman equation

$$\left(\frac{\eta_D^2 - 1}{\eta_D + 0.4}\right)\frac{1}{\rho} = \text{constant} \tag{23.9}$$

offers a more accurate means for checking the accuracy of experimental densities and refractive indices, and for calculating one from the other.

The molar refraction is equal to the specific refraction multiplied by the molecular weight. It is a more or less additive property of the groups or elements making up the compound. A set of atomic refractions will be found in Ref. 1.

23.9.2 Abbé Refractometer

The Abbé refractometer compares the angles at which light from a point source passes through the test liquid and into a prism whose refractive index is known. The refractive index of the test liquid is read from the dial (Fig. 23.15).

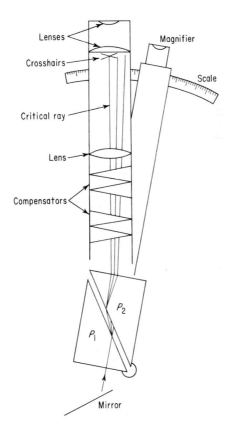

FIGURE 23.15 Schematic diagram of the Abbé refractometer.

To obtain a measurement, proceed as follows:

1. Adjust the constant-temperature bath to the desired temperature. Connect the refractometer to a small circulating pump.
2. Twist the knurled locking screw counter clockwise to unlock the prism assembly. Lower the bottom part of the hinged prism.

3. Clean the upper and lower prisms with a soft, lint-free cotton wetted with alcohol. Allow the prisms to dry.

4. Place a drop of a test solution of known refractive index at the temperature of the prism on the prism.

5. Close the prism assembly and lock by twisting the knurled knob.

6. Set the magnifier index on the scale to correspond with the known refractive index of the standard.

7. Look through the eyepiece and turn the compensator knob until the colored, indistinct boundary seen between the light and dark fields becomes a sharp line.

8. Adjust the knurled knob at the bottom of the magnifier arm until the sharp line exactly intersects the midpoint of the cross hairs in the image (Fig. 23.16). The refractometer is now standardized.

9. Repeat steps 2 through 5, placing a drop of the sample on the prisms.

10. Look through the eyepiece and move the magnifier arm until the sharp line exactly intersects the midpoint of the cross hairs on the image. Read the refractive index from the magnifier-index pointer.

11. Clean the prisms and lock them together.

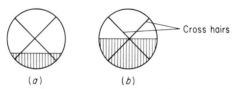

FIGURE 23.16   Adjustment of the refractometer: (*a*) incorrect and (*b*) correct.

23.9.3   Differential Refractometers

Differential refractometers have been described in Chap. 4; see especially Fig. 4.9. These instruments are useful for determining small differences in concentration.

23.10   CRYOSCOPIC MEASUREMENTS

The molecular weight of a substance can be found by observing the lowering of the freezing point of a solvent—the freezing-point depression—caused by the presence of a solute. This colligative property depends solely upon the number of solute particles dissolved in the solvent, not upon the kinds of particles.

Each solvent has its own characteristic molal freezing-point constant, that is, the depression is degrees Celsius of the freezing point when one mole of solute is dissolved in one kilogram of the solvent. A selection of cryoscopic constants K_f is given in Table 23.1. The method is applicable only to dilute solutions for which the number of moles of solute is negligible in comparison with the number of moles of solvent.

TABLE 23.1   Cryoscopic Constants

K_f	Melting point, °C	Compound
1.853	0.0	Water
5.12	5.533	Benzene
6.94	80.290	Naphthalene
18.26	57.88	Succinonitrile
39.3	6.544	Cyclohexanol

When a solvent is heated about 10°C above the melting point and then allowed to cool, the temperature will gradually fall until the melting point of the solvent is reached. At equilibrium the temperature will remain constant until the solvent has solidified completely, and then it will drop as it continues to cool. The melting point of the pure solvent is determined, and then the melting point of a solution of known concentration of the solute in the solvent is determined. The depression of the melting point T is often used for molecular weight determinations:

$$M_2 = \frac{1000 w_2 K_f}{w_1 \Delta T} \tag{23.10}$$

where w_1 is the weight of solvent and w_2 is the weight of solute whose molecular weight is M_2.

Proceed as follows:

1. Set up the apparatus as shown in Fig. 23.17.

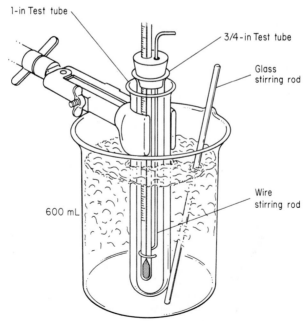

FIGURE 23.17   Freezing-point apparatus for cryoscopic measurements.

2. Measure out about 10 g (± 0.02 g) of selected solvent and transfer it quantitatively to the clean, dry test tube. If the solvent is a solid at room temperature, the bath must be heated to about 10°C above its melting point. If the solvent is a liquid, the bath must be a cooling bath to freeze the solvent. If the solvent is frozen, removing the cooling bath allows it to warm up slowly and melt, the temperature at equilibrium being the melting point.

3. Raise the test tube out of the bath and allow it to cool (or warm) slowly while mixing and record the equilibrium temperature (when the temperature remains constant). Special thermometers permit readings to be made to 0.001°C.

4. Measure out accurately (to 0.001 g) about 1 g of the powdered solute. Add it to the test tube containing the solvent. Repeat the heating (or cooling) step and record the freezing (or melting) point.

5. Calculate the molecular weight from Eq. 23.10.

23.11 EBULLIOSCOPIC MEASUREMENTS

A nonvolatile solute raises the boiling point of a solvent by an amount proportional to the concentration of the solute. The solute must be stable at the boiling point of the solvent. Molecular weights can be determined with the relation

$$M = K_b \frac{1000 w_2}{w_1 \Delta T_b} \qquad (23.11)$$

where T_b is the elevation of the boiling point brought about by the addition of w_2 grams of solute to w_1 grams of solvent and K_b is the ebullioscopic constant (Table 23.2). In the column headed "Barometric correction" is given the number of degrees for each millimeter of difference between the barometric reading and 760 mm Hg to be subtracted from K_b if the pressure is lower than 760 mm. In general, the effect is within experimental error if the pressure is within 10 mm of 760 mm.

TABLE 23.2 Ebullioscopic Constants

K_b	Boiling point, °C	Barometric correction	Compound
0.515	100.00	0.0008	Water
2.53	80.10	0.0007	Benzene
3.29	110.625	0.0008	Toluene
4.15	131.687	0.0011	Chlorobenzene
4.48	76.75	0.0013	Carbon tetrachloride
6.26	155.908	0.0016	Bromobenzene

Proceed as follows:

1. Arrange the experimental setup as shown in Fig. 23.18 for the micro-boiling point technique using the capillary-tube method.

2. Weigh out about 20 g (± 0.01 g) of solvent.

3. Heat the bath gently until a steady stream of bubbles issues from the capillary tube. Record the boiling point of the pure solvent. Discontinue the boiling so as not to lose solvent.

4. Remove the test tube from the bath and cool it. Discard the capillary tube.

5. Measure the mass of the test substance accurately to 0.01 g, and add it to the solvent. The substance must dissolve completely; ascertain its solubility experimentally beforehand.

6. Add a clean capillary tube and redetermine the boiling point of the solution.

7. Use Eq. 23.11 to calculate the molecular weight of the solute.

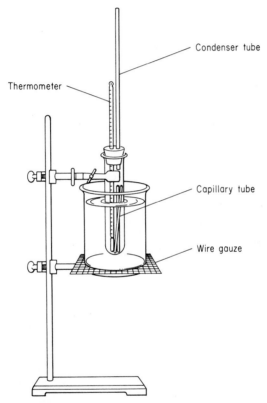

FIGURE 23.18   Apparatus for ebullioscopic measurements (micro method).

23.12 *VAPOR-PRESSURE OSMOMETRY*

At constant temperature and pressure, the vapor pressure of a pure liquid is lowered if a second substance is dissolved in it. At sufficiently low concentrations, the magnitude of the vapor pressure decrease is directly proportional to the molar concentration of solute.

23.12.1 Instrumentation

Two thermistors are placed in a closed chamber with a small amount of pure, volatile liquid in the bottom, as shown in Fig. 23.19. A built-in heater allows the chamber to be operated up to 130°C; however, the temperature of the chamber must be controlled to a few thousandths of a degree. A water circulation coil around the cylinder permits operation down to 25°C, and lower if cooled water is used.

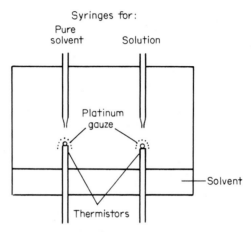

FIGURE 23.19   Schematic of chamber design for vapor pressure osmometry.

The thermistors are enclosed in glass, and the sensitive thermistor tip is covered with a small piece of fine platinum gauze. Solvent and solutions are dropped onto the thermistors through fine tubes leading from syringes on top of the instrument. Syringes can be heated independently of the measurement chamber to keep polymers in solution. Under these conditions both thermistors will be at exactly the same temperature and the chamber atmosphere will be saturated with solvent vapor. A small drop of solvent on each thermistor will not change this situation. Vapor saturation of the chamber atmosphere is achieved with cylindrical, porous paper wicks that are soaked by the solvent in the bottom of the chamber and extend up around the thermistors.

If the solvent dropped on one thermistor is replaced with a solution, the solvent vapor pressure in the solution will be lower than that of pure solvent. Condensation of solvent begins from the vapor-filled chamber. The heat of condensation warms the solution droplets until its vapor pressure rises to match that of pure solvent at the chamber temperature. A steady-state condition is established in 2 to 5 min with most solvents; readings can be taken at any time in the next 15 min with no more than 1 percent error.

The final temperature difference between the two thermistors is related to concentration and molecular weight by the expression

$$\Delta R = \frac{KC}{M} \tag{23.12}$$

where ΔR = resistance difference (proportional to the temperature difference) between two thermistors
C = solute concentration
M = solute molecular weight

The constant K is determined experimentally with a molecular weight standard.

The solvent selected must be very pure with a vapor pressure of 50 to 400 mmHg at the operating temperature. The solute must not have a vapor pressure more than 0.1 percent of that of the solvent at the operating temperature. If it does, it is necessary to use a solvent with a higher vapor pressure. Since the magnitude of the observed temperature change is inversely proportional to the solvent heat of vaporization, the best sensitivity is attained with solvents such as benzene and chloroform and the poorest with water.

A small tube in the measuring chamber permits removal of excess liquid. Thus, the chamber can be used indefinitely without opening it, if the solvent is not changed. Conversion from one solvent to another requires very careful cleaning of the chamber to remove all traces of the previous solvent. Interchangeable chambers are recommended, one for each solvent.

Molecular weights below 1000 can often be determined with a precision of 0.5 percent. The lower limit for molecular weight is around 100; the practical upper limit is about 25 000 for toluene as solvent and 5000 for water as solvent.

REFERENCES

1. J. A. Dean, *Handbook of Organic Chemistry*, McGraw-Hill, New York, 1986, p. 10-94.

CHAPTER 24
PRELIMINARY OPERATIONS OF ANALYSIS

24.1 HANDLING THE SAMPLE IN THE LABORATORY

Each sample should be completely identified, tagged, or labeled so that no question as to its origin or source can arise. Some of the information which may be on the sample is:

1. The number of the sample.
2. The notebook experiment identification number.
3. The date and time of day the sample was received.
4. The origin of the sample and cross-reference number.
5. Weight or volume (approximately) of the sample.
6. Identifying code of the container.
7. What is to be done with the sample, what determinations are to be made, or what analysis is desired?

A computerized laboratory data management system is the solution for these problems. Information as to samples expected, tests to be performed, people and instruments to be used, calculations to be performed, and results required are entered and stored directly in such a system. The raw experimental data from all tests can be collected by the computer automatically or can be entered manually. Status reports as to the tests completed, work in progress, priority work lists, statistical trends, and so on are always available automatically on schedule and on demand.

24.2 SAMPLING

Raw materials and products must be sampled in order to conduct analyses for components or to determine their purity. The size of the lot to be sampled can range from a few grams to tons, and yet the sample used in the analysis must represent as closely as possible the average composition of the total quantity being analyzed.

The gross sample of the lot being analyzed is supposed to be a miniature replica in composition and in particle-size distribution. If it does not truly represent the entire lot, all further work to reduce it to a suitable laboratory size and all laboratory procedures are a waste of time.

24.2.1 Basic Sampling Rules

The technique of sampling varies according to the substance being analyzed and its physical characteristics.

1. The size of the sample must be adequate, depending upon what is being measured, the type of measurement being made, and the level of contaminants.
2. The sample must be representative and reproducible; in static systems multi-level sampling must be made.

24.2.2 Sampling Gases

The size of the gross sample required for gases can be relatively small because any nonhomogeneity occurs at the molecular level. Relatively small samples contain tremendous quantities of molecules. The major problem is that the sample must be representative of the entire lot. This requires the taking of samples with a "sample thief" at various locations of the lot, and then combining the various samples into one gross sample.

24.2.3 Sampling Liquids

When liquids are pumped through pipes, a number of samples can be collected at various times and combined to provide the gross sample. Care should be taken that the samples represent a constant fraction of the total amount pumped and that all portions of the pumped liquid are sampled.

Homogeneous liquid solutions can be sampled relatively easily, provided that the material can be mixed thoroughly by means of agitators or mixing paddles. After adequate mixing, samples can be taken from the top and bottom and combined into one sample which is thoroughly mixed again; from this the final sample is taken for analysis.

24.2.4 Sampling Nonhomogeneous Solids

The task of obtaining a representative sample from a lot of nonhomogeneous solids requires that one proceeds as follows:

1. A gross sample is taken.
2. The gross sample is reduced to a representative laboratory-size sample. Two methods of doing this are described below.

24.2.4.1 Coning and Quartering. When very large lots are to be sampled, a representative sample can be obtained by coning (Fig. 24.1) and quartering (Fig.

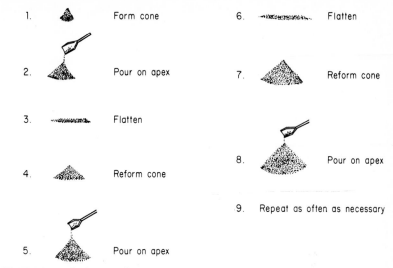

1. Form cone

2. Pour on apex

3. Flatten

4. Reform cone

5. Pour on apex

6. Flatten

7. Reform cone

8. Pour on apex

9. Repeat as often as necessary

FIGURE 24.1 Coning samples.

24.2). The first sample is formed into a cone, and the next sample is poured onto the apex of the cone. The result is then mixed and flattened, and a new cone is formed. As each successive sample is added to the re-formed cone, the total is mixed thoroughly and a new cone is formed prior to the addition of another sample.

After all the samples have been mixed by coning, the mass is flattened and a circular layer of material is formed. This circular layer is then quartered, and the alternate quarters are discarded. This process can be repeated as often as desired until a sample size suitable for analysis is obtained.

24.2.4.2 Rolling and Quartering. A representative cone of the sample is obtained by the coning procedure described in the preceding section. This sample is then placed on a flexible sheet of suitable size. The cone is flattened (Fig. 24.3), and then the entire mass of the sample is repeatedly rolled by pulling first one, then another, corner over to its opposite corner. The number of rollings required depends upon the size of the sample, the size of the particles, and the physical condition of the sample.

To collect the sample, raise all four corners of the sheet simultaneously and

FIGURE 24.2 Quartering samples. The cone is flattened, opposite quarters are selected, and the other two quarters are discarded.

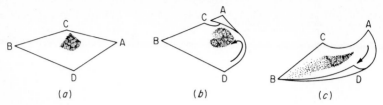

FIGURE 24.3   Rolling the sample. (*a*) Roll *A* to *B*, then (*b*) roll *C* to *D*. Do not roll *A* to *D* as mixing will be less effective.

collect the sample in the middle of the sheet. The resulting rolled sample is flattened into a circular layer and quartered until a suitably sized sample is obtained.

24.2.5 Sampling Metals

Metals can be sampled by drilling the piece to be sampled at regular intervals from all sides, being certain that each drill hole extends beyond the halfway point. Additional samples can be obtained by sawing through the metal and collecting the "sawdust."

A sample gun for molten metals is shown in Fig. 24.4. The sample is withdrawn into a glass holder. When the sample cools, the glass is broken to obtain the sample. Another design is shown in Fig. 24.5. The sampler is made of two

FIGURE 24.4   Sample gun for molten metals.

FIGURE 24.5   Sampler for molten or powdered mass.

concentric, slotted brass tubes. It is inserted into a molten or powdered mass, and the tube is rotated to secure a solid core representative of the lot.

24.3 CRUSHING AND GRINDING

In dealing with solid samples, a certain amount of crushing or grinding is sometimes required to reduce the particle size. Unfortunately, these operations tend to alter the composition of the sample, and for this reason the particle size should be reduced no more than is required for homogeneity and ready attack by reagents.

Ball or jar mills are jars or containers, fitted with a cover and gasket which can be securely fastened to the jar. The jar is half-filled with special balls, and then enough of the material which is to be ground is added to cover the balls and the voids between them. The cover is fastened securely, and the jar is revolved on a rotating assembly. The length of time for which the material is ground depends upon the fineness desired and the hardness of the material. The jar is then emptied on a coarse-mesh screen to separate the balls from the ground material.

Brittle materials that require shearing as well as impact are best handled with a Hammer-Cutter mill which is effective in grinding wool, paper, dried plants, wood, soft rocks, and similar materials. Flexible samples such as rubber and plastics may be chilled in liquid nitrogen and ground in freezer mills.

A serious error can arise during grinding and crushing as a consequence of mechanical wear and abrasion of the grinding surfaces. For this reason only the hardest materials such as hardened steel, agate, or boron carbide are employed for the grinding surface. Even with these, contamination of the sample is sometimes encountered.

Hardened steel is suitable for all-around grinding. Some iron and chromium contamination can be expected. Stainless steel is less subject to chemical attack but contributes nickel as well as iron and chromium.

Tungsten carbide is very versatile. Halide-releasing compounds which would corrode steel can be handled in tungsten carbide dishes or vessels.

Agate is harder than steel, and chemically inert to almost anything but hydrogen fluoride. It is brittle and must be handled carefully.

Alumina ceramic is ideal for extremely hard samples and in cases where steel and tungsten carbide contaminants are objectionable. It is lightweight and brittle, but very abrasion resistant.

24.3.1 Precautions in Grinding Operations

Several factors may cause appreciable alteration in the composition of the sample as a result of grinding. Among these is the heat that is inevitably generated. This can cause losses of volatile components in the sample. In addition, grinding increases the surface area of the solid and thus increases its susceptibility to reactions with the atmosphere. Water can be gained or lost from solids on grinding.

Another potential source of error arises from the difference in hardness of the sample components. The softer materials are converted to smaller particles more rapidly than the hard ones; any loss of the sample in the form of dust will thus cause an alteration in composition.

24.4 SCREENING

Intermittent screening of the material is often employed to increase the efficiency of grinding. In this operation the ground sample is placed upon a wire or cloth sieve (Fig. 24.6) that will pass particles of the desired size. The residual particles are then returned for further grinding; the operation is repeated until the entire sample passes through the screen. Screens are available in different sieve openings.

FIGURE 24.6   Typical sieve.

24.5 DRYING

It must be determined at the start if the analysis is to be reported on the "as received" basis or after drying to a constant weight by one of several methods described in Chap. 25. Most analytical results for solid samples should be expressed on a dry weight basis.

24.6 METHODS FOR DISSOLVING THE SAMPLE

Relatively few natural materials or organic materials are water soluble; they generally require treatment with acids or mixtures of acids, some combustion treatment, or even a fusion with some basic or acidic flux. The procedure adopted for the solution of a material will depend on the speed and convenience of the method and reagents employed and the desirability of avoiding the introduction of substances which interfere with the subsequent determinations or are tedious to remove. Consideration must also be given to the possible loss during the solution process of the constituents to be determined.

24.6.1 Volatilization Losses

Continual care must be exercised that some constituent to be determined is not lost by volatilization. Volatile weak acids are lost when materials are dissolved in stronger acids. Where these weak acids are to be determined, a closed apparatus must be used. Of course, volatile materials may be collected and determined. Volatile acids, such as boric acid, hydrofluoric acid, and the other halide acids, may be lost during the evaporation of aqueous solutions, and phosphoric acid may be lost when a sulfuric acid solution is heated to a high temperature.

Phosphorus may be lost as phosphine when a phosphide or a material containing phosphorus is dissolved in a nonoxidizing acid. A very definite loss of silicon as silicon hydride may occur by volatilization when aluminum and its alloys are dissolved in nonoxidizing acids.

Germanium tetrachloride, mercury(II) chloride, antimony(III) chloride, arsenic(III) chloride, and tin(IV) chloride are volatilized from hot hydrochloric acid solutions. Osmium, ruthenium, and rhenium may be lost by volatilization as the oxides from hot sulfuric acid or nitric acid solutions. In fact, under optimum conditions, the foregoing losses form the basis of quantitative volatilization procedures (*see* Chap. 30).

Relatively few things are lost by volatilization from alkaline fusions. Mercury may be reduced to the metal and lost, arsenic may be lost if organic matter is present, and of course any gases associated with the material are expelled. From acid fusions fluoride may be lost, carrying away with it some silicon or boron.

24.7 DECOMPOSITION OF INORGANIC SAMPLES

The electromotive force series (Table 16.1) furnishes a guide to the solution of metals in nonoxidizing acids such as hydrochloric acid, dilute sulfuric acid, or dilute perchloric acid, since this process is simply a displacement of hydrogen by the metal. Thus all metals below hydrogen in the series displace hydrogen, and thus dissolve in nonoxidizing acids with the evolution of hydrogen. Some exceptions to this may be found. The action of hydrochloric acid on lead, cobalt, nickel, cadmium, and chromium is slow, and lead is insoluble in sulfuric acid owing to the formation of surface film of lead sulfate.

Oxidizing acids must be used to dissolve the metals above hydrogen. The commonest of the oxidizing acids are nitric acid, hot concentrated sulfuric acid, hot concentrated perchloric acid, or some mixture which yields free chlorine or bromine. Addition of bromine or hydrogen peroxide to mineral acids is often useful.

24.7.1 Use of Liquid Reagents

24.7.1.1 Hydrochloric Acid. Concentrated hydrochloric acid (about 12 M) is an excellent solvent for many metal oxides as well as those metals which lie below hydrogen in the electromotive series. It is often a better solvent for the oxides than the oxidizing acids. After a period of heating in an open container, a constant-boiling 6 M solution remains (boiling point about 112°C).

24.7.1.2 Use of Nitric Acid. Concentrated nitric acid is an oxidizing solvent that finds wide use in attacking metals. It will dissolve most common metallic elements except aluminum and chromium, which become passive to the reagent. Many of the common alloys can also be decomposed by nitric acid. However, tin, antimony, and tungsten form insoluble oxides when treated with concentrated nitric acid. This treatment is sometimes employed to separate these elements from other sample components.

24.7.1.3   Use of Sulfuric Acid.   Hot concentrated sulfuric acid is often employed as a solvent. Part of its effectiveness arises from its high boiling point (about 340°C), at which temperature decomposition and solution of substances often proceed quite rapidly. Most organic compounds are dehydrated and oxidized under these conditions. Most metals and many alloys are attacked by the hot acid.

24.7.1.4   Use of Perchloric Acid.   Hot concentrated perchloric acid (72%) is a potent oxidizing agent and solvent. It attacks a number of ferrous alloys and stainless steels that are intractable to the other mineral acids. This acid also dehydrates and rapidly oxidizes organic materials; a treatment with hot nitric acid should precede perchloric acid treatment to avoid potential explosions.

Cold perchloric acid and hot dilute solutions are quite safe. However, all treatment of samples with perchloric acid should be done in specially designed hoods and behind explosion shields.

24.7.1.5   Use of Oxidizing Mixtures.   More rapid solvent action can sometimes be obtained by the use of mixtures of acids or by the addition of oxidizing agents to the mineral acids. Aqua regia, a mixture consisting of three volume units of concentrated hydrochloric acid and one of nitric acid, releases free chlorine to serve as oxidant. Addition of bromine to mineral acids serves the same purpose.

Addition of hydrogen peroxide (30% reagent) to mineral acids often increases their solvent action and hastens the oxidation of organic materials in the sample. Mixtures of nitric and perchloric acids are also useful for this purpose, as are mixtures of fuming nitric and concentrated sulfuric acids. Fume traps must be used to avoid venting noxious gases.

24.7.1.6   Use of Hydrofluoric Acid.   The primary use for hydrofluoric acid is the decomposition of silicate rocks and minerals where silica is not to be determined; the silicon escapes as silicon tetrafluoride. After decomposition is complete, the excess hydrofluoric acid is driven off by evaporation with sulfuric acid to fumes or with perchloric acid to virtual dryness. Sometimes residual traces of fluoride can be complexed with boric acid.

Caution   Hydrofluoric acid can cause serious damage and painful injury when brought in contact with the skin. Momentarily it acts like a "pain killer" while it penetrates the skin or works under fingernails.

24.7.2   Decomposition of Samples by Fluxes

Quite a number of common substances—such as silicates, some of the mineral oxides, and a few of the iron alloys—are attacked slowly, if at all, by the usual liquid reagents. Recourse to more potent fused-salt media, or fluxes, is then called for. Fluxes will decompose most substances by virtue of the high temperature required for their use and the high concentration of reagent brought in contact with the sample.

Where possible, the employment of a flux is usually avoided, for several dangers and disadvantages attend its use. In the first place, a relatively large quantity of the flux is required to decompose most substances—often 10 times the sample weight. The possibility of significant contamination of the sample by impurities in the reagent thus becomes very real.

Furthermore, the aqueous solution resulting from the fusion will have a high salt content, and this may lead to difficulties in the subsequent steps of the analysis. The high temperatures required for a fusion increase the danger of loss of pertinent constituents by volatilization. Finally, the container in which the fusion is performed is almost inevitably attacked to some extent by the flux; this again can result in contamination of the sample.

In those cases where the bulk of the substance to be analyzed is soluble in a liquid reagent and only a small fraction requires decomposition with a flux, it is common practice to employ the liquid reagent first. The undecomposed residue is then isolated by filtration and fused with a relatively small quantity of a suitable flux. After cooling, the melt is dissolved and combined with the rest of the sample.

24.7.2.1   Method of Carrying Out a Fusion.   In order to achieve a successful and complete decomposition of a sample with a flux, the solid must ordinarily be ground to a very fine powder; this will produce a high specific surface area. The sample must then be thoroughly mixed with the flux; this operation is often carried out in the crucible in which the fusion is to be done by careful stirring with a glass rod.

In general, the crucible used in a fusion should never be more than half-filled at the outset. The temperature is ordinarily raised slowly with a gas flame because the evolution of water and gases is a common occurrence at this point; unless care is taken there is the danger of loss by spattering. The crucible should be covered with its lid as an added precaution. The maximum temperature employed varies considerably and depends on the flux and the sample. It should be no greater than necessary in order to minimize attack on the crucible and decomposition of the flux. It is frequently difficult to decide when the heating should be discontinued. In some cases, the production of clear melt serves to indicate the completion of the decomposition. In others the condition is not obvious and the analyst must base the heating time on previous experience with the type of material being analyzed. In any event, the aqueous solution from the fusion should be examined carefully for particles of unattacked sample.

When the fusion is judged complete, the mass is allowed to cool slowly; then just before solidification the crucible is rotated to distribute the solid around the walls of the crucible so that the thin layer can be readily detached.

24.7.3   Types of Fluxes

The common fluxes used in analysis are listed in Table 24.1. Basic fluxes, employed for attack on acidic materials, include the carbonates, hydroxides, peroxides, and borates. The acidic fluxes are the pyrosulfates, the acid fluorides, and boric acid. Fluxes rich in lithium tetraborate are well suited for dissolving basic oxides, such as alumina. Lithium metaborate, on the other hand, is more basic and better suited for dissolving acidic oxides such as silica or titanium dioxide.

The lowest melting flux capable of reacting completely with a sample is usually the optimum flux. Accordingly, mixtures of lithium tetraborate, with the metaborate or carbonate, are often selected.

If an oxidizing flux is required, sodium peroxide can be used. As an alterna-

TABLE 24.1   The Common Fluxes

Flux	Melting point, °C	Types of crucible used for fusion	Types of substances decomposed
Na_2CO_3	851	Pt	For silicates, and silica-containing samples; alumina-containing samples; insoluble phosphates and sulfates
Na_2CO_3 + an oxidizing agent such as KNO_3, $KClO_3$, or Na_2O_2		Pt (not with Na_2O_2), Ni	For samples needing an oxidizing agent
NaOH or KOH	320–380	Au, Ag, Ni	For silicates, silicon carbide, certain minerals
Na_2O_2	Decomposes	Fe, Ni	For sulfides, acid-insoluble alloys of Fe, Ni, Cr, Mo, W, and Li; Pt alloys; Cr, Sn, Zn minerals
$K_2S_2O_7$	300	Pt, porcelain	Acid flux for insoluble oxides and oxide-containing samples
B_2O_3	577	Pt	For silicates and oxides when alkalis are to be determined
$CaCO_3$ + NH_4Cl		Ni	For decomposing silicates for determination of alkalis

tive, small quantities of the alkali nitrates or chlorates are mixed with sodium carbonate.

24.8 DECOMPOSITION OF ORGANIC COMPOUNDS

Analysis of the elemental composition of an organic substance generally requires drastic treatment of the material in order to convert the elements of interest into a form susceptible to the common analytical techniques. These treatments are usually oxidative in nature, involving conversion of the carbon and hydrogen of the organic material to carbon dioxide and water. In some instances, however, heating the sample with a potent reducing agent is sufficient to rupture the covalent bonds in the compound and free the element to be determined from the carbonaceous residue.

Oxidation procedures are sometimes divided into two categories. Wet ashing (or oxidation) makes use of liquid oxidizing agents. Dry ashing usually implies ignition of the organic compound in air or a stream of oxygen. In addition, oxidations can be carried out in certain fused-salt media, sodium peroxide being the most common flux for this purpose.

24.8.1 Oxygen Flask (Schöninger) Combustion

A relatively straightforward method for the decomposition of many organic substances involves oxidation with gaseous oxygen in a sealed container. The reaction products are absorbed in a suitable solvent before the reaction vessel is opened. Analysis of the solution by ordinary methods follows.

A simple apparatus for carrying out such oxidation has been suggested by Schöninger (Fig. 24.7). It consists of a heavy-walled flask of 300- to 1000-mL capacity fitted with a ground-glass stopper. Attached to the stopper is a platinum-gauze basket which holds from 2 to 200 mg of sample. If the substance to be analyzed is a solid, it is wrapped in a piece of low-ash filter paper cut in the shape shown in the illustration. Liquid samples can be weighed in gelatin capsules which are then wrapped in a similar fashion. A tail is left on the paper and serves as an ignition point (wick).

A small volume of an absorbing solution is placed in the flask, and the air in the container is then displaced by allowing tank oxygen to flow into it for a short period. The tail of the paper is ignited, and the stopper is quickly fitted into the flask. The container is then inverted to prevent the escape of the volatile oxidation products. Ordinarily the reaction proceeds rapidly, being catalyzed by the platinum gauze surrounding the sample. During the combustion, the flask is kept behind a safety shield to avoid damage in case of an explosion. Complete combustion takes place within 20 s.

After cooling, the flask is shaken thoroughly and disassembled; then the inner surfaces are rinsed down. The analysis is then performed on the resulting solution. This procedure has been applied to the determination of halogens, sulfur, phosphorus, and various metals in organic compounds. When sulfur is to be determined, any sulfur dioxide formed in the Schöninger combustion is subsequently oxidized to sulfur trioxide (actually sulfate) by treatment with hydrogen peroxide.

24.8.2 Peroxide Fusion

Sodium peroxide is a strong oxidizing reagent which, in the fused state, reacts rapidly and often violently with organic matter, converting carbon to the carbon-

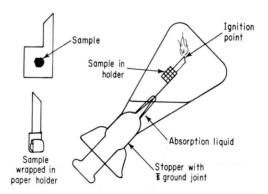

FIGURE 24.7 Schöninger (oxygen) combustion flask.

ate, sulfur to sulfate, phosphorus to phosphate, chlorine to chloride, and iodine and bromine to iodate and bromate. Under suitable conditions the oxidation is complete, and analysis for the various elements may be performed upon an aqueous solution of the fused mass.

Once started, the reaction between organic matter and sodium peroxide is so vigorous that a peroxide fusion must be carried out in a sealed, heavy-walled, steel bomb. Sufficient heat is evolved in the oxidation to keep the salt in the liquid state until the reaction is completed. Ordinarily the reaction is initiated by passage of current through a wire immersed in the flux or by momentary heating of the bomb with a flame.

The maximum size for a sample that is to be fused is perhaps 100 mg. The method is more suited to semimicro quantities of about 5 mg.

One of the main disadvantages of the peroxide-bomb method is the rather large ratio of flux to sample needed for a clean and complete oxidation. Ordinarily an approximate 200-fold excess is used. The excess peroxide is subsequently decomposed to hydroxide by heating in water. After neutralization, the solution necessarily has a high salt content.

24.8.3 Wet-Ashing Procedures

Solution in a variety of strong oxidizing agents will decompose organic samples. The main consideration associated with the use of these reagents is preventing volatility losses for the elements of interest. For the volatile elements, dissolution must be effected under reflux conditions. Use a 125-mL quartz flask fitted with a quartz condenser filled with quartz beads.

24.8.3.1 Use of Sulfuric Acid. One wet-ashing procedure is the Kjeldahl method for the determination of nitrogen in organic compounds; it is described in Sec. 22.4. Here concentrated sulfuric acid is the oxidizing agent. This reagent is also frequently employed for decomposition of organic materials where metallic constituents are to be determined. Commonly, nitric acid is added to the solution periodically to hasten the rate at which oxidation occurs. A number of elements may be volatilized, at least partially, by this procedure, particularly if the sample contains chlorine; these include arsenic, boron, germanium, mercury, antimony, selenium, tin, and the halogens.

Transfer a suitable sample to a quartz beaker, and treat it with 5 mL of water and 10 mL nitric acid. Digest tissues overnight with the beaker covered; other materials may take less digestion. Cool, add 5 mL of sulfuric acid and evaporate to fumes of sulfuric acid. Cover the beaker and add nitric acid dropwise to the hot solution to destroy any residual organic matter. Transfer the solution to a Teflon FEP beaker along with any remaining siliceous material. Add 1 mL of hydrofluoric acid and evaporate the solution to fumes of sulfuric acid.

24.8.3.2 Use of Perchloric Acid–Nitric Acid Mixtures. A good deal of care must be exercised in using a mixture of perchloric acid and nitric acid. Explosion can be avoided by starting with a solution in which the perchloric acid is well diluted with nitric acid and not allowing the mixture to become concentrated in perchloric acid until the oxidation is nearly complete. Properly carried out, oxidations with this mixture are rapid and losses of metallic ions negligible. *Caution*: This mixture should never be used to decompose pyridine and related compounds.

Transfer a suitable sample (usually 1 to 5 g) to a 100-mL Teflon FEP beaker and treat it with 25 mL of 1:1 nitric acid. Digest on a hot plate for 1 h, and cool and rinse the sides of the beaker with water. Add 10 mL of nitric acid and 5 mL of hydrofluoric acid and evaporate to approximately 10 mL. If the sample contains black particles of unreacted carbon, repeat the addition of 10 mL of nitric acid. Add 10 mL perchloric acid. Cover and raise the temperature gradually until fumes of perchloric acid appear. Continue heating until fumes of perchloric acid are no longer noted. Cool, wash down the sides of the beaker with distilled water, and heat again until fumes of perchloric acid are no longer noted. Continue the digestion until the sample volume is reduced to about 2 mL. Transfer to a suitable volumetric flask.

24.8.4 Dry-Ashing Procedure

The simplest method for decomposing an organic sample is to heat it with a flame in an open dish or crucible until all the carbonaceous material has been oxidized by the air. A red heat is often required to complete the oxidation. Analysis of the nonvolatile components is then made after solution of the residual solid. A great deal of uncertainty always exists with respect to the recovery of supposedly nonvolatile elements when a sample is treated in this manner. Some losses probably arise from the mechanical entrainment of finely divided particular matter in the hot convection currents around the crucible. Volatile metallic compounds may be formed during the ignition.

The dry-ashing procedure is the simplest of all methods for decomposing organic compounds. It is often unreliable and should not be employed unless tests have been performed that demonstrate its applicability to a given type of sample.

CHAPTER 25
MOISTURE AND DRYING

25.1 INTRODUCTION

The presence of water in a sample represents a common problem that frequently faces the analyst. Water may exist as a contaminant from the atmosphere or from the solution in which the substance was formed, or it may be bonded as a chemical compound, a hydrate. Regardless of its origin, water plays a part in determining the composition of the sample. Unfortunately, particularly in the case of solids, the water content is a variable quantity that depends upon such things as humidity, temperature, and the state of subdivision. Thus, the constitution of a sample may change significantly with environment and method of handling.

In order to cope with the variability in composition caused by the presence of moisture, the analyst may attempt to remove the water by drying prior to weighing samples for analysis. Alternatively the water content may be determined at the time the samples are weighed out for analysis; in this way results can be corrected to a dry basis.

25.2 FORMS OF WATER IN SOLIDS

It is convenient to distinguish among the several ways in which water can be held by a solid. The *essential water* in a substance is that water which is an integral part of the molecular or crystal structure of one of the components of the solid. It is present in that component in stoichiometric quantities. An example is $CaC_2O_4 \cdot 2H_2O$.

The second form is called *water of constitution*. Here the water is not present as such in the solid but rather is formed as a product when the solid undergoes decomposition, usually as a result of heating. This is typified by the processes

$$2KHSO_4 \rightarrow K_2S_2O_7 + H_2O$$
$$Ca(OH)_2 \rightarrow CaO + H_2O$$

Nonessential water is not necessary for the characterization of the chemical constitution of the sample and therefore does not occur in any sort of stoichiometric proportions. It is retained by the solid as a consequence of physical forces.

Adsorbed water is retained on the surface of solids in contact with a moist environment. The quantity is dependent upon humidity, temperature, and the specific surface area of the solid. Adsorption is a general phenomenon that is encountered in some degree with all finely divided solids. The amount of moisture adsorbed on the surface of a solid also increases with the humidity. Quite generally, the amount of adsorbed water decreases as the temperature increases, and in most cases approaches zero if the solid is dried at temperatures above 112°C. Equilibrium is achieved rather rapidly, ordinarily requiring only 5 or 10 min. This often becomes apparent to a person who weighs finely divided solids that have been rendered anhydrous by drying; a continuous increase in weight is observed unless the solid is contained in a tightly stoppered vessel.

A second type of nonessential water is called sorbed water. This is encountered with many colloidal substances such as starch, protein, charcoal, zeolite minerals, and silica gel. The amounts of sorbed water are often large compared with adsorbed moisture, amounting in some instances to as much as 20 percent or more of the solid. Solids containing even this much water may appear to be perfectly dry powders. Sorbed water is held as a condensed phase in the interstices or capillaries of the colloidal solids. The quantity is greatly dependent upon temperature and humidity.

A third type of nonessential moisture is *occluded water*. Here, liquid water is entrapped in microscopic pockets spaced irregularly throughout the solid crystals. Such cavities often occur naturally in minerals and rocks.

Water may also be dispersed in a solid in the form of a solid solution. Here the water molecules are distributed homogeneously throughout the solid. Natural glasses may contain several percent of moisture in this form.

25.2.1   Effect of Grinding on Moisture Content

Often the moisture content and thus the chemical composition of a solid is altered to a considerable extent during grinding and crushing. This will result in decreases in some instances and increases in others.

Decreases in water content are sometimes observed when solids containing essential water in the form of hydrates are ground. For example, the water content of calcium sulfate dihydrate is reduced from 20 to 5 percent by this treatment. Undoubtedly the change is a result of localized heating during the grinding and crushing of the particles.

Losses also occur when samples containing occluded water are reduced in particle size. Here, the grinding process ruptures some of the cavities and exposes the water so that it may evaporate.

More commonly, the grinding process is accompanied by an increase in moisture content, primarily because of the increase in surface area exposed to the atmosphere. A corresponding increase in adsorbed water results. The magnitude of the effect is sufficient to alter appreciably the composition of a solid.

25.3   DRYING SAMPLES

Samples may be dried by heating them to 110°C or higher if the melting point of the material is higher and the material will not decompose at that temperature. This procedure will remove the moisture bound to the surface of the particles. The arrangement for drying samples is shown in Fig. 25.1.

FIGURE 25.1   Arrangement for drying solid samples.

The procedure is as follows:

1. Fill the weighing bottle no more than half full with the sample to be dried.
2. Place a label on the beaker or loosely inside the beaker. Do not place a label on the weighing bottle as it will gradually char.
3. Place the weighing bottle in the beaker. Remove the cover from the weighing bottle, and place inside the beaker.
4. Cover the beaker with a watch glass supported on glass hooks.
5. Place the beaker with weighing bottle in a drying oven at the desired temperature for 2 h.
6. Remove from the oven. Cool somewhat before placing the weighing bottle, now covered with its cap, in a desiccator.

25.3.1   The Desiccator

A desiccator is a container (glass or aluminum) filled with a substance which absorbs water (a desiccant) (Fig. 25.2). Several desiccants and their properties are listed in Table 25.1. The ground-glass (or metal) rim is lightly greased with petroleum jelly or silicone grease. The desiccator provides a dry atmosphere for objects and substances. To use a desiccator, proceed as follows:

1. Remove the cover by sliding sideways.
2. Place the object to be dried on the porcelain platform plate.
3. Slide the lid back in position.

The desiccator's charge of desiccant must be frequently renewed to keep it effective. Surface caking signals the need to renew the desiccant. Some desiccants contain a dye which changes color when the desiccant can no longer absorb water.

Vacuum desiccators are equipped with side arms so that they may be connected to a vacuum. This type of desiccator should be used to dry crystals which are wet with organic solvents. Vacuum desiccators should not be used for substances which sublime readily.

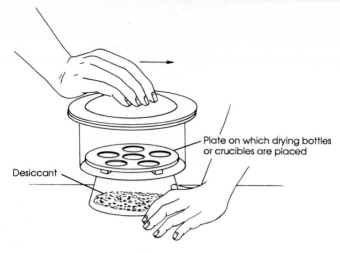

FIGURE 25.2 The desiccator, showing how the lid is removed.

TABLE 25.1 Drying Agents

Drying agent	Most useful for	Residual water, mg H_2O per liter of dry air (25°C)	Grams water removed per gram of desiccant
Al_2O_3	Hydrocarbons	0.002–0.005	0.2
$Ba(ClO_4)_2$	Inert gas streams	0.6–0.8	0.17
BaO	Basic gases, hydrocarbons, aldehydes, alcohols	0.0007–0.003	0.12
CaC_2	Ethers		0.56
$CaCl_2$	Inert organics	0.1–0.2	0.15 (1 H_2O)
CaH_2	Hydrocarbons, ethers, amines, esters, higher alcohols	1×10^{-5}	0.83
CaO	Ethers, esters, alcohols, amines	0.01–0.003	0.31
$CaSO_4$	Most organic substances	0.005–0.07	0.07
KOH	Amines	0.01–0.9	
$Mg(ClO_4)_2$	Gas streams	0.0005–0.002	0.24
$MgSO_4$	Most organic compounds	1–12	0.15–0.75
Molecular sieve 4×	Molecules with effective diameter > 4 Å	0.001	0.18
P_2O_5	Gas streams; not suitable for alcohols, amines, or ketones	2×10^{-5}	0.5
Silica gel	Most organic amines	0.002–0.07	0.2
H_2SO_4	Air and inert gas streams	0.003–0.008	Indefinite

25.4   DRYING COLLECTED CRYSTALS

Gravity-filtered crystals collected on a filter paper may be dried as follows:

1. Remove the filter paper from the funnel. Open up the filter paper and flatten it on a watch glass of suitable size or on a shallow evaporating dish. Cover the watch glass or dish with a piece of clean, dry filter paper and allow the crystals to air-dry.

Hygroscopic substances cannot be air-dried in this way.

2. Press out excess moisture from the crystals by laying filter paper on top of the moist crystals and applying pressure with a suitable object.
3. Use a spatula to work the pasty mass on a porous plate; then allow it to dry.
4. Use a portable infrared lamp to warm the sample and increase the rate of drying. Be sure that the temperature does not exceed the melting point of the sample.
5. Use a desiccator.

When very small quantities of crystals are collected by centrifugation, they can be dried by subjecting them to vacuum in the centrifuge tube while gently warming the tube.

25.5   DRYING ORGANIC SOLVENTS

Water can be removed from organic liquids and solutions by treating the liquids with a suitable drying agent to remove the water. The selection of drying agents must be carefully made. The drying agent selected should not react with the compound or cause the compound to undergo any reaction but should remove only the water. Table 25.1 lists drying agents.

25.5.1   Use of Solid Drying Agents

Solid drying agents are added to wet organic solvents. They remove the water, and then the hydrated solid is separated from the organic solvent by decantation and filtration. Proceed as follows:

1. Pour the organic liquid into a flask which can be stoppered. Add small portions of the drying agent, shaking the flask thoroughly after each addition. Add as much drying agent as required.
2. Allow to stand for a predetermined time.
3. Filter the solid hydrate from the liquid with a funnel and filter paper.

Several operations may be required. Repeat if necessary.

25.5.2   Efficiency of Drying Operations

The efficiency of a drying operation is improved if the organic solvent is repeatedly exposed to fresh portions of the drying agent. Some dehydrating agents are very powerful and dangerous, especially if the water content of the organic solvent is high. These should be used only after the wet organic solvent has been grossly predried with a weaker agent. Drying agents will clump together, sticking to the bottom of the flask when a solution is "wet." Wet solvent solutions appear to be cloudy; dry solutions are clear. If the solution is "dry," the solid drying agent will move about and shift easily on the bottom of the flask.

Molecular sieves are excellent drying agents for gases and liquids. In addition to absorbing water, they also absorb other small molecules. Several types have been described in Sec. 3.8.1. Regeneration of molecular sieves can be accomplished by heating to 250°C or applying a vacuum. In contrast to chemically acting drying agents, molecular sieves are unsuitable when the substance is to be dried in vacuo.

Calcium carbide, metallic sodium, and phosphorus(V) oxide remove moisture by chemical reaction with water. Do not use these drying agents where either the drying agent itself or the product that it forms will react with the compound or cause the compound itself to undergo reaction or rearrangement. These drying agents are useful in drying saturated hydrocarbons, aromatic hydrocarbons, and ethers. However, the compounds to be dried must not have functional groups, such as hydroxyl or carboxyl, which will react with the agents.

Some general precautions are:

1. Do not dry alcohols with metallic sodium.
2. Do not dry acids with basic drying agents.
3. Do not dry amines or basic compounds with acidic drying agents.
4. Do not use calcium chloride to dry alcohols, phenols, amines, amino acids, amides, ketones, or certain aldehydes and esters.

25.6 FREEZE DRYING

Some substances cannot be dried at atmospheric conditions because they are extremely heat sensitive, but they can be freeze-dried. Freeze drying is a process whereby substances are subjected to high vacuum after they have been frozen. Under these conditions, ice (water) will sublime and other volatile liquids will be removed. This leaves the nonsublimable material behind in a dried state.

Commercial freeze driers are self-contained units. They may consist merely of a vacuum pump, adequate vapor traps, and a receptacle for the material in solution (Fig. 25.3), or they may include refrigeration units to chill the solution plus more sophisticated instruments to designate temperature and pressure, plus heat and cold controls and vacuum-release valves.

Freeze drying differs from ordinary vacuum distillation in that the solution or substance to be dried is first frozen to a solid mass. It is under these conditions that the water is selectively removed by sublimation, the ice going directly to the water-vapor state. Proceed as follows:

1. Freeze the solution, spreading it out on the inner surface of the container to increase the surface area.
2. Apply high vacuum; the ice will sublime and leave the dried material behind.

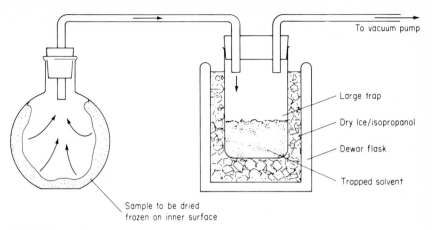

FIGURE 25.3 Setup for freeze drying in the laboratory.

3. Use dilute solutions in preference to concentrated solutions.
4. Protect the vacuum pump from water with a dry-ice trap, and insert chemical gas-washing towers to protect the pump from corrosive gases.

25.7 HYGROSCOPIC ION-EXCHANGE MEMBRANE

The Perma Pure driers utilize a hygroscopic, ion-exchange membrane in a continuous drying process to selectively remove water vapor from mixed gas streams. The membrane is a proprietary extrudable desiccant in tubular form. A single desiccant tube is fabricated in a shell-and-tube configuration and sealed into an impermeable shell which has openings adjacent to the sample inlet and product outlet (Fig. 25.4). If a wet gas stream flows through the tubes and a countercurrent dry gas stream purges the shell, water vapor molecules are transferred through the walls of the tubing. The wet gas is dried, and the dry purge gas becomes wet as it carries away the water vapor.

The efficiency and capacity of a drier at constant temperature and humidity are based on the drier's geometry (that is, internal volume, outside surface area, and shell volume), as well as the gas flows and pressures of the wet sample and

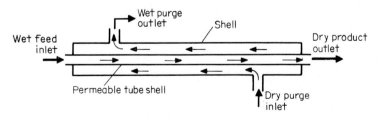

FIGURE 25.4 Perma Pure drier schematic diagram. (*Courtesy of Perma Pure Products, Inc.*)

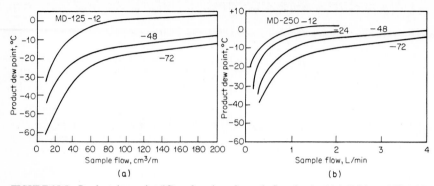

FIGURE 25.5 Product dew point (°C) as function of sample flow for the (a) 0.125-in and (b) 0.25-in. drier tube. Tube lengths are approximately 12, 24, 48, and 72 in. (*Courtesy of Perma Pure Products, Inc.*)

the dry purge. The reduction of water vapor in the product of a drier may be increased by reducing the sample flow or by increasing the drier volume (a longer tube length). Increasing the sample flow results in a higher dew point in the product. A bundle of tubes with a common header increases the volume of wet gas that can be handled. Units are available with shell diameters ⅛ or ¼ in; lengths range from 12 to 72 in. Product dew point is plotted against the product flow in Fig. 25.5 for ⅛- and ¼-in-diameter drier tubes.

The membrane is stable up to 160°C, as is the stainless-steel shell. Fluorocarbon and polypropylene shells are also available; their maximum use temperatures are 160 and 150°C, respectively. The chemical resistance of the driers to acid gases and liquids is shown in Table 25.2.

TABLE 25.2 Chemical Resistance of the Perma Pure Drier

Sample	Concentration, %	Stainless*	Polypropylene and fluorocarbon*
Chlorine	100	X	Y
HCl	10	X	Y
NO_2	0.02	Y	Y
NO_2	0.20	Y	Y
SO_2	0.50	Y	Y
SO_2	1	X	Y
EDC (liq)	100	Y	Y
Methylene chloride	100	Y	Y

*Y = usable; X = not usable.

25.8 MICROWAVE DRYING

Conventional microwave ovens have generally proved unsatisfactory for laboratory use because of the uneven distribution of microwave energy and the problem of excess reflected microwave energy. However, microwave driers, utilizing programmable computers are extremely versatile and easy to operate. The programmable aspect allows the microwave intensity to be varied during the drying or heating cycle.

A primary use is for rapid chemical digestions, using the microwave energy heat source. Digestion can be reduced from hours to minutes.

Samples can be analytically dried to less than 1 mg of water in several minutes. Water selectively absorbs the microwave energy and is removed through evaporation. In some systems, moisture determination is entirely automatic. The sample is placed on the balance pan, the door closed, and the start button depressed. The initial sample weight is stored in the computer, and the microwave system is actuated for the predetermined time. The final weight of the sample is ascertained when the oven is turned off. Weight loss and percentage moisture or solid residue are displayed.

25.9 CRITICAL POINT DRYING

Critical point drying offers a number of advantages over both air and freeze drying in the preparation of samples for examination under a transmission or a scanning electron microscope. Specimens treated by this method can be studied in the true original form without the physical deterioration and structural damage usually produced when water or other volatiles are removed from a sample by conventional drying techniques. Preparation time is measured in minutes.

The method takes advantage of the fact that at its critical point a fluid passes imperceptibly from a liquid to a gas with no evident boundary and no associated distortional forces. Heating a liquid to its critical point in a closed system causes the density of the liquid to decrease and the density of the vapor to increase until the two phases become identical at the critical point where the liquid surface vanishes completely.

To proceed, all the water in the sample is replaced by a carefully selected transitional fluid. Then, while the sample is completely immersed in the fluid, it is heated in a sealed bomb to slightly above the critical point. The vapor is then released while holding the bomb above the critical temperature. This procedure leaves a completely dry specimen which has not been subjected either to the surface tension forces of air drying or to the freezing and sublimation boundaries associated with freeze drying.

A suitable fluid must have these properties:

1. Be nonreactive with the specimen.

2. Have a critical temperature low enough to prevent damage to specimens which are temperature sensitive.

3. Have a critical pressure low enough that conventional equipment can be used, without requiring cumbersome and bulky pressure designs.

4. Be nontoxic and readily available.

Carbon dioxide and nitrous oxide were used in early studies. Freons are well suited; their critical and boiling temperatures are such that the preliminary steps can be conducted in open vessels for maximum control and visibility.

In most critical point drying procedures, an intermediate liquid is first used to displace the moisture present in the original specimen. Ethanol and acetone are

TABLE 25.3   Transitional and Intermediate Fluids for Critical Point Drying

Transitional fluid	Critical pressure, atm	Critical temperature, °C	Boiling point, °C	Intermediate liquid
CF_3CF_3 (Freon 116)	29.4	19.7	-78.2	Acetone
$CClF_3$ (Freon 13)	38.2	28.9	-81.4	Ethanol
CHF_3 (Freon 23)	47.7	25.9	-82.0	Ethanol
Carbon dioxide	72.85	31.04	Sublimes	Ethanol and pentyl acetate
Nitrous oxide	71.60	36.43	-88.47	Not required

two of the more popular reagents used for this purpose. Other liquids can be used if they are fully miscible with water and with the transitional fluid. The moisture in the sample is removed by passing the specimen stepwise through a graded series of solutions starting with a 10 percent concentration and moving up to a moisture-free, 100 percent liquid. The specimen is then removed from the intermediate liquid and transferred to the transitional fluid for treatment in the bomb. The several materials which have been successfully used as intermediate liquids, together with the principal transitional fluids, are listed in Table 25.3.

25.10 KARL FISCHER METHOD FOR MOISTURE

The determination of water is one of the most important and most widely practised analyses in industry. The field of application is so large that it is the subject of a three-volume series of monographs.[1] The Karl Fischer method relies on the specificity for water of the reagent devised by Fischer. The original reagent contained pyridine, sulfur dioxide, and iodine in an organic solvent (methanol). It reacts quantitatively with water.

$$C_5H_5N\!-\!I_2 + C_5H_5N\!-\!SO_2 + C_5H_5N + H_2O \rightarrow {}^1\!/_4 C_5H_5N\!-\!HI + C_5H_5N\!-\!SO_3$$
$$(25.1)$$

There is a secondary reaction with the solvent (methanol):

$$C_5H_5N\!-\!SO_3 + CH_3OH \rightarrow C_5H_5NH\!-\!O\!-\!SO_2\!-\!OCH_3 \qquad (25.2)$$

Various improvements have been suggested. The end point is usually ascertained by means of an amperometric titration using two indicator electrodes (see Sec. 19.6).

Iodine in the Karl Fischer reaction can be generated coulometrically with 100 percent efficiency. Now an absolute instrument is available, and the analysis requires no calibration or standardization (see Sec. 20.3.1).

An entirely automated titrimeter will determine moisture in the range from 1 $\mu g \cdot mL^{-1}$ to 100 percent water content. The instrument combines a buret, a sealed titration vessel, a magnetic stirrer, and a pump system for changing solvent. Liquid samples are injected through a septum; solid samples are inserted through a delivery opening in the titration head. The titration is kinetically controlled; the speed of titrant delivery is adjusted to the expected content of water.

Typically 50 mg of water are titrated in less than 1 min with a precision of better than 0.3 percent. An optional pyrolysis system allows the extraction of moisture from solid samples.

REFERENCES

1. J. Mitchell and D. M. Smith, *Aquammetry,* Wiley, New York, Vol. 1, 1977; Vol. 2, 1983; Vol. 3, 1980.

CHAPTER 26
EXTRACTION METHODS

Solutes have different solubilities in different solvents, and the process of selectively removing a solute from a mixture with a solvent is called extraction. The solute to be extracted may be in a solid or liquid medium, and the solvent used for the extraction process may be water, a water-miscible solvent, or a water-immiscible solvent. The selection of the solvent to be used depends on the solute and on the requirements of the experimental procedure.

26.1 SOLVENT EXTRACTION SYSTEMS

Extraction procedures based on the distribution of solutes between immiscible solvents are carried out for two purposes. *Exhaustive extraction* involves the quantitative removal of one solute; *selective extraction* involves the separation of two solutes.

26.1.1 Partition Coefficient

It is useful to consider first the behavior of a single solute between two immiscible liquids, which are usually water and an organic liquid. The distribution equilibrium in the simplest case involves the same molecular species in each phase.

$$A_{aq} \rightleftharpoons A_{org} \tag{26.1}$$

Included in this class are neutral covalent molecules which are not solvated by either of the solvents. The partition coefficient corresponds in value to the ratio of the saturated solubilities of the solute in each phase

$$K_d = \frac{[A]_{org}}{[A]_{aq}} \tag{26.2}$$

Ignoring any specific solute-solvent interactions, the solubility of such molecular species in organic solvents is generally at least an order of magnitude higher than in water; that is, $K_d \geq 10$. In each homologous series, increasing the alkyl chain length increases the partition coefficient by a factor of about four for each new methylene group incorporated into the molecule. Branching results in a lower K_d

as compared with the linear isomer. Incorporation of hydrophilic functional groups into the molecule lowers the partition coefficient as written.

26.1.2 Exhaustive Extraction

Water will extract inorganic salts, salts of organic acids, strong acids and bases, and low-molecular-weight (four carbons or less) carboxylic acids, alcohols, polyhydroxy compounds, and amines from any immiscible organic solvents which contain them. The completeness of the extraction may depend upon the pH of the aqueous phase, as will be discussed in a later section.

Consider an extraction of X moles of solute A dissolved in V_w milliliters of water, with V_o milliliters of organic solvent. If Y moles remain in the water phase after a single extraction, the fraction remaining unextracted is

$$\frac{Y}{X} = f = \frac{V_w}{V_w + K_d V_o} \tag{26.3}$$

The fraction remaining unextracted is independent of the initial concentration. Therefore, if n successive extractions are performed with fresh portions of solvent, the fraction remaining unextracted is

$$f = \left[1 + K_d \frac{V_o}{V_w} \right]^{-n} \tag{26.4}$$

The amount of solute remaining in the water phase after a single extraction is dependent upon two factors: (1) the partition coefficient (in later expression this is replaced by the distribution ratio D) and (2) the volume ratio of the phases. Exhaustive extraction will involve either repeated batch extractions or continuous extraction methods. After n extractions, the fraction of solute remaining in the water phase is

$$\frac{X}{Y} = \left[\frac{V_w}{K_d(V_o + V_w)} \right]^n \tag{26.5}$$

For a given amount of organic phase, the extraction is more efficient if carried out several times with equal portions of extracting phase. However, little is gained by dividing the volume of extractant into more than four or five portions. Equation 26.4 is useful in determining whether a given extraction is practicable, with a reasonable value of V_o/V_w, or whether an extractant with a more favorable partition coefficient (or distribution ratio) should be sought.

EXAMPLE 26.1 How might 99.0 percent of a substance with a partition coefficient of 4 be extracted into an organic phase?

Method 1. If a single extraction is to be done, it is necessary to determine the volume ratio needed. Proceed as follows:

$$0.01 = \left[1 + 4\left(\frac{V_o}{V_w}\right) \right]^{-1}$$

$$\frac{V_o}{V_w} = 25$$

The total amount of solute would determine whether 100 mL of organic phase would be used to extract 4 mL of aqueous phase, or some other ratio such as 500 mL and 20 mL, respectively. In any case, the volume ratio is just feasible. Even so, too large a volume of organic solvent is required.

Method 2. How many equilibrations would be required using fresh organic phase each time and equal volumes of each phase?

$$0.01 = (1 + 4)^{-n}$$

$$\log 0.01 = n \log 0.2 \quad \text{and} \quad -2.0 = -0.7n$$

$$n = 2.8 \text{ (or 3 extractions)}$$

This second approach requires much less organic solvent; and three successive equilibrations are not particularly time consuming.

If the partition coefficient or distribution ratio is unknown, it can be obtained by equilibrating known volumes of the aqueous phase and extracting solvent, then determining the concentration of distributing species in both phases. This should be performed over a range of concentrations.

The requirements of a separation will set the limits on the completeness of extraction required. If removal from the aqueous phase must be 99.9 percent, then D must be 1000 or greater when equal volumes of the two phases are used, but D must be 10 000 or greater when the aqueous phase is 10 times the volume of the organic phase. These limits assume one equilibration only. If additional equilibration steps are employed, D could be proportionally smaller. When performed manually, the use of liquid-liquid extraction methods is limited to values of D greater than, or equal to, one.

26.1.3 Selectivity of an Extraction

An extraction for separation purposes is more involved than the singular examples discussed in the preceding section. Now two or more components may be distributed between the two phases. Considering the requirements for the complete separation of one component, namely D_1 must be 1000 or greater when the phase volumes are equal, then D_2 must be 0.001 or less if not more than 0.1 percent of the second component is to coextract.

Do not be misled into thinking that a second equilibration will remedy an incomplete extraction when an insufficient difference exists between the two distribution ratios.

EXAMPLE 26.2 Assume that $D_1 = 10$ and $D_2 = 0.1$, and that the phase volumes are equal. A single extraction will remove 90.9 percent of component 1, but also 9.1 percent of component 2. Now a second extraction of the residual aqueous phase will remove an additional 8.3 percent of component 1 (for a total of 99.2 percent removal) but also remove an additional 8.3 percent of component 2 (for a total of 17.4 percent removal).

What is desired—a maximum removal of component 1 or less than quantitative removal of component 1 but in as pure a state as possible?

EXAMPLE 26.3 To continue with the conditions enumerated in the preceding example, but desiring pure component 1, a back-extraction should be tried. Equilibrate the organic extracting phase in step 1 with an equal volume of fresh aqueous phase.

True, only 82.6 percent of component 1 remains in the organic phase (8.3 percent is back-extracted), but only 0.7 percent of component 2 remains as contaminant (the remainder having been back-extracted). A second back-extraction with a second fresh aqueous phase would lower the contamination from component 2 to only 0.06 percent while providing a 75.0 percent recovery of component 1.

If the distribution ratios of two components are close in value, it becomes necessary to resort to chemical parameters, such as pH or masking agents, to improve the extraction conditions for the desired component. An alternative method would be countercurrent distribution methods in which distribution, transfer, and recombination of various fractions are performed a sufficient number of times to achieve separation.

26.2 EXTRACTION OF FORMALLY NEUTRAL SPECIES

Ionic compounds would not be expected to extract into organic solvents from aqueous solutions. However, through addition or removal of a proton or a masking ion, an uncharged extractable species may be formed. Included in this category are the neutral metal chelate complexes.

An example is furnished by the anion of a carboxylic acid and the influence of pH upon the distribution ratio of the neutral carboxylic acid molecule between an organic solvent and water. Only the neutral molecule partitions between the contacting phases. The partition coefficient is given by

$$K_d = \frac{[\text{RCOOH}]_{\text{org}}}{[\text{RCOOH}]_{\text{aq}}} \tag{26.6}$$

In the aqueous phase, an acid-base equilibrium is involved:

$$K_{\text{HA}} = \frac{[\text{RCOOH}]}{[\text{RCOO}^-][\text{H}^+]} \tag{26.7}$$

where K_{HA}, the acid association constant, is the reciprocal of the dissociation constant K_a. The overall distribution ratio is expressed by

$$D = \frac{[\text{RCOOH}]_{\text{org}}}{([\text{RCOOH}] + [\text{RCOO}^-])_{\text{aq}}} \tag{26.8}$$

Combining Eqs. 26.6, 26.7, and 26.8

$$D = \frac{K_d}{1 + 1/K_{\text{HA}}[\text{H}^+]} \tag{26.9}$$

or

$$\frac{1}{D} = \frac{1}{K_d} + \frac{1}{K_d K_{\text{HA}}[\text{H}^+]} \tag{26.10}$$

In a plot of $1/D$ versus $1/[\text{H}^+]$, the intercept is $1/K_d$ and the slope is $1/K_d K_{\text{HA}}$.

Figure 26.1 is a plot of Eq. 26.9. Two regions are apparent. When $1/K_{HA}[H^+]$ is much less than 1, $\log D = \log K_d K_{HA} - pH$, and the plot is a line of unit slope. At low pH values the uncharged molecular species dominate in the aqueous phase. In dilute hydrochloric acid (between 5 and 10% HCl), carboxylic acids, phenols, and other weakly ionized acids will be extracted from an aqueous phase by an immiscible organic solvent such as chloroform or carbon tetrachloride. Conversely, dilute NaOH will extract acidic solutes from an immiscible organic solvent by converting the acidic solute to the corresponding sodium salt.

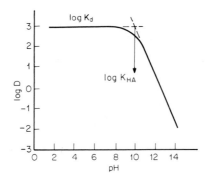

FIGURE 26.1 $\log D$ versus pH for a weak acid (RCOOH type) with $K_{HA} = 6.7 \times 10^9$ and $K_d = 720$.

Basic substances, such as organic amines, alkaloids, and cyclic nitrogen-containing ring compounds, follow just the reverse pattern from that of acidic materials. Dilute hydrochloric acid will convert basic substances into protonated species (positively charged) that will extract into the aqueous phase. Conversely, dilute NaOH will convert the protonated materials into neutral species that can be extracted from an aqueous phase into an immiscible organic phase. Control of the pH of the aqueous phase aids in separations. For example, phenols are not converted to the corresponding salt by $NaHCO_3$, whereas carboxylic acids are converted to the corresponding salt. An NaOH extraction will convert both phenols and carboxylic acids to the corresponding sodium salts and will extract both into the aqueous phase.

26.2.1 Association in the Organic Phase

Dimerization in the organic phase

$$2A_{org} \rightleftharpoons [(A)_2]_{org} \qquad (26.11)$$

increases the distribution ratio D

$$D = \frac{[A]_{org} + 2[(A_2)_{org}]}{[A]_{aq}} \qquad (26.12)$$

Incorporating the partition coefficient,

$$D = K_d(1 + 2K_2[A]_{org}) \tag{26.13}$$

where K_2 is the dimerization constant. Dimerization decreases the monomer concentration, the species which takes part directly in the phase partition, so that the overall distribution increases.

26.2.2   Metal Chelate Systems

To convert a metal ion in aqueous solution to an extractable species, the charge has to be neutralized and any waters of hydration have to be displaced. If the metal ions are effectively surrounded by hydrophobic ligands which are able to bring about charge neutralization and occupy all the positions in the coordination sphere of the metal ion, distribution will strongly favor an organic phase. Chelate systems offer one type of metal extraction system.

The equilibria involved in a metal chelate extraction system can be outlined in the following manner:

where M represents a metal ion of charge $n+$ and with m waters of hydration and HL represents a chelating agent. The chelating agent, usually a slightly dissociated, polyfunctional organic acid, distributes between the two phases. An additional acid-base dissociation system exists in the aqueous phase. Some of the free ligand ions in the aqueous phase react with the hydrated metal ions to form the extractable chelate ML_n which partitions between the two phases.

The expression for the distribution ratio of the metal between the two phases is

$$\frac{1}{D} = \frac{1}{(K_d)_c} + \frac{K_{HL}{}^n(K_d)_r{}^n[H^+]^n}{K_f(K_d)_c[HL]_{org}{}^n} \tag{26.14}$$

where $(K_d)_c$ = partition coefficient of metal chelate
K_{HL} = partition coefficient of chelating agent
K_f = formation constant of metal chelate in aqueous phase

The distribution ratio is dependent on the nature of the chelating reagent and its concentration, the pH of the aqueous phase, and the solubility of the chelating reagent in the organic solvent. The extent of extraction increases with the equilibrium (excess, unused) concentration of chelating reagent in the organic phase. Starting with a low pH, the logarithm of the distribution ratio increases with a slope of n, eventually reaching a constant, pH-independent value determined by the partition coefficient of the metal chelate. The distribution of zinc as a function of aqueous hydrogen-ion concentration for three different concentrations of chelating agent is shown in Fig. 26.2.

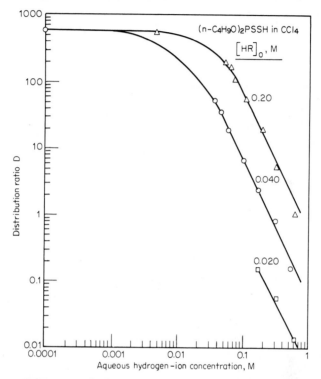

FIGURE 26.2   Distribution of zinc as a function of aqueous hydrogen ion concentration for three stated concentrations of chelating agent in CCl_4.   (*From work of T. H. Handley, R. H. Zucal, and J. A. Dean.*)

26.3   *SELECTIVE EXTRACTION*

Addition of a masking (complexing) agent moves the distribution curve to higher pH values, the shift being greater as the amount of masking agent is increased and as the magnitude of the formation constant of the masked metal increases. For example, copper(II) is extracted with 8-hydroxyquinoline in chloroform at a much lower pH than is uranium(IV); yet, if ethylenediaminetetraacetate (EDTA) is added to the system, the uranium extraction proceeds as before whereas the copper extraction is delayed until the pH is raised by more than five units.

Sometimes greater selectivity can be achieved by using a particular metal chelate, rather than the chelating reagent itself, as the ligand source. For example, copper(II) is selectively extracted in the presence of iron(III) and cobalt(II) using lead diethyldithiocarbamate, whereas all three metals would be coextracted with sodium diethyldithiocarbamate. The lead chelate is more stable (has a larger formation constant) than iron and cobalt chelates but less stable than the copper chelate.

26.4   ION ASSOCIATION SYSTEMS

26.4.1   Simple Ion Association

The simplest category of ion association systems involves large and bulky cations and anions whose size and structure are such that they do not have a primary hydration shell. Relatively simple, stoichiometric equilibria are sufficient to describe these ion association systems. There is dissociation of the ion pairs in the aqueous phase and association in the organic phase. An example involves the extraction of rhenium as the perrhenate(VII) anion from an aqueous phase with tetraphenylarsonium chloride contained in the organic phase as an ion pair (ion pair formation is indicated by a comma placed between the cation and anion which are in association with each other).

$$[(C_6H_5)_4As^+, Cl^-]_{org} + ReO_4^- \rightleftharpoons [(C_6H_5)_4As^+, ReO_4^-]_{org} + Cl^- \qquad (26.15)$$

26.5   CHELATION AND ION ASSOCIATION

Polyvalent cations can be extracted if the cation is made large and hydrocarbon-like. This can be accomplished with certain chelate reagents that have two uncharged coordinating atoms, such as 1,10-phenanthroline, which forms cationic complex chelates with metal ions, such as iron(II), iron(III), cobalt(II), and copper(II). In conjunction with a large anion such as perchlorate, tris(1,10-phenanthroline)iron(II) perchlorate extracts moderately well into chloroform and quite well into nitrobenzene. The extraction is improved when long-chain alkyl sulfonate ions replace the perchlorate ion.

Another general method for making small, polyvalent metal cations extractable is to transform them into large negatively charged chelate complexes. An example is the formation of an anionic complex of magnesium with excess 8-hydroxyquinoline ion (Ox) in a carbonate solution:

$$Mg^{2+} + 3Ox^- \rightleftharpoons MgOx_3^- \qquad (26.16)$$

and extraction with the large tetrabutylammonium cation:

$$[(C_4H_9)_4N^+, Cl^-]_{org} + MgOx_3^- \rightleftharpoons [(C_4H_9)_4N^+, MgOx_3^-]_{org} + Cl^- \qquad (26.17)$$

26.6   LIQUID ION EXCHANGERS

High-molecular-weight amines in acid solution form large cations capable of forming extractable ion pairs with a variety of anions. Tri(2-ethylhexyl)amine, methyldioctylamine, and tribenzylamine are useful and versatile in the extraction of mineral, organic, and complex metal acids. The anion attached to the organic ammonium cation is exchanged for other anionic species. This technique provides an ideal method for removing mineral acids from biologic preparations under mild conditions.

The removal of complex metal acids is illustrated by the extraction of zinc as $ZnCl_4^{2-}$ from an HCl solution using a benzene solution of tribenzylamine R_3N:

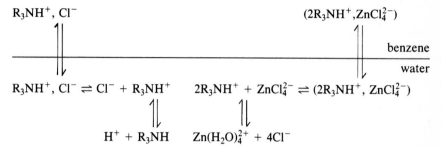

$$R_3NH^+, Cl^- \qquad\qquad (2R_3NH^+, ZnCl_4^{2-})$$

benzene

water

$$R_3NH^+, Cl^- \rightleftharpoons Cl^- + R_3NH^+ \qquad 2R_3NH^+ + ZnCl_4^{2-} \rightleftharpoons (2R_3NH^+, ZnCl_4^{2-})$$

$$H^+ + R_3NH \qquad Zn(H_2O)_4^{2+} + 4Cl^-$$

Obviously there are a number of interlocking equilibria. Sufficient chloride ion must be present to form the tetrachlorozincate anion, and hydrogen ion is needed to form the tributylammonium cation.

26.6.1 "Onium" Systems

Solvent molecules may participate in ionic reactions. Recognition of the hydrated hydronium ion $H_9O_4^+$ as the cation which pairs with halometallic anions in the extraction of complex halometallic acids, has clarified the role of the oxygen-containing solvent in oxonium extraction systems. This cation must be stabilized by hydrogen bonding to the organic solvent (denoted S). Thus, the coordinating ability of the solvent is important. The order of effectiveness is, roughly: isobutyl methyl ketone > butyl acetate > pentyl alcohol > diethyl ether. Solvents such as benzene and carbon tetrachloride are ineffective. The well known extraction of iron(III) from HCl solution is outlined with isobutyl methyl ketone as solvent:

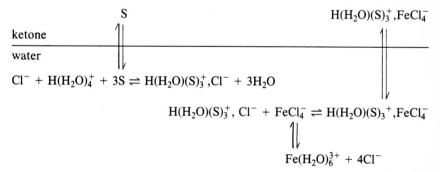

$$S \qquad\qquad\qquad H(H_2O)(S)_3^+, FeCl_4^-$$

ketone

water

$$Cl^- + H(H_2O)_4^+ + 3S \rightleftharpoons H(H_2O)(S)_3^+, Cl^- + 3H_2O$$

$$H(H_2O)(S)_3^+, Cl^- + FeCl_4^- \rightleftharpoons H(H_2O)(S)_3^+, FeCl_4^-$$

$$Fe(H_2O)_6^{3+} + 4Cl^-$$

Extraction of iron(III) is complete in one equilibrium from 6 M HCl when methyl isobutyl ketone is used as solvent in place of diethyl ether or other ethers. Ethers suffer from the formation of peroxides which at best are a nuisance and at worst may cause explosions.

The iron is easily stripped from the organic phase by washing it with a 0.1 M

HCl solution. The back-extraction is expedited by adding either a reducing agent [iron(II) is not extracted from HCl solutions] or 1 M phosphoric acid to the aqueous phase.

26.7 CONTINUOUS LIQUID-LIQUID EXTRACTIONS

When many extractions would be needed to remove a solute with a small distribution ratio, continuous liquid-liquid extractions may be used. However, to be feasible, the impurities must not be extractable and the solute must be stable at the temperature of the boiling solvent. Two approaches are described.

26.7.1 Higher-Density Solvent Extraction

In this method the extracting solvent has a higher density than the immiscible solution being extracted. The condensate from the total reflux of the extracting

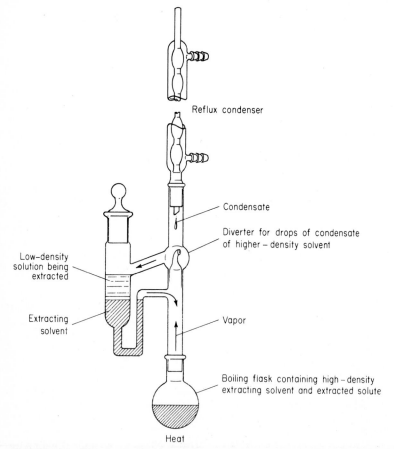

FIGURE 26.3   High-density liquid extractor.

heavier solvent is diverted through the solution to be extracted; passes through that solution, thus extracting the solute; and siphons back into the boiling flask (Fig. 26.3). Continuous heating vaporizes the higher-density solvent, and the process is continued as long as is necessary.

26.7.2 Lower-Density Solvent Extraction

The extracting solvent has a lower density than the immiscible solution being extracted. The condensate from the total reflux of the extracting lower-density solvent is caught in a tube (Fig. 26.4). As the tube fills, the increased pressure forces some of the lower-density solvent out through the bottom. It rises through the higher-density solvent, extracting the solute, and flows back to the boiling flask. Continuous heating vaporizes the low-density solvent, and the process is continued as long as is necessary.

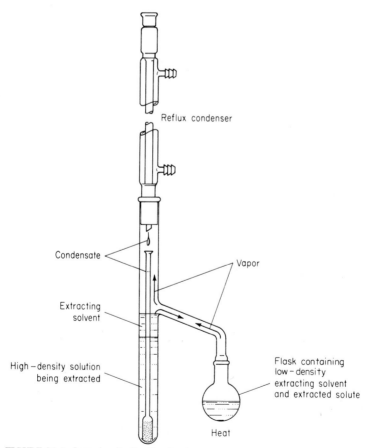

Reflux condenser

Condensate

Vapor

Extracting solvent

High-density solution being extracted

Flask containing low-density extracting solvent and extracted solute

Heat

FIGURE 26.4 Low-density liquid extractor.

26.8 EXTRACTION OF SOLIDS

A Soxhlet extractor (Fig. 26.5) can be used to extract solutes from solids, using any desired volatile solvent, which can be water-miscible or water-immiscible. The solvent is vaporized. When it condenses, it drops on the solid substance contained in a thimble and extracts soluble compounds. When the liquid level fills the body of the extractor, it automatically siphons into the flask. This process continues repeatedly as the solvent in the flask is vaporized and condensed.

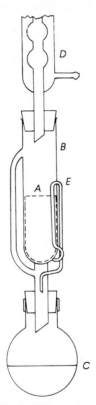

FIGURE 26.5 Soxhlet extractor. A = extraction thimble; B = body of extractor; C = extracting solvent; D = reflux condenser; and E = liquid return siphon.

Set up the apparatus, then proceed as follows:

1. Put the solid substance in the porous thimble, and place the partially filled thimble in the Soxhlet inner tube.
2. Fill the flask one-half full of extracting solvent.
3. Assemble the unit and turn on the cooling water.
4. Heat, using a heating mantle.

5. When the extraction is complete, pour the extraction solvent containing the solute into a beaker. Isolate the extracted components by evaporation or distillation of the solvent, provided the solute is nonvolatile and thermally stable. See Chap. 25 for additional methods applicable to solvent removal.

Soxhlet extractors are standard equipment in laboratories that analyze fats and oils in biologic samples. Separations can be achieved at low temperatures in inert atmospheres on a micro or macro scale by a discontinuous or continuous process.

26.9 LABORATORY WORK

The usual apparatus for batch extractions is a pear-shaped separatory funnel of the appropriate volume. The solute is extracted from one layer by shaking with a second immiscible phase in a separatory funnel (Fig. 26.6) until equilibrium has been attained (either from literature directions or by running a series of equilibrations at several time intervals to ascertain the time required to reach equilibrium). After equilibration the two layers are allowed to separate completely, and the layer containing the desired component is removed.

Simple inversions of the separatory funnel, repeated perhaps 50 times during 1 to 2 min, suffice to attain equilibrium in most extractions. After each of the first several inversions, the stopcock should be momentarily vented. Both the distribution ratio and the phase volumes are influenced by temperature changes.

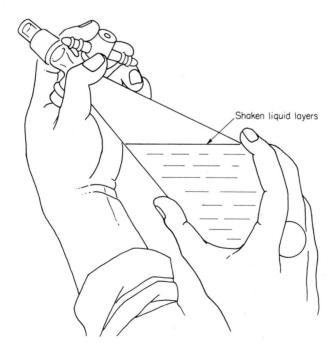

Shaken liquid layers

FIGURE 26.6 How to hold, swirl, and vent a separatory funnel.

Droplets of aqueous phase entrained in the organic phase can be removed by centrifugation or by filtering through a plug of glass wool. Whatman phase-separating papers, available in grade 1PS, are silicone-treated papers which serve as a combined filter and separator. The organic phase passes through the silicone-treated paper. The water phase will be held back at all times in the presence of an immiscible organic solvent. These papers function best when used flat on a Buchner- or Hirsch-type funnel.

The step-by-step procedure for extractions with a separatory funnel follow:

1. Use a clean separatory funnel, lubricating the barrel and plug of the funnel with a suitable lubricant.

2. Pour the solution to be extracted into the funnel, which should be large enough to hold at least twice the total volume of the solution and the extraction solvent.

3. Pour in the extraction solvent; close with the stopper.

4. Invert the funnel and swirl gently. Repeat these operations a number of times. Do not shake vigorously as emulsion formation should be avoided.

5. While the funnel is inverted, open the stopcock slowly to relieve the pressure that has built up. The disappearance of an audible "whoosh" of the escaping vapors indicates the cessation of pressure buildup.

6. Repeat step 4.

7. Place the funnel in a ring-stand support and allow the two layers of liquid to separate (Fig. 26.7). Remove the stopper closure.

8. Open the stopcock slowly and drain off the bottom layer.

26.9.1 Breaking Emulsions

Emulsions are more easily produced when the densities of the two solvents are similar. It is also desirable to select solvents that have a high interfacial tension for rapid separation of phases.

Suggestions for breaking emulsions follow.

1. Emulsions caused by too small a difference in the densities of the water and organic layer can be broken by the addition of a high-density organic solvent such as carbon tetrachloride.

2. Pentane can be added to reduce the density of the organic layer, if so desired, especially when the aqueous layer has a high density because of dissolved salts.

3. Sometimes troublesome emulsions can be broken by introducing a strong electrolyte. Saturated NaCl or Na_2SO_4 salt solutions will increase the density of aqueous layers.

4. Emulsions formed with ethereal solutions can be broken with the addition of small quantities of ethanol or 2-propanol.

5. Add a few drops of silicone defoamer.

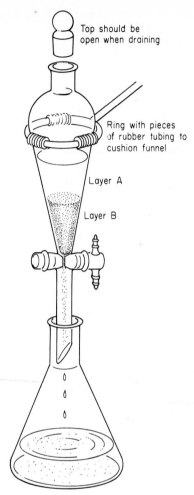

Top should be
open when draining

Ring with pieces
of rubber tubing to
cushion funnel

Layer A

Layer B

FIGURE 26.7 Separating immiscible liquids
by means of a separatory funnel.

26.9.2 Disposable Phase Separators

A separator, manufactured by Whatman, is a high-speed, disposable medium providing complete separation of immiscible aqueous solutions from organic solvents. The separator is low in cost so it can be discarded after use.

The phase separator can be used with solvents that are either lighter or heavier than water. If heavier than water, the solvent passes directly through the apex of the filter cone; if lighter than water, the solvent passes through the walls of the cone. The phase separator is unaffected by mineral acids to 4 M, and it will tolerate alkalies to 0.4 M.

Successful separations have been achieved at 90°C. However, surface tension is inversely proportional to temperature, so temperature limits for a given separation must be experimentally determined.

Proceed as follows:

1. Fold the phase-separator paper in the normal way, and place it in a conical filter funnel.
2. Pour the mixed phases directly into the funnel. It is not normally necessary to allow the phases to separate cleanly before pouring.
3. Allow the organic phase to filter completely through the paper.
4. If required, wash the retained aqueous phase with a small volume of clean organic solvent. This is normally necessary when separating lighter-than-water solvents in order to clean the meniscus of the aqueous phase.
5. Do not allow the aqueous phase to remain in the funnel after phase separation is complete, or it will begin to seep through.

BIBLIOGRAPHY

Lo, T. C., H. H. I. Baird, and C. Hanson, eds., *Handbook of Solvent Extraction*, Wiley-Interscience, New York, 1983.

Morrison, G. H., and H. Freiser, *Solvent Extraction in Analytical Chemistry*, Wiley, New York, 1957.

Stary, J., *Metal Chelate Solvent Extraction*, Pergamon, Oxford, 1965.

CHAPTER 27
DISTILLATION METHODS

Distillation is one of the principal methods used in the isolation, purification, and identification of volatile compounds. The process is sometimes suprisingly simple and usually gives a very complete separation. The only parameters which need to be varied are temperature and pressure. Separations may often be accomplished by a distillation of the constituent being determined. Occasionally the technique is used to remove a material whose presence interferes with another determination.

27.1 DISTILLATION AND VAPOR PRESSURE

Distillation is a process in which the liquid is vaporized, condensed, and collected in a receiver. The liquid which is not vaporized is called the residue. The resultant liquid, the condensed vapor, is called the *condensate* or *distillate*.

Distillation is used to purify liquids and to separate one liquid from another. It is based on the difference in the physical property of liquids called volatility. Volatility is a general term used to describe the relative ease with which the molecules escape from the surface of a pure liquid or a pure solid. The vapor pressure of a substance at a given temperature expresses this property.

A volatile substance is one which exerts a relatively high vapor pressure at room temperature. A nonvolatile substance is one which exerts a low vapor pressure. The more volatile a substance, the higher its vapor pressure and the lower its boiling point. The less volatile a substance, the lower its vapor pressure and the higher its boiling point.

All liquids and solids have a tendency to vaporize at all temperatures, and this tendency varies with temperature and the external pressure which is applied. When a solvent is enclosed, vaporization will take place until the partial pressure of the vapor above the liquid has reached the vapor pressure at that temperature. Further evaporation of the liquid can be accomplished by removing some of the vapor above it, which in turn reduces the vapor pressure over the liquid.

27.2 SIMPLE BATCH DISTILLATION

In simple batch distillation a batch of material is charged to a distillation flask, boiling is initiated, and the vapors are then continuously removed, condensed, and collected until their average composition has reached a desired value. An experimental setup for simple batch distillation is shown in Fig. 27.1. The glass equipment preferably should have ground-glass fitted joints. To be sure your setup is correct, follow the checklist below.

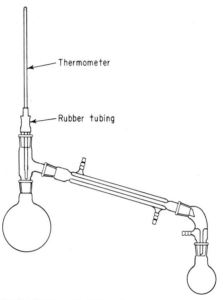

FIGURE 27.1   Apparatus for simple batch distillation at atmospheric pressure.

1. The distilling flask should accommodate twice the volume of the liquid to be distilled.
2. The thermometer bulb should be slightly below the sidearm opening of the flask. The boiling point of the corresponding distillate is normally accepted as the temperature of the vapor. If the thermometer is not positioned correctly, the temperature reading will be inaccurate. If the entire bulb of the thermometer is placed too high, above the side arm leading to the condenser, the entire bulb will not be heated by the vapor of the distillate and the temperature reading will be too low. If the bulb is placed too low, too near the surface of the boiling liquid, superheating may occur, and the thermometer reading will be too high.
3. All connections should be firm and tight.
4. The flask, condenser, and receiver should be clamped independently in their proper relative positions on a sturdy base.
5. The upper outlet for the cooling water exiting from the condenser should point upward to keep the condenser full of water.

Proceed as follows:

1. Pour the liquid into the distilling flask with a funnel which extends below the side arm.

2. Add a few boiling stones to prevent bumping.

3. Insert the thermometer.

4. Open the water valve for condenser cooling.

5. Heat the distilling flask until boiling begins. Adjust the heat input so that the rate of distillate is a steady 2 to 3 drops per second.

6. Collect the distillate in the receiver.

7. Continue distillation until only a small residue remains. Do not distill to dryness.

8. The distillate can be protected from atmospheric moisture by a drying tube.

27.3 INORGANIC APPLICATIONS

Distillation from solutions involves compounds in which covalent bonds prevail. Volatile inorganic substances form typical molecular lattices in which there exists an intimate association of a small number of atoms. This association is preserved on volatilization. Only small cohesive forces exist between individual molecules. Forces holding the individual atoms together within the molecule are much stronger. Thus, application of small amounts of energy to these systems easily disrupts the weak molecular bonds between these molecules and those of the solvent. Examples include the non-saltlike hydrides (H_2O, H_2S, NH_3), non-saltlike halides ($GeCl_4$, $AsCl_3$, $SeOCl_2$), a few oxides (CO_2, OsO_4), some of the metal carbonyls, and a few special cases among the carbon compounds.

27.3.1 Boron

As little as 0.1 $\mu g \cdot mL^{-1}$ of boron in aqueous solutions can be separated by distillation as the ester methyl borate $B(OCH_3)_3$ (bp 68.5°C) from methanol solutions at 75 to 80°C in all-silica glassware.

27.3.2 Fluoride

Fluoride ion is easily removed as HF by evaporation with sulfuric or perchloric acid. Separation is accomplished by distillation as H_2SiF_6. The sample is placed in a distilling flask with glass beads, sulfuric or perchloric acid is added, and the mixture steam distilled at 135 to 140°C.

27.3.3 Ge, As, Sb, and Sn

As their chlorides, the elements germanium, arsenic, antimony, and tin may be easily separated from other elements and from each other. $GeCl_4$ may be distilled from a 3 to 4 M HCl solution in an atmosphere of chlorine. Arsenic will be kept in the pentavalent state, which is not volatile. Tin is kept in solution as the stable hexachlorostannate(IV)(2−) ion.

After adding hydrazine as reductant, arsenic can be distilled quantitatively as $AsCl_3$ from 6 M HCl at 110 to 112°C. It may be carried over in a stream of carbon dioxide or by adding HCl from a dropping funnel into the distilling flask. After arsenic has distilled, phosphoric acid is added to the distilling flask to combine with the tin (and raise the boiling point) and the antimony is distilled as $SbCl_3$ at 160°C. Finally, at 140°C (in the presence of bismuth; otherwise 165°C) tin is distilled over as $SnBr_4$ by dropping in a mixture of HCl and HBr (3:1).

Simultaneous removal of arsenic, antimony, and tin is readily accomplished by dropping a mixture of HCl and HBr (3:1) into the sample that contains a reducing agent such as hydrazine.

27.3.4 Chromium

Chromium can be removed as chromyl(VI) chloride, CrO_2Cl_2, by evaporating the sample to fumes with perchloric acid, which converts the chromium to the hexavalent state; concentrated hydrochloric acid is then dropped into the boiling solution.

27.3.5 Osmium

Osmium is easily separated from the other platinum metals. The sample is treated in the distilling flask with 8 M HNO_3 or boiling $HClO_4$, and a slow stream of air passed through the boiling solution carries away the OsO_4 (bp 129°C). Absorb the osmium tetroxide in 6 M HCl saturated with sulfur dioxide to immediately reduce osmium to the nonvolatile lower valence states.

27.3.6 Sulfur

Where it is necessary to determine sulfur in the sulfide form, as distinct from total sulfur, the sample is treated with hydrochloric acid and the liberated hydrogen sulfide gas is absorbed in ammoniacal zinc sulfate solution.

27.3.7 Volatile Hydrides

Chemical vaporization has been applied to a series of elements (As, Bi, Ge, Sb, Se, Sn, and Te) that form volatile hydrides, and to the analysis of mercury. Gaseous hydrides may be generated with sodium borohydride, dispensed in pellet form, as the reducing agent added to an acid solution. A flow of inert gas transports the metallic hydride from the generation unit. The hydride-generating system finds considerable use as an adjunct to flame atomic absorption analyses (Chap. 10).

27.4 DISTILLATION OF A SOLUTION

This process effects the separation of nonvolatile dissolved solids because they remain in the residue and the volatile liquid is distilled, condensed, and collected.

1. The temperature of the distillate is constant throughout because it is pure.

2. The temperature of the boiling solution increases gradually throughout the distillation because the boiling solution becomes saturated with the nonvolatile solids.

3. When evaporating a solution to recover the solute or when using electric heat or burners to distill off large volumes of solvent to recover the solute, do not evaporate completely to dryness. The residue may be superheated and begin to decompose.

When it becomes necessary to distill off large volumes of solvent to recover very small quantities of the solute, it is advisable to use a large distilling flask at first. (Never fill any distilling flask more than half full.) When the volume has decreased, transfer the material to a smaller flask and continue the distillation. This minimizes losses caused by the large surface area of large flasks.

27.5 DISTILLATION OF A MIXTURE OF TWO LIQUIDS

Simple distillation of a mixture of two liquids will not effect a complete separation. If both are volatile, both will vaporize when the solution boils and both will appear in the condensate. The more volatile of the two liquids will vaporize and escape more rapidly and will form a larger proportion of the distillate initially. The less volatile constituent will concentrate in the liquid which remains in the distilling flask, and the temperature of the boiling liquid will rise.

In batch rectification the equipment consists of the distillation flask, a column (packed or plate), and a condenser. The system is first brought to steady state under total reflux, after which an overhead product is continuously withdrawn. The entire column thus operates as an enriching section. When the difference in the volatilities of the two liquids is large enough, the first distillate may be almost pure. As time proceeds, the composition of the material being distilled becomes less rich in the more volatile component, and the distillation must be stopped after a certain time to attain a desired average composition. The distillate collected between the first portion and the last portion will contain varying amounts of the two liquids (Fig. 27.2). The lower curve of the diagram gives the boiling points of all mixtures of these compounds. The upper curve gives the composition of the vapor in equilibrium with the boiling liquid phase.

27.6 AZEOTROPIC DISTILLATION

Azeotropic mixtures distill at constant temperature without change in composition. Obviously, one cannot separate azeotropic mixtures by normal distillation methods.

Azeotropic solutions are nonideal solutions. Some display a greater vapor pressure than expected; they are said to exhibit positive deviation. Within a certain composition range such mixtures boil at temperatures higher than the boiling temperature of either component; they are maximum-boiling azeotropes (see Table 27.1 and Fig. 27.3).

Mixtures which have boiling temperatures much lower than the boiling tem-

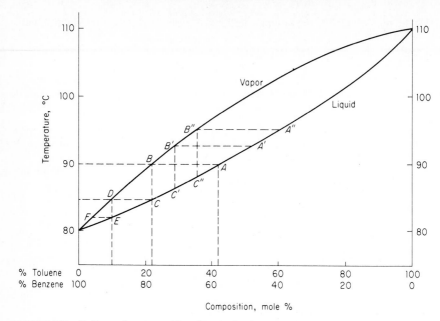

FIGURE 27.2   Boiling-point-composition diagram for the system benzene-toluene.

TABLE 27.1   Maximum-Boiling-Point Azeotropic Mixtures

Component A		Component B		BP of azeotropic mixture, °C	% B (by mass) in mixture
Substance	BP, °C	Substance	BP, °C		
Water	100.0	Formic acid	100.8	107.1	77.5
Water	100.0	HCl	19.4	120.0	37.0
Water	100.0	Nitric acid	86.0	120.5	68.0
Acetone	56.4	Chloroform	61.2	64.7	80.0
Acetic acid	118.5	Pyridine	115.5	130.7	65.0
Phenol	181.5	Aniline	184.4	186.2	58.0

perature of either component exhibit negative deviation; when such mixtures have a particular composition range, they act as though a third component were present. In Fig. 27.4 the minimum boiling point at Z is a constant boiling point because the vapor is in equilibrium with the liquid and has the same composition as the liquid. For example, pure ethanol cannot be obtained by fractional distillation of aqueous solutions which contain less than 95.57% ethanol because this is the azeotropic composition; the boiling point of this azeotropic mixture is 0.3°C lower than that of pure ethanol. Other examples are given in Table 27.2.

27.6.1   Use of an Entrainer

In azeotropic distillation a third solvent, called an entrainer, is added to the original azeotropic mixture, which will form a constant-boiling mixture with one of

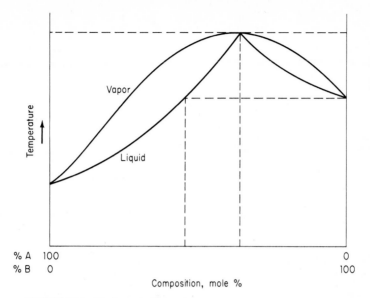

FIGURE 27.3 Maximum-boiling-point azeotrope.

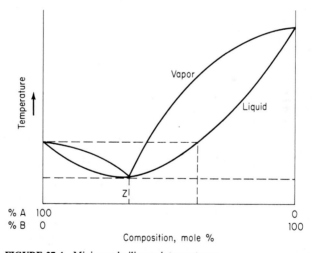

FIGURE 27.4 Minimum-boiling-point azeotrope.

the two components of the original mixture. Thus, a shift in the boiling point is often large enough to effect a further separation of close-boiling materials. The quantitative removal of one component of the original mixture is therefore dependent upon the addition of an excess of the entrainer. The second requirement is that the entrainer be easily removed by simple distillation.

Absolute ethanol can be obtained by distilling the azeotropic ethanol-water mixture with benzene. A new, lower-boiling (65°C) ternary azeotrope is formed

Table 27.2 Minimum-Boiling-Point Azeotropic Mixtures

Component A		Component B		BP of azeotropic mixture, °C	% A (by mass) in mixture
Substance	BP, °C	Substance	BP, °C		
Water	100.0	Ethanol	78.3	78.15	4.4
Water	100.0	2-Propanol	82.4	80.4	12.1
Water	100.0	Pyridine	115.5	92.6	43.0
Ethanol	78.3	Ethyl acetate	77.2	71.8	31.0
Water	100.0	Butanoic acid	163.5	99.4	18.4
Benzene	80.2	Cyclohexane	80.8	77.5	55.0
Ethanol	78.3	Benzene	80.2	68.2	32.4
Methanol	64.7	Chloroform	61.2	53.5	12.5
Ethanol	78.3	2-Butanone	79.6	74.8	40.0
Acetic acid	118.5	Toluene	110.6	105.4	28.0

(74% benzene, 7.5% water, and 18.5% ethanol). Distillation of the ternary azeotrope accomplishes the quantitative removal of water from the system but leaves the ethanol contaminated with benzene. However, the benzene-ethanol azeotrope (67.6% benzene and 32.4% ethanol) boils at 68.3°C and distills over by simple fractional distillation from the anhydrous ethanol which can be collected at 78.5°C.

Other examples are the use of acetic acid to separate ethylbenzene from vinylbenzene and butyl acetate for the dehydration of acetic acid. In fact, the judicious selection of entrainer liquids can facilitate the separation of complex mixtures.

27.7 FRACTIONAL DISTILLATION

The separation and purification of a mixture of two or more liquids, present in appreciable amounts, into various fractions by distillation is called *fractional distillation*. Essentially it consists of the systematic redistillation of distillates (fractions of increasing purity). Fractionations can be carried out with an ordinary distilling flask; but, where the components do not have widely separated boiling points, it is a very tedious process. A fractionating column of some type is needed (Fig. 27.5). The column is essentially an apparatus for performing a large number of successive distillations without the necessity of actually collecting and redistilling the various fractions.

The separation of mixtures by this means is a refinement of ordinary separation by distillation. Thus a series of distillations involving partial vaporization and condensation concentrates the more volatile component in the first fraction of the distillate and leaves the less volatile component in the last fraction or in the residual liquid. The vapor leaves the surface of the liquid and passes up through the packing of the column. There it condenses on the cooler surfaces and redistills many times from the heat of the rising vapors before entering the condenser. Each minute distillation causes a greater concentration of the more volatile liquid in the rising vapor and an enrichment of the residue which drips down through

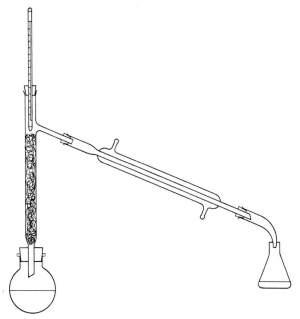

FIGURE 27.5   Fractional distillation apparatus for use under vacuum
or at atmospheric pressure. A packed column is shown.

the column. By means of long and efficient columns, two liquids may be completely separated.

Sometimes a fractionating column must be heated in order to achieve the most efficient fractionation of distillates. This may be accomplished by wrapping the column with heating tape.

27.7.1   Efficiency of Fractionating Columns

The enrichment of finite samples can be achieved by using a distillation column so designed that excellent contact is made between vapor rising through the column and condensate falling through the column. The efficiency of such columns is usually expressed in terms of the number of theoretical plates in the column. A *theoretical plate* is that length of column required to give the same change in the composition of liquid as that brought about by one equilibrium stage in the temperature-composition diagram. The more of these plates (or stages) there are in a column, the higher the efficiency or separating power. Theoretical plate numbers are useful in predicting the number of liquid-vapor equilibrations necessary to separate a given pair of components.

The number of theoretical plates (n) can be determined from one form of the Fenske equation.

$$n - 1 = \frac{\log\left[(X_a/X_b)(Y_b/Y_a)\right]}{\log \alpha} \tag{27.1}$$

where X_a = percent of low boiler in head
$\quad$ X_b = percent of low boiler in distillation flask
$\quad$ Y_a = percent of high boiler in head
$\quad$ Y_b = percent of high boiler in distillation flask
$\quad$ α = ratio of vapor pressures of two components

All that is necessary to make this determination is to distill a mixture of normal solvents whose relative volatility α is known and then analyze the distillate and residue by withdrawing small samples while the column is operating under total reflux (that is, no distillate taken off). Analysis of the samples by gas chromatography will provide the data needed. An approximate value of α can be calculated from the expression

$$\log \alpha = \frac{\Delta T}{85} \tag{27.2}$$

where ΔT is the difference in boiling points of the components. This approximation is valid for materials with boiling points near 100°C. Column efficiency is often evaluated experimentally with a mixture of hexane and methylcyclohexane. The theoretical plates required to obtain a 99.6:0.4 separation of a 30:70 starting mixture for various boiling point differences are shown in Table 27.3.

TABLE 27.3   Theoretical Plates Required for Separation in Terms of Boiling Point Difference and Alpha

30:70 v/v in flask and 99.6:0.4 v/v in distillate

ΔT, °C	α	Plates required
40	3.00	5
25	2.00	8
15	1.50	12
7	1.20	26
3.5	1.10	50
1.8	1.05	97
1.1	1.03	159
0.4	1.01	470

EXAMPLE 27.1   A mixture of 21.5% dibutyl phthalate and 78.5% dibutyl nonanedioate was distilled at a takeoff rate of 3 drops · min^{-1}. Analysis of an early takeoff gave a head concentration of 98.4% dibutyl phthalate and 1.6% of the other ester. The vapor pressure of dibutyl phthalate at 342°C is 760 mm and the vapor pressure of dibutyl nonanedioate at 342°C is 723.5 mm. Thus,

$$\alpha = \frac{760}{723.5} = 1.047$$

Using the preceding information in the Fenske equation:

$$n = 1 + \frac{\log \frac{98.4}{21.5} \left(\frac{78.5}{1.6}\right)}{\log 1.047} = 118$$

for a total of 118 plates.

As a general rule, to obtain maximum efficiency, the reflux ratio must closely approximate the number of plates to perform the separation. In the example, the reflux ratio should be at least 100:1 and preferably somewhat larger.

27.8 COLUMN DESIGNS

Distillation columns provide the contact area between streams of descending liquids and ascending vapor and thereby furnish an approach toward vapor-liquid equilibrium. Fractionation occurs in the packing. Depending upon the type of distillation which is used, the contact may occur in discrete steps, called plates or trays, or in a continuous differential contact on the surface of the packing. The column can be filled with a variety of packings (Fig. 27.6), or in some columns, the packings are built in, such as in the Vigreux column. Spinning band columns, to be discussed later, offer another option.

The following discussion centers around the several basic column designs usually employed in the laboratory. In the final selection of a column a compromise must be found which combines high separating power per unit length of column and high capacity with low holdup. The holdup should be less than 10 percent of

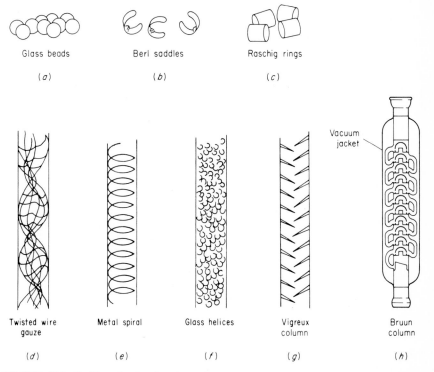

Glass beads   Berl saddles   Raschig rings

(*a*)   (*b*)   (*c*)

Twisted wire gauze   Metal spiral   Glass helices   Vigreux column   Bruun column   Vacuum jacket

(*d*)   (*e*)   (*f*)   (*g*)   (*h*)

FIGURE 27.6 Packings for fractionating columns. Some (*a, b, c*) are shown loose; others (*d, e, f*) are shown in place. In the (*g*) Vigreux and (*h*) Brunn columns the packings are built in.

the amount of material to be separated, but must not be of the same order of magnitude as the material to be resolved. There is not much point in using columns with more than 40 to 50 theoretical plates for normal laboratory work because of the unusually long time required for the column to come to equilibrium and the difficulty of operating such columns efficiently.

27.8.1 Plate Columns

Plate fractionating columns have a definite number of trays or plates and are fitted with either bubble caps or sieve perforations, or modifications of these two, which bring about intimate vapor-liquid contact. In the crossflow plate, the liquid flows across the plate and from plate to plate. In the counterflow plate, the liquid flows down through the same orifices as the rising vapor. There are three designs for counterflow plates: the sieve plate, the valve plate, and the bubble-cap plate (Fig. 27.7).

The sieve plate has contacting holes that measure 3.2 to 50 mm in diameter. Valve plates have valves that open or close to expose the holes for vapor passage. The latter arrangement will maintain efficient operation over a wider oper-

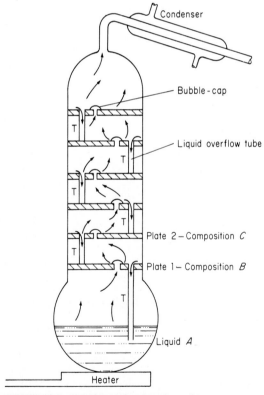

FIGURE 27.7 Bubble-cap fractionating column.

ating range. In the bubble-cap plate the vapor flows through a hole in the plate through a riser, reverses direction in the dome of the cap, flows downward in an annular area between the rise and the cap, and exits through slots in the cap.

A column 1 m in length containing 30 trays or plates can possess as many as 15 to 18 theoretical plates. Operating conditions with such a column are highly reproducible, but the holdup per plate is greater than that in a packed column.

27.8.2 Packed Columns

A practical method for increasing vapor-liquid contact inside the fractionating equipment is to fill the column with a material that offers a high surface area on which the condensed liquid can form a layer that the ascending vapors contact. Commonly, glass beads, broken glass, short lengths of tubing, saddle-shaped devices, or wire spirals are used as packing (see Fig. 27.6). The primary objective is the formation of an extremely tortuous vapor path to provide the maximum amount of rectification. Although high efficiency can be achieved in this way, the holdup is extremely large. Higher pot temperatures are required to overcome the pressure drop of the tightly packed column; this may be destructive to many compounds. It is possible to separate materials with boiling points as close as 0.5 to 1.0°C, but only at takeoff rates of one drop every 4 h.

Glass, single-turn helices, 4 mm in diameter, are perhaps the most generally useful material for packed columns. Glass helices have high throughput capacity and low liquid holdup. Because of the low pressure drop of these columns, they are well adapted for either atmospheric or vacuum fractionations.

For small diameter columns of 5 mm or less, the spiral packing is useful. One type consists of a precision-made Monel rod, 1.5 mm in diameter, which is wound with 2.2 turns per centimeter of the same metal. The center rod is removable for vacuum fractionation.

27.8.3 Spinning Band Systems

Spinning band systems have a motor-driven band which rotates inside the column (Fig. 27.8). This configuration combines efficiencies exceeding those of the best packed columns with the low holdup and low pressure drop of the most basic unpacked columns. Metal mesh columns are constructed of spirally wound wire mesh, permanently fastened to a central shaft, which is connected to a motor. As the tightly fitted band is rotated, the exposed wire at the side of the band "combs" through the layer of liquid on the walls, continuously forming hundreds of vapor-liquid interfaces in each small section of the column. The band is fashioned so that it is constantly pumping downward. Since the spinning band is constantly whirling through the ascending vapors, any tendency to form vapor channels is eliminated. Mesh spinning columns can be used to obtain complete separations of materials with boiling point differences of 5°C with relatively high throughputs.

Annular Teflon spinning bands provide superior performance as compared with mesh spinning bands. The tightly fitted Teflon spiral contacts the inner glass walls of the column, wiping the descending refluxed liquid into a thin film only several molecular layers deep. This provides the maximum amount of exposure of all liquid molecules to the heated vapor which is being hurled into it as a result of the spinning action of the band. Rotation speeds can be higher than with metal

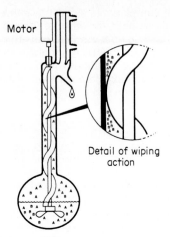

Motor

Detail of wiping
action

FIGURE 27.8 Teflon spinning band column. (*Courtesy of Perkin-Elmer Corporation.*)

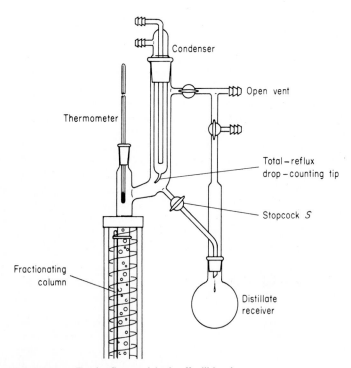

Condenser

Open vent

Thermometer

Total – reflux
drop – counting tip

Stopcock *S*

Fractionating
column

Distillate
receiver

FIGURE 27.9 Total-reflux partial-takeoff still head.

bands since the natural "slipperiness" of Teflon allows constant, high-speed rotation against the column walls without wear of the band or the glass wall.

Annular Teflon stills with a column length of 1 m provide a separating capability in excess of 200 theoretical plates. One example of a difficult separation is the separation of an equal percentage mixture of *m*-xylene and *p*-xylene, where boiling point difference is 1°C. With a boil-up rate of 60 drops $\cdot$ min^{-1} and a 100:1 reflux ratio, distillate can be collected at a rate of 1.7 mL $\cdot$ h^{-1}.

27.9   TOTAL-REFLUX PARTIAL-TAKEOFF HEADS

For truly effective fractionation, use a total-reflux partial-takeoff distilling head, as shown in Fig. 27.9. With the stopcock *S* completely closed, all condensed vapors are returned to the distilling column, a total reflux condition. With the stopcock partially opened, the number of drops of condensate falling from the condenser which return to the fractionating column can be adjusted. The ratio of the number of drops of reflux to the number of drops of distillate allowed to pass through stopcock *S* into the receiver is called the *reflux ratio*. With an efficient column, reflux ratios as high as 100:1 can be used to separate effectively compounds which have very close boiling points.

27.10   VACUUM DISTILLATION

Many substances cannot be distilled satisfactorily at atmospheric pressure because they are sensitive to heat and decompose before the boiling point is reached. Vacuum distillation—distillation under reduced pressure—makes it possible to distill at much lower temperatures. The boiling point of the material is affected by the pressure in the system. The lower the pressure, the lower the boiling point. For each twofold pressure reduction, the boiling point of most compounds is lowered by about 15°C. Thus, by lowering the normal 760-mm atmospheric pressure to about 3 mm, it is possible to subtract roughly 120°C from the boiling point of a mixture.

A setup for vacuum fractionation is shown in Fig. 27.10. It is necessary that all glass joints, stopcocks, and hose connections be thoroughly lubricated with a vacuum type lubricant. In addition, the system should be tested for leaks before the sample is placed in the distilling flask. When the system will hold a vacuum of less than 1 mm of mercury, the assembly is ready to use. For many purposes, a vacuum maintained at 10 mm of mercury is satisfactory for laboratory fractional distillation.

Vacuum fractionation is recommended only for those liquids which are affected by oxidation or are decomposed at their boiling points. One disadvantage of vacuum fractionation is that the throughput capacity of any column is considerably less than it is at atmospheric pressure. Consequently, much more time is required to fractionate at a desired reflux ratio. Of course, more precautions are necessary to maintain a vacuum-tight system at the desired reduced pressure.

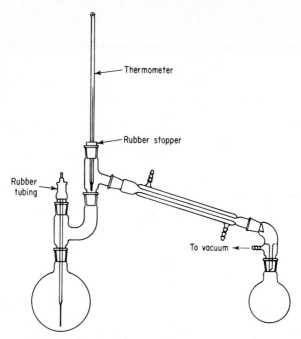

FIGURE 27.10   Vacuum distillation with gas-capillary bubbler.

27.11   STEAM DISTILLATION

Steam distillation is a simple distillation where vaporization of the charge is achieved by flowing live steam directly through it. The practice has special value as a means of separating and purifying heat-sensitive organic compounds because the distillation temperature is lowered. The organic compounds must be essentially immiscible with water. In these cases, both the compound and water will exert their full vapor pressure upon vaporization from the immiscible two-component liquid.

When steam is passed into a mixture of the compound and water, the compound will distill with the steam. In the distillate, this distilled compound separates from the condensed water because it is insoluble (or nearly so) in water. Both the temperature and the pressure of the distillation can be chosen at any desired value (except under a condition where liquid water forms in the distillation flask). If the steam is superheated and remains so during its travel through the liquid, then

$$\frac{p_s}{P} = \frac{L_s}{L_T} \tag{27.3}$$

where p_s = partial pressure of steam
P = system total pressure
L_s = moles of steam issuing from liquid
L_T = total moles of vapor generated from liquid

Both L_s and L_T may be independently varied by varying the rate of steam supplied and its degree of superheat, and, in some cases, by supplying heat from an external source.

Most compounds, regardless of their normal boiling point, will distill by steam distillation at a temperature below that of pure boiling water. For example, naphthalene is a solid with a boiling point of 218°C. It will distill with steam and boiling water at a temperature of less than 100°C.

Some high-boiling compounds decompose at their boiling point. Such substances can be successfully distilled at a lower temperature by steam distillation. Steam distillation can also be used to rid substances of contaminants because some water-insoluble substances are steam-volatile and others are not. Commercial applications of steam distillation include the distilling of turpentine and certain essential oils.

If a source of piped steam is not readily available, the steam can be generated in an external steam generator and then passed into the mixture to be steam-distilled.

27.12 MOLECULAR DISTILLATION

Many organic substances cannot be distilled by any of the ordinary distilling methods because (1) they are extremely viscous, so any condensed vapors would plug up the distilling column, side arm, or condenser or (2) their vapors are extremely heat sensitive or susceptible to condensation reactions. For the distillation of high-molecular-weight substances (molecular weights around 1300 for hydrocarbons and around 5000 for silicones and halocarbons), molecular stills are used.

In molecular distillation the vapor path is unobstructed and the condenser is separated from the evaporator by a distance less than the mean free path (the average distance traveled by a molecule between collisions) of the evaporating molecules. Thus, molecular distillation differs from other distillation in operating procedure (mainly the pressure, or vacuum, employed).

1. Distillation is conducted at low pressures, typically 0.001 mmHg or lower. At this low pressure the boiling point of high-molecular-weight high-boiling substances may be reduced as much as 200 to 300°C. It is the degree of thermal exposure that is so markedly less in molecular distillation.

2. The distance from the surface of the liquid being vaporized to the condenser is arranged to be less than the mean free path of the vapor at the operating pressure and temperature. For air at 25°C and 0.001 mmHg, the mean free path is 5.09 cm.

3. All condensed vapor flows to the distillate receiver with very few vapor molecules ever returning to the original material.

4. No equilibrium exists between the vapor and condensed phases. Ideal distillation conditions are attained when the rate of evaporation is equal to the rate of condensation.

A simple form of the apparatus (Fig. 27.11) is that in which a cooled condensing surface is supported a few centimeters (or even millimeters) above a thin, heated layer of sample and the whole enclosed in a highly evacuated chamber.

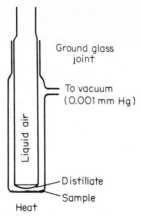

FIGURE 27.11   Schematic diagram of a simple molecular still.

For many organic materials the distance between condenser and evaporator surface is usually only several millimeters. The condenser is kept at liquid air or dry-ice temperatures to minimize the rebound of vapor molecules from the condenser surface.

A useful expression for the distilling rate W in grams per second per square meter of liquid surface is

$$W = \frac{0.0583PT^{1/2}}{M^{1/2}} \tag{27.4}$$

where P is the saturation vapor pressure in millimeters of mercury at the solution temperature T in kelvins, M is the molecular weight of the distilling substance, and the constant 0.0583 is the result of combining the molar gas constant and various conversion factors.

The degree of separation effected by molecular distillation is comparable to that produced by a simple batch distillation and is greatest when components differ in boiling point by 50° or more. Of course, when fractionation is not complete, a greater degree of separation can be attained by redistillation of the distillate.

Since no laboratory pump can achieve the reduced pressures required in one stage, two or more pumps are employed in series.

In a rotating still, materials are fed at slightly above their melting point into the still. They are distributed evenly and thinly over the heated evaporating surface. Since the fractionating power of the molecular still is restricted to the preferential evaporation of the most volatile constituents from the surface layer of liquid, it is imperative that the surface layer be continually replenished.

27.13   SUBLIMATION

Some solids can go from the solid to the vapor state without passing through the liquid state. This phenomenon is called sublimation. The vapor can be resolidified by reducing the temperature. The process can be used for purifying solids if the impurities in the solids have a much lower vapor pressure than that of the desired compound. A sublimation point is a constant property like a melting or boiling point. Examples of materials which sublime under ordinary conditions are iodine, naphthalene, and dry ice (solid carbon dioxide).

Many solids do not develop enough vapor pressure at atmospheric pressure to sublime, but they do develop enough vapor pressure to sublime at reduced pressure. For this reason, most sublimation equipment is constructed with fittings making it adaptable for vacuum connections.

27.13.1 Methods of Sublimation

There are two basic approaches to the practical sublimation of substances. (1) Evacuate the system while holding the temperature below the melting point of the substance. (2) For substances which sublime at atmospheric pressure, the system is heated but the temperature is held below the melting point.

A simple laboratory setup for conducting sublimation at atmospheric pressure is shown in Fig. 27.12. Gently heat the sublimable compound in a container which has a loosely fitting cover that is chilled with cold water or ice.

Apparatus based on the use of cold finger as a condenser is shown in Fig. 27.13. The coolant can be a mixture of ice and water.

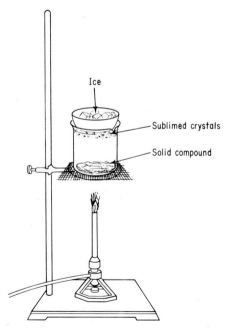

FIGURE 27.12   Simple laboratory setup for sublimation.

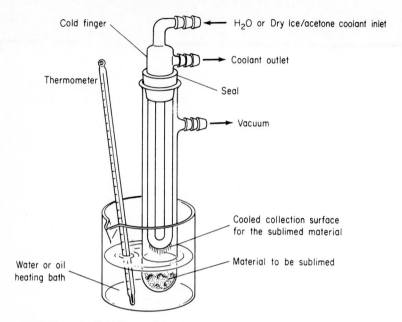

FIGURE 27.13   Cold-finger sublimation apparatus.

27.13.2   Advantages of Sublimation

The advantages of sublimation as a purification method are:

1. No solvent is used.
2. It is faster than crystallization. It may not be as selective if the vapor pressures of the sublimable solids are very close together.
3. More volatile impurities can be removed from the desired substance by subliming off the impurities.
4. Substances which contain occluded molecules of solvation can sublime to form the nonsolvated product, losing the water or other solvent during the sublimation process.
5. Nonvolatile or less volatile solids can be separated from the more volatile ones.

CHAPTER 28

PREPARATION OF SOLUTIONS

Solutions must be prepared for the various procedures performed in the laboratory. It is most important that directions be followed exactly and that all cautions be observed. Always recheck the label of the chemical that you are using. Use of the wrong chemical can ruin a determination.

28.1 GRADES OF PURITY OF CHEMICALS

Chemicals are manufactured in varying degrees of purity. Select the grade of chemical that meets the needs of the work to be done.

28.1.1 Technical Grade

This grade is used industrially, but it is generally unsuitable for laboratory reagents because of the presence of many impurities.

28.1.2 Practical Grade

This grade does contain impurities, but it is usually pure enough for most organic preparations. It may contain some of the intermediates resulting from its preparation.

28.1.3 USP

USP grade chemicals are pure enough to pass certain tests prescribed by the U.S. Pharmacopoeia (edition specified) and are acceptable for drug use, although there may be some impurities for which tests have not been made. This grade is generally acceptable for most laboratory purposes.

28.1.4 CP

CP stands for chemically pure. Chemicals of this grade are almost as pure as reagent-grade chemicals, but the intended use for the chemicals determines whether the purity is adequate for the purpose. The classification is an ambiguous one; read the label and use caution when substituting for reagent-grade chemicals.

28.1.5 Spectroscopic Grade

Solvents of special purity are required for spectrophotometry in the ultraviolet, infrared, or near-infrared regions, as well as in NMR spectrometry and fluorometry. Specifications of the highest order in terms of absorbance characteristics, water content, and evaporation residues are given. The principal requirement of a solvent for these procedures is that the background absorption be as low as possible. Most of these chemicals are accompanied by a label which states the minimum transmission at given wavelengths. Residual absorption within certain wavelengths is mainly due to the structure of the molecule.

28.1.6 Chromatography Grade

These chemicals have a minimum purity level of 99+ mol % as determined by gas chromatography. Each is accompanied by its own chromatogram indicating the column and parameters of the analysis. No individual impurity should exceed 0.2%.

28.1.7 Reagent Analyzed (Reagent Grade)

Reagent-grade chemicals are those which have been certified to contain impurities in concentrations below the specifications of the committee on Analytical Reagents of the American Chemical Society (ACS). Each bottle is identified by batch number. Use only reagent-analyzed chemicals in chemical analysis.

Be certain that the bottle has not been contaminated in previous use. Impurities may not have been tested for, and the manufacturer's analysis may possibly have been in error.

28.1.8 Primary Standard

Substances of this grade are sufficiently pure and of the stated molecular composition that they may serve as reference standards in analytical procedures. They may be used directly to prepare standard solutions by dissolving weighed amounts in solvents and then diluting them to known volumes.

Many compounds form hydrates; some form more than one hydrate. When using tables from reference handbooks or those accompanying this chapter, check carefully to confirm that the formula is the one sought and not another hydrate.

One example will serve to point out potential pitfalls. Potassium hydrogen phthalate is available in several grades. For the preparation of standard buffer solutions, the reagent-grade chemical is satisfactory. However, only the primary standard is suitable when one is standardizing a solution of sodium hydroxide. The primary standard chemical guarantees a 1:1:1 ratio of potassium:hydrogen:phthalate as well as minimal impurities.

Occasionally, a trace impurity in one manufacturer's batch will be detrimental for the intended analytical use. Try another manufacturer's reagent because the raw chemicals used in the preparation of the primary standard may not contain the same level of trace impurity.

28.2 CHEMICAL SAFETY PRACTICES

Anyone who works with hazardous chemicals, toxic substances, or enclosed liquid samples should take every precaution possible to avoid exposure to these agents. *Always think of the chemical laboratory as a hazardous environment in which you must maintain a high standard of vigilance.*

Consult Tables 28.1, 28.2, and 28.3 for a guide in storing chemicals that require refrigeration.

TABLE 28.1 Chemicals Recommended for Refrigerated Storage Because of Possible Chemical Decomposition or Polymerization

Acetaldehyde	Isoprene
Acrolein	Lecithin
Adenosine triphosphoric acid	Mercaptoacetic acid
Bromoacetaldehyde, diethyl acetal	Methyl acrylate
Bromosuccinimide	2-Methyl-1-butene
3-Buten-2-one	Methylenedi-*p*-phenylene
	diisocyanate
tert-Butyl hydroperoxide	4-Methyl-1-pentene
2-Chlorocyclohexanone	α-Methylstyrene
Cupferron	1-Naphthyl isocyanate
1,3-Cyclohexadiene	1-Pentene
1,3-Dihydroxy-2-propanone	Isopentyl acetate
Divinylbenzene	Pyruvic acid
Ethyl methacrylate, monomer	Styrene, stabilized
Glutathione	Tetramethylsilane
Glycidol	Thioacetamide
Histamine, base	Veratraldehyde
Hydrocinnamaldehyde	Vitamin E (and the acetate)

TABLE 28.2 Refrigerated Storage Recommended Because of Flammability and High Volatility

Acetaldehyde	Isoprene
Bromoethane	Isopropylamine
tert-Butylamine	Methylal
Carbon disulfide	2-Methylbutane
1-Chloropropane	2-Methyl-2-butene
3-Chloropropane	Methyl formate
Cyclopentane	Pentane
Diethyl ether	Propylamine
2,2-Dimethylbutane	Propylene oxide
Dimethyl sulfide	Trichlorosilane
Furan	
Iodomethane	

TABLE 28.3   Chemicals Which Polymerize or Decompose as a
Result of Freezing on Extended Storage at Low Temperature

Formaldehyde	Sodium methoxide
Hydrogen peroxide	Sodium nitrate
Sodium chlorite [sodium chlorate(IV)]	Sodium peroxide
Sodium chromate(VI)	Strontium nitrate
Sodium dithionite	Urea
Sodium ethoxide	

The following are some general rules on how to handle chemicals safely.

1. Store chemicals in a safe place. Reactive chemicals (Table 28.4) must be segregated from other materials. Furthermore, chemicals that are dangerous in combination should not be stored together; examples of incompatible chemicals are given in Table 28.5.

2. Clean up spills and refuse promptly. Never leave volatile, combustible, or dangerous liquids exposed on counters, benches, or other work surfaces.

3. Make certain that all chemical containers are properly labeled and classified and that especially hazardous materials are appropriately designated with clearly understood warnings.

4. Never taste or inhale unmarked chemicals.

5. Use special care when handling sealed-glass samples that are rapidly heated or cooled. The rapid heating of some samples may result in the formation of a solid bolus in the sample tube and cause the tube to explode.

6. Know the temperatures at which chemicals evaporate and their boiling points.

7. Dispose of the contents of all unlabeled containers via the so-called Labpack route where many individual bottles are packed with a suitable absorbing material in one steel drum. Unknown material must be identified, minimally, as to the chemical class. If possible, all disposed substances should be completely identified.

TABLE 28.4   Reactive Chemicals Which Must Be Segregated

Ammonium nitrate	Magnesium perchlorate
Ammonium perchlorate	Nitroethane
Benzoyl peroxide	Nitromethane
2-Butanone peroxide	2-Nitropropane
tert-Butyl hydroperoxide	2-Nitrotoluene
tert-Butyl peroxide	4-Nitrotoluene
Calcium hydride	Phosphorus, red
Cesium metal	Picric acid
Chromium(III) nitrate	Potassium metal
Chromium(VI) oxide (trioxide)	Potassium *tert*-butoxide
1,3-Dinitrobenzene	Potassium nitrate
Hydrazine	Potassium perchlorate
Hydrazine hydrate	Sodium metal (any form)
Hydrogen peroxide	Sodium azide
Lauroyl peroxide	Sodium chlorate

TABLE 28.5   Examples of Incompatible Chemicals

Chemical	Keep out of contact with
Acetic acid	Chromic acid, nitric acid, hydroxyl compounds, ethylene glycol, perchloric acid, peroxides, permanganates
Acetylene	Bromine, chlorine, copper, fluorine, mercury, silver
Alkali metals, aluminum, magnesium	Water, chlorinated hydrocarbons, carbon dioxide, halogens
Ammonia, anhydrous	Mercury, chlorine, calcium hypochlorite, bromine, iodine, hydrofluoric acid (anhydrous)
Ammonium nitrate	Acids, metal powders, flammable liquids, chlorates, nitrites, sulfur, finely divided organic or combustible materials
Aniline	Nitric acid, hydrogen peroxide
Bromine	Same as for chlorine
Carbon, activated	Calcium hypochlorite, all oxidizing agents
Chlorates	Ammonium salts, acids, metal powders, sulfur, finely divided organic or combustible materials
Chromic acid	Acetic acid, alcohols, camphor, glycerol, turpentine, flammable liquids in general
Chlorine	Ammonia, acetylene, alkanes, benzene, butadiene, hydrogen, sodium carbide, finely divided metals, turpentine
Chlorine dioxide	Ammonia, hydrogen sulfide, methane, phosphine
Copper	Acetylene, hydrogen peroxide
Cumene hydroperoxide	Acids (inorganic or organic)
Flammable liquids	Ammonium nitrate, chromic acid, the halogens, hydrogen peroxide, nitric acid
Fluorine	Isolate from everything
Hydrocarbons	Bromine, chlorine, fluorine, chromic acid, sodium peroxide
Hydrocyanic acid	Nitric acid, alkali metals
Hydrofluoric acid, anhydrous	Ammonia (aqueous or anhydrous)
Hydrogen peroxide	Copper, chromium, iron, most metals or their salts, alcohols, acetone, organic materials, flammable liquids, combustible materials
Hydrogen sulfide	Fuming nitric acid, oxidizing gases
Iodine	Acetylene, ammonia (aqueous or anhydrous), hydrogen
Mercury	Acetylene, fulminic acid, ammonia
Nitric acid (concentrated)	Acetic acid, aniline, chromic acid, hydrocyanic acid, hydrogen sulfide, flammable liquids, flammable gases
Oxalic acid	Silver, mercury
Perchloric acid	Acetic anhydride, bismuth and its alloys, alcohol, paper, wood, many organic materials
Potassium	Carbon dioxide, water, tetrachloromethane
Potassium chlorate	Sulfuric and other acids
Potassium perchlorate	Sulfuric and other acids
Potassium permanganate	Benzaldehyde, ethylene glycol, glycerol, sulfuric acid
Silver	Acetylene, ammonium compounds, oxalic acid, tartaric acid
Sodium	Same as for potassium
Sodium peroxide	Ethanol or methanol, glacial acetic acid, acetic anhydride, benzaldehyde, carbon disulfide, glycerol, ethylene glycol, ethyl acetate, methyl acetate, furfural
Sulfuric acid	Chlorates, perchlorates, permanganates

28.2.1 Waste Types Defined

28.2.1.1 Corrosive. A corrosive hazardous waste has a pH of 3 or less or a pH of 12 or more. A substance may also be classed as corrosive if it corrodes steel at a rate greater than 6.4 mm · year^{-1} at a temperature of 54°C.

28.2.1.2 Ignitable. An ignitable hazardous waste is defined as:

1. A solid liable to cause fire through friction, absorption of moisture, spontaneous chemical change, or situations encountered during its disposal.
2. A liquid with a flash point less than 60°C.
3. An ignitable compressed gas or an oxidizer as defined by the Federal Department of Transportation.

28.2.1.3 Radioactive. Waste designated as radioactive has either (1) a mean concentration of 5 picocurie per gram or more solid waste or 50 picocurie per liter for liquid waste.

28.2.1.4 Reactive. A reactive hazardous waste will possess one of these characteristics:

1. It is commonly unstable and readily undergoes violent chemical reaction without detonating.
2. It reacts violently with water or generates toxic vapors or fumes when mixed with water.
3. It is a cyanide- or sulfide-bearing waste that can generate toxic fumes when exposed to mildly acidic or alkaline conditions.
4. It is capable of detonating or exploding with strong initiating force at ordinary temperatures and pressures.
5. It is a forbidden explosive or Class A or B explosive as defined by the Federal Department of Transportation.

28.2.1.5 Toxic. A toxic hazardous waste is defined in terms of a prescribed extraction procedure. If the extract obtained from a representative sample of the waste exceeds the minimum allowable concentration of any of a series of inorganic or organic substances, then the waste is considered toxic. Additional classifications exist if the waste meets the definition of hazardous waste as found in the Federal Resource Conservation and Recovery Act.

A toxic organic waste contains any organic substance that has a calculated human lethal dose (LD_{50}) of less than 800 mg · kg^{-1}.

28.3 WATER FOR LABORATORY USE

Water is needed in the laboratory in various grades of purity depending on the procedures for which it is to be used. The specific conductance (see Chap. 20) is the characteristic used to specify the purity of water. Depending upon the particular processes used in its purification and the efficiency of those processes, average city water normally contains dissolved inorganic substances, microorganisms, dissolved organic compounds of vegetable and plant origin, and particulate matter. This tap water contains sufficiently high levels of these impurities so that it cannot be used for testing procedures and analytical evaluations. Water is also characterized as "hard" or "soft." Hard water contains appreciable quantities of calcium and magnesium ions plus lesser amounts of other minerals.

28.3.1 Softening Hard Water

Some waters are temporarily hard because they contain the hydrogen carbonates of calcium, magnesium, and/or iron. These hydrogen carbonates can be removed simply by heating, thus converting the soluble hydrogen carbonate to the insoluble carbonate, which can be filtered out.

When water contains sulfates or chlorides of calcium, magnesium, or iron, it is permanently hard water and is unsuitable for most laboratory work.

28.3.2 Ion-Exchange Resins

Soft water can be obtained by passing hard water through a bed of special ion-exchange resins. The "bed" may be a mixture of cation- and anion-exchange resins in the sodium and chloride form, respectively, or it may be a tandem arrangement of the cation and anion resin beds. In any case, the sodium ions will be exchanged for the calcium, magnesium, iron, and other multivalent cations.

$$Ca^{2+} + 2Na^+R^- \rightleftharpoons Ca^{2+}(R^-)_2 + 2Na^+ \qquad (28.1)$$

and the chloride exchanges with the sulfate and carbonate anions.

$$SO_4^{2-} + 2R_4N^+Cl^- \rightleftharpoons (R_4N^+)_2SO_4^{2-} + 2Cl^- \qquad (28.2)$$

In the equations, R represents an exchange site in the resin. The effluent from the ion-exchange resins will contain essentially sodium chloride. See Sec. 4.9 for further discussion of these resins.

For a complete removal of all ionic material in solution, the cation exchanger must be in the hydrogen form and the anion exchanger in the hydroxide form. Now the effluent will be free of ionic material. However, pathogens and nonelectrolytes will not be removed.

28.3.3 Reverse Osmosis

High-quality water of consistent purity can be obtained by the process of reverse osmosis whereby the dissolved solids are separated from feed water by applying a pressure differential across a semipermeable membrane. This membrane allows the water to flow through it but prevents dissolved ions, molecules, and solids from passing through. Rejection of salts is as high as 98 percent while the rejection of most organic molecules is 100 percent.

28.3.4 Distilled Water

Water purified by distillation is called distilled water. Electrically heated or steam-heated water stills distill raw water to give high-grade distilled water and vent volatile impurities. High-quality distilled water has a specific conductivity of less than 1 μS $\cdot$ cm^{-1}, which corresponds to about 0.5 mg $\cdot$ L^{-1} of a dissolved salt. Absolutely pure water has a conductivity of 0.55 μS $\cdot$ cm^{-1} at 25°C.

For the highest grade water, the distilling unit should be fabricated from quartz and the distillate collected in a tin-faced receptacle. Triple-distilled water prepared under these conditions is considered high-quality water.

28.3.5 Reagent-Grade Water

Reagent-grade water is water of the highest purity that is available from a practical standpoint. It is even purer than triple-distilled water. Reagent-grade water is obtained from pretreated water, that is, water which has been distilled, deionized, or subjected to reverse osmosis. It is then passed through an activated carbon cartridge to remove the dissolved organic materials, through two deionizing cartridges to remove any dissolved inorganic substances, and through membrane filters to remove microorganisms and any particulate matter with a diameter larger than 0.22 μm.

Bacteria may still be a problem. Organic-free water for HPLC and fluorescent spectrophotometry can be prepared and preserved for 30 days. An immersion-type ultraviolet unit kills bacteria present in the water and prevents bacteria from forming. In general, the organics associated with reagent-grade water is 100 to 300 μg $\cdot$ L^{-1}, expressed as total organic carbon; organic-pure water has less than 5 μg $\cdot$ L^{-1}.

28.3.6 Storage of Pure Water

After purification, water must be stored in containers that will not degrade the purity of the water from either metals leaching from the container or organic materials dissolving from plastics. Freezing offers the least chance for contaminants to enter the system.

Pure tin is recognized as the most practical material for storage of high purity distilled water because tin is chemically inert in the presence of pure water and will not contaminate the water in any way. High-quality water stills and storage tanks are always faced with pure tin inside, and it is logical to use the same safeguard through to the point of use (i.e., any piping).

The use of plastic materials must be considered carefully. The simplest plastic is the polyethylene family. It has a simple chain, with no plasticizers or additive atoms; therefore it is a good choice among the plastics. However, there is no way of identifying specific organic molecules or residual organic molecules that might leach into the water. These materials can be used in any application when organic contamination, below measurable levels, is not critical. The use of these materials in biomedical applications is not recommended.

Most other plastics, particularly the polyvinyl chlorides, use plasticizers or stabilizers which are usually soluble in pure water and are therefore likely to cause organic contamination.

Glass containers, particularly soft glass, will contaminate solutions stored in them through leaching of ions from the glass.

Bacteria may still be a problem. Organic-free water for HPLC and fluorescent spectrophotometry can be prepared and preserved for 30 days. An immersion-type ultraviolet unit kills bacteria present in the water and prevents bacteria from forming. In general, the organics associated with reagent-grade water are 100 to

$300 \ \mu g \cdot L^{-1}$, expressed as total organic carbon; organic-pure water has less than $5 \ \mu g \cdot L^{-1}$.

28.4 SOLUBILITY RULES FOR INORGANIC COMPOUNDS

Table 28.6 gives general solubility rules for inorganic compounds, and Table 28.7 shows some soluble complex ions.

TABLE 28.6   General Solubility Rules for Inorganic Compounds

Nitrates	All nitrates are soluble.
Acetates	All acetates are soluble; silver acetate is moderately soluble.
Chlorides	All chlorides are soluble except AgCl, $PbCl_2$, and Hg_2Cl_2. $PbCl_2$ is soluble in hot water, slightly soluble in cold water.
Sulfates	All sulfates are soluble except barium and lead. Silver, mercury(I), and calcium are slightly soluble.
Hydrogen sulfates	The hydrogen sulfates are more soluble than the sulfates.
Carbonates, phosphates, chromates, silicates	All carbonates, phosphates, chromates, and silicates are insoluble, except those of sodium, potassium, and ammonium. An exception is $MgCrO_4$ which is soluble.
Hydroxides	All hydroxides (except lithium, sodium, potassium, cesium, rubidium, and ammonium) are insoluble; $Ba(OH)_2$ is moderately soluble; $Ca(OH)_2$ and $Sr(OH)_2$ are slightly soluble.
Sulfides	All sulfides (except alkali metals, ammonium, magnesium, calcium, and barium) are insoluble. Aluminum and chromium sulfides are hydrolyzed and precipitate as hydroxides.
Sodium, potassium, ammonium	All sodium, potassium, and ammonium salts are soluble. Exceptions: $Na_4Sb_2O_7$, $K_2NaCo(NO_2)_6$, K_2PtCl_6, $(NH_4)_2PtCl_6$, and $(NH_4)_2NaCl(NO_2)_6$.
Silver	All silver salts are insoluble. Exceptions: $AgNO_3$ and $AgClO_4$; $AgC_2H_3O_2$ and Ag_2SO_4 are moderately soluble.

If a solution retains more than the equilibrium concentration of the dissolved solute, which happens under certain conditions, the solution is said to be supersaturated. *Supersaturation* is an unstable condition which can revert back to a stable one as a result of physical shock, decreased temperature, merely standing for a period of time, or some indeterminate factor.

Although a solvent will only dissolve a limited quantity of solute at a definite temperature, the rate at which the solute dissolves can be speeded up by the following methods:

TABLE 28.7 Soluble Complex Ions

Cation	NH_3	CNS^-	Cl^-	CN^-	OH^-
Aluminum					$Al(OH)_4^-$
Silver	$Ag(NH_3)_2^+$			$Ag(CN)_2^-$	
Cadmium	$Cd(NH_3)_4^{2+}$		$CdCl_4^{2-}$	$Cd(CN)_4^{2-}$	
Cobalt	$Co(NH_3)_6^{3+}$			$Co(CN)_4^{2-}$	
Copper	$Cu(NH_3)_4^{2+}$		$CuCl_4^{2-}$	$Cu(CN)_2^-$	
Iron(II)				$Fe(CN)_6^{4-}$	
Iron(III)		$Fe(CNS)^{2+}$	$FeCl_4^-$	$Fe(CN)_6^{3-}$	
Mercury(II)	$Hg(NH_3)_4^{2+}$	$Hg(CNS)_4^{2-}$	$HgCl_4^{2-}$	$Hg(CN)_4^{2-}$	
Nickel	$Ni(NH_3)_6^{2+}$			$Ni(CN)_4^{2-}$	
Lead			$PbCl_4^{2-}$		$Pb(OH)_3^-$
Zinc	$Zn(NH_3)_4^{2+}$			$Zn(CN)_4^{2-}$	$Zn(OH)_4^{2-}$

1. Pulverizing or grinding the solid increases the surface area of the solid in contact with the liquid.
2. Heating the solvent, except in a few instances, increases the rate of solution.
3. Stirring.

28.4.1 Colloidal Solutions

Colloidal dispersions are dispersions of particles which are hundreds and thousands of times larger than molecules or ions. Yet these particles are not large enough to settle rapidly out of solution. These particles, which make up the dispersed phase in colloidal systems, are intermediate in size between those of a true solution and those of a coarse suspension.

Colloidal dispersions, like true solutions, may be perfectly clear to the naked eye, but when they are examined at right angles to a beam of light, they will appear turbid. This phenomenon is caused by the scattering of light by the colloidal particles. The dispersed phase will pass through ordinary filter paper but not through semipermeable membranes.

28.5 LABORATORY SOLUTIONS

The solutions of chemical reagents used in the laboratory are prepared so that their concentrations or compositions are known and can be used in appropriate calculations. Chemical calculations and computations for the preparation of laboratory solutions follow.

28.5.1 Weight Percent

Weight percent implies grams of solute per 100 grams of solution. For example, 25 g of NaCl in 100 g of water is a 20% by weight solution.

$$\text{Weight percent} = \frac{\text{weight of solute}}{\text{weight of total solution}} \times 100$$

$$= \frac{25\text{gNaCl}}{25\text{gNaCl} + 100\text{gH}_2\text{O}} \times 100 = 20\%$$

To prepare 1000 g of a 10% NaCl solution by weight, proceed as follows:

$$10\%, \text{ or } \frac{1}{10} \times 1000 \text{ g} = 100 \text{ g NaCl needed}$$

1000 g of solution − 100 g NaCl = 900 g of water needed

Therefore, dissolve 100 g of NaCl in 900 g of water.

28.5.2 Volume Percent

Volume percent expresses the milliliters of solute per 100 mL of solution. For example, 10 mL of ethanol plus 90 mL of water is a 10% by volume solution of ethanol. This rule generally holds true, but the volume percent cannot always be calculated directly from the volumes of the components mixed because the final volume may not equal the sum of the separate volumes.

To prepare volume percent solutions, decide how much solution you want to make and of what strength (expressed as a fraction or decimal instead of a percent). For example, to make 1000 mL of a 5% by volume solution of ethylene glycol in water, take 50 mL (that is, 0.05 × 1000 mL) of ethylene glycol and dilute with water to 1000 mL.

28.5.3 Conversion of Weight Percent to Volume Percent

Using as an example a 10% by weight solution of ethanol in water, which contains 10 g of ethanol for each 90 g of water, the volume percent is determined as follows:

From the density of ethanol ($0.794 \text{ g} \cdot \text{mL}^{-1}$) and the density of the solution ($0.983 \text{ g} \cdot \text{mL}^{-1}$), both obtained from reference tables:

$$\text{Volume of ethanol} = \frac{10 \text{ g}}{0.794 \text{ g} \cdot \text{mL}^{-1}} = 12.6 \text{ mL}$$

$$\text{Volume of solution} = \frac{100 \text{ g}}{0.983 \text{ g} \cdot \text{mL}^{-1}} = 101.3 \text{ mL}$$

$$\text{Volume percent} = \frac{\text{volume of ethanol}}{\text{total volume of solution}} = \frac{12.6}{101.8} \times 100 = 12.4\%$$

Reverse this process to convert volume percent to weight percent.

28.5.4 Molal and Molar Solutions

Molality is the gram-molecular weight of solute per 1000 g of solvent. For example, a solution containing 58.44 g of NaCl (the molecular weight expressed in grams) per 1000 g of water is a 1 molal (1 m) solution.

Molarity is the gram-molecular weight of solute per liter of solution. If 58.44 g of NaCl is dissolved in sufficient water to prepare a solution whose volume is 1000 mL, it is a 1 molar (1 M) solution.

28.5.5   Equivalent Weight and Normal Solutions

The equivalent weight of an element or compound is that weight which in a given reaction has the total reactive power equal to that of one atomic weight of hydrogen (1.008 g).

1. For acids, the equivalent weight is the molecular weight divided by the number of hydrogen ions donated per molecular unit.
2. For bases, it is the molecular weight divided by the number of hydrogen ions accepted per molecular unit.
3. Since one electron will convert a hydrogen ion into a hydrogen atom, the equivalent weight for oxidation-reduction reactions will be the molecular weight divided by the number of electrons added to (or subtracted from) the reacting entity.
4. In precipitation and complexation reactions, the equivalent weight is that weight which reacts with the equivalent of that amount of hydrogen. In most cases the equivalent weight is found by dividing the formula weight by the net number of charges on the constituent actually taking part in the reaction.

The following examples will express the equivalent weight for each particular reaction.

Equivalent weight	Reaction	Comments
HCl/1	$HCl + NaOH = NaCl + H_2O$	Also, NaOH/1
$H_2C_2O_4$/1	$H_2C_2O_4 + NaOH = NaHC_2O_4 + H_2O$	Also, NaOH/1
$H_2C_2O_4$/2	$H_2C_2O_4 + 2NaOH = Na_2C_2O_4 + 2H_2O$	Also, NaOH/1
H_2CrO_4/2	$H_2CrO_4 + 2NaOH = Na_2CrO_4 + 2H_2O$	As an acid
H_2CrO_4/3	$H_2CrO_4 + 3e^- + 6H^+ = Cr^{3+} + 4H_2O$	As an oxidant
H_2CrO_4/2	$H_2CrO_4 + 2Ag^+ = Ag_2CrO_4 + 2H^+$	As precipitant
NH_3/1	$NH_3 + H^+ = NH_4^+$	As a base
$2NH_3$	$Ag^+ + 2NH_3 = Ag(NH_3)_2^+$	As complexer*

*The equivalent of 2 mol of ammonia is 1 mol of silver, or 0.5 mol of silver is equivalent to 1 mol of ammonia.

Another example that illustrates the need for balanced chemical equations to determine the reactants and the products is the reaction of phosphoric acid with sodium hydroxide. Two inflection points are observed on a potentiometric titration plot. The first occurs around pH 5 after the first hydrogen has completely reacted; the second occurs at about pH 9 after the second hydrogen has reacted. The equivalent weight of phosphoric acid based on the first end point would be H_3PO_4/1 whereas that based on the second end point would be H_3PO_4/2. If we use only the titrant volume between the first and second end points, the equivalent weight would be H_3PO_4/1.

28.5.6 Method for Preparing Solutions of Definite Molarity

Use the following formula to prepare solutions of definite molarity.

$$\frac{\text{Molecular weight of compound} \times}{\text{molarity wanted} \times \text{number of liters}} = \frac{\text{grams of compound needed}}{\text{to make up the solution}}$$

For example, to prepare 50 mL (0.050 L) of a 0.5 M NaCl solution.

$$58.45 \text{ g} \cdot \text{mol}^{-1} \times 0.5 \; M \; (\text{mol} \cdot \text{L}^{-1}) \times 0.050 \text{ L} = 1.46 \text{ g NaCl}$$

In a 50-mL volumetric flask, dissolve 1.46 g of NaCl in a small amount of water and then add water until the mark is reached.

28.5.7 Preparing Concentrated Molar Solutions

After deciding on the molarity and amount of the solution to be prepared, use the following formula to calculate how much of the more concentrated solution you need.

$$\frac{\text{Volume of concentrated solution} \times}{\text{molarity of concentrated solution}} = \frac{\text{volume of final solution} \times}{\text{molarity of final solution}}$$

For example, to prepare 100 mL of an exactly 1 M HCl solution from stock 12.1 M HCl,

$$100 \text{ mL} \times 1.00 \; M = ? \times 12.1 \; M$$

Pipet exactly 8.3 mL of stock 12.1 M HCl, and dilute to a total volume of 100 mL in a volumetric flask.

28.5.8 Calculating Molarity from Gram Concentrations

This type of calculation is often used when preparing more dilute solutions from stock reagents, especially acids and bases. Freshly opened bottles of these reagents are generally of the concentrations indicated in Table 28.8. This may not be true of bottles that are long opened, and this is especially true of ammonium hydroxide, which rapidly loses its strength.

A commercial CP reagent usually comes to the laboratory in a bottle having a label which states its molecular weight w, its density (or its specific gravity) d, and its percentage assay p. When such a reagent is used to prepare an aqueous solution of desired molarity M, a convenient formula to employ is:

$$V = \frac{100wM}{pd}$$

where V is the number of milliliters of concentrated reagent required for 1 L of the dilute solution. For example, sulfuric acid has the molecular weight of 98.08.

TABLE 28.8   Concentrations of Commonly Used Acids and Bases

Reagent	Formula weight	Density, g · mL^{-1} (20°C)	Weight % (approx.)	Molarity	V, mL*
Acetic acid	60.05	1.05	99.8	17.45	57.3
Ammonium hydroxide	35.05	0.90	56.6	14.53	60.0
(as NH$_3$)	17.03		28.0		
Ethylenediamine	60.10	0.899	100.0	15.0	66.7
Formic acid	46.03	1.20	90.5	23.6	42.5
Hydrobromic acid	80.92	1.49	48.0	8.84	113.0
Hydrochloric acid	36.46	1.19	37.2	12.1	82.5
Hydroiodic acid	127.91	1.70	57.0	7.6	132.0
Nitric acid	63.01	1.42	70.4	15.9	63.0
Perchloric acid	100.47	1.67	70.5	11.7	85.5
Phosphoric acid	97.10	1.70	85.5	14.8	67.5
Sulfuric acid	98.08	1.84	96.0	18.0	55.8
Triethanolamine	149.19	1.124	100	7.53	132.7

*V, mL = volume in milliliters needed to prepare 1 L of 1 M solution.

If the concentrated acid assays as 95.5% and has a specific gravity of 1.84, the volume required for 1 L of a 0.100 M solution is:

$$V = \frac{100 \times 98.08 \times 0.100}{95.5 \times 1.84} = 5.58 \text{ mL}$$

28.6 PREPARATION OF STANDARD LABORATORY SOLUTIONS

A series of tables now follow which describe the preparation of laboratory solutions frequently used in the laboratory. Standard stock solutions containing 1000 μg · mL^{-1} as the element in a final volume of 1 L and the procedure for preparation are listed in Table 28.9.

The preparation of standard volumetric (titrimetric) solutions is described in Table 28.10. Acid-base indicators are given in Table 28.11.

Table 28.12 describes the preparation of general laboratory reagents whose strengths need only to be known approximately. Those reagents listed in Table 28.10 are not repeated in this table.

28.6.1   Relation of Concentration Units

$$1 \, \mu g \cdot mL^{-1} = 1 \, mg \cdot L^{-1} = 1 \, ppm \, (w/v)$$

$$1 \, meq \cdot L^{-1} = (\text{weight in grams per equivalent} \div 1000) \text{ in } 1 \, L$$

$$1 \, mg \text{ atom per liter} = (\text{atomic weight in grams} \div 1000) \text{ in } 1 \, L$$

$$1 \, \mu g \cdot mL^{-1} = \frac{(1 \, meq \cdot L^{-1})(1000)}{\text{weight in grams per equivalent}}$$

TABLE 28.9   Standard Stock Solutions

1000 µg · mL^{-1} as the element in a final volume of 1 L

Element	Procedure
Aluminum	Dissolve 1.000 g Al wire in minimum amount of 2 *M* HCl, and dilute to volume.
Antimony	Dissolve 1.000 g Sb in (1) 10 mL HNO$_3$ plus 5 mL HCl, and dilute to volume when dissolution is complete, or (2) 18 mL HBr plus 2 mL liquid Br$_2$; when dissolution is complete add 10 mL HClO$_4$, heat in a well-ventilated hood while swirling until white fumes appear and continue for several minutes to expel all HBr, then cool and dilute to volume.
Arsenic	Dissolve 1.3203 g of As$_2$O$_3$ in 3 mL 8 *M* HCl and dilute to volume, or treat the oxide with 2 g NaOH and 20 mL water. After dissolution dilute to 200 mL, neutralize with HCl (pH meter), and dilute to volume.
Barium	(1) Dissolve 1.7787 g BaCl$_2$ · 2H$_2$O (fresh crystals) in water, and dilute to volume. (2) Treat 1.4367 g BaCO$_3$ with 300 mL water, slowly add 10 mL of HCl, and after the CO$_2$ is released by swirling, dilute to volume.
Beryllium	Dissolve 1.000 g Be in 25 mL 2 *M* HCl, then dilute to volume.
Bismuth	Dissolve 1.000 g Bi in 8 mL of 10 *M* HNO$_3$, boil gently to expel brown fumes, and dilute to volume.
Boron	Dissolve 5.720 g fresh crystals of H$_3$BO$_3$, and dilute to volume.
Bromine	Dissolve 1.489 g KBr (or 1.288 g NaBr) in water, and dilute to volume.
Cadmium	Dissolve 2.282 g 3CdSO$_4$ · 8H$_2$O in water, and dilute to volume.
Calcium	Place 2.4973 g CaCO$_3$ in volumetric flask with 300 mL water, carefully add 10 mL HCl; after CO$_2$ is released by swirling, dilute to volume.
Cerium	Dissolve 4.515 g (NH$_4$)$_4$Ce(SO$_4$)$_4$ · 2H$_2$O in 500 mL water to which 30 mL H$_2$SO$_4$ had been added, cool, and dilute to volume. Advisable to standardize against As$_2$O$_3$.
Cesium	Dissolve 1.267 g CsCl, and dilute to volume. Standardize: Pipet 25 mL of final solution to Pt dish, add 1 drop H$_2$SO$_4$, evaporate to dryness, and heat to constant weight at 800°C. Cs (in µg · mL^{-1}) = (40)(0.734)(wt of residue).
Chlorine	Dissolve 1.648 g NaCl, and dilute to volume.
Chromium	Dissolve 2.829 g K$_2$Cr$_2$O$_7$ in water, and dilute to volume.
Cobalt	Dissolve 1.000 g Co in 10 mL of 2 *M* HCl, and dilute to volume.
Copper	Dissolve 3.929 g fresh crystals of CuSO$_4$ · 5H$_2$O, and dilute to volume.
Fluorine	Dissolve 2.210 g NaF in water, and dilute to volume.
Gallium	Dissolve 1.000 g Ga in 50 mL of 2 *M* HCl, and dilute to volume.
Germanium	Dissolve 1.4408 g GeO$_2$ with 50 g oxalic acid in 100 mL of water, and dilute to volume.
Gold	Dissolve 1.000 g Au in 10 mL of hot HNO$_3$ by dropwise addition of HCl, boil to expel oxides of nitrogen and chlorine, and dilute to volume. Store in amber container away from light.
Indium	Dissolve 1.000 g in 50 mL of 2 *M* HCl, and dilute to volume.
Iodine	Dissolve 1.308 g KI in water, and dilute to volume.
Iridium	Dissolve 2.465 g Na$_3$IrCl$_6$ in water, and dilute to volume.
Iron	Dissolve 1.000 g Fe wire in 20 mL of 5 *M* HCl; dilute to volume.
Lanthanum	Dissolve 1.1717 g La$_2$O$_3$ (dried at 110°C) in 50 mL of 5 *M* HCl, and dilute to volume.
Lead	Dissolve 1.5985 g Pb(NO$_3$)$_2$ in water plus 10 mL HNO$_3$, and dilute to volume.
Lithium	Dissolve a slurry of 5.3228 g Li$_2$CO$_3$ in 300 mL of water by addition of 15 mL HCl; after release of CO$_2$ by swirling, dilute to volume.

TABLE 28.9   Standard Stock Solutions (*Continued*)

1000 μg · mL^{-1} as the element in a final volume of 1 L

Element	Procedure
Magnesium	Dissolve 1.000 g Mg in 50 mL of 1 M HCl, and dilute to volume.
Manganese	Dissolve 1.000 g Mn in 10 mL HCl plus 1 mL HNO$_3$, and dilute to volume.
Mercury	Dissolve 1.000 g Hg in 10 mL of 5 M HNO$_3$, and dilute to volume.
Molybdenum	Dissolve 1.5003 g MoO$_3$ in 100 mL of 2 M ammonia, and dilute to volume.
Nickel	Dissolve 1.000 g Ni in 10 mL hot HNO$_3$, cool, and dilute to volume.
Niobium	Transfer 1.000 g Nb (or 1.4305 g Nb$_2$O$_5$) to Pt dish, add 20 mL HF, and heat gently to complete dissolution. Cool, add 40 mL H$_2$SO$_4$, and evaporate to fumes of SO$_3$. Cool and dilute to volume with 8 M H$_2$SO$_4$.
Osmium	Dissolve 1.3360 g OsO$_4$ in water and dilute to volume. Prepare only as needed because solution loses strength on standing.
Palladium	Dissolve 1.000 g Pd in 10 mL of HNO$_3$ by dropwise addition of HCl to hot solution, and dilute to volume.
Phosphorus	Dissolve 4.260 g (NH$_4$)$_2$HPO$_4$ in water, and dilute to volume.
Platinum	Dissolve 1.000 g Pt in 40 mL of hot aqua regia, evaporate to incipient dryness, add 10 mL HCl, and again evaporate to moist residue. Add 10 mL HCl, and dilute to volume.
Potassium	Dissolve 1.9067 g KCl (or 2.8415 g KNO$_3$) in water, and dilute to volume.
Rhenium	Dissolve 1.000 g Re in 10 mL of 8 M HNO$_3$ in an ice bath until initial reaction subsides, and dilute to volume.
Rubidium	Dissolve 1.4148 g RbCl in water. Standardize as described under cesium. Rb (in μg · mL^{-1}) = (40)(0.320)(wt of residue).
Ruthenium	Dissolve 1.317 g RuO$_2$ in 15 mL of HCl, and dilute to volume.
Selenium	Dissolve 1.4050 g SeO$_2$ in water, and dilute to volume.
Silicon	Fuse 2.1393 g SiO$_2$ with 4.60 g Na$_2$CO$_3$, maintaining melt for 15 min in Pt crucible. Cool, dissolve in warm water, and dilute to volume.
Silver	Dissolve 1.5748 g AgNO$_3$ in water, and dilute to volume.
Sodium	Dissolve 2.5421 g NaCl in water, and dilute to volume.
Strontium	Dissolve a slurry of 1.6849 g SrCO$_3$ in 300 mL of water by careful addition of 10 mL of HCl; after release of CO$_2$ by swirling, dilute to volume.
Sulfur	Dissolve 4.122 g (NH$_4$)$_2$SO$_4$ in water, and dilute to volume.
Tantalum	Transfer 1.000 g Ta to Pt dish, add 20 mL HF, and heat gently to complete the dissolution. Cool, add 40 mL of H$_2$SO$_4$ and evaporate to heavy fumes of SO$_3$. Cool and dilute to volume with 50% H$_2$SO$_4$.
Tellurium	Dissolve 1.2508 g TeO$_2$ in 10 mL of HCl, and dilute to volume.
Thallium	Dissolve 1.3034 g TlNO$_3$ in water, and dilute to volume.
Thorium	Dissolve 2.3794 g Th(NO$_3$)$_4$ · 4H$_2$O in water, add 5 mL HNO$_3$, and dilute to volume.
Tin	Dissolve 1.000 g Sn in 15 mL of warm HCl, and dilute to volume.
Titanium	Dissolve 1.000 g Ti in 10 mL of H$_2$SO$_4$ with dropwise addition of HNO$_3$, and dilute to volume with 5% H$_2$SO$_4$.
Tungsten	Dissolve 1.7941 g of Na$_2$WO$_4$ · 2H$_2$O in water, and dilute to volume.
Uranium	Dissolve 2.1095 g UO$_2$(NO$_3$)$_2$ · 6H$_2$O in water, and dilute to volume.
Vanadium	Dissolve 2.2963 g NH$_4$VO$_3$ in 100 mL of water plus 10 mL of HNO$_3$, and dilute to volume.
Yttrium	Dissolve 1.2692 g Y$_2$O$_3$ in 50 mL of 2 M HCl, and dilute to volume.
Zinc	Dissolve 1.000 g Zn in 10 mL of HCl, and dilute to volume.
Zirconium	Dissolve 3.533 g ZrOCl$_2$ · 8H$_2$O in 50 mL of 2 M HCl, and dilute to volume. Solution should be standardized.

TABLE 28.10   Standard Laboratory Solutions

Weight (or volume) to be dissolved and diluted to 1 L with distilled water

Substance	Procedure
Alkaline arsenite, 0.1000 N As(III) to As(V)	Dissolve 4.9460 g of As_2O_3 in 40 mL of 30% NaOH solution. Dilute with 200 mL of water. Acidify the solution with 6 M HCl to the acid color of methyl red indicator. Add to this solution 40 g of $NaHCO_3$, and dilute to 1 L. One milliliter of this solution is equivalent to 0.01269 g of iodine.
Ammonium thiocyanate, 0.1 N	Dissolve 7.6120 g of NH_4SCN in water, and dilute to 1 L. Standardize against standard silver nitrate solution.
Cerium(IV) sulfate, 0.1 N Ce(IV) to Ce(III)	Dissolve 63.26 g of cerium(IV) ammonium sulfate dihydrate in 500 mL of 1 M sulfuric acid. Dilute to 1 L, and standardize against the alkaline arsenite solution using o-phenanthroline indicator. To 30–40 mL of arsenite solution, measured accurately, in a flask, and slowly add 20 mL of 2 M sulfuric acid and 2 drops of 0.01 M osmium tetroxide solution. Titrate to a faint blue endpoint.
Hydrochloric acid, 0.1 N	Add 8 mL of concentrated HCl to 1 L of distilled water. Standardize against weighed portions of sodium carbonate.
Iodine, 0.1000 N I(0) to I(1−)	Dissolve 12.690 g of resublimed iodine in 25 mL of a solution containing 15 g of KI which is free from iodate. When dissolved after swirling, dilute to 1 L.
Iron(II) ammonium sulfate, 0.1000 N	Dissolve 39.2139 g of $FeSO_4 \cdot (NH_4)_2SO_4 \cdot 6H_2O$ in 500 mL of 0.5 M sulfuric acid, and dilute to 1 L. Standardize against standard permanganate or dichromate solution for more accurate normality.
Potassium cyanide for nickel titration, 0.1 N = 0.4 × KCN = 26.0479 g · L	Dissolve this amount of KCN in water, add 10 g of KOH to prevent hydrolysis, and dilute to 1 L. For silver titration, use 13.0240 g of KCN. *Wear protective gloves and work in a hood* when handling cyanide solids or solutions. *Never pipet solutions by the mouth.*
Potassium dichromate(VI), 0.1000 N as Cr(VI) to Cr(III)	Weigh out 4.9030 g of $K_2Cr_2O_7$ that has been oven-dried and cooled, and diluted with water to 1 L.
Potassium hydroxide, 0.1 N	This solution should contain 5.61 g of KOH per liter. Dissolve about 6 g in 500 mL of water, add enough barium chloride to precipitate any carbonate that may be present, allow the precipitate to settle, filter into a liter flask, and dilute to the mark with CO_2-free water. Standardize against potassium hydrogen phthalate. Use a guard tube containing Ascarite.
Potassium iodate, 0.1000 N as I(V) to I(1−)	Weigh out 3.5667 g of KIO_3 and about 15 g KI, dissolve both in water, and dilute to 1 L.
Potassium permanganate, 0.1 N as Mn(VII) to Mn(II)	Dissolve about 3.3 g in a liter of water. Allow to stand 2–3 days, then filter through a Gooch crucible and into the container in which it is to be stored. Discard the first 25 mL of filtrate, and allow the last inch to remain in the original bottle. Never allow the solution to come into contact with rubber, filter paper, or any organic matter. Standardize against sample of sodium oxalate.

TABLE 28.10 Standard Laboratory Solutions (*Continued*)

Weight (or volume) to be dissolved and diluted to 1 L with distilled water

Substance	Procedure
Silver nitrate, 0.1000 N	Dissolve 16.9875 g of silver nitrate in exactly 1 L of water.
Sodium chloride, 0.1000 N	Dissolve 5.8443 g of NaCl in water and dilute to 1 L.
Sodium EDTA, 0.01 N	Dissolve about 3.8 g of the disodium dihydrate salt, and dilute to 1 L. Standardize against a magnesium solution of known strength.
Sodium hydroxide, 0.1 N	Dissolve about 4 g of NaOH pellets. Proceed as described for the potassium hydroxide solution.
Sodium thiosulfate, 0.1000 N	Dissolve 15.811 g of $Na_2S_2O_3$ crystals in water and dilute almost to 1 L. Add 0.5 g of sodium carbonate or 0.5 mL of chloroform as preservative, and dilute to volume.
Sulfuric acid, 0.1 N	Pour slowly 3 mL of concentrated sulfuric acid into 200 mL of water. Cool and dilute to 1 L. Standardize.

TABLE 28.11 Preparation of Acid-Base Indicator Solutions

To prepare solutions of the indicators, grind 0.1 g of the dye in a clean mortar with the quantities of 0.05 M sodium hydroxide given in the table, and dilute with distilled water to 200 mL for a 0.05% stock solution

Indicator	Molecular weight	pH range	mL 0.05 M NaOH per 0.1 g of indicator
Cresol red (acid range)	382	0.0–1.0	5.3
Thymol blue (acid range)	456	1.2–2.8	4.3
m-Cresol purple (acid range)	382	1.2–2.8	5.3
Bromophenol blue	670	3.0–4.6	3.0
Bromocresol green	698	3.8–5.4	2.9
Methyl red	269	4.2–6.2	4.7
Chlorophenol red	423	5.4–6.8	4.7
Bromocresol purple	540	5.2–6.8	5.3
Bromothymol blue	624	6.0–7.6	3.2
Phenol red	354	6.8–8.2	5.7
Cresol red	404	7.0–8.8	5.3
m-Cresol purple	382	7.6–9.2	5.3
Thymol blue	456	8.0–9.2	4.3
Phenolphthalein	318	8.0–10.0	*
Thymolphthalein	431	8.8–10.5	*

*Phenolphthalein and thymolphthalein will dissolve in ethanol.

Table 28.12 General Laboratory Solutions

Substance	Formula	Molecular weight	Conc'n, M	Weight (or volume) in 1 L
Aluminum chloride	$AlCl_3 \cdot 6H_2O$	241.43	0.05	12.1 g
Aluminum nitrate	$Al(NO_3)_3 \cdot 9H_2O$	375.13	0.1	37.5 g
Aluminum sulfate	$Al_2(SO_4)_3 \cdot 18H_2O$	666.42	0.083	55 g
Ammonium acetate	$NH_4C_2H_3O_2$	77.08	1	77 g
Ammonium chloride	NH_4Cl	53.49	0.01	0.54 g
Ammonium molybdate	$(NH_4)_2MoO_4$	196.01	0.5	Dissolve 72 g MoO_3 in 200 mL water, add 60 mL NH_4OH, filter into 270 mL HNO_3, dilute to 1 L.
Ammonium nitrate	NH_4NO_3	80.04	0.1	8 g
Ammonium oxalate	$(NH_4)_2C_2O_4 \cdot H_2O$	142.11	0.25	35.5 g
Ammonium peroxodisulfate	$(NH_4)_2S_2O_8$	228.18	0.1	22.8 g
Ammonium sulfate	$(NH_4)_2SO_4$	132.14	0.1	13 g
Antimony trichloride	$SbCl_3$	228.11	0.1	22.7 g in 200 mL HCl, and dilute to 1 L.
Aqua regia				Mix 1 part HNO_3 with 3 parts HCl.
Barium chloride	$BaCl_2 \cdot 2H_2O$	244.28	0.1	24.4 g
Barium nitrate	$Ba(NO_3)_2$	261.35	0.1	26 g
Bismuth chloride	$BiCl_3$	315.34	0.1	31.5 g in 200 mL HCl; dilute to 1 L.
Boric acid	H_3BO_3	61.83	0.1	6.2 g
Bromine in CCl_4	Br_2	159.82	0.1	1 g liquid Br_2 in 63 mL CCl_4
Bromine water				Saturate 1 L water with 15 g liquid Br_2.
Cadmium chloride	$CdCl_2$	183.31	0.1	18.3 g
Cadmium sulfate	$CdSO_4 \cdot 4H_2O$	280.5	0.25	70 g
Calcium chloride	$CaCl_2 \cdot 6H_2O$	219.1	0.1	22 g
Chromium(III) nitrate	$Cr(NO_3)_3 \cdot 9H_2O$	400.15	0.1	40 g
Cobalt(II) sulfate	$CoSO_4 \cdot 7H_2O$	281.10	0.25	70 g
Copper(II) nitrate	$Cu(NO_3)_2 \cdot 6H_2O$	295.64	0.1	29.5 g
Copper(II) sulfate	$CuSO_4 \cdot 5H_2O$	249.68	0.1	25 g
Fehling's solution A				35 g $CuSO_4 \cdot 5H_2O$ in 500 mL water.

Table 28.12   General Laboratory Solutions (*Continued*)

Substance	Formula	Molecular weight	Conc'n, M	Weight (or volume) in 1 L
Fehling's solution B				Dissolve 173 g $KNaC_4H_4O_6 \cdot 4H_2O$ in 200 mL water, add 50 g NaOH in 200 mL water, and dilute to 500 mL with water.
Hydrogen peroxide	H_2O_2	34.01	0.2	Dilute 23 mL 30% H_2O_2 to 1 L
Iron(III) ammonium sulfate	$Fe(NH_4)_2(SO_4)_2 \cdot 12H_2O$	482.19	0.2	96 g plus 10 mL H_2SO_4
Lead nitrate	$Pb(NO_3)_2$	331.2	0.1	33 g
Magnesium chloride	$MgCl_2 \cdot 6H_2O$	203.31	0.25	51 g
Magnesium sulfate	$MgSO_4 \cdot 7H_2O$	246.48	0.25	62 g
Magnesium uranyl acetate				
Solution A				Dissolve 100 g $UO_2(C_2H_3O_2)_2 \cdot 7H_2O$ in 60 mL acetic acid, and dilute to 500 mL.
Solution B				Dissolve 330 g $Mg(C_2H_3O_2)_2 \cdot 4H_2O$ in 60 mL acetic acid at 70°C. Dilute to 200 mL with water. Mix solutions A and B, cool to exactly 20°C, and filter.
Manganese chloride	$MnCl_2 \cdot 4H_2O$	197.91	0.25	50 g
Manganese sulfate	$MnSO_4 \cdot 7H_2O$	277.11	0.25	69 g
Mercury(II) chloride	$HgCl_2$	271.5	0.25	68 g
Mercury(I) nitrate	$Hg_2(NO_3)_2 \cdot 2H_2O$	561.22	0.1	56 g plus 50 mL HNO_3; dilute to 1 L.
Nickel sulfate	$NiSO_4 \cdot 6H_2O$	262.86	0.25	66 g
Oxalic acid	$H_2C_2O_4 \cdot 2H_2O$	126.07	0.1	12.6 g
Potassium bromide	KBr	119.01	0.1	11.9 g
Potassium carbonate	K_2CO_3	138.21	1.0	138 g
Potassium chloride	KCl	74.56	0.1	7.45 g
Potassium chromate	K_2CrO_4	194.20	0.1	19.4 g
Potassium dihydrogen phosphate	KH_2PO_4	136.09	0.1	13.6 g
Potassium ferricyanide	$K_3Fe(CN)_6$	329.26	0.167	55.0 g
Potassium ferrocyanide	$K_4Fe(CN)_6 \cdot 3H_2O$	422.41	0.1	42.3 g
Potassium hydrogen phosphate	K_2HPO_4	174.18	0.1	17.4 g
Potassium hydrogen phthalate	$KHC_8H_4O_4$	204.22	0.100	20.42 g

Table 28.12 General Laboratory Solutions (*Continued*)

Substance	Formula	Molecular weight	Conc'n, M	Weight (or volume) in 1 L
Potassium hydrogen sulfate	$KHSO_4$	136.17	0.1	13.6 g
Potassium hydrogen sulfite	$KHSO_3$	120.17	0.2	24.0 g
Potassium iodide	KI	166.01	0.1	16.6 g (protect from light)
Potassium nitrate	KNO_3	101.11	0.1	10.1 g
Potassium sulfate	K_2SO_4	174.27	0.1	17.4 g
Potassium thiocyanate	KSCN	97.18	0.1	9.7 g
Sodium acetate	$NaC_2H_3O_2 \cdot 3H_2O$	136.08	0.5	68.0 g
Sodium hydrogen carbonate	$NaHCO_3$	84.01	0.1	8.4 g
Sodium bromide	NaBr	102.9	0.1	10.3 g
Sodium carbonate	Na_2CO_3	105.99	0.5	53.0 g
Sodium cobaltinitrite	$Na_3Co(NO_2)_6$	404	0.08	Dissolve 25 g $NaNO_3$ in 75 mL water, add 2 mL acetic acid, then 2.5 g of $Co(NO_2)_2 \cdot 6H_2O$, dilute to 100 mL with water.
Sodium hydrogen phosphate	$Na_2HPO_4 \cdot 12H_2O$	358.14	0.1	35.8 g
Sodium hydrogen sulfate	$NaHSO_4$	120.0	1	120 g
Sodium iodide	NaI	149.9	0.1	15 g
Sodium nitrate	$NaNO_3$	85.00	0.2	17 g
Sodium nitroprusside	$Na_2Fe(CN)_5NO \cdot 2H_2O$	197.95	10%	100 g
Sodium polysulfide	Na_2S_x			Dissolve 480 g $Na_2S \cdot 9H_2O$ in 500 mL water, add 40 g NaOH and 16 g sulfur, mix well, and dilute to 1 L.
Sodium sulfate	Na_2SO_4	142.04	0.1	14.2 g
Sodium sulfite	Na_2SO_3	126.04	0.1	12.6 g
Sodium tetraborate	$Na_2B_4O_7 \cdot 10H_2O$	381.37	0.0025	9.5 g
Tin(II) chloride	$SnCl_2 \cdot 2H_2O$	225.63	0.15	Dissolve 33.8 g in 75 mL HCl, and dilute to 1 L. Keep Sn granules in container.
Zinc chloride	$ZnCl_2$	136.3	0.1	13.6 g
Zinc nitrate	$Zn(NO_3)_2 \cdot 6H_2O$	297.5	0.1	30 g
Zinc sulfate	$ZnSO_4 \cdot 7H_2O$	287.6	0.25	72 g

28.7 PREPARATION OF STANDARD (NIST) BUFFER SOLUTIONS

To prepare the standard buffer solutions recommended by the National Institute of Standards and Technology (formerly National Bureau of Standards) (U.S.), the indicated weights of the pure materials in Table 28.13 should be dissolved in a good grade of water, either freshly distilled or deionized, and diluted to 1 L. The excess of solid potassium hydrogen tartrate and calcium hydroxide must be removed, the latter by filtration through a sintered glass filter (No. 3 porosity). Water for preparation of alkaline solutions should be boiled and protected from carbon dioxide while cooling or should be purged with CO_2-free air. The entrance of CO_2 into borax and calcium hydroxide buffers must be avoided. The buffer solutions may be stored in polyethylene or Pyrex containers. The solutions should be replaced every 2 months, or sooner if formation of mold is noticed. A crystal of thymol may be added as preservative.

The tartrate, phthalate, carbonates, and phosphates may each be dried for 2 h at 110°C. Potassium tetroxalate and calcium hydroxide need not be dried. Fresh-looking crystals of sodium tetraborate should be used. Calcium hydroxide supplied by the Bureau of Standards should be used; alternatively, directions for its preparation should be followed carefully.

28.8 STANDARDS FOR pH MEASUREMENT OF BLOOD AND BIOLOGIC MATERIAL

Blood is a well-buffered medium. Several standard solutions are recommended by the National Institute of Standards and Technology (U.S.) (NIST). The NIST phosphate standard of 0.025 M (Table 28.13) has pH_s = 6.840 at 38°C.

TABLE 28.13 Compositions of Standard Buffer Solutions, National Institute of Standards and Technology (U.S.)

Standard	Weight of salt,* g
$KHC_2O_4 \cdot 2H_2O$, 0.05 M	12.70
Potassium hydrogen tartrate, about 0.034 M	Saturated at 22–28°C
Potassium hydrogen phthalate, 0.05 M	10.21
Phosphate (solution 1):	
KH_2PO_4, 0.025 M	3.40
Na_2HPO_4, 0.025 M	3.55
Phosphate (solution 2:	
KH_2PO_4, 0.008665 M	1.179
Na_2HPO_4, 0.03032 M	4.33
$Na_2B_4O_7 \cdot H_2O$, 0.01 M	3.81
Carbonate:	
$NaHCO_3$, 0.025 M	2.10
Na_2CO_3, 0.025 M	2.65
$Ca(OH)_2$, about 0.0203 M	Saturated at 25°C

*Air weights of salt per liter of buffer solution.

Another phosphate reference solution containing the same salts, but in the molal ratio 1:4, has an ionic strength of 0.13. It is prepared by dissolving 1.360 g of KH_2PO_4 and 5.677 g of Na_2HPO_4 (air weights) in carbon dioxide–free water to make 1 L of solution. The pH_s is 7.416 at 37.5 and 38°C.

The phosphate-succinate system gives these pH_s values at 25°C.

Molality* KH_2PO_4 = molality $Na_2C_6H_5O_7$	pH_s	$(pH_s)/dt$, deg^{-1}
0.005	6.251	− 0.000 86
0.010	6.197	− 0.000 71
0.015	6.162	
0.020	6.131	
0.025	6.109	− 0.000 4

*The molecular weight of disodium succinate is 162.05, and that of potassium dihydrogen phosphate is 136.09.

The tris(hydroxymethyl)aminomethane system covers the pH range 7.0 to 9.0. To 50 mL of 0.1 M tris(hydroxymethyl)aminomethane (12.114 g · L^{-1}), add x mL of 0.1 M HCl. Dilute to 100 mL. At 25°C the pH values are listed. pH/dt is approximately −0.028. The ionic strength equals 0.001x.

pH	x
7.00	46.6
7.20	44.7
7.40	42.0
7.60	38.5
7.80	34.5
8.00	29.2
8.20	22.9
8.40	17.2
8.60	12.4
8.80	8.5
9.00	5.7

28.9 OTHER BUFFER SOLUTIONS

Premixed, ready-to-use buffer solutions are available commercially in pint and quart containers. Also available are vials or packets of concentrate which can be quantitatively transferred to specified volumetric flasks and diluted to the mark with distilled water. Thus, several stock solutions of various pH values can be kept in a very small space.

28.9.1 Buffers with pH Values at Narrow Intervals

Sometimes one has need of buffers over a narrow interval of pH values. The range of the buffering effect of a single weak acid group is approximately one pH

TABLE 28.14   pH Ranges of Selected Buffer Systems

Materials	pH range
Glycine and HCl	1.0–3.7
Citrate and HCl	1.3–4.7
Formate and HCl	2.8–4.6
Succinic acid and sodium tetraborate (borax)	3.0–5.8
Acetate and acetic acid	3.7–5.6
Succinate and succinic acid	4.8–6.3
2-(N-Morpholino)ethanesulfonic acid, "MES," and NaOH	5.2–7.1
2,2-Bis(hydroxymethyl)-2,2′,2″-nitrilotriethanol and HCl	5.8–7.2
Potassium dihydrogen phosphate and sodium tetraborate	5.8–9.2
KH_2PO_4 and Na_2HPO_4	6.1–7.5
N-Tris(hydroxymethyl)methyl-2-aminoethanesulfonic acid, "TES," and NaOH	6.8–8.2
N-2-Hydroxyethylpiperazine-N′-2-ethanesulfonic acid, "HEPES," and NaOH	6.9–8.3
Triethanolamine and HCl	6.9–8.5
Diethylbarbiturate (Veronal) and HCl	7.0–8.5
Tris(hydroxymethyl)aminomethane, "Tris," and HCl	7.2–9.0
N-Tris(hydroxymethyl)methylglycine, "Tricine," and HCl	7.2–9.0
N,N-Bis(2-hydroxyethyl)glycine, "Bicine," and HCl	7.4–9.2
Sodium tetraborate and HCl	7.6–8.9
Glycine and NaOH	8.2–10.1
Ethanolamine and HCl	8.6–10.4
Sodium carbonate and sodium hydrogen carbonate	9.2–11.1
Disodium hydrogen phosphate and NaOH	11.0–12.0

TABLE 28.15   Preparation of Buffer Solutions* (25°C)

Desired pH	Solution 1, mL†	Solution 2, mL‡	Desired pH	Solution 1, mL†	Solution 2, mL‡
2.0	97.5	2.5	7.5	46.0	54.0
2.5	92.0	8.0	8.0	42.5	57.5
3.0	88.0	12.0	8.5	39.0	61.0
3.5	83.0	17.0	9.0	34.5	65.5
4.0	77.5	22.5	9.5	30.0	70.0
4.5	72.0	28.0	10.0	27.0	73.0
5.0	67.0	33.0	10.5	24.5	75.5
5.5	63.0	37.0	11.0	22.0	78.0
6.0	59.0	41.0	11.5	16.5	83.5
6.5	54.5	45.5	12.0	8.5	91.5
7.0	49.5	50.5			

*To prepare 100 mL of any buffer listed in the table, pipet the exact amounts of each solution into a dry container.

†Solution 1. Dissolve 12.37 g of anhydrous boric acid, H_3BO_3, and 10.51 g of citric acid monohydrate in distilled water, and dilute to 1 L in a volumetric flask. This makes a 0.200 M boric acid and a 0.0500 M citric acid solution.

‡Solution 2. Dissolve 38.01 g of $Na_3PO_4 \cdot 12H_2O$ in distilled water, and dilute to 1 L in a volumetric flask. This makes a 0.100 M tertiary sodium phosphate solution.

unit on either side of pK_a. The ranges of some useful buffer systems are collected in Table 28.14. After the components have been brought together, the pH of the resulting solution should be determined with a pH meter. Minor adjustments can be made by further addition of the acidic or basic component.

When there are two or more acid groups per molecule, or a mixture is composed of several overlapping systems, the useful buffering range is larger. Some of these "universal" buffer systems are included in Table 28.15.

BIBLIOGRAPHY

Berlin, R. E., and C. C. Stanton, *Radioactive Waste Management,* Wiley, New York, 1989.

Fawcett, H., ed., *Hazardous and Toxic Materials,* 2d ed., Wiley, 1989.

Sax, N. I., and R. J. Lewis, Sr., *Hazardous Chemicals Desk Reference,* Van Nostrand Reinhold, New York, 1987.

CHAPTER 29
VOLUMETRIC ANALYSIS

A volumetric method is one in which the analysis is completed by measuring the volume of a solution of established concentration needed to react completely with the substance being determined. It is customary to divide the reactions of volumetric analysis into four groups:

1. Neutralization methods (acidimetry and alkalimetry)
2. Oxidation and reduction (redox) methods
3. Precipitation methods
4. Complex formation (ion combination) methods

In Sec. 28.5.5 the principles underlying the use of equivalents and normal solutions were taken up in a general way.

29.1 TITRATION THEORY

The concentration of a species in a sample can be accurately determined by a chemical reaction. Progressive measured addition of a reagent is known as a titration. The amount of added reagent is measured until the titration reaction is quantitatively complete. At this point, the amount of reagent is chemically equivalent to the amount of the species in the sample. The unknown concentration of a known volume of sample, or the unknown weight of the species, can be calculated from the known volume and concentration of the titrant (reagent). Of course, some means must be available for ascertaining the end point.

29.1.1 Equivalent Weight in Neutralization Methods

The fundamental reaction of neutralization methods is

$$H^+ + OH^- \rightarrow H_2O \tag{29.1}$$

The equivalent weight of a substance acting as an acid is that weight of it which is equivalent in total neutralizing power to one gram atom (1.008 g) of hydrogen ions. The equivalent weight of a substance acting as a base is that weight of it which will neutralize one gram atom of hydrogen ions (or its equivalent in total neutralizing power to 17.008 g of hydroxyl ions).

A normal solution of an acid or base contains one equivalent weight (in grams) of the acid or base in exactly one liter of solution; this can also be stated as one milliequivalent (meq) in one milliliter of solution. Total neutralizing power should not be confused with degree of ionization; equivalent weight is based on total neutralizing power. It follows that the product of the number of milliliters of a given solution and the normality of the solution gives the number of milliequivalents of solute present (either in the titrating solution or in the species reacting with the titrant), or

$$\text{mL} \times N(= \text{meq} \cdot \text{mL}^{-1}) = \text{number of milliequivalents} \qquad (29.2)$$

and we have the following relationship between two reaction solutions:

$$\text{mL}_A \times N_A = \text{mL}_B \times N_B \qquad (29.3)$$

A solution can therefore be standardized by determining what volume of the solution will react exactly with a definite volume of another solution the normality of which is already known, such as a primary standard solution. When dealing with a known weight of a species:

$$\frac{\text{Weight in grams}}{\text{Milliequivalent weight}} = \text{number of milliequivalents of solid} \qquad (29.4)$$

where the milliequivalent weight of the solid is stated in grams.

29.1.2 Definition of Terms

29.1.2.1 Primary Standard. A primary standard should possess these characteristics:

1. High equivalent weight
2. Correct stoichiometry (see Sec. 28.1.8) and high purity
3. Stability to drying at 120°C without decomposition, and should not react with components in the atmosphere.

29.1.2.2 Equivalence Point. The point in a titration at which the volume of titrant added is chemically equivalent to the amount of substance being titrated is called the equivalence point. It is a theoretical concept. We estimate its position by observing physical changes associated with the end point.

29.1.2.3 End Point. The end point occurs when there is an abrupt change of some property of the system. This might be one of the following:

1. Indicator color change or change in fluorescence
2. Potential changes across a pair of electrodes as in potentiometric titration methods (Chap. 18)
3. Change in the diffusion current, as in amperometric titrations (Sec. 19.6)
4. Changes in conductivity, as in conductometric titrations (Sec. 20.4.4)
5. Temperature changes, as in thermometric titration methods (Sec. 21.6.1).

29.1.2.4 Titration Error. The titration error is the difference between the end point and the equivalence point. This can be largely eliminated by standardizing the titrant in the same manner as the titration will be conducted; in other words, standardize as you will analyze. If a *blank titration* is run, the volume of titrant used is subtracted from the total amount of titrant consumed when titrating a sample or standard. The blank should contain all the reagents in approximately the same volume of solution as will the unknown, except the the actual material being titrated is absent.

29.2 TOOLS OF VOLUMETRIC ANALYSIS

Pipets, burets, and volumetric flasks are standard volumetric equipment. Volumetric apparatus calibrated *to contain* a specified volume is designated *TC*, and apparatus calibrated *to deliver* a specified amount, *TD*. Calibration of volumetric ware is usually done at 20°C; therefore, the ware should be used at approximately this temperature. *Never* subject volumetric ware to hot or cold liquids; following heating or cooling it takes many hours for volumetric ware to contract or expand to its correct volume.

Before use, volumetric ware should be washed with detergent and, if necessary, cleaning solution. Then it should be carefully and repeatedly rinsed with distilled water. Drying is done by drawing clean, filtered (to remove oil and dust particles), compressed air through the units.

29.2.1 Volumetric Flasks

Volumetric flasks are calibrated to contain a specified volume when filled to the line etched on the neck (Fig. 29.1). Types of ware include:

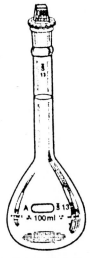

FIGURE 29.1 Volumetric flask with standard taper stopper.

Class A: Manufactured to the highest tolerances from Pyrex, borosilicate, or Kimax glass.

Class B (economy ware): Possesses tolerances about twice those of Class A. Tolerances are given in Table 29.1.

Nalgene polypropylene ware: Is autoclavable and meets Federal Specifications NNN-F-289 (Class B).

TABLE 29.1   Tolerances of Volumetric Flasks

Capacity, mL	Tolerances,* ±mL		Capacity, mL	Tolerances,* ±mL	
	Class A	Class B		Class A	Class B
5	0.02	0.04	200	0.10	0.20
10	0.02	0.04	250	0.12	0.24
25	0.03	0.06	500	0.20	0.40
50	0.05	0.10	1000	0.30	0.60
100	0.08	0.16	2000	0.50	1.00

*Accuracy tolerances for volumetric flasks at 20°C are given by ASTM standard E288.

After introducing the sample through a pipet or a funnel whose tip extends below the line etched on the neck, bring the liquid level almost to the mark with solvent. Then use a medicine dropper to make such final additions of solvent as necessary.

29.2.2   Volumetric Pipets

Pipets are designed for the transfer of known volumes of liquid from one container to another. Pipets which deliver a fixed volume are called *volumetric* or *transfer pipets* (Fig. 29.2). Other pipets, known as *measuring* or *serological* pipets, are calibrated in convenient units so that any volume up to maximum capacity can be delivered (Fig. 29.3). Micropipets are discussed later. Pipet capacity tolerances are given in Table 29.2.

The following instructions pertain specifically to the manipulation of transfer pipets, but with minor modifications they may be used for other types as well. Liquids are usually drawn into pipets through the application of a slight vacuum. *Never* pipet directly by mouth. For manually controlled filling and discharge of pipets, a rubber suction bulb (Fig. 29.4) is useful. It consists of a strong rubber bulb with three corrosion resistant ball valves which, operated by finger pressure, permit delivery within approximately 0.01 mL. An autoclavable unit is constructed of polypropylene, with rubber aspiration bulb (taken apart for autoclaving) and polyolefin check valve; it can be used with pipets of 20-mL capacity or smaller. Also available is a disposable pipet mouthpiece protector which acts as a hydrophobic barrier to prevent passage of aqueous solutions when pipetting by mouth. It consists of a nylon-reinforced acrylic filter in an acrylic housing attached to a 40-cm length of flexible plastic tubing; it does not prevent the passage of vapors or organic solvents.

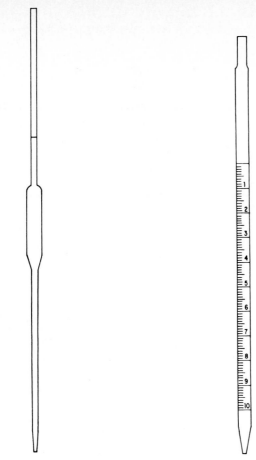

FIGURE 29.2 Volumetric or transfer pipet.   **FIGURE 29.3** Measuring pipet.

TABLE 29.2 Pipet Capacity Tolerances

Volumetric transfer pipets			Measuring and serological pipets	
Capacity, mL	Tolerances,* ±mL		Capacity, mL	Tolerances,† ±mL
	Class A	Class B		Class B
0.5	0.006	0.012	0.1	0.005
1	0.006	0.012	0.2	0.008
2	0.006	0.012	0.25	0.008
3	0.01	0.02	0.5	0.01
4	0.01	0.02	0.6	0.01
5	0.01	0.02	1	0.02
10	0.02	0.04	2	0.02
15	0.03	0.06	5	0.04
20	0.03	0.06	10	0.06
25	0.03	0.06	25	0.10
50	0.05	0.10		
100	0.08	0.16		

*Accuracy tolerances for volumetric transfer pipets are given by ASTM standard E969 and Federal Specification NNN-P-395.

†Accuracy tolerances for measuring pipets are given by Federal Specification NNN-P-350 and for serological pipets by Federal Specification NNN-P-375.

FIGURE 29.4   Pipet filler used to transfer liquids safely.

29.2.2.1 Pipetting Protocol.   When pipetting, proceed as follows, consulting Fig. 29.5:

1. Keep the tip of the pipet below the surface of the liquid.
2. Draw the liquid up the pipet using one of the devices described above. Avoid holding the pipet by the glass bulb. Fill until the liquid is slightly above the etched mark on the upper stem.
3. Release the pressure to allow the meniscus to approach the calibration mark.
4. At the mark, stop the liquid flow and drain the drop on the tip by touching it to the wall of the liquid-holding container.
5. Tilt the pipet slightly, and wipe away any drops on the outside.
6. Transfer the pipet to the container to be used, and release pressure on the rubber bulb. If a transfer pipet, allow the solution to drain completely with the tip touching the side of the container. Allow a time lapse of 10 s, or the period specified on the pipet.
7. If a measuring pipet, drain the solution only to the desired calibration mark. Remove the last drop by touching the wall of the container (below the etch mark if a volumetric flask). *Do not* blow out the pipet unless it is a serological pipet (calibrations down to the tip).

After use the pipet should be thoroughly rinsed with distilled water or an appropriate solvent and dried. Store in a protective can.

29.2.3 Micropipets

Micropipets are available in various styles. Many are repetitive. One type of high-precision micropipet comes with a tip ejector (Eppendorf). The air displacement design provides superior pipetting performance and eliminates carryover and contamination problems often associated with positive displacement systems (those with a plunger to displace the liquid). Some have a fixed capacity while

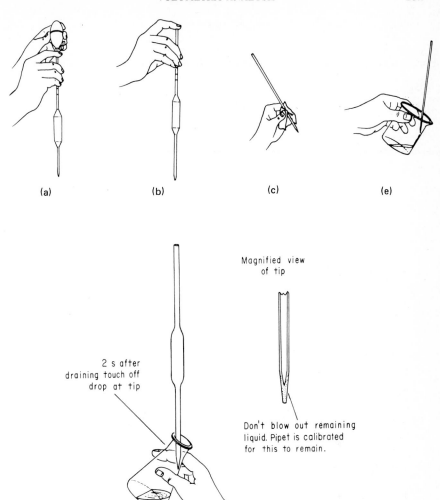

(a)   (b)   (c)   (e)

Magnified view
of tip

2 s after
draining touch off
drop at tip

Don't blow out remaining
liquid. Pipet is calibrated
for this to remain.

(f)

FIGURE 29.5 Techniques for using a volumetric pipet. (*a*) Draw liquid past the graduation mark. (*b*) Use forefinger to maintain liquid level above the graduation. (*c*) Tilt pipet slightly and wipe away any drops on the outside surface. (*d*) (not illustrated) Drain liquid until the meniscus lies at the graduation mark. (*e*) Touch off any drop at the tip into waste. (*f*) Allow pipet to drain freely into the intended container. (*g*) Don't blow out remaining liquid in the tip because pipet is calibrated for this to remain.

others have multivolume capabilities which feature an adjustable, spring-loaded rotating plunger button at the top of the pipet for selection of different volumes. With digital pipets the volume is adjusted by means of a ratchet mechanism built into the control button. Each twist of the button produces an audible click, assuring that a new volume setting is locked into place. Each of the micropipets described use disposable polypropylene pipet tips; some feature automatic tip ejection. Typical accuracy and precision for micropipets is shown in Table 29.3.

TABLE 29.3   Tolerances of Micropipets (Eppendorf)

Capacity, μL	Accuracy, %	Precision, %	Capacity, μL	Accuracy, %	Precision, %
10	1.2	0.4	100	0.5	0.2
40	0.6	0.2	250	0.5	0.15
50	0.5	0.2	500	0.5	0.15
60	0.5	0.2	600	0.5	0.15
70	0.5	0.2	700	0.5	0.15
80	0.5	0.2	1000	0.5	0.15

29.2.4   Burets

Burets, like measuring pipets, deliver any volume up to their maximum capacity. Burets of the conventional type (Fig. 29.6) must be manually filled; usual sizes are 10 and 50 mL. They are made from precision-bore glass tubing. Permanent graduation marks are etched into the glass and filled with fused-on enamel of some color. A blue stripe on the outside back is a distinct aid in meniscus reading. Stopcocks vary from the ground-glass barrel and glass plug (or plastic analog) to a separable polyethylene and Teflon valve assembly. One option involves automatic self-zeroing buret with a dual stopcock plug, one to control filling and the other, dispensing. Tolerances of burets are given in Table 29.4.

29.2.4.1   Cleaning the Buret.   Before being placed in service, a buret must be scrupulously clean. In addition, it must be established that the stopcock is liquid-tight. Grease films that appear unaffected by cleaning solution may yield to treatment with such organic solvents as acetone or benzene. Thorough washing with detergent should follow such treatment. After filling with the titrating solution, no drops of solution should adhere to the inner wall. If they do, reclean the buret.

29.2.4.2   Filling the Buret.   To fill the buret, make certain that the stopcock is closed. Add 5 to 10 mL of solution and carefully rotate the buret to wet the walls completely; allow the liquid to drain through the tip. Repeat this procedure if the buret was not dry initially. Then fill the buret above the zero mark (at top of buret). Free the tip of any air bubbles by allowing small quantities of solution to pass. Finally, lower the level of the solution to, or somewhat below, the zero mark. After allowing about a minute for drainage, take an initial volume reading. It is not necessary to adjust the initial reading to exactly zero.

29.2.4.3   Holding the Stopcock.   Always push the plug into the barrel while rotating the plug during a titration. A right-handed person points the handle of the

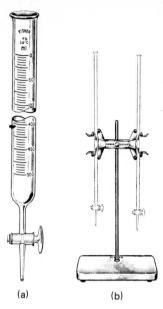

(a)   (b)

FIGURE 29.6 (*a*) Single-dispensing buret with graduated etched scale, standard taper, and stopcock; (*b*) titrating assembly and stand with white base for easy observation of color changes.

TABLE 29.4   Buret Accuracy Tolerances

Capacity, mL	Subdivision, mL	Accuracy, ±mL	
		Class A* and precision grade	Class B and standard grade
10	0.05	0.02	0.04
25	0.10	0.03	0.06
50	0.10	0.05	0.10
100	0.20	0.10	0.20

*Class A conforms to specifications in ASTM E694 for standard taper stopcocks and to ASTM E287 for Teflon or polytetrafluoroethylene stopcock plugs. The 10-mL size meets the requirements for ASTM D664.

stopcock to the right, operates the plug with the left hand, and grasps the stopcock from the left side (see Fig. 29.10).

29.2.4.4 Greasing the Stopcock. Glass plugs and barrels must be lightly greased. First, remove the stopcock and clean the barrel and plug with a swab soaked in benzene. Lightly regrease with a line of grease around the circumference of the plug on both sides of the plug hole (Fig. 29.7).

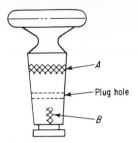

FIGURE 29.7 How to grease a stopcock.

Teflon plugs and polyethylene barrels do not require any lubrication, a distinct advantage. Water adequately lubricates the surfaces.

29.2.5 Reading the Meniscus

Volumetric flasks, burets, pipets, and graduated cylinders are calibrated to measure volumes of liquids. When a liquid is confined in a narrow tube, such as a buret or pipet, the surface is found to exhibit a marked curvature, called a meniscus. It is common practice to use the bottom of the meniscus in calibrating and using volumetric ware (Fig. 29.8). Special care must be used in reading this meniscus.

If a blue line running vertically up the buret or measuring pipet is not present, the bottom of the meniscus, which is transparent, is made more distinct by positioning a black-striped white card behind the glass wall (see Fig. 29.9). Magnification of the stripe is different above and below the meniscus.

Location of the eye in reading the meniscus must be on the level of the me-

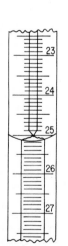

FIGURE 29.8 Reading a meniscus with a blue stripe.

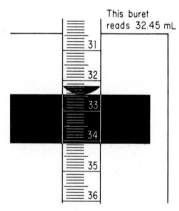

This buret reads 32.45 mL

The black portion of the card is adjusted to make the meniscus plainly visible.

FIGURE 29.9 Useful technique for reading a meniscus. The black portion of the card is adjusted to make the meniscus plainly visible.

niscus (not above, when too small a volume is observed, nor below when too large a volume is observed).

29.3 PERFORMING A TITRATION

To perform a titration, proceed as follows:

1. Use a setup such as that shown in Figs. 29.6 and 29.10. Fill the buret with the titrant as described in Sec. 29.2.4.2.
2. Add the titrant to the titration flask slowly, swirling the flask with the right hand (if right-handed) until the end point is obtained. To avoid overstepping the end point, and with the buret tip well within the titration vessel, introduce solution from the buret in increments of a milliliter or so. Swirl the sample constantly to assure efficient mixing.
3. Rinse the walls of the flask several times during the titration.
4. Reduce the volume of the additions as the titration progresses. In the immediate vicinity of the end point, the titrant should be added a drop at a time. When it is judged that only a few more drops are needed, rinse down the walls of the titration flask. Allow a minute to elapse between the last addition of titrant and the reading of the buret.
5. Near the end point, the trail of color from each drop is quite long. The end point is reached when the color change does not disappear after 30 s.
6. For precision work, volumes of less than one drop can be rinsed off the tip of the buret with wash water.

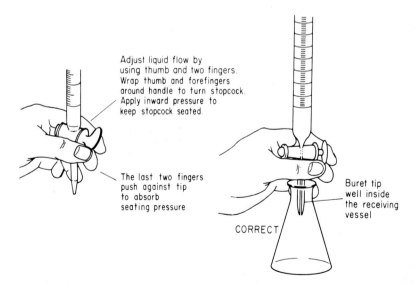

FIGURE 29.10 Preferred method for manipulating a stopcock and using a buret during a titration.

29.4   NEUTRALIZATION TITRATIONS

pH changes in neutralization titrations can be followed continuously with the equipment described in Sec. 18.5. When using color indicators, the correct indicator for a given titration is the one that changes color when the solution has the same pH value as a solution obtained by dissolving the indicator in the same volume of water as formed by the salt after neutralization.

29.4.1   Indicator Selection

Indicator selection may be made in several ways.

1. Base the selection on a potentiometric titration curve (pH versus titrant volume): Choose an indicator whose pK_a value most nearly coincides with the inflection point on the potentiometric titration plot.
2. Arrange several small beakers in pairs; in each beaker place about 10 mL of a solution of the salt formed by the neutralization. Add two drops of one indicator to each beaker of a pair. Now add one drop of titrant to the first beaker (let us say the strong acid titrant) and one drop of strong base titrant to the second beaker. Repeat this operation for successive pairs of beakers, using a different indicator. An appropriate color indicator is one which exhibits a sharp color change when one drop either of acid or base is added.
3. Calculate the pH value at the equivalence point as shown in subsequent sections. Neutralization titrations can be grouped into these categories: (a) strong acid–strong base, (b) weak acid–strong base or strong acid–weak base, and (c) weak acid–weak base.

29.4.2   Strong Acid Titrated with Strong Base

For the titration of a strong acid, i.e., one that is completely ionized in the solvent, with a strong base, the fundamental reaction is

$$H^+ + OH^- \rightarrow H_2O \qquad (29.5)$$

If exactly 25 mL of a 0.1000 M HCl solution is diluted to 100 mL, the pH of the resulting solution is calculated as follows. Since HCl is considered to be 100% ionized, the number of moles of hydrogen ions in solution is

$$(0.025 \text{ L})(0.1000 \text{ mol} \cdot \text{L}^{-1}) = 2.50 \times 10^{-3} \text{ mol}$$

which is now dissolved in a total volume of 100 mL (0.100 L).

$$[H^+] = \frac{2.50 \times 10^{-3} \text{ mol}}{0.100 \text{ L}} = 2.50 \times 10^{-2} \text{ mol} \cdot \text{L}^{-1}$$

The pH is calculated as follows:

$$pH = -\log [H^+] = -\log (2.50 \times 10^{-2}) = -\log 2.50 - (-2) = 1.68$$

Now as the titration progresses, the pH of the solution after 20 mL of 0.110 M NaOH has been added is

$$\underbrace{(0.025 \text{ L})(0.100 \text{ mol} \cdot \text{L}^{-1})}_{\text{Original HCl}} - \underbrace{(0.020 \text{ L})(0.110 \text{ mol} \cdot \text{L}^{-1})}_{\text{NaOH added}} = \underbrace{0.0003 \text{ mol}}_{\text{HCl remaining}}$$

The 0.0003 mol of HCl is now dissolved in 120 mL of solution. Then

$$[\text{H}^+] = 0.0003 \text{ mol}/0.120 \text{ L} = 0.0025 \text{ mol}$$

$$\text{pH} = -\log 0.0025 = -\log 2.5 - (-3) = 2.60$$

At the equivalence point, pH = pOH = 7.00.

The graph of this titration curve is shown in Fig. 29.11. At the right of the illustration are shown the approximate pH values at which three common indicators change color. Either would be suitable for this titration.

In general, the interval between the initial pH of the solution to be titrated and the point at which 99.9 percent neutralization has occurred is three pH units. Similarly, the point at which 0.1 percent excess of unreacted NaOH exists in the titration flask is three pH units less than the pH of the strong base titrant. In our example the initial pH was 1.68; ignoring any dilution effects, the pH at which 99.9 percent has been neutralized would be 4.68. Again ignoring any dilution effect due to added titrant, the pH corresponding to 0.1 percent excess NaOH would be roughly 10 (actually 9.3). Any indicator changing color in the interval from pH 4.7 to pH 9.3 would be suitable.

The minimum initial concentration that can be titrated within 0.1 percent of completion is thus seen to be roughly 0.0001 M. Actually it must be slightly larger because of dissolved carbon dioxide (from the air) in solutions (pH about 6.7).

When titrating a strong base with a strong acid, similar considerations apply. The titration curve shown in Fig. 29.11 would be traced from the end to the be-

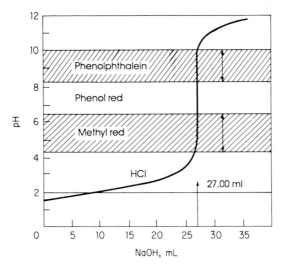

FIGURE 29.11 Titration curve: strong acid–strong base.

ginning. The pH range suitable for a color indicator would again be roughly from pH 9.3 to 4.7.

29.4.3 Weak Acid Titrated with a Strong Base

The titration curve for a weak acid, acetic acid, titrated with a strong base is shown in Fig. 29.12. All calculations for the pH at any point during the titration and up to the end point center around the ionization constant for the equilibrium.

$$HOAc \rightleftharpoons H^+ + OAc^- \qquad (29.6)$$

$$\frac{[H^+][OAc^-]}{[HOAc]} = K_a = 1.86 \times 10^{-5} \qquad (29.7)$$

Now we shall consider various points on the titration curve. Initially

$$[H^+] = [OAc^-] \quad \text{and} \quad [HOAc] = C_{initial} - [H^+]$$

For a 0.100 M solution of HOAc,

$$[H^+]^2 = (0.100 - [H^+])(1.86 \times 10^{-5})$$

Solving, $[H^+] = 1.36 \times 10^{-3}$ and pH = 2.87.

After some titrant has been added and until the equivalence point,

$$[OAc^-] = [OH^-]_{added} \quad \text{and} \quad [HOAc] = C_{initial} - [OH^-]_{added}$$

If we assume that 60 percent HOAc remains untitrated, then 40 percent is [OAc⁻].

$$[H^+] = \frac{[HOAc]}{[OAc^-]}K_a = \frac{0.60}{0.40}1.86 \times 10^{-5} = 2.79 \times 10^{-5}$$

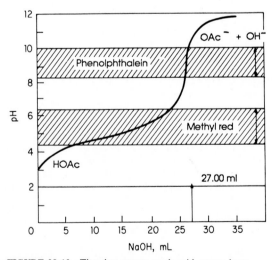

FIGURE 29.12 Titration curve: weak acid–strong base.

and the solution pH = 4.55. Note that the original concentration and the solution volume in the numerator and denominator cancel out. At the equivalence point, the salt formed, sodium acetate, hydrolyzes.

$$OAc^- + H_2O \rightleftharpoons HOAc + OH^- \tag{29.8}$$

Equal amounts of unreacted acetic acid and hydroxyl ion coexist. Although these entities have been brought together in stoichiometric amounts, the neutralization reaction is slightly incomplete (and becomes more so as the acid becomes weaker). Writing the equilibrium expression

$$\frac{[HOAc][OH^-]}{[OAc^-]} = K_{hydrolysis} \tag{29.9}$$

Multiplying numerator and denominator by $[H^+]$

$$\frac{[HOAc][H^+][OH^-]}{[H^+][OAc^-]} = \frac{K_w}{K_a} = K_{hydrolysis} \tag{29.10}$$

This shows that the hydrolysis constant is just the ratio of the ion product of water and the ionization constant of the weak acid. Now reverting to the initial expression for the hydrolysis at the equivalence point,

$$[OH^-]^2 = \frac{K_w}{K_a}[OAc^-] \tag{29.11}$$

If 25.00 mL of 0.1000 M acetic acid is titrated with 0.1000 M NaOH, the final volume will be 50.00 mL and the acetate concentration will be 0.0500 M (actually slightly less because of the hydrolysis, which will be ignored). Substituting data into Eq. 29.11:

$$[OH^-] = \frac{1.0 \times 10^{-14}}{1.86 \times 10^{-5}}(0.050) \quad \text{and} \quad [OH^-] = 5.18 \times 10^{-6}$$

$$pOH = -\log[OH^-] = 5.29 \quad \text{and} \quad pH = 8.71$$

Suitable indicators are very restricted and limited to the pH range from 8.71 to approximately 10 (the point when 0.1 percent excess titrant is present). Phenolphthalein is suitable as shown in Fig. 29.12.

It is also possible to titrate a weak base with a strong acid (Fig. 29.13).

29.4.4 Other Neutralization Systems

We will not calculate the pH values for titration of a weak acid–weak base. Such titrations are of no value in general analytical work. There is no sudden inflection of the titration curve at the equivalence point and no indicator has a sharp enough change in color to indicate the equivalence point with satisfactory precision.

29.4.5 Summary of Acid-Base Titrations

The previous sets of calculations can be summarized as follows:

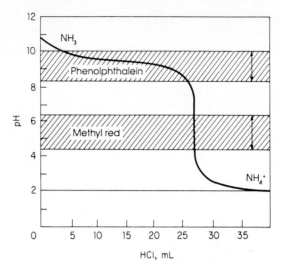

FIGURE 29.13 Titration curve: weak base–strong acid.

Strong acid: $pH = - \log [\text{acid}]$
Strong base: $pH = 14.00 + \log [\text{base}]$
Weak acid: $pH = \frac{1}{2} pK_a - \frac{1}{2} \log [\text{acid}]$
Weak base: $pH = 14.00 - \frac{1}{2} pK_b + \frac{1}{2} \log [\text{base}]$
Salt formed by a weak acid and a strong base:

$$pH = 7.00 + \frac{1}{2} pK_a + \frac{1}{2} \log [\text{salt}]$$

Salt formed by a weak base and a strong acid:

$$pH = 7.00 + \frac{1}{2} pK_b + \frac{1}{2} \log [\text{salt}]$$

Acid salts of a dibasic acid:
$$pH = \frac{1}{2} pK_1 + \frac{1}{2} pK_2 - \frac{1}{2} \log [\text{salt}] + \frac{1}{2} \log (K_1 + [\text{salt}])$$

Mixture of a weak acid and its salt (buffer solution):

$$pH = pK_a + \log \frac{[\text{salt}] + [\text{H}_3\text{O}^+] - [\text{OH}^-]}{[\text{acid}] - [\text{H}_3\text{O}^+] + [\text{OH}^-]}$$

From the calculations of the pH during the course of the titration of a strong acid, note that an interval of three pH units is involved from the initial pH to the pH at which 99.9 percent of the acid has been neutralized. This imposes a limit on the initial acid concentration; it cannot be less than 0.0001 M.

The same limitations apply to the titration of a strong base; the initial concentration cannot be less than 0.0001

For the case of a weak acid titrated with a strong base, the pH at the equivalence point cannot exceed the pH produced by an excess (unused) strong base that corresponds to 0.100 mL (two drops) of a 0.1 M titrant in 100 mL total volume of solution. This pH is 10.0. The detection of the end point may present a problem.

Similar limitations apply to the titration of a weak base when titrated with a strong acid. For a 0.1 M titrant and when 0.1 mL excess is present in 100 mL, the pH will be 4.00.

In summary, when titrating weak acids (or bases), adhere to this simple relationship: $CK_a > 10^{-7}$. Between the midpoint and end point (assuming 99.9 percent completion) of a weak acid, the interval is three pH units. Since 0.1 percent excess titrant (when 0.1 M) represents pH 10, K_a must be 10^{-6} or larger. This leaves only one pH unit gap for an indicator color change. Certain instrumental methods permit the successful titration of somewhat weaker acids.

The preparation of standard volumetric solutions can be found in Chap. 28.

29.4.6 Indicators

With acid-base indicators (Table 17.5), the acid form has one color (or is colorless) and the base has a different one. Acid-base indicators are weak acids or bases themselves and follow the same equilibria as outlined for acids (or bases) during their titration. Their colors are so intense that only a few drops are needed. However, when titrating very dilute solutions, it will be necessary to compute a correction for the consumption of titrant by the indicator. This can be done by running a blank titration using the same total volume of solution and the same number of drops of indicator. Alternatively, the indicator can be adjusted beforehand to the color at the end point. Blank titrations are preferable.

The indicator cation (or anion) is usually visible to the human eye when about 10 percent is present. Considering the equilibrium,

$$HIn \rightleftharpoons H^+ + In^- \tag{29.12}$$

$$pH = pK_{HIn} + \log [In^-] - \log [HIn] \tag{29.13}$$

the usual indicator color change occurs over an interval of two pH units; that is, over the interval pH = $pK_{ind} \pm 1$. Review Figs. 29.11, 29.12, and 29.13, which illustrate the salient facts discussed.

29.4.7 Fluorescent Acid-Base Indicators

Some fluorescent substances are so sensitive to pH that they can be used as indicators in acid-base titrations (Table 29.5). The merit of such indicators is that they can be employed in turbid or intensely colored systems such as foodstuffs, plant extracts, animal liquors, and essential oils. The solution to be titrated is placed in a dark box illuminated by a mercury lamp equipped with a black glass envelope; the progress of the titration is observed visually through a viewing port.

29.5 PRECIPITATION TITRATIONS

Precipitation titrations are used for the determination of chloride, bromide, and iodide—their insoluble silver salts are formed—and, conversely, for the determi-

TABLE 29.5   Fluorescent Acid-Base Indicators

Name	pH range	Color change acid to base	Indicator solution*
3,6-Dihydroxyphthalimide	0.0–2.4	Blue to green	1
	6.0–8.0	Green to yellow-green	
Salicylic acid	2.5–4.0	Non-fl to blue	3
Erythrosin B (tetraiodofluorescein)	2.5–4.0	Non-fl to green	3(0.2%)
Chromotropic acid	3.1–4.4	Non-fl to lt blue	3(5%)
Fluorescein	4.0–4.5	Pink/green to green	3(1%)
5,6-Benzoquinoline	4.4–6.3	Blue to non-fl	2
Umbelliferone	6.5–8.0	Non-fl to blue	
Coumaric acid	7.2–9.0	Non-fl to green	1
Naphthol AS	8.2–10.3	Non-fl to yellow-green	3

*Indicator solutions: (1) 1% solution in ethanol, (2) 0.05% solution in 90% ethanol, (3) sodium or potassium salt in distilled water.

nation of silver—as silver thiocyanate. For these rather simple systems, at the equivalence point,

$$[Ag^+] = [X^-] = \sqrt{K_{sp}} \tag{29.14}$$

where X is the anion of any insoluble silver salt. The more insoluble is the silver salt, the more distinct will be the end point in a titration. While it may seem desirable to stop the titration only at the equivalence point, this is not necessary for the less soluble salts. Silver chloride represents a borderline situation at the equivalence point. The chloride ion remaining in solution is $1.8 \times 10^{-5} M$, which is slightly larger than the $1 \times 10^{-6} M$ quantity usually accepted for a quantitative reaction.

The titration will not be precise if the solubility product exceeds about 10^{-8} when 0.1 M solutions of titrant and 0.01 M solutions of the unknown are used. Obviously the situation will be improved by keeping the unknown concentration high or by decreasing the solubility product. If no alternative system is available, the solubility may be decreased (often by a factor of 10) by adding ethanol or some other water-miscible organic solvent. Coprecipitation is always a danger.

29.5.1   Indicators for Precipitation Titrations

Several different methods are used in determining end points in this branch of volumetric analysis. Of course, potentiometric titration with a silver indicator electrode and a reference electrode separated from the titration system by a double salt bridge (to avoid chloride contamination) remains a possibility. A glass electrode makes a fine reference electrode; in the usual acidic solutions it possesses a constant (although unknown) potential.

Selective ion electrodes are available for the halides and sulfide, as well as for silver. The sodium electrode's extreme sensitivity to silver permits its use in monitoring many argentometric (silver) titrations, such as chloride (Fig. 29.14), bromide, iodide, cyanide, and thiocyanate.

The common color indicators are (1) the blood-red color of the tetrathiocyanatoiron(III)(-1) anion (the Volhard method); (2) the appearance of the red, insoluble silver chromate molecule (the Mohr method); and (3) various adsorption indicators.

29.5.1.1 Volhard Method.

In the titration of the halides (not fluoride) by the Volhard method, a small excess of standard silver nitrate is added. Add indicator [ammonium iron(III) bisulfate] and back titrate with standard KSCN. It is necessary to filter off the coagulated silver chloride precipitate, wash the precipitate, and acidify the filtrate with nitric acid, before back titrating with KSCN because silver thiocyanate is less soluble than silver chloride. The first faint brownish red tinge appearing throughout the liquid phase is taken as the end point. Subtract the milliequivalents of KSCN from the total milliequivalents of silver added to arrive at the milliequivalents of halide present. The Volhard method is also used to determine silver by direct titration with KSCN.

29.5.1.2 Mohr Method.

In the Mohr method the chloride solution should be neutral (pH about 6.3). Addition of excess calcium carbonate is an easy way to adjust the solution pH. Potassium chromate solution is added, and the titration commenced. The end point is signaled by the appearance of the brownish tinge of silver chromate. If the titration is done rapidly, the method can be applied to bromide. However, iodide reacts with chromate to form iodine and chromium(III).

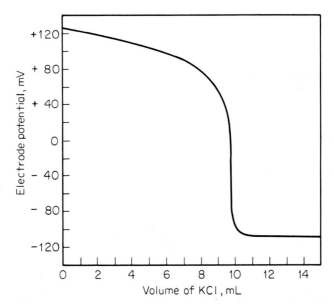

FIGURE 29.14 Silver nitrate titrated with potassium chloride using a sodium glass electrode (which responds to silver ions) as the indicator electrode.

29.5.1.3 Adsorption Indicators.   When present during a titration, certain types of dyestuff ions are adsorbed more strongly on the surface of colloidal silver halide precipitation when on one side of the equivalence point than the other. Dyestuff anions are more strongly adsorbed when an excess of the positive ion of the precipitate is present in the solution than at the equivalence point or when an excess of the negative ion is present. For example, dyes of the fluorescein series (particularly dichlorofluorescein) are used in the form of their sodium salts; the anions of these salts are more strongly adsorbed on a silver halide when a slight excess of silver ion is present in the solution. After the equivalence point in a halide titration, the dye changes sharply to a pink color which appears to be uniformly distributed through the solution but is in reality at the surface of the colloidal particles. Conversely, cations of certain basic dyestuffs, like the rhodamines, are more strongly adsorbed on a colloidal silver halide when a slight excess of the halide ion is present. It is important to remember that the precipitate must be colloidal, and for that purpose a protective colloid, such as dextrin (potato starch), must be added to the system before the titration is started.

29.6 OXIDATION-REDUCTION TITRATIONS

29.6.1 Theory

First, let us consider an ideal oxidation-reduction system. Stating the Nernst expression in terms of concentration and assuming the temperature is 25°C, the electrode potential is given by

$$E = E^{\circ\prime} - \frac{0.059\ 16}{n} \log \frac{[red]}{[ox]} \tag{29.15}$$

Remembering that an oxidation is considered to be quantitatively complete when 0.1 percent or less of the oxidant remains (and 99.9 percent is now in the form of reductant), the potential range encompassed during a titration will be

from $E^\circ + 3(0.059/n)$
through E°
to $E^\circ - 3(0.059/n)$

Any other oxidation-reduction system infringing upon this region constitutes an interference.

If $n = 1$ and E° is 0.77 V, the range extends from 0.95 V (all oxidant) to 0.59 V (all reductant), a span of 0.36 V. If $n = 2$, the span would be only 0.18 V.

For the reductant to be suitable for use as a titrant, it should have an E° value that is at least $0.36/n$ V less than the E° of the oxidant. This is the minimum difference because a potentiometric titration curve would only show an inflection at the end point. To use colored indicators, the difference would have to be increased at least 0.12 V; this assumes an indicator (Table 29.6) is available that fits exactly into the potential gap between 0.59 and 0.47 V when E° of the oxidant is 0.77 V. The E° of the indicator would have to be 0.53 V. A little more room on the potential scale is seen to be desirable, either to provide an adequate visual color change with an indicator (Table 29.6) or to provide a "jump" (noticeable break) on the potential axis in potentiometry.

TABLE 29.6   General Oxidation-Reduction Indicators

Name	Reduction potential, V (30°C)		Color change upon oxidation
	pH = 0	pH = 7	
Bis(5-bromo-1,10-phenanthroline) ruthenium(II) dinitrate	1.41		Red to faint blue
Tris(5-nitro-1,10-phenanthroline) iron(II) sulfate	1.25		Red to faint blue
Tris(4,7-diphenyl-1,10-phenanthroline) iron(II) disulfate	1.13 (4.6 M H_2SO_4)		Red to faint blue
4-Nitrodiphenylamine	1.06		Colorless to violet
Erioglaucine A	1.00*		Green yellow to bluish-red
Diphenylamine-4-sulfonate (Na salt)	0.85		Colorless to violet
4'-Ethoxy-2,4-diamino-azobenzene	0.76		Red to pale yellow
2,6-Dichloroindophenol (Na salt)	0.67	0.22	Colorless to blue
Methylene blue	0.53	0.01	Colorless to blue
Indigo-5-monosulfonic acid (Na salt)	0.26	−0.16	Colorless to blue
Safranine T	0.24	−0.29	Colorless to violet-blue

*First noticeable color transition often 60 mV less than $E°$.

29.6.2   Prior Oxidation and Reduction

In many analyses it is necessary to carry out oxidation or reduction steps prior to the actual determination in order to bring the sample constituent quantitatively to a particular oxidation state. Several conditions must be fulfilled:

1. The prior oxidation or reduction must be quantitative and reasonably rapid for practical application.
2. The excess oxidant or reductant must be easy to remove.
3. The prior oxidation or reduction must be selective to avoid interference from other sample components.

29.6.3   Oxidizing Agents for Prior Oxidations

Selected oxidizing agents are listed in Table 29.7. In place of the free halogens used directly, these species can be generated at the point of mixing in an acid medium by using a mixture of chlorate plus chloride (heated) for the generation of chlorine, bromate and bromide for bromine, and iodate plus iodide for iodine. Care must be taken when using concentrated hydrogen peroxide that its concentration never exceed 30 percent. Solid compounds containing the peroxo group, such as potassium peroxodisulfate, offer an alternative.

TABLE 29.7   Oxidizing Agents Useful for Prior Oxidation

Name	Standard potential, V (25°C)	Comments
Ozone	2.07	Special generator needed
Peroxydisulfate	2.0	Silver ion acts as catalyst; excess oxidant removed by boiling
Hydrogen peroxide (30%)	1.77 (acid)	Decomposed by boiling
	0.88 (base)	Good for alkaline solution
Sodium bismuthate	1.59	Excess removed by filtration through glass frit or asbestos
Permanganate	1.51	Excess removed by sodium azide followed by boiling the solution
Hydrogen peroxide (30%) plus HCl	~ 1.4	Free chlorine formed; excess peroxide removed by boiling
Perchloric acid	1.34	Hot, concentrated (72%); after cooling and dilution, boil solution to remove chlorine
Bromine (aqueous)	1.08	Excess boiled away

Coulometry offers a convenient method for producing oxidants whose solutions are not stable on standing. These include bromine, iodine, and chlorine.

29.6.4   Reducing Agents for Prior Reductions

Table 29.8 contains a listing of selected reducing agents. When present in aque-

TABLE 29.8   Reducing Agents for Prior Reductions

Name	Reduction potential, V (25°C)	Comments
Sodium metal	−2.7	
Sodium borohydride in 1–4 M HCl	unk	Forms volatile hydrides with As, Ge, Sb, Se, Sn, and Te
Zinc (amalgamated)	−0.76	Solution passed through column of zinc granules
Hypophosphorous acid	−0.51	
Chromium(II) chloride	−0.4	Excess removed by air oxidation
Raney nickel	−0.25	
Hydrogen (Pt catalyst)	0.0	
Titanium(III) chloride	0.1	
Tin(II) chloride	0.14	
Sulfur dioxide	0.17	Excess removed in stream of carbon dioxide
Ascorbic acid	~0.17	
Hydrazine hydrochloride	0.17	
Silver (with HCl)	0.22	Solution passed through column of silver granules

ous solution, the stronger reductants often bring about the liberation of hydrogen upon standing. Oxygen must be strictly excluded; a blanket of nitrogen gas or carbon dioxide gas (conveniently released by pellets of dry ice) should cover the system. Coulometry offers a convenient method to produce reductants in situ without problems of storage; these reductants include chromium(II), tin(II), titanium(III), and copper(I).

29.7 COMPLEXOMETRIC TITRATIONS

Of the reagents used in complexometric titrations, the field is dominated by ethylenediaminetetraacetic acid, abbreviated EDTA or H_4Y. Coordinating agents that occupy a single coordination position, like NH_3 or CN^-, are invariably added in a succession of steps (Chap. 24). Unless one particular step happens to be extraordinarily stable, there will be no extended range of concentration of complexing agent over which a single species of complex is formed except for the highest complex. By contrast, EDTA, a polydentate ligand (one that has several coordination sites occupied simultaneously), forms very stable 1:1 complexes with many metal ions.

The successive pK_a values of H_4Y are: 1.99, 2.67, 6.16, and 10.26 at 20°C and an ionic strength of 0.1. The fraction α_4 of Y^{4-} present at various pH values is given in Table 29.9.

TABLE 29.9 Values for α_4 for EDTA in Solutions of Various pH

pH	α_4	$-\log \alpha_4$	pH	α_4	$-\log \alpha_4$
2.0	3.7×10^{-14}	13.44	8.0	5.4×10^{-3}	2.29
3.0	2.5×10^{-11}	10.60	9.0	0.052	1.29
4.0	3.6×10^{-9}	8.48	10.0	0.35	0.46
5.0	3.5×10^{-7}	6.45	11.0	0.85	0.07
6.0	2.2×10^{-5}	4.66	12.0	0.98	0.00
7.0	4.8×10^{-4}	3.33			

29.7.1 Formation of Metal-EDTA Complexes

The formation constants of the metal-EDTA complexes, which are equilibrium constants of reactions of the type

$$M^{n+} + Y^{4-} \rightleftharpoons MY^{(n-4)+} \tag{29.16}$$

$$K_{MY} = \frac{[MY^{(n-4)+}]}{[M^{n+}][Y^{4-}]} \tag{29.17}$$

are listed in Table 29.10. There are three categories of metal-EDTA complexes. (1) Those with very large formation constants can be titrated in strongly acid solutions (pH 0 to 2). (2) Those with constants (log K values) that lie between 14 and 19 can be titrated in weakly acid media (pH 4 to 6). (3) The remainder are best titrated in strongly ammoniacal solutions. The minimum pH for EDTA titration of various metal ions is shown in Fig. 29.15.

To work with a large fraction of Y^{4-} and at the same time prevent the precipitation of metals as hydroxides (or basic salts), an auxiliary masking agent is

TABLE 29.10   Formation Constants of EDTA Complexes

Metal ion	log K_{MY}	Metal ion	log K_{MY}
Cr(III)	36	Ni(II)	18.56
Co(III)	36	Pb(II)	18.3
In(III)	25	Zn(II)	16.4
Fe(III)	24.23	Cd(II)	16.4
Th(IV)	23.2	La(III)	16.34
Bi(III)	22.8	Al(III)	16.11
Hg(II)	21.80	Fe(II)	14.33
Zr(IV)	19.40	Mn(II)	13.8
Cu(II)	18.7	Ca(II)	11.0
Co(II)	16.3	Mg(II)	8.64
		Sr(II)	8.80

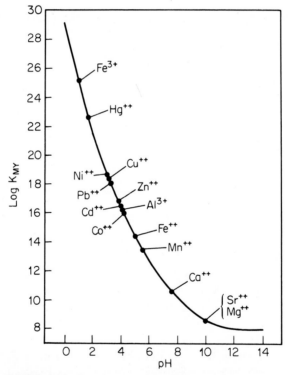

FIGURE 29.15   The minimum pH for EDTA titration of various metal ions to give a satisfactory end point.

added to the titration system. Often the auxiliary masking agent is ammonia, although citrate is useful when working in acidic media.

29.7.2 Titration Curves

The titration of calcium ion with EDTA does not involve any auxiliary masking agent, although the ammonium-ammonia buffer system is used to adjust the solution to about pH 9.2. Figure 29.16 shows the titration curves at various pH values. No distinct potential break occurs until pH = 9 or greater. Just as with the other types of titrations, the pCa must change by three units between the initial value and the end point value. Additional space must be provided for the color indicator's reaction before the pCa curve begins to bend toward the volume axis.

The titration of nickel(II) with EDTA illustrates the interaction between masking agent and pH (Fig. 29.17). Over the range from pH 4 to 7, only nickel(II) aquo ions exist in the solution until they react with EDTA. However, beginning at pH > 7, increasing amounts of $Ni(NH_3)_4^{2+}$ are formed, which lifts the curve until auxiliary complexation is complete. This shifts the end point to higher values of pNi. At the same time, the fraction of Y^{4-} is increasing. Suitable end points lie between pH 6 and 11; the low end might be an acetate buffer and the high end an

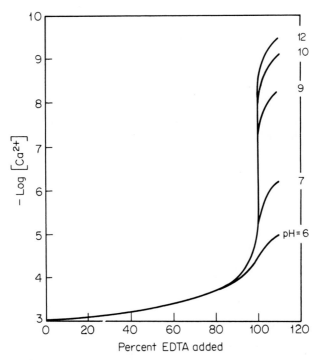

FIGURE 29.16   Titration curves of calcium with EDTA at various values of solution pH.

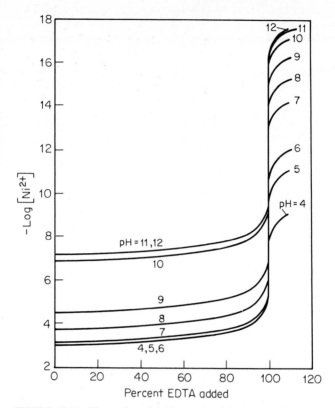

FIGURE 29.17 Illustration of the interaction between masking agent and pH. The total nickel concentration is 0.001 M. The sum of free ammonia and ammonium ion is 0.1 M.

ammoniacal buffer. The minimum pH for a satisfactory end point is shown in Fig. 29.15 for the EDTA titration of various metal ions.

29.7.3 Buffers

Except when working in acidic solutions of pH > 2.5, it is necessary to have a buffer present to remove the proton displaced from EDTA by the metal. An acetate buffer is suitable for those metals with log K_{MY} between 14 and 18. For less stable complexes, an ammoniacal buffer is used.

29.7.4 Color Indicator Systems

1. Eriochrome black T is a widely used metallochromic indicator. It exhibits a blue color in the uncomplexed form which changes to red when complexed with the metal. With direct titration, the indicator is suitable for Ba, Ca, Cd, In, Pb, Mg, Mn, rare earths, Sr, and Zn. Some elements are best handled by adding an excess of EDTA, then back titrating the unused EDTA with a metal ion which forms a less stable EDTA complex, including Al, Bi, Co, Ga, Fe, Hg, Ni, Pd, and Tl.

2. Murexide finds use in the titration of Ca, Co, Cu, and Ni.

3. 1-(2-Pyridylazo)-2-naphthol (PAN) has been used for the direct titration of Cd, Cu, In, Sc, and Zn, and the back titration of Bi, Co, Cu, Ga, Fe, Pb, Ni, Sc, and Zn.

4. Pyrocatechol violet has been used in the direct titration of Al, Cd, Co, Cu, Ga, Fe, Pb, Mg, Mn, Ni, Th, and Ti.

29.7.5 Potentiometric Indicator Electrodes

Ion-selective electrodes, when available, are very suitable indicator electrodes for complexometric titrations (see Table 18.9).

A small stationary mercury electrode (see Chap. 19) or gold amalgam wire in contact with a solution to which a small quantity of mercury(II) chelonate, HgY^{2-}, has been added, establishes the half-cell:

$$Hg \mid HgY^{2-}, MY^{(n-4)+}, M^{n+}$$

where the electrode potential is given by

$$E = E^\circ + \frac{0.0592}{2} \log \frac{[M^{n+}][HgY^{2-}]}{[MY^{(n-4)+}]} \tag{29.18}$$

Because a fixed amount of HgY^2 is present, the potential is dependent upon the ratio $[M^{n+}]/[MY^{(n-4)+}]$. The species HgY^2 must be considerably more stable than $MY^{(n-4)+}$.

CHAPTER 30
FILTRATION

30.1 INTRODUCTION

Filtration is the process of removing material, often but not always a solid, from a substrate in which it is suspended. This process is a physical one; any chemical reaction is inadvertent and normally unwanted. Filtration is accomplished by passing the mixture to be processed through one of the many available sieves called filter media. These are of two kinds: surface filters and depth filters.

With the surface filter, filtration is essentially an exclusion process. Particles larger than the filter's pore or mesh dimensions are retained on the surface of the filter; all other matter passes through. Examples are filter papers, membranes, mesh sieves, and the like. These are frequently used when the solid is to be collected and the filtrate is to be discarded. Depth filters, however, retain particles both on their surface and throughout their thickness; they are more likely to be used in industrial processes to clarify liquids for purification.

In the laboratory, filtration is generally used to separate solid impurities from a liquid or a solution or to collect a solid substance from the liquid or solution from which it was precipitated or recrystallized. This process can be accomplished with the help of gravity alone (Fig. 30.1) or it can be speeded up by using vacuum techniques (discussed later).

The efficiency of filtration depends on the correct selection of the method to be used, the availability of various pieces of apparatus. the use of the filter medium most appropriate for the particular process, and the use of correct laboratory technique in performing the manipulations involved. Although the carrier liquid is usually relatively nonreactive, it is sometimes necessary to filter materials from high alkaline or acidic carrier liquids or to perform filtration under other highly reactive conditions. A variety of filter media exists from which it is possible to select one that best fits the particular objectives and conditions of a given process.

30.2 FILTER MEDIA

30.2.1 Paper

There are several varieties or grades of filter paper (Table 30.1) for special purposes. There are qualitative grades, low-ash or ashless quantitative grades, hardened grades, and even glass-fiber papers. For a given filtration, you must select

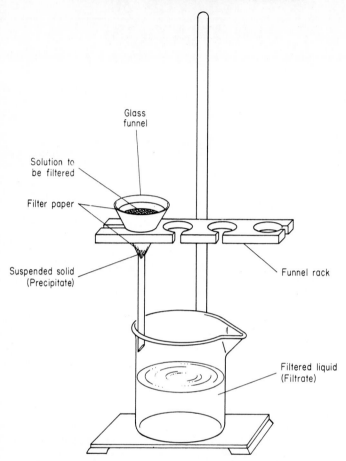

FIGURE 30.1 A gravity-filtration setup using filter paper and a funnel.

the proper filter paper with regard to porosity and residue (or ash). Qualitative-grade paper will leave an appreciable amount of ash upon ignition (on the order of 0.7 to 1 mg from a 9-cm circle) and is therefore unsuitable for applications in quantitative analysis where precipitates are to be ignited on the paper and weighed. They are widely used for clarification of solutions, filtration of precipitates which will later be dissolved, and general nonquantitative separations of precipitates from solution.

Low-ash or ashless quantitative-grade papers can be ignited without leaving an ash. The residue left by an 11-cm circle of low-ash paper may be as low as 0.06 mg and that left by an ashless-grade paper is 0.05 mg or less. This small residue can be considered negligible for most work.

Hardened-grade papers are designed for use in vacuum filtrations and are processed to have great wet strength and hard, lintless surfaces. They are available in low-ash and ashless as well as regular grades.

All grades of filter paper are manufactured in a variety of sizes and in several degrees of porosity. Select the proper porosity for a given precipitate. If too

TABLE 30.1   Commonly Used Filter Papers*

S-W	W	S&S	Porosity†	Speed	Used for
			Qualitative or regular grade papers		
500	4	604	Coarse	Very rapid	Coarse or gelatinous precipitates
501	1	595	Medium	Medium	Ordinary crystalline precipitates
503	3	602	Medium	Slow	Fine precipitates; used with Buchner funnel
			Quantitative grade papers (less than 0.1 mg ash)		
800	41	589 black	Coarse	Very rapid	Coarse or gelatinous precipitates
850	40	589 white	Medium	Medium	Ordinary crystalline precipitates
900	42	589 blue	Fine	Slow	Fine crystalline precipitates

*Code: S-W, Sargent-Welch; W, Whatman; S&S, Schleicher & Schuell.
†Relative retention for porosity: coarse, 20–25 μm; medium, 11 μm; fine, 2.5 μm.

coarse a paper is used, very small crystals may pass through, while use of too fine a paper will make filtration unduly slow. The main objective is to carry out the filtration as rapidly as possible, retaining the precipitate on the paper with a minimum loss.

30.2.2 Glass-Fiber Filters

Glass-fiber filters are produced from borosilicate glass. Their soft texture and high water absorption results in excellent sealing in most types of filter holders. These filters give a combination of fine retention, rapid filtration, high particulate loading capacity, and inertness not found in any cellulose paper. They must always be used flat; folding can rupture the glass-fiber matrix, leading to cracks and poor retention.

These filters can be used as a prefilter, which is placed on top of the filter to be protected. The prefilter removes all coarse particles, allowing only fine particles to pass and be trapped by the second filter. For continuous filtration procedures, the life of the second filter may be markedly extended.

30.2.3 Membrane Filters

There are two types of membrane filters—surface and depth. Surface membrane filters are used for final filtration or prefiltration, whereas a depth membrane filter is generally used in clarifying applications, where quantitative retention is not required, or as a prefilter to prolong the life of a downstream surface membrane filter.

Membrane filters are thin polymeric (plastic) structures with extraordinarily fine pores. Such filters are distinctive in that they remove from a gas or liquid passing through them all particulate matter or microorganisms larger than the filter pores. With proper filter selection, they yield a filtration that is ultraclean and/or sterile. Membrane filters are available in a wide variety of pore sizes in a number of different polymeric materials. The range of pore sizes and the uniformity of pore size in a typical filter are shown in Table 30.2. For sterilization, membrane pore sizes are usually 0.2 μm. For clarification of solutions a pore size of 0.45 μm is usually used. Separation of foreign particles in sampling waters and gases utilizes a pore size of 3 to 5 μm.

TABLE 30.2   Membrane Filters

Filter pore size, μm	Maximum rigid particle to penetrate, μm	Filter pore size, μm	Maximum rigid particle to penetrate, μm
14	17	0.65	0.68
10	12	0.60	0.65
8	9.4	0.45	0.47
7	9.0	0.30	0.32
5	6.2	0.22	0.24
3	3.9	0.20	0.25
2	2.5	0.10	0.108
1.2	1.5	0.05	0.053
1.0	1.1	0.025	0.028
0.8	0.95		

Membrane systems are operated in a tangential flow mode. Feed material sweeps tangentially across the upstream surface of the membrane as filtration occurs. Membranes can be repeatedly regenerated with strong cleaning agents.

30.2.4 Fritted Ware

Fritted ware includes funnels and crucibles with fritted-glass disks sealed permanently into the lower portion of the unit. These units are useful in instances where a paper filter might be attacked. Porosity is controlled in manufacture, and disks are graded into several classifications (Table 30.3). The extra coarse and coarse porosities are held toward the maximum pore diameter listed, whereas the medium, fine, very fine, and ultrafine are held toward the minimum pore diameter listed. With care fritted glass crucibles can be used for quantitative analyses requiring ignition to a temperature as high as 500°C.

30.2.4.1 Care and Cleaning of Fritted Ware. A new fritted filter should be washed by suction with hot HCl and then rinsed with water before it is used. This treatment removes loose particles of foreign matter. Clean again immediately after use. Many precipitates can be removed from the filter surface simply by rinsing from the reverse side with water under pressure not exceeding 15 lb · in.$^{-1}$. Alternatively, draw water through the filter from the reverse side with a vacuum pump. When the pores remain clogged, chemical treatment will be needed. Cleaning solutions for fritted ware are given in Table 30.4.

TABLE 30.3   Porosities of Fritted Glassware

Porosity	Nominal maximum pore size, μm	Principal uses
Extra coarse	170–220	Filtration of very coarse materials. Gas dispersion, gas washing, and extractor beds. Support of other filter materials
Coarse	40–60	Filtration of coarse materials. Gas dispersion, gas washing, gas absorption. Mercury filtration. For extraction apparatus
Medium	10–15	Filtration of crystalline precipitates. Removal of "floaters" from distilled water
Fine	4–5.5	Filtration of fine precipitates. As a mercury valve. In extraction apparatus
Very fine	2–2.5	General bacteria filtrations
Ultrafine	0.9–1.4	General bacteria filtrations

TABLE 30.4   Cleaning Solutions for Fritted Glassware

Material	Cleaning solution
Fatty materials	Carbon tetrachloride
Organic matter	Hot concentrated sulfuric acid plus a few drops of sodium or potassium nitrate
Albumen	Hot aqueous ammonia or hot hydrochloric acid
Glucose	Hot mixed acid (sulfuric plus nitric acids)
Copper or iron oxides	Hot hydrochloric acid plus potassium chlorate
Mercury residue	Hot nitric acid
Silver chloride	Aqueous ammonia or sodium thiosulfate
Aluminous and siliceous residues	Hydrofluoric acid, 2%, followed by concentrated sulfuric acid; rinse immediately with distilled water followed by a few milliliters of acetone. Repeat rinsing until all trace of acid removed.

30.3   FILTERING ACCESSORIES

30.3.1   Filter Supports

Some solutions tend to weaken the filter paper, and at times, the pressure on the cone of the filter will break the filter paper, ruining the results of the filtration. Thin, woven textile disks (Fig. 30.2) can be used to support the tip of the filter-paper cone. They are approximately the same thickness as the filter paper and therefore ensure close contact of the reinforced paper with the funnel walls. They are folded along with the filter paper when it is formed into the normal conical shape, and they can easily be removed from the wet filter paper after the filtration has been completed, if ashing is desired.

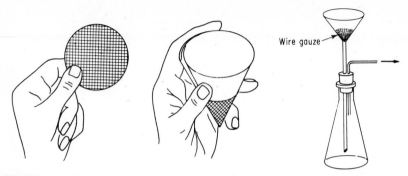

FIGURE 30.2   Woven filter-paper support and funnel with a wire-gauze cone as support for the filter paper.

Cones constructed of platinum mesh are available. They are completely inert and are inserted into a funnel, and the folded filter paper is placed within the cone.

30.3.2   Filter Aids

During filtration certain gummy, gelatinous, flocculent, semicolloidal, or extremely fine particulates often quickly clog the pores of a filter paper. Filter aids consist of diatomaceous earth and are sold under the trade names of Celite or FilterAid. They are pure and inert powderlike materials which form a porous film or cake on the filter medium. In use, they are slurried or mixed with the solvent to form a thin paste and then filtered through the paper. An alternative procedure involves the addition of the filter aid directly to the problem slurry with thorough mixing.

Filter aids cannot be used when the object of the filtration is to collect a solid product because the precipitate collected also contains the filter aid.

For quantitative purposes, filtration of gelatinous materials can be assisted by adding shredded ashless paper tablets to the solution before filtration. The dispersed fibers form passages through the precipitate for flow of liquid.

30.4   MANIPULATIONS ASSOCIATED WITH THE FILTRATION PROCESS

Whether one uses gravity or vacuum filtration, three operations must be performed: decantation, washing, and transfer (Fig. 30.3).

30.4.1   Decantation

When a solid readily settles to the bottom of a liquid and shows little or no tendency to remain suspended, it can be separated easily from the liquid by carefully pouring off the liquid so that no solid is carried along. Proceed as follows:

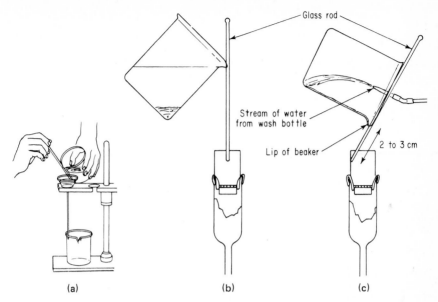

FIGURE 30.3 (*a*) Gravity filtering operation; (*b*) decantation; and (*c*) transfer of the last portions of precipitate. Also illustrates vacuum filtration with a Gooch crucible.

1. Hold the container which has the mixture in it in one hand and have a glass stirring rod in the other.
2. Incline the container until the liquid has almost reached the lip.
3. Touch the center of the glass rod to the lip of the container and the end of the rod to the side of the container into which you wish to pour the liquid.
4. Continue the inclination of the original container until the liquid touches the glass rod and flows along into the second container. The glass rod enables you to pour the liquid from the original container slowly enough that the solid is not carried along and also prevents the liquid from running back along the outside of the original container.

30.4.2 Washing

The objective of washing is to remove the excess liquid phase and any soluble impurities which may be present in the precipitate. Use a solvent which is miscible with the liquid phase but does not dissolve the precipitate.

Solids can be washed in the beaker after decantation of the supernatant liquid phase. Add a small amount of the wash liquid, and thoroughly mix it with the precipitate. Allow the solid to settle. Decant the wash liquid through the filter. Allow the precipitate to settle, with the beaker tilted slightly so that the solid accumulates in the corner of the beaker under the spout. Repeat this procedure several times. Several washings with small volumes of liquid are more effective in removing soluble contaminants than a single washing using the total volume.

30.4.3 Transfer of the Precipitate

Remove the bulk of the precipitate from the beaker to the filter by using a stream of wash liquid from a wash bottle. Use the stirring rod to direct the flow of liquid into the filtering medium. The last traces of precipitate are removed from the walls of the beaker by scrubbing the surfaces with a rubber (tip) policeman attached to the stirring rod. All solids collected are added to the main portion of the filter paper. If the precipitate is to be ignited, use small fragments of ashless paper to scrub the sides of the beaker; then add these fragments to the bulk of the precipitate in the filter with the collected solid.

30.4.4 Gravity Filtration

During gravity filtration the filtrate passes through the filter medium under the forces of gravity and capillary attraction between the liquid and the funnel stem. The most common procedure involves the use of filter paper and a conical funnel. The procedure is slow, but it is highly favored for gravimetric analysis over the more rapid vacuum filtration because there is better retention of fine particles of precipitate and less rupturing or tearing of the paper. Moreover, gravity filtration is generally the fastest and more preferred method for filtering gelatinous precipitates.

Avoid accumulating precipitate on the filter paper during the early stages of the filtration process. This is necessary for rapid filtering. Remove as much liquid as possible by decantation through the filter. Washing the precipitate while it is still in the beaker and decanting the liquid speeds filtration. The bulk of the precipitate is not added to the filter until the last stages of the filtration as part of the washing process.

Optimum filtering speed is achieved in gravity filtration by proper folding and positioning of the paper in the funnel (Fig. 30.4). Follow these suggestions.

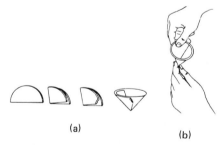

(a)

(b)

FIGURE 30.4 (a) Steps in folding paper for use in filtering with a regular funnel. The second fold is not exactly at a right angle. Note the tear, which makes the paper stick better to the funnel; (b) seating of a filter paper.

1. Take maximum advantage of capillary attraction to assist in drawing the liquid phase through the paper. Use a long-stemmed funnel and maintain a continuous column of water from the tip of the funnel stem to the undersurface of the paper. The tip of the funnel should touch the side of the vessel which receives

the filtrate. Accurate fit of the folded paper helps maintain an airtight seal between the funnel and the top edge of the wet filter paper.

2. Expose as much of the paper as possible to provide free flow of liquid through the paper. If you fold the paper as shown in Fig. 30.4, the paper will not coincide exactly with the walls of the funnel and the liquid will be able to flow between the paper and the glass. This type of fold will also help to maintain an airtight seal between the top edge of the filter paper and the funnel.

3. Fluting the paper (Fig. 30.5) permits free circulation of air and is advantageous when filtering hot solutions. Prepleated and prefolded fluted filters can be purchased commercially.

4. Many precipitates will spread over a wetted surface against the force of gravity; this behavior is known as creeping, and it can cause loss of precipitate. For this reason, be sure to pour the solution into the filter until the paper cone is no more than three-quarters filled. This provides an area near the top of the paper which is free of precipitate. By grasping this "clean" portion, you can remove the cone from the funnel and fold it for ignition.

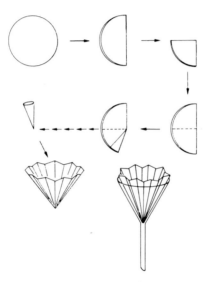

FIGURE 30.5 Folding a fluted filter. Fold the filter paper in half, then fold this half into eight equal sections, like an accordion. The fluted filter paper is then opened and placed in a funnel.

30.5 VACUUM FILTRATION

Vacuum filtration is a very convenient way to speed up the filtration process, but the filter medium must retain the very fine particles without clogging. The vacuum is normally provided by a water aspirator, although a vacuum pump, protected by suitable traps, can be used. Because of the inherent dangers of flask collapse from the reduced pressure, thick-walled filter flasks should be used.

Vacuum filtration is advantageous when the precipitate is crystalline. It should not be employed for gelatinous precipitates as clogging will occur. When performing vacuum filtration with filter paper, the folded paper is inserted into a small porous metal liner in the apex of the funnel.

Solutions of very volatile liquids and hot solutions are not filtered conveniently with suction. The suction may cause excessive evaporation of the solvent, which cools the solution enough to cause precipitation of the solute.

30.5.1 Arrangement of Equipment

A typical setup for carrying out a vacuum filtration is shown in Fig. 30.6. This illustration shows the use of a Büchner funnel in which the wetting filter paper must be seated before the suction is applied. The funnel or crucible is fitted to a suction flask. The sidearm of the flask is connected to a source of vacuum. A water-trap bottle is inserted between the flask and the source of vacuum.

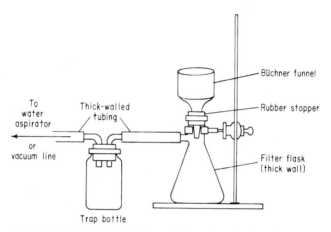

FIGURE 30.6   Vacuum filtration assembly using a Büchner funnel.

30.5.2 Equipment and Filter Media

30.5.2.1 Büchner Funnels.   Büchner funnels (Fig. 30.7) are often used for vacuum filtration. They have a flat, perforated bottom. A filter-paper circle of a diameter sufficient to cover the perforations is placed on the flat bottom, moistened, and tightly sealed against the bottom by applying a slight vacuum. In use, the precipitate is allowed to settle and the liquid phase is first decanted by pouring it down a stirring rod aimed at the center of the filter paper, applying only a light vacuum until sufficient solid has built up over the paper to protect it from breaking. The vacuum is then increased, and the remainder of the precipitate is added.

The precipitate is washed by adding small amounts of wash liquid over the surface of the precipitate, allowing the liquid to be drawn through the solid slowly with the vacuum. Precipitates cannot be ignited and weighed in Büchner funnels.

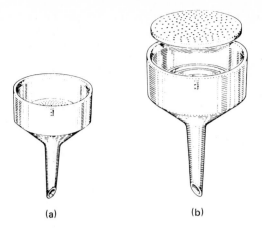

(a)   (b)

FIGURE 30.7 Büchner funnels without and with a removable plate.

Büchner funnels are useful in synthetic work. Precipitates can be air-dried by allowing them to stand in the funnel and drawing a current of air from the room through the precipitate with the source of vacuum.

30.5.2.2 Porous Porcelain and Monroe Crucibles. Porcelain crucibles with porous ceramic disks permanently sealed in the bottom are used in the same way as sintered-glass crucibles except that they may be ignited at high temperatures. The Monroe crucible is made of platinum with a burnished platinum mat serving as the filter medium. The latter possesses a high degree of chemical inertness.

30.5.2.3 Gooch Crucible. The Gooch crucible (Fig. 30.8) is a porcelain thimble with a perforated base. The filter medium is either a mat of asbestos or a fiberglass paper disk. The mat is prepared by pouring a slurry of asbestos fiber sus-

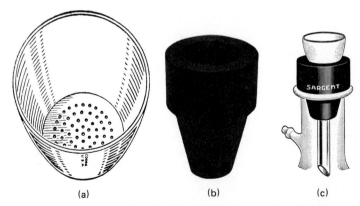

(a)   (b)   (c)

FIGURE 30.8 (*a*) Gooch crucible with perforated bottom; (*b*) rubber adapter for suction filtration; (*c*) adapter in use with a Gooch crucible in a suction flask.

pended in water into the crucible and applying light suction. The mat is sufficiently thick when no light can be seen when looking through the perforated base.

Asbestos mats permit precipitates to be quantitatively ignited to extremely high temperatures without danger of reduction by carbon from paper. With fiberglass mats, ignition temperatures must not exceed 500°C. Both filter media are resistant to attack by most chemicals.

The filter media used in the Gooch crucible are quite fragile. Exercise extreme care when adding the liquid so that the asbestos or glass paper will not be disturbed or broken, allowing the precipitate to pass through.

30.5.2.4 *Platinum Crucible.*

Platinum is useful in crucibles for specialized purposes. The chemically valuable properties of platinum include its resistance to attack by most mineral acids, including hydrofluoric acid; its inertness with respect to many molten salts; its resistance to oxidation, even at elevated temperatures; and its very high melting point.

With respect to limitations, platinum is readily dissolved on contact with aqua regia and with mixtures of chlorides and oxidizing agents generally. At elevated temperatures it is also dissolved by fused alkali oxides, peroxides, and, to some extent, hydroxides. When heated strongly, it readily alloys with such metals as gold, silver, copper, bismuth, lead, and zinc. Because of this predilection toward alloy formation, contact between heated platinum and other metals or their readily reduced oxides must be avoided. Slow solution of platinum accompanies contact with fused nitrates, cyanides, alkali chlorides, and alkaline-earth chlorides at temperatures above 1000°C. Hydrogen sulfates attack the metal slightly at temperatures above 700°C. Surface changes result from contact with ammonia, chlorine, volatile chlorides, sulfur dioxide, and gases possessing a high percentage of carbon. At red heat, platinum is readily attacked by arsenic, antimony, and phosphorus, the metal being embrittled as a consequence. A similar effect occurs upon high-temperature contact with selenium, tellurium, and to a lesser extent, sulfur and carbon. Finally, when heated in air for prolonged periods at temperatures greater than 1500°C, a significant loss in weight due to volatilization of the metal must be expected.

There are a number of rules governing the use of platinum ware:

1. Use platinum equipment only in those applications which will not affect the metal. When the nature of the system is in doubt, demonstrate the absence of potentially damaging components before committing platinum ware to use.

2. Avoid violent changes in temperature; deformation of a platinum container can result if its contents expand upon cooling.

3. Supports made of clean, unglazed ceramic materials, fused silica, or platinum itself may be safely used in contact with incandescent platinum; tongs of nichrome or stainless steel may be employed only after the platinum has cooled below the point of incandescence.

4. Clean platinum with an appropriate chemical agent immediately following use. Recommended cleaning agents are hot hydrogen chromate(VI) solution for removal of organic materials, boiling hydrochloric acid for removal of carbonates and basic oxides, and fused potassium hydrogen sulfate for the removal of silica, metals, and their oxides. A bright surface should be maintained by burnishing with sea sand.

5. Avoid heating platinum under reducing conditions, particularly in the presence of carbon. Specifically, (a) do not allow the reducing portion of the

burner flame to contact a platinum surface, and (b) char filter papers under the mildest conditions possible and with free access of air.

30.5.2.5 Hirsch Funnel. Hirsch funnels, used for collecting small amounts of solids, are usually made of porcelain. The inside bottom of the funnel is a flat plate with holes, which supports the filter paper. Funnels are also made of glass with sintered-glass disks. When using these funnels, a rubber ring forms the seal between the funnel and the filter disk. To use, the funnel is connected to the vacuum line or to the aspirator.

30.6 GRAVIMETRIC ANALYSIS

One of the most important applications of filtration is in gravimetric analysis. The substances which are weighed are obtained in two ways. (1) An insoluble precipitate is formed from the desired component; from the weight of the precipitate, the percentage calculations can be done, given the mass of the original sample. (2) A volatile component can be distilled off, and the volatile and nonvolatile portions can be determined by weighing.

30.6.1 Suitable Characteristics of a Precipitate

1. It should be relatively insoluble, to the extent that any loss due to its solubility would not significantly affect the result.
2. It should be readily filterable. Particle size should be large enough to be retained by a filter medium.
3. Its crystals should be reasonably pure, with easily removed solid contaminants.
4. It should have a known chemical composition or be easily converted to a substance which does have a known composition.
5. It should not be hygroscopic.
6. It should be stable.

30.6.2 Aging and Digestion of Precipitates

Aging and digestion of precipitates frequently help to make them suitable for analytical procedures. Freshly formed precipitates are aged by leaving them in contact with the supernatant liquid at room temperature for a period of time. There are frequently changes in the surface, such as a decrease in the total surface area or removal of strained and imperfect regions. Both effects are due to recrystallization because small particles tend to be more soluble than large ones, and ions located in imperfect and strained regions are less tightly held than is normal and therefore tend to return to the solution. On aging they are deposited again in more perfect form. These changes cause a beneficial decrease in adsorbed foreign ions and yield a more filterable as well as a purer precipitate.

Heating during the aging process is called *digestion*. Increasing the temperature greatly enhances digestion. The precipitate is kept in contact with the supernatant at a temperature near boiling for a period of time. Flocculated colloids usually undergo rapid aging, particularly on digestion, and a major portion of the adsorbed contaminants may often be removed.

30.6.3 Filtration and Ignition

Eventually, the substances precipitated and prepared for filtration must be filtered, dried, and weighed. Some precipitates are collected in tared crucibles and oven-dried to constant weight; others are ignited and ashed before being brought to constant weight.

Empty crucibles are brought to constant weight at the temperature to be used in the final drying or ashing of the precipitate. Agreement with 0.2 mg between consecutive measurements is considered constant weight. Store the prepared crucible in a desiccator until it is needed.

When filtration and washing are completed, transfer the filter paper and its contents from the funnel to the prepared crucible. Flatten the cone along its upper edge, then fold the corners inward. Next, fold the top over. Finally, ease the paper and contents into the crucible so that the bulk of the precipitate is near the bottom (see Fig. 30.9).

If a heat lamp is available, place the crucible on a clean, nonreactive surface. Position the lamp about 6 mm from the top of the crucible and turn it on. Charring of the paper will take place without further intervention. Removal of the remaining carbon is accomplished with a burner.

Considerably more attention must be paid to the charring process when a burner is employed to ash a filter paper. Since the burner can produce much higher temperatures, the danger exists of expelling moisture so rapidly in the initial stages of heating that mechanical loss of the precipitate occurs. A similar possibility arises if the paper is allowed to flame. (A clean crucible cover is placed over the mouth of the crucible until the flame is extinguished.) Finally, as long as carbon is present, there is also the possibility of chemical reduction of the precipitate. This is a serious problem where reoxidation following ashing of the paper is not convenient. In order to minimize these difficulties, the crucible is placed as illustrated in Fig. 30.10. The tilted position of the crucible allows for the ready access to air.

Always place the hot cover or crucible on a wire gauze—never on the bench surface.

Heating is commenced with a small burner flame. This is gradually increased as moisture is evolved and the paper begins to char. The smoke that is given off serves as a guide to the intensity of heating that can be safely tolerated. If the smoke does flame, it should be snuffed out immediately with the crucible cover.

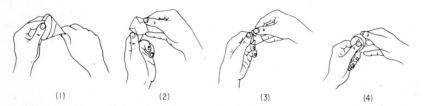

     (1)   (2)   (3)   (4)

FIGURE 30.9 Technique for transferring a filter paper with precipitate to a crucible.

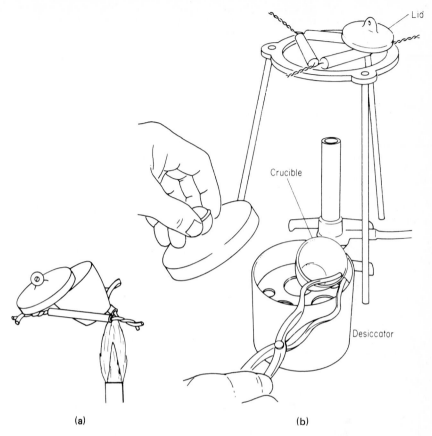

(a)   (b)

FIGURE 30.10 (*a*) Ignition of a precipitate with access to air. Start heating slowly from the side. Do not let the flame enter the crucible. Move the flame over the bottom when the paper has been charred. (*b*) Using tongs, transfer the cooled crucible to a dessicator.

Finally, when no further smoking can be detected, the residual carbon is removed by gradually lowering the crucible into the full oxidizing flame of the burner. Reposition the crucible to expose fresh portions to the crucible to the highest temperature of the flame.

When ignition is complete, cool the crucible almost to room temperature and then transfer to a desiccator. This avoids heating the desiccator and hastens the final attainment of room temperature. Weigh the crucible and its contents. Repeat the ignition until constant weight is attained.

30.6.4 Gravimetric Calculations

In most cases the desired constituent is not weighed directly, but is precipitated and weighed as some other compound. It is then necessary to convert the weight obtained to the weight in the desired form by using a *gravimetric factor*. For ex-

TABLE 30.5　Selected Gravimetric Factors

Sought	Weighed	Factor
Ba	$BaSO_4$	0.5885
C	CO_2	0.2729
Ca	$CaSO_4$	0.2944
Cu	CuO	0.7988
	$CuCNS$	0.5224
Fe	Fe_2O_3	0.6994
H	H_2O	0.1119
K	$KClO_4$	0.2822
	K_2SO_4	0.4487
Mg	$Mg_2P_2O_7$	0.2184
Na	$NaOAc \cdot Mg(OAc)_2 \cdot UO_2(OAc)_2 \cdot 6.5H_2O$	0.01527
Na	Na_2SO_4	0.3237
Ni	Ni dimethylglyoxime	0.2032
P	$P_2O_5 \cdot 24MoO_3$	0.01723
Pb	$PbCrO_4$	0.6411
Si	SiO_2	0.4674
S	CdS	0.2220
Sn	SnO_2	0.7876
Zn	$Zn_2P_2O_7$	0.4291
ZnO	$Zn_2P_2O_7$	0.5341

ample, a molecule of silver chloride comprises one atom of silver and one atom of chlorine. The ratio of silver to silver chloride is as $Ag/AgCl$; similarly, the ratio of chlorine to silver chloride is as $Cl/AgCl$. Substituting the atomic weights into the ratios, the gravimetric factors become:

$$Ag/AgCl = 107.87/(107.87 + 35.45) = 0.7527$$

and　　　　　　$$Cl/AgCl = 35.45/(107.87 + 35.45) = 0.2473$$

A number of useful gravimetric factors are listed in Table 30.5.

CHAPTER 31
THE ANALYTICAL BALANCE

31.1 INTRODUCTION

Weighing is the most common and most fundamental procedure in chemical work. Today's laboratory balances incorporate the latest advancements in electronics, precision mechanics, and materials science. Gains to the chemist are unprecedented ease of use, versatility, and accuracy. A balance user should pay particular attention to the following aspects, which form some of the main topics of this chapter.

1. Select the proper balance for a given application. Understand the technical specifications of balances.
2. Understand the functions and features of the instrument, and use them correctly to obtain the performance that the particular balance was designed to provide.
3. Know how to ascertain the accuracy and functionality of a balance through correct installation, care, and maintenance.
4. Use proper and efficient techniques in weighing operations.
5. Apply proper judgment in interpreting weighing results. High accuracy may require corrections for air buoyancy.
6. Be aware that most electronic balances can be interfaced to printers, computers, and specialized application devices. It is usually more reliable and efficient to process weighing records and associated calculations electronically.

31.1.1 Mass and Weight

The terms mass and weight are both legitimately used to designate a quantity of matter as determined by weighing. However, the following scientific terminology should be adhered to in any technical context.

Mass: An invariant measure of the quantity of matter in an object. The basic unit of mass is the kilogram, which is embodied in a standard kept in Paris. Masses are more practically expressed in grams or milligrams.

Apparent mass: Apparent mass is the mass of an object minus the mass of the air that the object displaces. Air buoyancy corrections are discussed in a later section. The distinction between apparent mass and absolute mass is insignificant in chemical work for all but a few special applications.

Weight: The force exerted on a body by the gravitational field of the earth, measured in units of force (Newton, abbreviation N). The weight of a body varies with geographic latitude, altitude above sea level, density of the earth at the location, and, to a very minute degree, with the lunar and solar cycles. Electronic balances must be calibrated on location against a mass standard.

Balance: A laboratory instrument used for the precise measurement of small masses.

31.1.2   Classification of Balances

To guide the user in selecting the correct equipment for a given application, balances are classified according to the graduation step or division of the reading device (scale dial or digital display) and according to their weighing capacity. Balances are classified in Table 31.1. Various types of balances are discussed in the following sections.

TABLE 31.1   Classification of Balances by Weighing Range

Nomenclature	Smallest division	Capacity (typical)
Ultramicroanalytical	0.1 μg	3 g
Microanalytical	0.001 mg	3 g
Semimicroanalytical	0.01 mg	30 g
Macroanalytical	0.1 mg	160 g
Precision	1 mg	160 g to 60 kg

31.2   GENERAL PURPOSE LABORATORY BALANCES

31.2.1   Top-Loading Balances

Top-loading balances are economical and easy to use for routine weighing, and for educational and quality assurance applications. Because of their design, top-loaders generally sacrifice at least one order of magnitude of readability. Models are available for many tasks; readabilities range from 0.001 to 0.1 g and capacities from 120 to 12 000 g. The latter also represent tare ranges. Balances with ranges above 5000 g are for special applications. Operating temperature is usually from 15 to 40°C. Typical specifications of single range models are given in Table 31.2.

Many top loading balances are dual- or poly-range balances which offer variable readability throughout their capacities for high resolution at each weight. Dual-range balances offer two levels of readability within their capacity range; poly-range balances offer a series of incremental adjustments in readability.

To use the balance, proceed as follows:

1. Slide all poises or riders to zero.
2. Zero the balance with balance-adjustment nuts.
3. Place the specimen on the pan of the balance.

4. Move the the heaviest poise or rider to the first notch that causes the indicating pointer to drop; then move the rider back one notch, causing the pointer to rise.

5. Repeat procedure 4 with the next highest rider.

6. Repeat this procedure with the lightest rider, adjusting the rider position so that the indicator points to zero.

7. The weight or mass of the specimen is equal to the sum of the values of all the rider or poise positions, which are read directly from the position of the riders on the marked beams.

Electronic top-loading balances may have additional features described in Sec. 31.4.3.

TABLE 31.2 Specifications of Balances

Capacity, g	Readability, mg	Stabilization time, s	Tare range, g
40	0.01	5	0–40
60	0.1	3	0–60
160	0.1	3	0–160
400	1.0	2	0–400
800	10	3	0–800
2200	10	3	0–2200
5000	100	3	0–5000
12000	100	5	0–12000

31.2.2 The Triple-Beam Balance

The classic triple-beam balance (Fig. 31.1a) provides a modest capacity of 111 g (201 g with an auxiliary weight placed in the 100-g notch). The three tiered scales are: 0 to 1 g by 0.01 g, 0 to 10 g by 1 g, and 0 to 100 g by 10 g. The 1-g scale is notchless and carries a rider; the others are notched and carry suspended weights.

Triple beam platform balances (Fig. 31.1b) have a sensitivity of 0.1 g and a total weighing capacity of 2610 g when used with the auxiliary weight set, or 610 g without it. Three tiered scales (front to back) are 0 to 10 g by 0.1 g, 0 to 500 g by 100 g, and 0 to 100 g by 10 g. One 500-g and two 1000-g auxiliary weights fit in a holder on the base. The aluminum beam is magnetically damped and has a spring-loaded zero adjust.

31.2.3 Dial-O-Gram Balances

A dial mechanism is used to obtain the weights from 0 to 10 g in 0.1-g intervals. In use the dial is rotated to 10.0 g. After moving the 200-g poise on the rear beam to the first notch which causes the pointer to drop, and then moving it back a notch, the same procedure is repeated with the 100-g poise. Finally, the dial knob is rotated until the pointer is centered. As shown in Fig. 31.2, the vernier scale provides readings to the nearest 0.1 g.

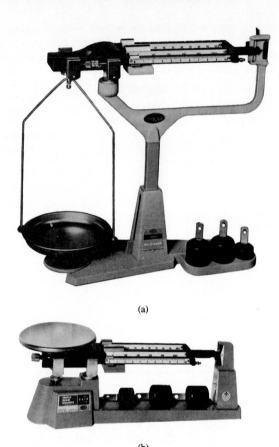

(a)

(b)

FIGURE 31.1   Two types of triple-beam balance.

31.3 MECHANICAL ANALYTICAL BALANCES

31.3.1 Equal-Arm Balance

The classical equal-arm balance consists of a symmetrical level balance beam, two pans suspended from its ends, and a pivotal axis (fulcrum) at its center (Fig. 31.3). Ideally, the two pan suspension pivots (E) are located in a straight line with the fulcrum (F) and the two lever arms are of exactly equal length. A rigid, truss-shaped construction of the beam minimizes the amount of bending when the pans are loaded. The center of gravity of the beam (C) is located just slightly below the center fulcrum, which gives the balance the properties of a physical pendulum. With a slight difference in pan loads, the balance will come to rest at an inclined position, the angle of inclination being proportional to the load differential. By reading the pointer position on a graduated angular scale, it is possible to determine fractional amounts of mass in between the even step values of a standard mass set of weights.

(a)

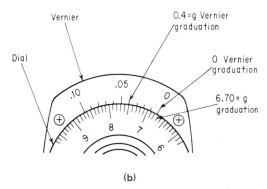

(b)

FIGURE 31.2 Dial-O-Gram control knob. (*a*) Harvard trip balance; (*b*) reading is 6.74 g. (*Courtesy of Ohaus Scale Co.*)

Variations and refinements of the equal-arm balance include (1) agate or synthetic sapphire knife edge pivots, (2) air damping or magnetic damping of beam oscillations, (3) sliding poises or riders, (4) built-in mass sets operated by dial knobs, (5) microprojector reading of the angle of beam inclination, (6) arrestment devices to disengage and protect pivots, and (7) pan brakes to stop the swing of the balance pans.

31.3.2 Single Pan Substitution Balance

Substitution balances (Fig. 31.4) have only one hanger assembly which incorporates both the load pan and a built-in set of weights on a holding rack. The hanger assembly is balanced by a counterpoise which is rigidly connected to the other side of the beam and whose weight equals the maximum capacity of the particular

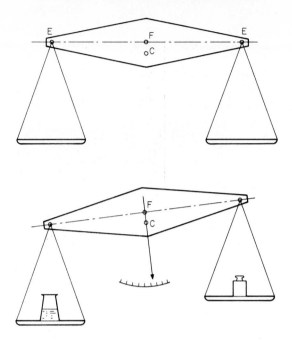

FIGURE 31.3   Critical design aspects in an equal-arm balance.

balance. The weight of an object is determined by lifting weights off the holding rack until enough weights have been removed to equal almost the weight of the object. In this condition the balance returns to an equilibrium position within its angular, differential weighing range. Small increments of weight in between the the discrete dial weight steps (usually in gram increments) are read from the projected screen image of a graduated optical reticle which is rigidly connected to the balance beam.

With only two pivot axes, this design disposes of the problem of keeping pivots in a straight line with and equidistant from the fulcrum. Furthermore, the elastic deformation of the beam stays unchanged as the loads on the beam remain essentially constant. Although these balances are completely mechanical in operation, the optical readout system was usually assisted with a light bulb.

While single pan substitution balances are no longer manufactured, there are a number of these products still in use. Electronic balances possess superior accuracy and operating convenience.

31.3.3   Top-Loading Balances

Top-loading balances are based on the same principles as equal-arm balances. As the pans move up and down, they remain horizontal, stabilized by a parallelogram linkage, as shown in Fig. 31.5. With the pans stabilized by lever linkages, top-loading equal-arm balances are more convenient to operate and give unobstructed access for tall or wide weighing loads.

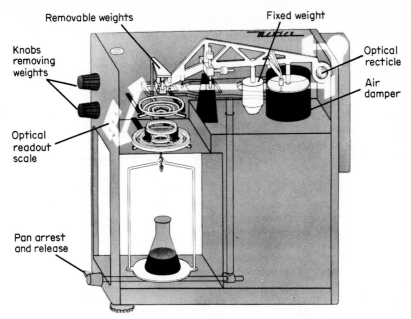

FIGURE 31.4 Cut-away side view of a single-pan substitution analytical balance. (*Courtesy of Mettler Instrument Co.*)

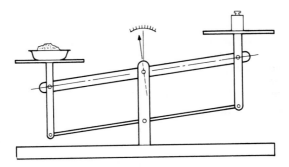

FIGURE 31.5 Parallelogram linkage of top-loading equal-arm balance.

31.4 ELECTRONIC BALANCES

Today there are two dominant types of electronic balances in use—the hybrid and the electromagnetic force balance. The hybrid balance uses a mix of mechanical and electronically generated forces, whereas the electromagnetic force balance uses electronically generated forces entirely.

31.4.1 Design and Operating Principles

In every electromechanical weighing system, there are three basic functions (Fig. 31.6):

1. The load transfer mechanism, composed of the weighing platform or pan, levers, and guides, receives the weighing load on the pan as a randomly distributed pressure force P and translates it into a measurable single force F. The platform is stabilized by flexure-pivoted guides; pivots are formed by elastically flexible sections E in the horizontal guide members. These features are shown in Fig. 31.7.

2. The electromechanical force transducer, often called the load cell, converts the mechanical input force into an electrical output (Fig. 31.8). A direct current i generates a static force on the coil which, in turn, counterbalances the weight force from the object on the balance pan (usually aided by one or more force reduction levers). The amount of coil current is controlled by a closed loop servo circuit which monitors the vertical deflections of the pan support through a photoelectric sensor and adjusts the coil current as required to maintain equilibrium between weighing load and compensation force.

3. The servo system is the electronic signal processing part of the balance. It receives the output signal of the load cell, converts it to numbers, performs computations, and displays the final weight data on the readout. In one method a continuous current is driven through the servomotor coil and in the other method the current is pulsed.

31.4.2 The Hybrid Balance

The hybrid balance is identical to the substitution balance except the balance beam is never allowed to swing through large angular displacements when the

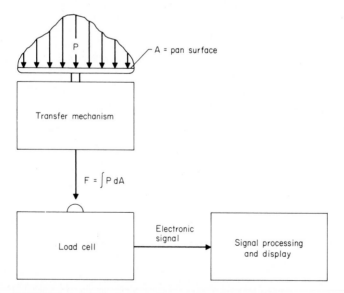

FIGURE 31.6 Schematic block diagram of an electronic balance.

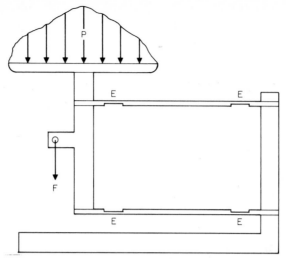

FIGURE 31.7 Schematic showing the operation of the transfer mechanism of an electronic balance.

applied loading changes. Instead the motion is very limited and when in equilibrium the beam is always restored to a predetermined reference position by a servo-controlled electromagnetic force applied to the beam (Fig. 31.9). The most salient features that distinguish the electronic hybrid are the balance beam and the built-in weights utilized in conjunction with the servo restoring force to hold the beam at the null position.

31.4.3 The Electromagnetic Force Balance

A magnetic force balances the entire load either by direct levitation or through a fixed-ratio lever system. The loading on the electromechanical mechanism that constitutes the balance is not constant but varies directly with the applied load. With this design the sensitivity and response are largely controlled by the servo system characteristics. The force associated with the sample being weighed is mechanically coupled to a servomotor that generates the opposing magnetic force (Fig. 31.10). When the two forces are in equilibrium the error detector is at the reference position and the average electric current in the servomotor coil is proportional to the resultant force that is holding the mechanism at the reference position. When the applied load changes, a slight motion occurs between the fixed and moving portions of the error-detector components, resulting in a very rapid change in current through the coil.

31.4.4 Special Routines for Electronic Balances

Signal processing in electronic balances usually involves special computation routines:

1. *Programmable stability control:* Variable integration permits compensation for unstable weight readings due to environmental vibration or air currents.

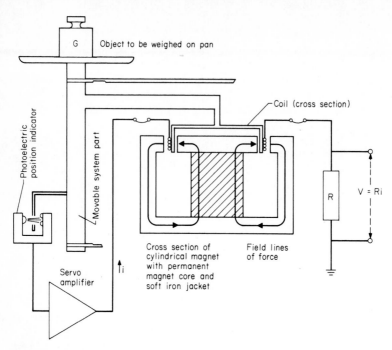

FIGURE 31.8 Electromagnetic force compensation principle of an electronic balance.

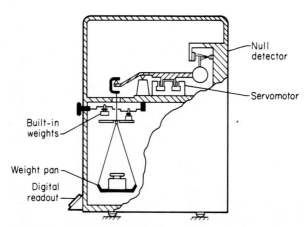

FIGURE 31.9 The hybrid electronic balance. [*Reprinted with permission from R. M. Schoonover, Anal. Chem., 54:974A (1982). Copyright 1982 American Chemical Society.*]

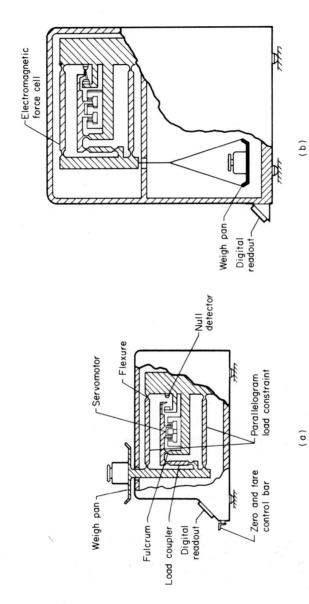

FIGURE 31.10 The electromagnetic force balance cell: (*a*) with pan and (*b*) top-loading.   [Reprinted with permission from R. M. Schoonover, *Anal. Chem.*, *54:976A (1982). Copyright 1982 American Chemical Society.*]

31.11

Preprogrammed filters minimize noise due to air currents and vibration. Four settings vary integration time, update rate of display, vibration filtering, and damping.

The adjustable stability range includes nine different settings (from 0.25 to 64 counts in the last significant digit) that control the tolerance range within which the stability indicator appears (a symbol which appears when actual sample weight is displayed within preset stability-range tolerances).

2. *Autotracking:* Eliminates distracting display flicker. Automatically rezeros balance during slight weight changes. Once the balance stabilizes and displays the weight, the autotracking feature takes over. Can be turned off to weigh hygroscopic samples or highly volatile substances.

3. *Serial interface:* Allows two-way communication with printers, computers, and other devices at baud rates up to 9600.

4. *Autocalibration:* When pushed, a calibration button activates a built-in test weight (up to four on some models). The balance calibrates itself to full accuracy and returns to weighing mode within seconds.

5. *Full-range taring:* Just set a container on the pan and touch the rezero button and the display automatically resets to zero.

6. *Overload protection:* Full-range taring and overload protection prevent damage to the balance if excess weight is placed on the pan.

31.4.5 Precautions and Procedures

With analytical balances the following guidelines should be observed:

1. Level the balance using the air-bubble float.
2. Observe the required warmup period or leave the balance constantly under power. Follow manufacturer's recommendations.
3. Use any built-in calibration mass and microprocessor controlled calibration cycle at the beginning of every work day.
4. Handle weighing objects with forceps. Fingerprints on glassware or the body heat from an operator's hand can influence results.
5. Always close the sliding doors on the balance when weighing, zeroing, or calibrating.
6. Accept and record the displayed result as soon as the balance indicates stability; observe the motion detector light in the display. Never attempt to mentally average the displayed numbers.

31.5 THE WEIGHING STATION

The finer the readability of a balance, the more critical is the choice of its proper location and environment. Observe these precautions:

1. Avoid air currents. Locate away from doors, windows, heat and air conditioning outlets.
2. Avoid having radiant heat sources, such as direct sunlight, ovens, baseboard heaters, nearby.

3. Avoid areas with vibrations. Locate away from elevators and rotating machinery. Special vibration-free work tables may be needed. To test, the displayed weight should not change if the operator shifts his or her weight, leans on the table, or places a heavy object next to the balance.

4. Choose an area free from abnormal radio-frequency and electromagnetic interference. The balance should not be on the same circuit as equipment that generates such interference, such as electric arcs or sparks.

5. Maintain the humidity in the range from 15 to 85 percent. Dry air can cause weighing errors through electrostatic charges on the weighing object and the balance windows. A room humidifier might help. Humid air causes problems because of moisture absorption by samples and container surfaces. A room dehumidifier would help.

6. Maintain even temperature of room and object weighed. If an object is warm relative to the balance, convection currents cause the pan to be buoyed up, and the apparent mass is less than the true mass.

7. Weigh materials that take up water or carbon dioxide from the air during the weighing process in a closed system.

8. Weigh volatile materials in a closed system.

9. Sit down when doing precise weighing. The operator should be able to plant elbows on the work table, thus allowing a steady hand in handling delicate samples.

31.6 AIR BUOYANCY[1]

In the fields of practical chemistry and technology, there is an unspoken agreement that no correction for air buoyancy is made in the weighing results. Therefore, the apparent mass of a body of unit density determined by weighing is about 0.1 percent smaller than its true mass. The exact mass on a balance is indicated only when the object to be weighed has the same density as the calibration standards used and air density is the same as at the time of calibration. For interested persons, conversions of weighings in air to those in vacuo are discussed in Ref. 2.

REFERENCES

1. R. M. Schoonover and F. E. Jones, "Air Buoyancy Correction in High Accuracy Weighing on Analytical Balance," *Anal. Chem.,* **53**:900 (1981).

2. J. A. Dean, ed., *Lange's Handbook of Chemistry,* 13th ed., McGraw-Hill Book Company, New York, 1985, p. 2-71.

BIBLIOGRAPHY

Ewing. G. W., "Electronic Laboratory Balances," *J. Chem. Educ.,* **53**:A252, (1976); **53**: A292 (1976).

Hirsch, R. F., "Modern Laboratory Balances," *J. Chem. Educ.,* **44**:A1023 (1967); **45**:A7 (1968).

Leonard, R. O., "Electronic Laboratory Balances," *Anal. Chem.,* **48**:879A (1976).

Schoonover, R. M., "A Look at the Electronic Analytical Balance," *Anal. Chem.,* **54**:973A (1982).

CHAPTER 32
PRESSURE AND VACUUM

32.1 EQUIPMENT FOR CONTROL AND REGULATION OF GASES

32.1.1 Gas Regulators

A gas-pressure regulator is a precision instrument designed to reduce a high source pressure to a safe value. Each regulator will control a chosen delivery pressure within the bounds of the regulator's delivery-pressure range. This prevents the overpressurization of any apparatus downstream of the regulator and permits stable flow rates to be established. *Pressure regulators are not flow regulators*. Use flow controllers to control flow if the downstream pressures are subject to variations.

A regulator reduces gas pressure by causing a counteraction of gas pressure on a diaphragm against the compression of a spring which can be adjusted externally with the pressure-adjusting screw (Fig. 32.1). In operation the pressure-adjusting screw is turned to exert force on the spring and diaphragm. This force is transmitted to the valve assembly, pushing the valve away from the seat. The high-pressure gas will flow past the valve into the low-pressure chamber. When the force of gas pressure on the diaphragm equals the force of the spring, the

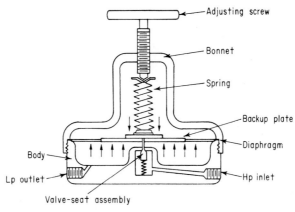

FIGURE 32.1 Schematic of a single-stage regulator.

valve and seat assemblies close, preventing the flow of additional gas into the low-pressure chamber.

32.1.2 Flowmeters

Flowmeters (Fig. 32.2) are used to indicate the rate of flow of the gas (or liquid) by registering the scale graduation at the center of the spherical float. Flowmeters do not control the rate of flow of the gas unless they are specifically equipped with control valves or flow controllers. Flowmeters must be installed in a vertical position. The gas or liquid enters the bottom and exits the top of the meter. A condition of dynamic equilibrium results when the upward force, due to fluid pressure, and the buoyant force due to the float, balance the weight of the float.

Flowmeters are designed for specific gases and for varying amounts of flow. The center of the float is the register of the fluid flow. The higher the float rises, the greater the flow rate. Parallax can affect the accuracy and reproducibility of measurements.

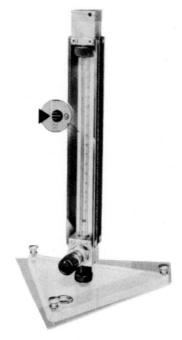

FIGURE 32.2   Flowmeter.

32.1.3 Safety Devices

It is necessary to provide further supplementary safety devices to prevent overpressuring of lines and to prevent suckback of materials in the the cylinder controls. The danger of suckback can be eliminated by providing a trap (Fig. 32.3) which will hold all material that can possibly be sucked back, or by using a

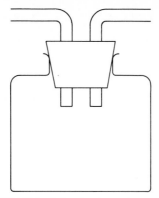

FIGURE 32.3 Safety bottle used to trap materials that are sucked back.

check valve or suitable vacuum break. Check valves prevent the return flow of gas and thus keep foreign matter out of gas lines, regulators, and cylinders ahead of the valve. The valves are spring-loaded.

Pressure increases due to uncontrolled reactions or unexpected surges of pressure can be relieved by means of a safety relief valve installed in the gas line.

Quick couplers permit regulators, needle valves, and other components of a gas system to be connected and interconnected to cylinder outlets quickly and safely.

32.2   HUMIDITY

The relative humidity refers to how much water vapor a volume of air is actually holding relative to (compared with) how much water vapor it is capable of holding. When it is holding the maximum amount of water vapor for a given temperature and pressure, it it said to be saturated. Relative humidity is usually expressed in percent.

A saturated aqueous solution in contact with an excess of the solute when kept in an enclosed space will maintain a constant humidity at a given temperature. A series of such systems are tabulated in Table 32.1.

32.2.1   Measurement of Humidity

The dew point of wet air is measured directly by observing the temperature at which moisture begins to form on an artificially cooled polished surface. It is usual practice to take the dew point as the average of the temperature when the fog first appears on cooling and disappears on heating.

Humidity and temperature can be measured quickly and accurately at atmospheric humidities by using the wet- and dry-bulb technique. The wet-bulb temperature is measured by contacting the air with a thermometer whose bulb is covered by a wick saturated with water. Traditionally, the unit is whirled manually to give the desired gas velocity across the bulb. Battery-operated units measure

TABLE 32.1   Humidity (%) Maintained by Saturated Solutions of Various Salts at Specified Temperatures

Solid phase	Temperature, °C						
	10	20	25	30	40	60	80
K_2SO_4	98	97	97	96	96	96	—
KNO_3	95	93	92.5	91	88	82	—
KCl	88	85	84	84	82	81	79.5
NaCl	76	75.7	75.3	74.9	74.7	74.9	76.4
$Mg(NO_3)_2 \cdot 6H_2O$	57	55	53	52	49	43	—
$K_2CO_3 \cdot 2H_2O$	47	44	43	42	40	—	—
$MgCl_2 \cdot 6H_2O$	34	33	33	33	32	30	—
$KF \cdot 2H_2O$	—	—	—	27.4	22.8	21.0	22.8
KOH	13	9	9	7	6	5	—
Aqueous tension at 100% humidity, torr	9.21	17.54	23.76	31.82	55.32	149.4	355.1

these values by having a battery-operated fan draw the air past the bulbs. When the wet- and dry-bulb temperatures are known, the humidity is readily obtained from tables, such as those in Ref. 1.

32.3   SOURCES OF VACUUM

There are several kinds of pumping equipment used to produce vacuum in the laboratory; the choice depends on the pressure range desired. Most commonly used are the hydroaspirator or filter pump, the rotary vane pump, the diffusion pump, the turbomolecular pump, the cryopump, and the ion or chemical pump.

32.3.1   Aspirator

The aspirator uses a water stream for entrainment of gases. It is often used for vacuum filtration and similar purposes. This device can reduce pressures to the vapor pressure of the pumping liquid. In the case of the hydroaspirator at room temperature, an ultimate pressure of 2.5 mm of mercury can be obtained.

32.3.2   Rotary Vane Pump

For rough vacuum in the range of a few millimeters of mercury, a rotary vane pump is frequently used; in fact, these pumps are used as the basis for central vacuum systems. The rotary vane pump relies on the sweeping action of multiple vanes mounted in slots in an eccentrically mounted rotor turning within a cylindrical housing. As the rotor turns, gas is swept up at the intake port, compressed as the confined volume is decreased, and expelled at the exhaust port, as shown in Fig. 32.4. The pump operates within a reservoir of oil which provides necessary sealing as well as lubrication. These pumps are composed of either one or

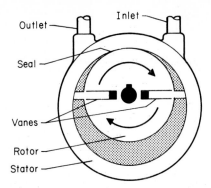

FIGURE 32.4 Cross section of an oil-sealed rotary vane pump.

two stages. Single-stage pumps are limited to system pressures of about 0.15 torr and above; two-stage pumps should be used for pressures below 0.15 torr.

A convenient feature found on some pumps is the vented exhaust. This device comprises an external needle valve and air passage which permits introduction of a controlled amount of air into the exhaust stage of the pump at the point of maximum compression. It thus provides a means of sweeping out condensable vapors which might otherwise remain in the pump and thus reduce its efficiency. This feature is useful where the pump is often exposed to water and solvent vapors.

For work in the very high or ultrahigh vacuum ranges, the rotary vane pump is used in tandem with a diffusion pump, turbomolecular pump, or a cryosorption or ion pump.

32.3.3 Diffusion Pump

In principle, the vapor diffusion pump resembles an aspirator. A stream of vapor is generated by evaporating a pumping fluid (Fig. 32.5). The molecules of the fluid, while passing through the space where the gas molecules of the system are being evacuated, strike those gas molecules and push them toward the outlet. The gas molecules are then replaced by the vapor molecules from the pump until practically all have been exhausted. The vaporized pumping fluid is then condensed and returned to the still.

Diffusion pumps yield a vacuum of 10^{-3} to 10^{-6} torr. A diffusion pump is used in conjunction with a rotary vane pump and takes over when the pressure is reduced by the latter unit. Diffusion pump oils are esters, hydrocarbons, polyphenyl ethers, or silicones that possess excellent stability toward heat and extremely low vapor pressures.

32.3.4 Turbomolecular Pump

The turbomolecular pump is the only purely mechanical vacuum pump that attains pressures of 10^{-9} torr without utilizing cold traps and oil traps. It creates a hydrocarbon-free vacuum which is desirable when a clean "backing" vacuum for

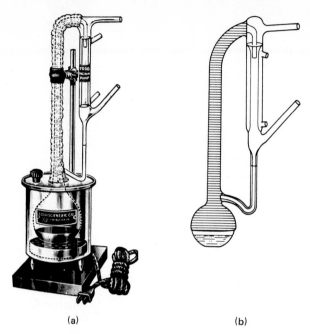

(a)   (b)

FIGURE 32.5   Schematic diagram of a mercury diffusion pump.

ion, sorption, or cryopumps is needed. The turbomolecular pump has a rotor, composed of a series of axial-flow bladed disks, driven at high speed and inter- leaved with stators. Each pair constitutes a pumping stage. Molecules of gases in the system to be purged randomly enter the pump. Some of the molecules are struck by the first rotor blades and tend to rebound in a favorable axial direction entering the blade passages of the first stator disk. The molecules rebound again from the stators in such a direction as to increase the probability of their being favorably impelled by the second rotor disk. This process is repeated through all the stages of the pump.

The maximum pressure ratio and pumping speed are attained when the gas in all stages is in the molecular flow pressure range. A rotary vane pump is required for roughing and backing purposes.

32.3.5   Cryopump

A cryopump operates by freezing gases into solids, thereby reducing vapor pres- sures to such a point that a high vacuum is created. It utilizes a closed-cycle he- lium refrigeration system and a set of surfaces cooled to cryogenic temperatures. Random molecular motion brings gas molecules into contact with these cold sur- faces, where they are removed by condensation or adsorption. All condensed gases are reduced to a vapor pressure below 10^{-12} torr. Helium, hydrogen, and neon are pumped by adsorption in a bed of charcoal cooled by helium. Cryopumps must be periodically recycled by liquefying and pumping off the fro- zen gases.

32.3.6 Ion Pump

An ion pump functions in the molecular flow range. Gas molecules enter the pump by random diffusion. Once in the pump the gas molecules are bombarded, ionized, and collected at a charged surface. The pump must be recycled periodically to remove the accumulated gases.

32.3.7 Sorption Pump

The sorption pump removes gases by immobilizing them via a chemical reaction or simple adsorption. It is also capable of lowering pressures to the ultrahigh vacuum range of 10^{-12} torr. The reaction or adsorbing surface must be periodically regenerated.

32.3.8 Care and Maintenance of Vacuum Pumps

1. The correct and full voltage must be present at the motor to secure rated performance. Be sure that the circuit is not overloaded and that the proper size electrical wiring is used.
2. Care must be taken to prevent foreign material from entering the pump. A small particle of glass or metal will impair the normal life expectancy of any pump.
3. Gases must not be allowed to condense in the pump oil. Corrosion of the finely machined parts is likely to occur. Adequate gas traps must be used. Gases are commonly dried with calcium sulfate (Drierite) or magnesium perchlorate dihydrate (Anhydrone) packed in heavy glass towers.
4. A single-stage pump should not be operated continually above 5 torr, and a double-stage pump should not be operated continually above 1 torr.
5. Leaks and cracks in vacuum systems are easily detected and located by the use of a high-frequency Tesla coil. When the Tesla coil is passed over the surface of evacuated glassware assemblies, an imperfect joint is pinpointed by a yellow glow at the point where the electric discharge enters the system.
6. All hoses should be clamped securely; use an average torque with a medium-sized screwdriver (about 15 to 30 in · lb).

32.4 DEVICES TO MEASURE PRESSURE OR VACUUM

32.4.1 U-Tube Manometers

An open-end or U-tube manometer (Fig. 32.6) consists of a U-tube filled with mercury or some other liquid, such as water, alcohol, or oil, which has one end open to the atmosphere and the other end attached to the system to be measured. When the pressure in the system is equal to the atmospheric pressure, the heights of the liquids are the same. Any difference in the levels of the two arms is a measure of the difference in pressure.

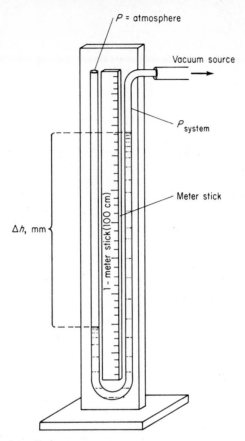

FIGURE 32.6   Open-end manometer.

32.4.1.1   *Hints and Precautions* .

1. The open-end manometer is limited to the measurement of pressure moderately above atmospheric.

2. The length of the columns must be able to show (theoretically) zero pressure when used for measurements of vacuum. When mercury is used as the liquid, the arms should be about 100 cm tall.

3. Low-density liquids should be used to measure small differences in pressure. Under the same pressure or vacuum, manometers filled with low-density liquids will show greater differences in height than manometers filled with a high-density liquid, such as mercury. Inclined manometers accentuate small differences in height.

4. A correction must be made for the vapor pressure of the liquid used in the manometer.

5. The ambient temperature must be carefully controlled because changes in temperature change the density of the liquid.

6. Viscosity of the liquid is very significant when used in an inclined manometer.

Sufficient time must elapse to allow the true level to be attained before taking a reading.

32.4.2 Closed-End Manometer

The closed-end manometer (Fig. 32.7) is normally filled with mercury and used to measure pressure less than atmospheric. It is connected to the system with a Y connector. As the pressure is reduced, the mercury will rise in the tube which is connected to the vacuum system. The pressure is determined by reading the difference in the heights of the mercury columns in the two tubes.

When the vacuum is interrupted and the system is brought back to atmospheric pressure, care must be exercised. A sudden change in pressure may cause the mercury to break the closed end of the tube by striking it sharply as it rushes back. A needle valve provides control of the rate of air intake.

32.4.3 Gauges

Bourdon-type gauges consist of a length of thin-walled metal tubing flattened into an elliptical cross section and then rolled into a C shape. When pressure is ap-

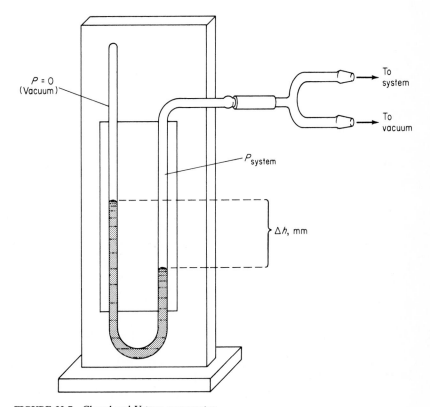

FIGURE 32.7 Closed-end U-type manometer.

plied, the tube tends to straighten out, and that motion is transmitted by levers and gears to a pointer which indicates the pressure (Fig. 32.8). These gauges can also be used to actuate electronic control equipment.

The McLeod gauge (Fig. 32.9) is the primary standard for the absolute measurement of pressure. A chamber (part of the gauge) of known volume is evacuated and filled with mercury. This chamber terminates in a sealed capillary which is calibrated in micrometers of mercury (absolute pressure). When the mercury fills the chamber, the gas therein is compressed to approximately atmospheric pressure and any trapped vapors are liquefied and have no significant volume. In other words, this gauge does not measure the pressure caused by any vapors present. Its reading represents only the total pressure of gases. This type of gauge is not considered suitable for ordinary use because it does not read continuously and is usually fragile.

The tilting McLeod gauge (Fig. 32.10) is very compact and portable. When the gauge is tilted, mercury traps the sample of gas in the measuring tube.

A *Pirani gauge* measures the pressure of the gas by indicating the ability of the gas to conduct heat away from a hot filament. The greater the pressure of a gas,

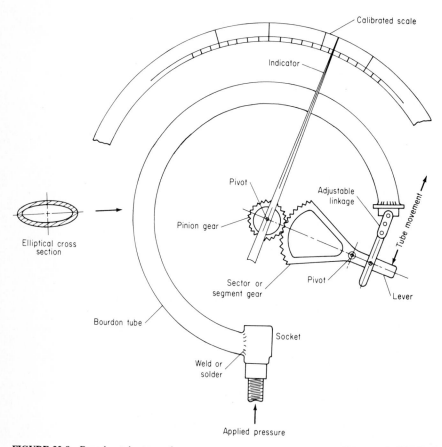

FIGURE 32.8 Bourdon tube type of pressure gauge.

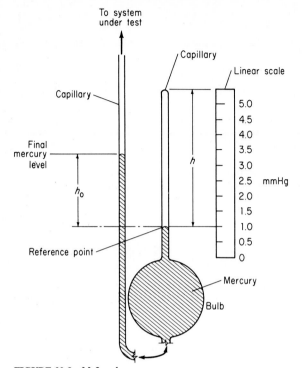

FIGURE 32.9 McLeod gauge.

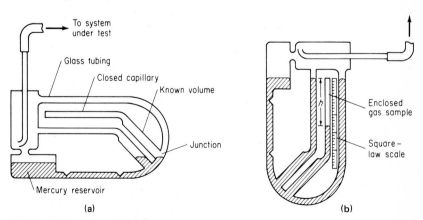

FIGURE 32.10 Tilting McLeod gauge.

the greater the conduction of heat from the filament. Also, different gases have different thermal conductivities. Usually a Wheatstone bridge circuit is employed with a microammeter calibrated to read in micrometers of mercury (Fig. 32.11). These gauges measure the presence of vapors as well as gases. They are usually used in the pressure range from 10^{-3} to 1 torr.

A *thermocouple gauge* (Fig. 32.12) resembles a Pirani gauge. The basic difference is that a thermocouple measures the filament temperature which will be a function of the pressure of the gases and vapors. This gauge is more rugged than the Pirani; its range is approximately the same.

Ionization gauges (Fig. 32.13) form ions of the gas molecules present. The amount of current carried by this ionized gas depends upon the amount of gas present. There are two types: the thermionic forms the gas ions by electrons emitted from a hot filament and the cold-cathode type uses a collimating magnetic field which forces the electrons to traverse a greatly increased path length. Depending on the exact type and design, these gauges will measure down to 10^{-6} torr.

32.5 VACUUM REGULATORS OR MANOSTATS

Manostats are inserted into vacuum systems to maintain a particular pressure. They open or close an orifice or capillary as necessary to compensate for leaks into or from the system.

The *cartesian type* (diver) is a glass vessel which floats on a pool of mercury in a chamber connected to the controlled system (Fig. 32.14). A sample of gas is trapped in the chamber at the desired pressure. Changes in the surrounding pres-

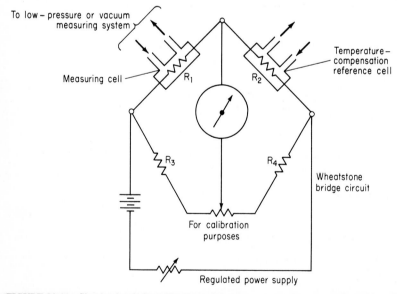

FIGURE 32.11 Simple circuit for a Pirani gauge.

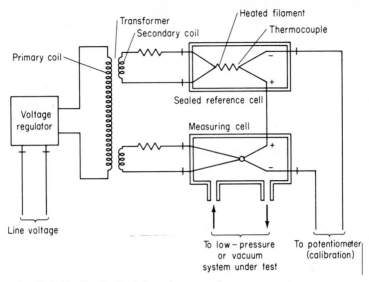

FIGURE 32.12 Simple circuit for a thermocouple vacuum gauge.

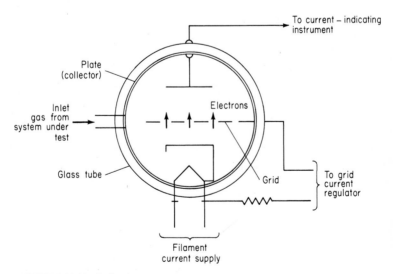

FIGURE 32.13 Ionization gauge.

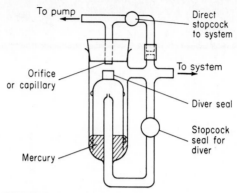

FIGURE 32.14   A cartesian diver manostat.

sure affect the buoyancy of the float, causing it to rise or sink vertically, thereby sealing or opening a capillary or an orifice.

The *Lewis* manostat uses mercury itself to open and seal the system by the rise of mercury against a fritted glass insert (Fig. 32.15). The operation is easier than the one using the cartesian diver.

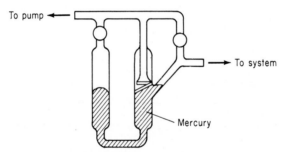

FIGURE 32.15   A modified Lewis manostat.

REFERENCES

1. J. A. Dean, Lange's *Handbook of Chemistry*, McGraw-Hill, New York, 13th ed., pp. 10-82 to 10-83.

CHAPTER 33
PRACTICAL LABORATORY INFORMATION

33.1 HEATING, COOLING, AND TEMPERATURE MEASUREMENT

33.1.1 Electrically Heated Oil Baths

Electrically heated oil baths using ordinary mineral oil can be used safely up to 200°C. Mineral oil eventually oxidizes and darkens. Any of the liquids listed in Table 33.1 can be substituted for mineral oil, each having its particular advantages and disadvantages. The bath is controlled with a variable-voltage transformer, and the heating element is an immersion coil, which minimizes danger of fire. Because it requires a fairly long time for oil baths to reach the desired temperature, it is advisable to preheat the unit partially before it is actually needed. Baths reduce the possibility of charring, which can occur when reaction vessels are heated by means of a flame.

The oil bath unit may be very heavy; a jack can be used to adjust the height of

TABLE 33.1 Substances Which Can Be Used for Heating Baths

Medium	Melting point, °C	Boiling point, °C	Useful range, °C	Flash point, °C	Comments
Water	0	100	0–100	None	Ideal
Silicone oil	−50	—	30–250	315	Somewhat viscous at low temperature
Triethylene glycol	−7	285	0–250	165	Noncorrosive
Glycerol	18	290	−20–260	160	Water-soluble, nontoxic
Paraffin	50	—	60–300	199	Flammable
Dibutyl o-phthalate	−35	340	150–320	171	Generally used

the bath. Water should be kept away from oil baths. Water in hot oil will cause it to splatter, and hot oil burns are painful and dangerous.

33.1.2 Cooling Baths

When low temperatures are required, the salt-ice mixtures listed in Table 33.2 can be used to attain the desired temperatures. Factors which contribute to achieving the stated final temperatures include:

1. Rate of mixing.
2. Heat transfer of the container. Thermos flasks or foam-insulated containers minimize heat transfer.
3. Fineness of the crushed ice.

Ice baths are actually ice-water baths. Pieces of ice make poor contact with the walls of the vessel; therefore, liquid water is needed to make the cooling medium efficient.

Dry ice (solid carbon dioxide) provides an easy way to obtain very low temperatures. The dry ice is crushed and mixed with the solvent. *Caution*: Dry ice is

TABLE 33.2   Cooling Mixtures

Substance	Quantity of substance, g	Quantity of water, mL	Temperature of mixture, °C
Ammonium nitrate	100	94	−4.0
Sodium nitrate	75	100	−5.3
Sodium thiosulfate 5-water	110	100	−8.0
Sodium chloride	36	100	−10.0
Sodium nitrate	50	100 (ice)	−17.8
Sodium chloride	33	100 (ice)	−21.3
Sodium bromide	66	100 (ice)	−28
Magnesium chloride	85	100 (ice)	−34
Calcium chloride 6-water	100	81 (ice)	−40.3
Calcium chloride 6-water	100	70 (ice)	−55

With these organic substances, dry ice (−78°C) in small lumps is added to the solvent until a slight excess of dry ice remains, or liquid nitrogen (−196°C) can be poured into the solvent until a slush is formed that consists of the solid-liquid mixture at its melting point.

Substance	Temperature, °C	Substance	Temperature, °C
Ethylene glycol	−13	Ethyl acetate	−84
1,2-Dichlorobenzene	−17	2-Butanone	−87
Carbon tetrachloride	−22.9	Hexane	−95
Bromobenzene	−31	Methanol	−98
Methoxybenzene	−37	Carbon disulfide	−112
Chlorobenzene	−45	Bromoethane	−119
N-Methylaniline	−57	Pentane	−130
p-Cymene	−68	2-Methylbutane	−160
Acetone	−77		

dangerous and must be handled with care. Never handle with bare hands or fingers as severe cold burns will result.

33.1.3 Temperature Measurements

With glass-mercury thermometers, the depth of immersion is very important, because the amount of mercury in the stem is significant when compared to the amount in the bulb. For this reason, thermometers are marked "full immersion" or "partial immersion," with definite markings to indicate the required depth of immersion for correct readings. If thermometers are not immersed to the specified mark, errors in the readings will occur, and they must be compensated for with a stem correction (see Fig. 33.1).

In melting-point or boiling-point determinations, the entire mercury column is not completely immersed in the vapor or liquid. Therefore, corrections must be made. At temperatures 0 to 100°C the error is negligible. However, around 200°C the error may be 3 to 5°C.

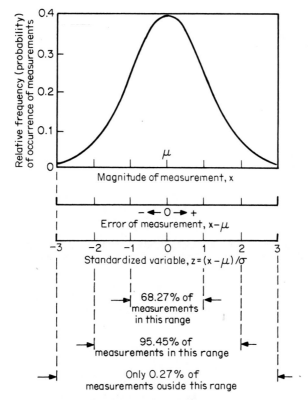

FIGURE 33.1   The normal distribution curve.

33.1.3.1 Stem Correction. The stem correction, in degrees Celsius, is given by

$$KN(T_o - T_m) = \text{degrees Celsius} \qquad (33.1)$$

where N = length of exposed thermometer, °C (that is, the length not in contact with vapor or liquid)
T_o = observed temperature on thermometer
T_m = mean temperature of exposed column (obtained by placing auxiliary thermometer alongside with its bulb midpoint)
K = constant, characteristic of particular kind of glass and temperature (see Table 33.3)

EXAMPLE 33.1 The temperature reads 240°C on a heat-resistant glass thermometer whose exposed scale reads from 110 to 360°C, which is a 250° interval. The temperature of the auxiliary thermometer reads 50°C.
Substituting into Eq. 33.1

$$(0.000\ 17)(250)(240 - 50) = 8.1$$

The corrected temperature should be 240 + 8 = 248°C.

TABLE 33.3 Values of K for Stem Correction of Thermometers

Temperature, °C	Soft glass	Heat-resistant glass
0–150	0.000 158	0.000 165
200	0.000 159	0.000 167
250	0.000 161	0.000 170
300	0.000 164	0.000 174
350		0.000 178
400		0.000 183
450		0.000 188

33.2 STATISTICAL TREATMENT OF DATA

Mistakes can be considered as incorrect tabulations of data or miscalculations of numbers. The only guarantee against mistakes is a check on each item and computation.
Errors are usually random and can be determinate or indeterminate. The precision of measurements depends upon many factors, such as the kind of apparatus used, the method employed, and the experience of the operator.

33.2.1 Determinate Errors

Determinate errors are traceable to a specific source.

1. Instrument and apparatus errors, such as calibration of gauges, precision of instrument scale readings, and accuracy of balances.

2. Operational errors, caused by carelessness of the operator, material loss, or misreading instruments.

3. Inherent error, arising from a basic practical or theoretical error in the procedure.

4. Personal errors, caused by personal prejudices or physical disabilities such as color blindness.

33.2.2 Indeterminate Errors

Indeterminate errors occur because of the laws of probability and chance; they are sometimes called random errors. Positive and negative errors occur with the same frequency. Small errors occur more often than large errors.

33.2.3 Error-Distribution Curve

The error-distribution curve (Fig. 33.1) suggests that the average of a large number of measurements (usually considered to be 30 or more) will be the correct value of the measurement, provided that the measurements are affected solely by random errors. This is the basis of the statistical evaluation of experimental measurements.

The breadth or spread of the curve indicates the precision of the measurements. A narrow-width curve implies good precision. Precision is determined by and related to the standard deviation. Before proceeding further, it is necessary to state that English letters are used to denote estimates (that is, data from a limited number of samples), whereas statisticians use the corresponding Greek letters for the parameters (that is, when a very large population is involved).

33.2.4 Measures of Dispersion

The standard deviation σ is the square root of the average squared difference between the individual observations x_i and the population mean, denoted by μ:

$$\sigma = \sqrt{\frac{\sum\limits_{i=1}^{N}(x_i - \mu)^2}{N}} \tag{33.2}$$

From a small sample set, the standard deviation s may be estimated by substituting the arithmetic mean of the set for the population mean. Since two pieces of information, namely s and $\bar{x}$, have been extracted from the data, we are left with $N - 1$ independent data points available for measurement of precision. Under these restraints, the standard deviation is

$$s = \sqrt{\frac{\sum\limits_{i=1}^{N}(x_i - \bar{x})^2}{N - 1}} \tag{33.3}$$

EXAMPLE 33.2   Find the standard deviation of these numbers from a measurement set: 12, 6, 7, 15, 10, and 8.

First, the arithmetic mean is calculated.

$$\bar{x} = \frac{12 + 6 + 7 + 15 + 10 + 8}{6} = 9.7$$

Now substitute the appropriate numbers into Eq. 33.3:

$$s = \sqrt{\frac{(12 - 9.7)^2 + (6 - 9.7)^2 + (7 - 9.7)^2 + (15 - 9.7)^2 + \cdots}{6 - 1}} = 3.39$$

With reference to Fig. 33.1, we can predict that 68 percent of the numbers from additional sets of measurements would lie within the confidence interval $\bar{x} \pm s$ or 9.7 ± 3.4 and 95 percent will lie within the confidence interval $\bar{x} \pm 2s$ or 9.7 ± 6.8.

33.2.5 Variance

For many purposes one uses the variance, which for the sample of the total population is s^2 and for the entire population is σ^2. The variance s^2 of a finite sample is an unbiased estimate of σ^2, whereas the standard deviation s is not an unbiased estimate of σ. From Example 33.2, the variance for the given set of numbers is 11.97.

When a series of observations can be logically arranged into k subgroups, the variance is calculated by summing the squares of the deviations for each subgroup, and then adding all the k sums and dividing by $N - k$ because one degree of freedom is lost in each subgroup.

33.2.6 Tests of Significance

If the data contained only random (or chance) errors, the cumulative estimates of the arithmetic mean and s would gradually approach the limits μ and σ. Were the true mean of the infinite population known, it would be expected that the averaged means for each group of data would also have some symmetrical type of distribution centered around μ. However, it would be expected that the spread of this dispersion about the mean would depend on the sample size.

33.2.7 Student's Distribution (or t Test)

The standard deviation of the distribution of means (equal to σ/N) is called the "Student's distribution", and the corresponding test of significance, a measure of error between μ and x, the t test. The distribution of the statistic is

$$\pm t = \frac{\bar{x} - \mu}{s/\sqrt{N}} \quad \text{or} \quad \mu = \bar{x} \pm \frac{ts}{\sqrt{N}}$$

This distribution is symmetrical about zero, and its dispersion is a function of the degrees of freedom $N - 1$. Its limits are called *confidence limits*. The percentage probability that μ lies within this interval is called the confidence level. Values of t are in Table 33.4 for any desired degrees of freedom and various confidence levels. For example, a confidence level of 0.95 implies that there would be only a 5 percent likelihood that μ would lie outside the confidence limits.

TABLE 33.4 Ordinates and Areas between Abscissa Values −z and +z of the Normal Distribution Curve

z	X	Y	A	1 − A	z	X	Y	A	1 − A
0	μ	0.399	0.0000	1.0000	±1.50	μ ± 1.50σ	0.1295	0.8664	0.1336
±0.05	μ ± 0.05σ	0.398	0.0399	0.9601	±1.55	μ ± 1.55σ	0.1200	0.8789	0.1211
±0.10	μ ± 0.10σ	0.397	0.0797	0.9203	±1.60	μ ± 1.60σ	0.1109	0.8904	0.1096
±0.15	μ ± 0.15σ	0.394	0.1192	0.8808	±1.65	μ ± 1.65σ	0.1023	0.9011	0.0989
±0.20	μ ± 0.20σ	0.391	0.1585	0.8415	±1.70	μ ± 1.70σ	0.0940	0.9109	0.0891
±0.25	μ ± 0.25σ	0.387	0.1974	0.8026	±1.75	μ ± 1.75σ	0.0863	0.9199	0.0801
±0.30	μ ± 0.30σ	0.381	0.2358	0.7642	±1.80	μ ± 1.80σ	0.0790	0.9281	0.0719
±0.35	μ ± 0.35σ	0.375	0.2737	0.7263	±1.85	μ ± 1.85σ	0.0721	0.9357	0.0643
±0.40	μ ± 0.40σ	0.368	0.3108	0.6892	±1.90	μ ± 1.90σ	0.0656	0.9426	0.0574
±0.45	μ ± 0.45σ	0.361	0.3473	0.6527	±1.95	μ ± 1.95σ	0.0596	0.9488	0.0512
±0.50	μ ± 0.50σ	0.352	0.3829	0.6171	±2.00	μ ± 2.00σ	0.0540	0.9545	0.0455
±0.55	μ ± 0.55σ	0.343	0.4177	0.5823	±2.05	μ ± 2.05σ	0.0488	0.9596	0.0404
±0.60	μ ± 0.60σ	0.333	0.4515	0.5485	±2.10	μ ± 2.10σ	0.0440	0.9643	0.0357
±0.65	μ ± 0.65σ	0.323	0.4843	0.5157	±2.15	μ ± 2.15σ	0.0396	0.9684	0.0316
±0.70	μ ± 0.70σ	0.312	0.5161	0.4839	±2.20	μ ± 2.20σ	0.0355	0.9722	0.0278
±0.75	μ ± 0.75σ	0.301	0.5467	0.4533	±2.25	μ ± 2.25σ	0.0317	0.9756	0.0244
±0.80	μ ± 0.80σ	0.290	0.5763	0.4237	±2.30	μ ± 2.30σ	0.0283	0.9786	0.0214
±0.85	μ ± 0.85σ	0.278	0.6047	0.3953	±2.35	μ ± 2.35σ	0.0252	0.9812	0.0188
±0.90	μ ± 0.90σ	0.266	0.6319	0.3681	±2.40	μ ± 2.40σ	0.0224	0.9836	0.0164
±0.95	μ ± 0.95σ	0.254	0.6579	0.3421	±2.45	μ ± 2.45σ	0.0198	0.9857	0.0143
±1.00	μ ± 1.00σ	0.242	0.6827	0.3173	±2.50	μ ± 2.50σ	0.0175	0.9876	0.0124
±1.05	μ ± 1.05σ	0.230	0.7063	0.2937	±2.55	μ ± 2.55σ	0.0154	0.9892	0.0108
±1.10	μ ± 1.10σ	0.218	0.7287	0.2713	±2.60	μ ± 2.60σ	0.0136	0.9907	0.0093
±1.15	μ ± 1.15σ	0.206	0.7499	0.2501	±2.65	μ ± 2.65σ	0.0119	0.9920	0.0080
±1.20	μ ± 1.20σ	0.194	0.7699	0.2301	±2.70	μ ± 2.70σ	0.0104	0.9931	0.0069
±1.25	μ ± 1.25σ	0.183	0.7887	0.2113	±2.75	μ ± 2.75σ	0.0091	0.9940	0.0060
±1.30	μ ± 1.30σ	0.171	0.8064	0.1936	±2.80	μ ± 2.80σ	0.0079	0.9949	0.0051
±1.35	μ ± 1.35σ	0.160	0.8230	0.1770	±2.85	μ ± 2.85σ	0.0069	0.9956	0.0044
±1.40	μ ± 1.40σ	0.150	0.8385	0.1615	±2.90	μ ± 2.90σ	0.0060	0.9963	0.0037
±1.45	μ ± 1.45σ	0.139	0.8529	0.1471	±2.95	μ ± 2.95σ	0.0051	0.9968	0.0032

TABLE 33.4 Ordinates and Areas between Abscissa Values $-z$ and $+z$ of the Normal Distribution Curve (Continued)

z	X	Y	A	1 − A	z	X	Y	A	1 − A
±1.50	μ ± 1.50σ	0.130	0.8664	0.1336	±3.00	μ ± 3.00σ	0.0044	0.9973	0.0027
					±4.00	μ ± 4.00σ	0.0001	0.99994	0.00006
±0.000	μ	0.3989	0.0000	1.0000	±5.00	μ ± 5.00σ	0.000001	0.9999994	0.0000006
±0.126	μ ± 0.126σ	0.3958	0.1000	0.9000	±1.036	μ ± 1.036σ	0.2331	0.7000	0.3000
±0.253	μ ± 0.253σ	0.3863	0.2000	0.8000	±1.282	μ ± 1.282σ	0.1755	0.8000	0.2000
±0.385	μ ± 0.385σ	0.3704	0.3000	0.7000	±1.645	μ ± 1.645σ	0.1031	0.9000	0.1000
±0.524	μ ± 0.524σ	0.3477	0.4000	0.6000	±1.960	μ ± 1.960σ	0.0584	0.9500	0.0500
±0.674	μ ± 0.674σ	0.3178	0.5000	0.5000	±2.576	μ ± 2.576σ	0.0145	0.9900	0.0100
±0.842	μ ± 0.842σ	0.2800	0.6000	0.4000	±3.291	μ ± 3.291σ	0.0018	0.9990	0.0010
					±3.891	μ ± 3.891σ	0.0002	0.9999	0.0001

Source: Perry, Chilton, and Kirkpatrick, *Chemical Engineers' Handbook*, 4th ed., McGraw-Hill, New York, 1963; by permission.

33.8

EXAMPLE 33.3 From the following data, do the two methods actually give concordant results?

Sample	Method A	Method B	Difference
1	33.27	33.04	$d_1 = 0.23$
2	51.34	50.96	$d_2 = 0.38$
3	23.91	23.77	$d_3 = 0.14$
4	47.04	46.79	$d_4 = 0.25$
			$\overline{d} = 0.25$

$$s_d = \sqrt{\frac{\Sigma(d - \overline{d})^2}{N - 1}} = 0.099$$

$$t = \frac{0.25}{0.099}\sqrt{4 - 1} = 4.30$$

From the table of t values, $t_{.975} = 3.18$ (at 95 percent probability) and $t_{.995} = 5.84$ (at 99 percent probability). Since the difference is significant at the 99 percent probability level (although not at the 95 percent probability), there is reason to believe the difference between the two methods is significant.

33.3 CHEMICAL RESISTANCE OF POLYMERS AND RUBBERS

Table 33.5 lists the resistance of various polymers and rubbers to many chemicals. However, the table is intended to be a general guide only. Tests should be run under proposed working conditions. The reactive combination of compounds from two or more classes could cause an undesirable chemical effect. As the temperature increases at which the material is used, the resistance to attack decreases.

Never store strong oxidizing agents in plastic labware except that constructed from fluorinated ethylene-propylene resin. Silver solutions will be reduced by some plastics. Never place plastic labware on a hot plate or in a flame.

For most applications, plastics may be washed in a mild nonalkaline detergent, followed by rinses with tap water and then distilled water. Most plastics, particularly the polyolefins, have nonwetting surfaces which resist attack and are easy to clean. Never use abrasive cleaners or scouring pads on any plastic labware. Do not use strong alkaline cleaning agents with polycarbonate polymers.

TABLE 33.5 Resistance of Selected Polymers and Rubbers to Various Chemicals at 20°C

The information in this table is intended to be used only as a general guide. The chemical resistance classifications are E = excellent (30 days of exposure causes no damage), G = good (some damage after 30 days), F = fair (exposure may cause crazing, softening, swelling, or loss of strength), N = not recommended (immediate damage may occur).

Polymers	Acids, dilute or weak	Acids, strong and concentrated	Alcohols, aliphatic	Aldehydes	Alkalies, concentrated	Esters	Ethers	Glycols	Hydrocarbons, aliphatic	Hydrocarbons, aromatic	Hydrocarbons, halogenated	Ketones	Oxidizing agents, strong
Acetals	F	N	F	N	N	N	N	G	N	N	N	N	N
Acrylics: poly(methyl methacrylate)	G	N	E	—	N	N	E	E	G	N	N	N	N
Allyls: diallyl phthalate	G	—	—	—	N	—	—	—	E	G	G	N	—
Cellulosics: cellulose-acetate-butyrate and cellulose-acetate-propionate polymers	F	N	N	N	N	N	N	G	F	N	N	N	—
Fluorocarbons	E	E	E	E	E	E	E	E	E	E	E	E	E
Polyamides	N	N	G	E	E	G	—	G	G	F	F	G	N
Polycarbonates	G	N	G	F	N	N	N	G	N	N	F	N	N
Polyesters	G	G	N	—	N	N	F	G	G	F	F	N	F
Poly(methyl pentene)	E	E	G	G	E	G	N	E	F	G	N	F	F

Material												
Low-density polyethylene	E	E	E	E	G	N	E	F	F	N	G	F
High-density polyethylene	E	E	E	E	G	N	G	G	G	N	G	F
Polybutadiene	G	F	E	—	—	—	—	—	E	E	E	E
Polypropylene and polyallomer	E	E	E	E	G	N	G	F	F	F	E	G
Polystyrene	N	—	E	N	N	N	N	N	N	N	N	N
Styrene-acrylonitrile copolymers	—	—	N	—	—	—	—	—	—	—	—	—
Styrene-acrylonitrile-butadiene copolymers	—	N	G	N	F	G	N	F	N	N	N	N
Sulfones: polysulfone	G	N	F	E	F	N	F	F	N	N	N	N
Vinyls: poly(vinyl chloride)	E	G	E	G	E	N	G	G	N	N	N	N

Rubbers

Material												
Natural rubber	—	E	—	—	N	N	E	N	N	N	N	N
Nitrile rubber	—	E	—	—	G	G	E	E	E	E	N	N
Polychloroprene	—	E	—	F	F	F	F	F	N	N	N	N
Polyisobutylene	—	E	—	E	F	E	E	N	F	F	N	N
Polysulfide rubbers: Thiokol	—	E	—	E	E	E	E	E	E	F	N	N
Styrene-butadiene rubber	—	E	—	N	N	N	E	N	N	N	N	N

Source: J. A. Dean, ed., *Handbook of Organic Chemistry*, McGraw-Hill, New York, 1986, pp. 10-85 to 10-86.

BIBLIOGRAPHY

Currie, L., "Sources of Error and the Approach to Accuracy in Analytical Chemistry," in: *Treatise on Analytical Chemistry,* 2d ed., I. Kolthoff and P. Elving, eds., Part I, Vol. 1, Chap. 4, Wiley, New York, 1978.

Mandel, J., "Accuracy and Precision: Evaluation and Interpretation of Analytical Results," in: *Treatise on Analytical Chemistry,* 2d ed., I. Kolthoff and P. Elving, eds., Part I, Vol. 1, Chap. 5, Wiley, New York, 1978.

McGee, T. D., *Principles and Methods of Temperature Measurement,* Wiley, New York, 1988.

Miller, J., and J. Miller, *Statistics for Analytical Chemistry,* Halsted, New York, 1984.

INDEX

1

ABOUT THE EDITORS:

Gershon J. Shugar is Professor of Engineering Technologies with Essex County College, Newark, New Jersey. In 1947 he established a chemical manufacturing business which became the largest exclusive pearlescent pigment manufacturing company in the United States. After selling his interest in the company in 1968, Dr. Shugar was appointed assistant professor and taught chemistry at Rutgers University in Newark, New Jersey. After one year in this position, he was appointed to the Essex County College faculty. He is the author of *Chemical Technicians' Ready Reference Handbook*, Third Edition.

John A. Dean is Professor Emeritus of Chemistry, University of Tennessee at Knoxville. His research interests, reflected in over 100 publications, include flame spectrometric methods, chromatographic and solvent extraction methods, and polarography. He is a member of the American Chemical Society, Society for Applied Spectroscopy (SAS), and many other learned societies. For six years he has been the newsletter editor for SAS. Dr. Dean is the editor of *Lange's Handbook of Chemistry*, Thirteenth Edition, and *The Handbook of Organic Chemistry*.